Handbook of Cloud Computing

Borko Furht · Armando Escalante
Editors

Handbook of Cloud Computing

 Springer

Editors
Borko Furht
Department of Computer and Electrical
    Engineering and Computer Science
Florida Atlantic University
777 Glades Road
Boca Raton, FL 33431, USA
bfurht@fau.edu

Armando Escalante
LexisNexis
6601 Park of Commerce Boulevard
Boca Raton, FL 33487, USA
armando.escalante@lexisnexis.com

ISBN 978-1-4899-9456-1   ISBN 978-1-4419-6524-0 (eBook)
DOI 10.1007/978-1-4419-6524-0
Springer New York Dordrecht Heidelberg London

Printed on acid-free paper

Springer is part of Springer Science+Business Media (www.springer.com)

Preface

Cloud computing has become a great solution for providing a flexible, on-demand, and dynamically scalable computing infrastructure for many applications. Cloud computing also presents a significant technology trends, and it is already obvious that it is reshaping information technology processes and the IT marketplace.

This Handbook is a carefully edited book – contributors are 65 worldwide experts in the field of cloud computing and their applications. The Handbook Advisory Board, comprised of nine researchers and practitioners from academia and industry, helped in reshaping the Handbook and selecting the right topics and creative and knowledgeable contributors. The scope of the book includes leading-edge cloud computing technologies, systems, and architectures; cloud computing services; and a variety of cloud computing applications.

The Handbook comprises four parts, which consist of 26 chapters. The first part on *Technologies and Systems* includes articles dealing with cloud computing technologies, storage and fault tolerant strategies in cloud computing, workflows, grid computing technologies, and the role of networks in cloud computing.

The second part on *Architectures* focuses on articles on several specific architectural concepts applied in cloud computing, including enterprise knowledge clouds, high-performance computing clouds, clouds with vertical load distribution, and peer-to-peer based clouds.

The third part on *Services* consists of articles on various issues relating to cloud services, including types of services, service scalability, scientific services, and dynamic collaborative services.

The forth part on *Applications* describes various cloud computing applications from enterprise knowledge clouds, scientific and statistical computing, scientific data management, to medical applications.

With the dramatic growth of cloud computing technologies, platforms and services, this Handbook can be the definitive resource for persons working in this field as researchers, scientists, programmers, engineers, and users. The book is intended for a wide variety of people including academicians, designers, developers, educators, engineers, practitioners, researchers, and graduate students. This book can also be beneficial for business managers, entrepreneurs, and investors. The book can have a great potential to be adopted as a textbook in current and new courses on Cloud Computing.

The main features of this Handbook can be summarized as:

1. The Handbook describes and evaluates the current state-of-the-art in a new field of cloud computing.
2. It also presents current systems, services, and main players in this explosive field.
3. Contributors to the Handbook are the leading researchers from academia and practitioners from industry.

We would like to thank the authors for their contributions. Without their expertise and effort, this Handbook would never come to fruition. Springer editors and staff also deserve our sincere recognition for their support throughout the project.

Boca Raton, Florida   Borko Furht
                                                                                Armando Escalante

Contents

Part II Architectures

Part III Services

Part IV Applications

Contributors

Ankur Agarwal Department of Computer Science and Engineering, FAU, Boca Raton, FL, USA, ankur@cse.fau.edu

Tim Bell Department of Computer Science and Software Engineering, University of Canterbury, Christchurch, New Zealand, tim.bell@canterbury.ac.nz

Paolo Bientinesi AICES, RWTH, Aachen, Germany, pauldj@aices.rwth-aachen.de

Norman Bobroff IBM Watson Research Center, Hawthorne, NY, USA, bobroff@us.ibm.com

Juan Cáceres Telefónica Investigación y Desarrollo, Madrid, Spain, caceres@tid.es

David Chapman Computer Science and Electrical Engineering Department, University of Maryland, Baltimore County, MD, USA, dchapm2@umbc.edu

Jinjun Chen Faculty of Information and Communication Technologies, Swinburne University of Technology, Hawthorn, Melbourne, Australia 3122, jchen@swin.edu.au

Lei Chen Hong Kong University of Science and Technology, Clear Water Bay, Hong Kong, leichen@cse.ust.hk

Karim Chine Cloud Era Ltd, Cambridge, UK, karim.chine@polytechnique.org

David Cohen Cloud Infrastructure Group, EMC Corporation, Cambridge, MA, USA, david.cohen@emc.com

Vincenzo D. Cunsolo University of Messina, Contrada di Dio, S. Agata, Messina, Italy, vdcunsolo@unime.it

Kemal A. Delic Hewlett-Packard Co., New York, NY, USA, kemal.delic@hp.com

Mac Devine IBM Corporation, Research Triangle Park, NC, USA, wdevine@us.ibm.com

Salvatore Distefano University of Messina, Contrada di Dio, S. Agata, Messina, Italy, sdistefano@unime.it

Patrick Dreher Renaissance Computing Institute, Chapel Hill, NC, 27517 USA, dreher@renci.org

Kathleen Ericson Department of Computer Science, Colorado State University, Fort Collins, CO, USA, ericson@cs.colostate.edu

Liana Fong IBM Watson Research Center, Hawthorne, NY, USA, llfong@us.ibm.com

Ian Foster Argonne National Laboratory, Argonne, IL, USA; The University of Chicago, Chicago, IL, USA, foster@anl.gov

Marc Eduard Frîncu Institute e-Austria, Blvd. Vasile Parvan No 4 300223, Room 045B, Timisoara, Romania, mfrincu@info.uvt.ro

Borko Furht Department of Computer & Electrical Engineering and Computer Science, Florida Atlantic University, Boca Raton, FL, USA, bfurht@fau.edu

Wei Gao Services Computing Technology and System Lab; Cluster and Grid Computing Lab, Huazhong University of Science and Technology, Wuhan, China, gaowei715@gmail.com

Giulio Giunta Department of Applied Science, University of Napoli Parthenope, Napoli, Italy, giulio.giunta@uniparthenope.it

Milt Halem Computer Science and Electrical Engineering Department, University of Maryland, Baltimore County, MD, USA, halem@umbc.edu

Mohammad Mehedi Hassan Department of Computer Engineering, Kyung Hee University, Global Campus, South Korea, hassan@khu.ac.kr

Nathan Henehan Senior Software Developer, NACS Solutions, Oberlin, OH, USA, nhenehan@gmail.com

Juan J. Hierro Telefónica Investigación y Desarrollo, Madrid, Spain, jhierro@tid.es

Dachuan Huang Services Computing Technology and System Lab; Cluster and Grid Computing Lab, Huazhong University of Science and Technology, Wuhan, China, hdc1112@gmail.com

Eui-Nam Huh Department of Computer Engineering, Kyung Hee University, Global Campus, South Korea, johnhuh@khu.ac.kr

Roman Iakymchuk AICES, RWTH, Aachen, Germany, iakymchuk@aices.rwth-aachen.de

Shadi Ibrahim Services Computing Technology and System Lab; Cluster and Grid Computing Lab, Huazhong University of Science and Technology, Wuhan, China, shadi@hust.edu.cn

Hai Jin Services Computing Technology and System Lab; Cluster and Grid Computing Lab, Huazhong University of Science and Technology, Wuhan, China, hjin@hust.edu.cn

Karuna P. Joshi Computer Science and Electrical Engineering Department, University of Maryland, Baltimore County, MD, USA, kjoshi1@umbc.edu

Burton S. Kaliski Jr. Office of the CTO, EMC Corporation, Hopkinton, MA, USA, burt.kaliski@emc.com

Hari Kalva Department of Computer Science and Engineering, FAU, Boca Raton, FL, USA, hari.kalva@fau.edu

Giuliano Laccetti Department of Mathematics and Applications, University of Napoli Federico II, Napoli, Italy, giuliano.laccetti@unina.it

Rui Li Hong Kong University of Science and Technology, Clear Water Bay, Hong Kong, cslr@cse.ust.hk

Wen-Syan Li SAP Technology Lab, Shanghai, China, wen-syan.li@sap.com

Geng Lin IBM Alliance, Cisco Systems, San Francisco, CA, USA, gelin@cisco.com

Xiao Liu Faculty of Information and Communication Technologies, Swinburne University of Technology, Hawthorn, Melbourne, Australia 3122, xliu@groupwise.swin.edu.au

Yanbin Liu IBM Watson Research Center, Hawthorne, NY, USA, ygliu@us.ibm.com

Anthony M. Middleton LexisNexis Risk Solutions, Boca Raton, FL, USA, tony.middleton@lexisnexis.com

Raffaele Montella Department of Applied Science, University of Napoli Parthenope, Napoli, Italy, raffaele.montella@uniparthenope.it

Jeff Napper Vrije Universiteit, Amsterdam, Netherlands, jnapper@cs.vu.nl

Phuong Nguyen Computer Science and Electrical Engineering Department, University of Maryland, Baltimore County, MD, USA, phuong3@umbc.edu

Jeffrey M. Nick Office of the CTO, EMC Corporation, Hopkinton, MA, USA, jeff.nick@emc.com

Sangyoon Oh WISE Research Lab, School of Information and Communication Engineering, Ajou University, Suwon, South Korea, syoh@ajou.ac.kr

Shrideep Pallickara Department of Computer Science, Colorado State University, Fort Collins, CO, USA, shrideep@cs.colostate.edu

Sangmi Lee Pallickara Department of Computer Science, Colorado State University, Fort Collins, CO, USA, sangmi@cs.colostate.edu

A.S. Pandya Department of Computer Science and Engineering, FAU, Boca Raton, FL, USA, pandya@fau.edu

Manish Parashar NSF CAC, Rutgers University, Piscataway, NJ, USA, parashar@rutgers.edu

Thomas Phan Microsoft Corporation, Washington, DC, USA, thomas.phan@acm.org

Marlon Pierce Community Grids Lab, Indiana University, Bloomington, IN, USA, mpierce@cs.indiana.edu

Álvaro Polo Telefónica Investigación y Desarrollo, Madrid, Spain, apv@tid.es

Antonio Puliafito University of Messina, Contrada di Dio, S. Agata, Messina, Italy, apuliafito@unime.it

Jeff A. Riley Hewlett-Packard Co., New York, NY, USA, jeff.riley@hp.com

Ivan Rodero NSF CAC, Rutgers University, Piscataway, NJ, USA, irodero@rutgers.edu

Luis Rodero-Merino INRIA-ENS, INRIA, Lyon, France, luis.rodero-merino@ens-lyon.fr

S. Masoud Sadjadi CIS, Florida International University, Miami, FL, USA, sadjadi@cis.fiu.edu

Marco Scarpa University of Messina, Contrada di Dio, S. Agata, Messina, Italy, mscarpa@unime.it

Eric Sills North Carolina State University, Raleigh, NC 27695, USA, edsills@ncsu.edu

Mikael Fernandus Simalango WISE Research Lab, Ajou University, Suwon, South Korea, mikael@ajou.ac.kr

Vivek Somashekarappa Armellini Inc., Palm City, FL, USA, vsomashekar@gmail.com

Luis M. Vaquero Telefónica Investigación y Desarrollo, Madrid, Spain, lmvg@tid.es

David Villegas CIS, Florida International University, Miami, FL, USA, dvill013@cis.fiu.edu

Mladen A. Vouk Department of Computer Science, North Carolina State University, Box 8206, Raleigh, NC 27695, USA, vouk@ncsu.edu

Song Wu Services Computing Technology and System Lab; Cluster and Grid Computing Lab, Huazhong University of Science and Technology, Wuhan, China, wusong@mail.hust.edu.cn

Yun Yang Faculty of Information and Communication Technologies, Swinburne University of Technology, Hawthorn, Melbourne, Australia 3122, yyang@groupwise.swin.edu.au

Yelena Yesha Computer Science and Electrical Engineering Department, University of Maryland, Baltimore County, MD, USA, yeyesha@csee.umbc.edu

Yaacov Yesha Computer Science and Electrical Engineering Department, University of Maryland, Baltimore County, MD, USA, yayesha@umbc.edu

Dong Yuan Faculty of Information and Communication Technologies, Swinburne University of Technology, Hawthorn, Melbourne, Australia 3122, dyuan@groupwise.swin.edu.au

Gaofeng Zhang Faculty of Information and Communication Technologies, Swinburne University of Technology, Hawthorn, Melbourne, Australia 3122, gzhang@groupwise.swin.edu.au

Jinzy Zhu IBM Cloud Computing Center, China, jinzyzhu@cn.ibm.com

About the Editors

Borko Furht is a professor and chairman of the Department of Electrical & Computer Engineering and Computer Science at Florida Atlantic University (FAU) in Boca Raton, Florida. He is also Director of recently formed NSF-sponsored

Industry/University Cooperative Research Center on Advanced Knowledge Enablement. Before joining FAU, he was a vice president of research and a senior director of development at Modcomp (Ft. Lauderdale), a computer company of Daimler Benz, Germany; a professor at University of Miami in Coral Gables, Florida; and a senior researcher in the Institute Boris Kidric-Vinca, Yugoslavia. Professor Furht received a Ph.D. degree in electrical and computer engineering from the University of Belgrade. His current research is in multimedia systems, video coding and compression, 3D video and image systems, wireless multimedia, and Internet and cloud computing. He is presently Principal Investigator and Co-PI of several multiyear, multimillion-dollar projects, including NSF PIRE project and NSF High-Performance Computing Center. He is the author of numerous books and articles in the areas of multimedia, computer architecture, real-time computing, and operating systems. He is a founder and editor-in-chief of the *Journal of Multimedia Tools and Applications* (Springer). He has received several technical and publishing awards, and has consulted for many high-tech companies including IBM, Hewlett-Packard, Xerox, General Electric, JPL, NASA, Honeywell, and RCA. He has also served as a consultant to various colleges and universities. He has given many invited talks, keynote lectures, seminars, and tutorials. He served on the Board of Directors of several high-tech companies.

Armando J. Escalante is Senior Vice President and Chief Technology Officer of Risk Solutions for the LexisNexis Group, a division of Reed Elsevier. In this position, Escalante is responsible for technology development, information systems and operations. Previously, Escalante was Chief Operating Officer for Seisint, a privately owned company, which was purchased by LexisNexis in 2004. In this position, he was responsible for Technology, Development and Operations. Prior to 2001, Escalante served as Vice President of Engineering and Operations for Diveo Broadband Networks, where he led world-class Data Centers located in the U.S. and Latin America. Before Diveo Broadband Networks, Escalante was VP for one of the fastest growing divisions of Vignette Corporation, an eBusiness software leader. Escalante earned his bachelor's degree in electronic engineering at the USB in Caracas, Venezuela and a master's degree in computer science from Stevens Institute of Technology, as well as a master's degree in business administration from West Coast University.

Part I
Technologies and Systems

Chapter 1
Cloud Computing Fundamentals

Borko Furht

1.1 Introduction

In the introductory chapter we define the concept of cloud computing and cloud services, and we introduce layers and types of cloud computing. We discuss the differences between cloud computing and cloud services. New technologies that enabled cloud computing are presented next. We also discuss cloud computing features, standards, and security issues. We introduce the key cloud computing platforms, their vendors, and their offerings. We discuss cloud computing challenges and the future of cloud computing.

Cloud computing can be defined as a new style of computing in which dynamically scalable and often virtualized resources are provided as a services over the Internet. Cloud computing has become a significant technology trend, and many experts expect that cloud computing will reshape information technology (IT) processes and the IT marketplace. With the cloud computing technology, users use a variety of devices, including PCs, laptops, smartphones, and PDAs to access programs, storage, and application-development platforms over the Internet, via services offered by cloud computing providers. Advantages of the cloud computing technology include cost savings, high availability, and easy scalability.

Figure 1.1, adapted from Voas and Zhang (2009), shows six phases of computing paradigms, from dummy terminals/mainframes, to PCs, networking computing, to grid and cloud computing.

In phase 1, many users shared powerful mainframes using dummy terminals. In phase 2, stand-alone PCs became powerful enough to meet the majority of users' needs. In phase 3, PCs, laptops, and servers were connected together through local networks to share resources and increase performance. In phase 4, local networks were connected to other local networks forming a global network such as the Internet to utilize remote applications and resources. In phase 5, grid computing provided shared computing power and storage through a distributed computing

B. Furht (✉)
Department of Computer & Electrical Engineering and Computer Science, Florida Atlantic University, Boca Raton, FL, USA
e-mail: bfurht@fau.edu

B. Furht, A. Escalante (eds.), *Handbook of Cloud Computing*,
DOI 10.1007/978-1-4419-6524-0_1, © Springer Science+Business Media, LLC 2010

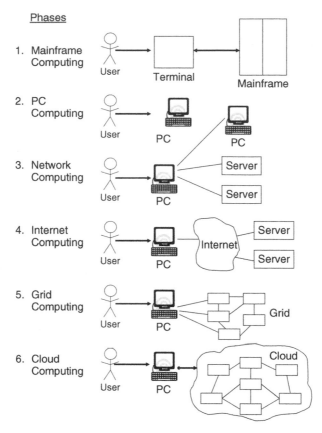

Fig. 1.1 Six computing paradigms – from mainframe computing to Internet computing, to grid computing and cloud computing (adapted from Voas and Zhang (2009))

system. In phase 6, cloud computing further provides shared resources on the Internet in a scalable and simple way.

Comparing these six computing paradigms, it looks like that cloud computing is a return to the original mainframe computing paradigm. However, these two paradigms have several important differences. Mainframe computing offers finite computing power, while cloud computing provides almost infinite power and capacity. In addition, in mainframe computing dummy terminals acted as user interface devices, while in cloud computing powerful PCs can provide local computing power and cashing support.

1.1.1 Layers of Cloud Computing

Cloud computing can be viewed as a collection of services, which can be presented as a layered cloud computing architecture, as shown in Fig. 1.2 [Jones XXXX]. The

Fig. 1.2 Layered architecture of Cloud Computing (adapted from Jones)

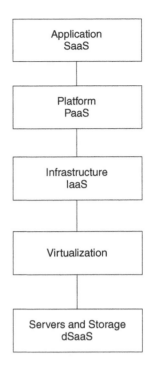

services offered through cloud computing usually include IT services referred as to SaaS (Software-as-a-Service), which is shown on top of the stack. SaaS allows users to run applications remotely from the cloud.

Infrastructure-as-a-service (IaaS) refers to computing resources as a service. This includes virtualized computers with guaranteed processing power and reserved bandwidth for storage and Internet access.

Platform-as-a-Service (PaaS) is similar to IaaS, but also includes operating systems and required services for a particular application. In other words, PaaS is IaaS with a custom software stack for the given application.

The data-Storage-as-a-Service (dSaaS) provides storage that the consumer is used including bandwidth requirements for the storage.

An example of Platform-as-aService (PaaS) cloud computing is shown in Fig. 1.3 ["Platform as a Service," http://www.zoho.com/creator/paas.html]. The PaaS provides Integrated Development Environment (IDE) including data security, backup and recovery, application hosting, and scalable architecture.

According to Chappell (2008) there are three categories of cloud services, as illustrated in Fig. 1.4. Figure 1.4a shows the cloud service SaaS, where the entire application is running in the cloud. The client contains a simple browser to access the application. A well-known example of SaaS is salesforce.com.

Figure 1.4b illustrates another type of cloud services, where the application runs on the client; however it accesses useful functions and services provided in the cloud. An example of this type of cloud services on the desktop is Apple's iTunes.

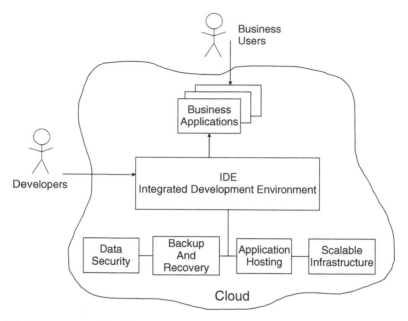

Fig. 1.3 The concept of Platform-as-a-Service, Zoho Creator (adapted from "Platform as a Service," http://www.zoho.com/creator/paas.html)

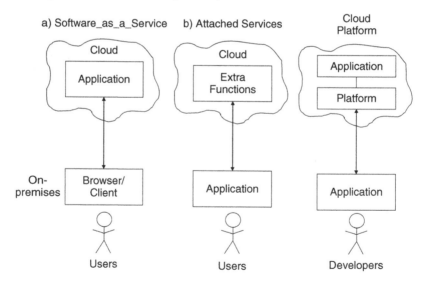

Fig. 1.4 The categories of cloud services (adopted from Chappell (2008))

The desktop application plays music, while the cloud service is used to purchase a new audio and video content. An enterprise example of this cloud service is Microsoft Exchange Hosted Services. On-premises Exchange Server is using added services from the cloud including spam filtering, archiving, and other functions.

Finally, Fig. 1.4c shows a cloud platform for creating applications, which is used by developers. The application developers create a new SaaS application using the cloud platform.

1.1.2 Types of Cloud Computing

There are three types of cloud computing ("Cloud Computing," Wikipedia, http://en.wikipedia.org/wiki/Cloud_computing): (a) public cloud, (b) private cloud, and (c) hybrid cloud, as illustrated in Fig. 1.5.

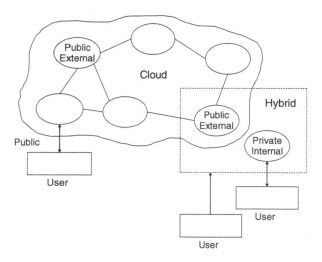

Fig. 1.5 Three types of cloud computing

In the public cloud (or external cloud) computing resources are dynamically provisioned over the Internet via Web applications or Web services from an off-site third-party provider. Public clouds are run by third parties, and applications from different customers are likely to be mixed together on the cloud's servers, storage systems, and networks.

Private cloud (or internal cloud) refers to cloud computing on private networks. Private clouds are built for the exclusive use of one client, providing full control over data, security, and quality of service. Private clouds can be built and managed by a company's own IT organization or by a cloud provider.

A hybrid cloud environment combines multiple public and private cloud models. Hybrid clouds introduce the complexity of determining how to distribute applications across both a public and private cloud.

1.1.3 Cloud Computing Versus Cloud Services

In this section we present two tables that show the differences and major attributes of cloud computing versus cloud services (Jens, 2008). Cloud computing is the IT foundation for cloud services and it consists of technologies that enable cloud services. The key attributes of cloud computing are shown in Table 1.1.

In Key attributes of cloud services are summarized in Table 1.2 (Jens, 2008).

Table 1.1 Key Cloud Computing Attributes (adapted from Jens (2008))

Attributes	Description
Infrastructure systems	It includes servers, storage, and networks that can scale as per user demand.
Application software	It provides Web-based user interface, Web services APIs, and a rich variety of configurations.
Application development and deployment software	It supports the development and integration of cloud application software.
System and application management software	It supports rapid self-service provisioning and configuration and usage monitoring.
IP networks	They connect end users to the cloud and the infrastructure components.

Table 1.2 Key Attributes of Cloud Services (adapted from Jens (2008))

Attributes	Description
Offsite. Third-party provider	In the cloud execution, it is assumed that third-party provides services. There is also a possibility of in-house cloud service delivery.
Accessed via the Internet	Services are accessed via standard-based, universal network access. It can also include security and quality-of-service options.
Minimal or no IT skill required	There is a simplified specification of requirements.
Provisioning	It includes self-service requesting, near real-time deployment, and dynamic and fine-grained scaling.
Pricing	Pricing is based on usage-based capability and it is fine-grained.
User interface	User interface include browsers for a variety of devices and with rich capabilities.
System interface	System interfaces are based on Web services APIs providing a standard framework for accessing and integrating among cloud services.
Shared resources	Resources are shared among cloud services users; however via configuration options with the service, there is the ability to customize.

1.2 Enabling Technologies

Key technologies that enabled cloud computing are described in this section; they include virtualization, Web service and service-oriented architecture, service flows and workflows, and Web 2.0 and mashup.

1.2.1 Virtualization

The advantage of cloud computing is the ability to virtualize and share resources among different applications with the objective for better server utilization. Figure 1.6 shows an example Jones]. In non-cloud computing three independent platforms exist for three different applications running on its own server. In the cloud, servers can be shared, or virtualized, for operating systems and applications resulting in fewer servers (in specific example two servers).

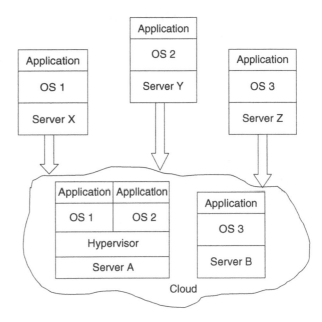

Fig. 1.6 An example of virtualization: in non-cloud computing there is a need for three servers; in the cloud computing, two servers are used (adapted from Jones)

Virtualization technologies include virtual machine techniques such as VMware and Xen, and virtual networks, such as VPN. Virtual machines provide virtualized IT-infrastructures on-demand, while virtual networks support users with a customized network environment to access cloud resources.

1.2.2 Web Service and Service Oriented Architecture

Web Services and Service Oriented Architecture (SOA) are not new concepts; however they represent the base technologies for cloud computing. Cloud services are typically designed as Web services, which follow industry standards including WSDL, SOAP, and UDDI. A Service Oriented Architecture organizes and manages Web services inside clouds (Vouk, 2008). A SOA also includes a set of cloud services, which are available on various distributed platforms.

1.2.3 Service Flow and Workflows

The concept of service flow and workflow refers to an integrated view of service-based activities provided in clouds. Workflows have become one of the important areas of research in the field of database and information systems (Vouk, 2008).

1.2.4 Web 2.0 and Mashup

Web 2.0 is a new concept that refers to the use of Web technology and Web design to enhance creativity, information sharing, and collaboration among users (Wang, Tao,

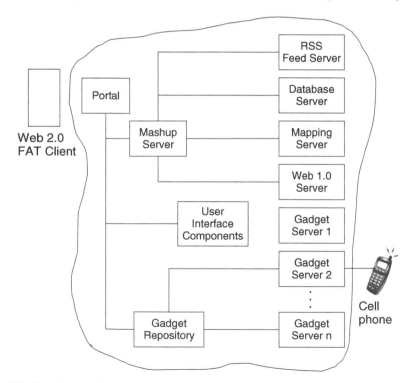

Fig. 1.7 Cloud computing architecture uses various components at different levels (adapted from Hutchinson and Ward (2009))

& Kunze, 2008). On the other hand, Mashup is a web application that combines data from more than one source into a single integrated storage tool. Both technologies are very beneficial for cloud computing.

Figure 1.7 shows a cloud computing architecture, adapted from Hutchinson and Ward (2009), in which an application reuses various components. The components in this architecture are dynamic in nature, operate in a SaaS model, and leverage SOA. The components closer to the user are smaller in nature and more reusable. The components in the center contain aggregate and extend services via mashup servers and portals.Data from one service (such as addresses in a database) can be mashed up with mapping information (such as Yahoo or Google maps) to produce an aggregated view of the information.

1.3 Cloud Computing Features

Cloud computing brings a number of new features compared to other computing paradigms (Wang et al., 2008; Grossman, 2009). There are briefly described in this section.

- Scalability and on-demand services
  Cloud computing provides resources and services for users on demand. The resources are scalable over several data centers.

- User-centric interface
  Cloud interfaces are location independent and can be accesses by well established interfaces such as Web services and Internet browsers.

- Guaranteed Quality of Service (QoS)
  Cloud computed can guarantee QoS for users in terms of hardware/CPU performance, bandwidth, and memory capacity.

- Autonomous system
  The cloud computing systems are autonomous systems managed transparently to users. However, software and data inside clouds can be automatically reconfigured and consolidated to a simple platform depending on user's needs.

- Pricing
  Cloud computing does not require up-from investment. No capital expenditure is required. Users pay for services and capacity as they need them.

1.3.1 Cloud Computing Standards

Cloud computing standards have not been yet fully developed; however a number of existing typically lightweight, open standards have facilitated the growth

Table 1.3 Cloud computing standards ("Cloud Computing," Wikipedia, http://en.wikipedia.org/wiki/Cloud_computing)

	Communications: HTTP, XMPP
	Security: OAuth, OpenID, SSL/TLS
Applications	Syndication: Atom
Client	Browsers: AJAX
	Offline:' HTML5
Implementations	Virtualization: OVF
Platform	Solution stacks: LAMP
Service	Data: XML, JSON
	Web services: REST

of cloud computing ("Cloud Computing," Wikipedia, http://en.wikipedia.org/wiki/Cloud_computing). Table 1.3 illustrates several of these open standards, which are currently used in cloud computing.

1.3.2 Cloud Computing Security

One of the critical issues in implementing cloud computing is taking virtual machines, which contain critical applications and sensitive data, to public and shared cloud environments. Therefore, potential cloud computing users are concerned about the following security issues ("Cloud Computing Security," Third Brigade, www.cloudreadysecurity.com.).

- Will the users still have the same security policy control over their applications and services?
- Can it be proved to the organization that the system is still secure and meets SLAs?
- Is the system complaint and can it be proved to company's auditors?

In traditional data centers, the common approaches to security include perimeter firewall, demilitarized zones, network segmentation, intrusion detection and prevention systems, and network monitoring tools.

The security requirements for cloud computing providers begins with the same techniques and tools as for traditional data centers, which includes the application of a strong network security perimeter. However, physical segmentation and hardware-based security cannot protect against attacks between virtual machines on the same server. Cloud computing servers use the same operating systems, enterprise and Web applications as localized virtual machines and physical servers. Therefore, an attacker can remotely exploit vulnerabilities in these systems and applications. In addition, co-location of multiple virtual machines increases the attack surface and risk to MV-to-VM compromise. Intrusion detection and prevention systems need to be able to detect malicious activity in the VM level, regardless of the location of the

VM within the virtualized cloud environment ("Cloud Computing Security," Third Brigade, www.cloudreadysecurity.com.).

In summary, the virtual environments that deploy the security mechanisms on virtual machines including firewalls, intrusion detection and prevention, integrity monitoring, and log inspection, will effectively make VM cloud secure and ready for deployment.

1.4 Cloud Computing Platforms

Cloud computing has great commercial potential. According to market research firm IDC, IT cloud services spending will grow from about $16B in 2008 to about $42B in 2012 and to increase its share of overall IT spending from 4.2% to 8.5%.

Table 1.4 presents key players in cloud computing platforms and their key offerings.

Table 1.4 Key Players in Cloud Computing Platforms (adapted from Lakshmanan (2009))

Company	Cloud computing platform	Year of launch	Key offerings
Amazon. com	AWS (Amazon Web Services)	2006	Infrastructure as a service (Storage, Computing, Message queues, Datasets, Content distribution)
Microsoft	Azure	2009	Application platform as a service (.Net, SQL data services)
Google	Google App. Engine	2008	Web Application Platform as a service (Python run time environment)
IBM	Blue Cloud	2008	Virtualized Blue cloud data center
Salesforce.com	Force.com	2008	Proprietary 4GL Web application framework as an on Demand platform

Table 1.5 compares three cloud computing platforms, Amazon, Google, and Microsoft, in terms of their capabilities to map to different development models and scenarios ("Which Cloud Platform is Right for You?," www.cumulux.com.).

1.4.1 Pricing

Pricing for cloud platforms and services is based on three key dimensions: (i) storage, (ii) bandwidth, and (iii) compute.

Storage is typically measured as average daily amount of data stored in GB over a monthly period.

Table 1.5 Cloud Computing Platforms and Different Scenarios (adapted from "Which Cloud Platform is Right for You?," www.cumulux.com.)

(1) Scenario	On-premise application unchanged in the cloud
Characteristics	Multiple red legacy, java or .NET based application
Amazon	Threat the machine as another server in the data center and do the necessary changes to configuration
Google	Needs significant refactoring of application and data logic for existing Java application
Microsoft	If existing app is ASP.NET application, then re-factor data, otherwise refactoring effort can be quite significant depending on the complexity
(2) Scenario	Scalable Web application
Characteristics	Moderate to high Web application with a back-end store and load balancing
Amazon	Threat the machine instance as another server in the data center and do the necessary changes to configuration. But scalability and elasticity is manual configuration
Google	Use dynamically scalable features of AppEngine and scripting technologies to build rich applications
Microsoft	Build scalable Web applications using familiar .NET technologies. Scaling up/down purely driven by configuration.
(3) Scenario	Parallel processing computational application
Characteristics	Automated long running processing with little to no user interaction.
Amazon	Need to configure multiple machine instances depending on the scale needed and manage the environments.
Google	Platform has minimal built-in support for building compute heavy applications. Certain application scenarios, such as image manipulation, are easier to develop with built-in platform features.
Microsoft	With worker roles and storage features like Queues and blobs, it is easy to build a compute heavy application that can be managed and controlled for scalability and elasticity.
(4) Scenario	Application in the cloud interacts with on-premise data
Characteristics	Cloud based applications interacting with on-premise apps for managing transactions of data
Amazon	Applications in EC2 server cloud can easily be configured to interact with applications running on premise.
Google	No support from the platform to enable this scenario. Possible through each app using intermediary store to communicate.
Microsoft	From features like Service Bus to Sync platform components it is possible to build compelling integration between the two environments.
(5) Scenario	Application in the cloud interacts with on-premise application
Characteristics	On-premise applications
Amazon	Applications in EC2 server cloud can easily be configured to interact with applications running on premise.
Google	No support from the platform to enable this scenario. Possible through each app using intermediary store to communicate.
Microsoft	From features like Service Bus to Sync platform components it is possible to build compelling integration between the two environments.

Bandwidth is measured by calculating the total amount of data transferred in and out of platform service through transaction and batch processing. Generally, data transfer between services within the same platform is free in many platforms.

Compute is measured as the time units needed to run an instance, or application, or machine to servicing requests. Table 6 compares pricing for three major cloud computing platforms.

In summary, by analyzing the cost of cloud computing, depending on the application characteristics the cost of deploying an application could vary based on the selected platform. From Table 1.6, it seems that the unit pricing for three major platforms is quite similar. Besides unit pricing, it is important to translate it into monthly application development, deployments and maintenance costs.

Table 1.6 Pricing comparison for major cloud computing platforms (adapted from "Which Cloud Platform is Right for You?," www.cumulux.com.)

Resource	UNIT	Amazon	Google	Microsoft
Stored data	GB per month	$0.10	$0.15	$0.15
Storage transaction	Per 10 K requests	$0.10		$0.10
Outgoing bandwidth	GB	$0.10 – $0.17	$0.12	$0.15
Incoming bandwidth	GB	$0.10	$0.10	$0.10
Compute time	Instance Hours	$0.10 – $1.20	$0.10	$0.12

1.4.2 Cloud Computing Components and Their Vendors

The main elements comprising cloud computing platforms include computer hardware, storage, infrastructure, computer software, operating systems, and platform virtualization. The leading vendors providing cloud computing components are shown in Table 1.7 ("Cloud Computing," Wikipedia, http://en.wikipedia.org/wiki/Cloud_computing).

Table 1.7 The leading vendors of cloud computing components

Cloud computing components	Vendors
Computer hardware	Dell, HP, IBM, Sun
Storage	Sun, EMC, IBM
Infrastructure	Cisco, Juniper Networks, Brocade Communication
Computer software	3tera. Eucalyptus. G-Eclipse. Hadoop
Operating systems	Solaris, AIX, Linux (Red Hat, Ubuntu)
Platform virtualization	Citrix, VMWare, IBM, Xen, Linux KVM, Microsoft, Sun xVM

1.5 Example of Web Application Deployment

In this section we present an example how the combination of virtualization and on of self service facilitate application deployment (Sun Microsystems, 2009). In this example we consider a two-tier Web application deployment using cloud, as illustrated in Fig. 1.8.

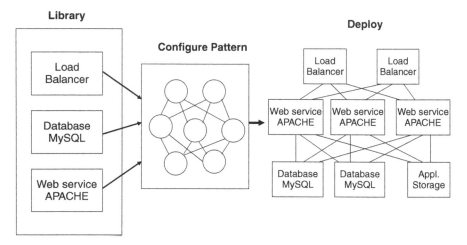

Fig. 1.8 Example of the deployment of an application into a two-tier Web server architecture using cloud computing (adapted from Sun Microsystems (2009))

The following steps comprise the deployment of the application:

- The developer selects a load balancer, Web server, and database server appliances from a library of preconfigured virtual machine images.
- The developer configures each component to make a custom image. The load balancer is configured, the Web server is populated with its static content by uploading it to the storage cloud, and the database server appliances are populated with dynamic content for the site.
- The developer than layers custom code into the new architecture, in this way making the components meet specific application requirements.
- The developer chooses a pattern that takes the images for each layer and deploys them, handling networking, security, and scalability issues.

The secure, high-availability Web application is up and running. When the application needs to be updated, the virtual machine images can be updated, copied across the development chain, and the entire infrastructure can be redeployed.

In this example, a standard set of components can be used to quickly deploy an application. With this model, enterprise business needs can be met quickly, without the need for the time-consuming, manual purchase, installation, cabling, and configuration of servers, storage, and network infrastructure.

Small and medium enterprises were the early adopters to cloud computing. However, there are recently a number of examples of cloud computing adoptions in the large enterprises. Table 1.8 illustrates three examples of cloud computing use in the large enterprises (Lakshmanan, 2009).

Table 1.8 Cloud computing examples in large enterprises

Enterprise	Scenario	Usage	Benefits
Eli Lilly	R&D High Performance Computing	Amazon server and storage cluster for drug discovery analysis and modeling.	Quick deployment time at a lower cost.
New York Times	Data Conversion	Conversion of archival articles (3 million) into new data formats using Amazon elastic compute services.	Rapid provisioning and higher elasticity on the infrastructure resources.
Pitney Bowes	B2B Application	Hosted model mail printing application for clients. Uses MS Azure.net and SQL services for the hosted model option (2009 Go live).	Flexibility at a lower cost and new business opportunity.

1.6 Cloud Computing Challenges

In summary, the new paradigm of cloud computing provides a number of benefits and advantages over the previous computing paradigms and many organizations are adopting it. However, there are still a number of challenges, which are currently addressed by researchers and practitioners in the field (Leavitt, 2009). They are briefly presented below.

1.6.1 Performance

The major issue in performance can be for some intensive transaction-oriented and other data-intensive applications, in which cloud computing may lack adequate performance. Also, users who are at a long distance from cloud providers may experience high latency and delays.

1.6.2 Security and Privacy

Companies are still concerned about security when using cloud computing. Customers are worried about the vulnerability to attacks, when information and

critical IT resources are outside the firewall. The solution for security assumes that
that cloud computing providers follow standard security practices, as described in
Section 1.3.2.

1.6.3 Control

Some IT departments are concerned because cloud computing providers have a
full control of the platforms. Cloud computing providers typically do not design
platforms for specific companies and their business practices.

1.6.4 Bandwidth Costs

With cloud computing, companies can save money on hardware and software; however they could incur higher network bandwidth charges. Bandwidth cost may be
low for smaller Internet-based applications, which are not data intensive, but could
significantly grow for data-intensive applications.

1.6.5 Reliability

Cloud computing still does not always offer round-the-clock reliability. There were
cases where cloud computing services suffered a few-hours outages.

In the future, we can expect more cloud computing providers, richer services,
established standards, and best practices.

In the research arena, HP Labs, Intel, and Yahoo have launched the distributed
Cloud Research Test Bad, with facilities in Asia, Europe, and North America, with
the objective to develop innovations including cloud computing specific chips. IBM
has launched the Research Computing Cloud, which is an on-demand, globally
accessible set of computing resources that support business processes.

1.7 Cloud Computing in the Future

In summary, cloud computing is definitely a type of computing
paradigm/architecture that will remain for a long time to come. In the near
future, cloud computing can emerge in various directions. One possible scenario
for the future is that an enterprise may use a distributed hybrid cloud as illustrated
in Fig. 1.9.

According to this scenario, the enterprise will use the core applications on its
private cloud, while some other applications will be distributed on several private
clouds, which are optimized for specific applications.

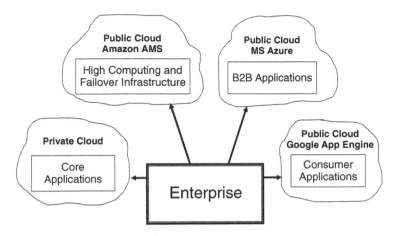

Fig. 1.9 Distributed hybrid cloud architecture (adapted from Lakshmanan (2009))

References

Chappell, D. (August 2008). *A short introduction to cloud platforms: An enterprise-oriented view.* San Francisco, CA: Chappel and Associates.

Grossman, R. L. (March/April 2009). The case for cloud computing. *IEEE ITPro*, 23–27.

Hutchinson, C., & Ward, J. (March/April 2009). Navigation the next-generation application architecture. *IEEE ITPro*, 18–22.

Jens, F. (September 2008). *Defining cloud services and cloud computing.* http://blogs.idc.com/ie/?p=190.

Jones, M. T. *Cloud computing with linux.* www.ibm.com/developerworks/linux/library/l-cloud-computing.

Lakshmanan, G. (April 2009). *Cloud computing – Relevance to enterprise.* Infosys White Paper.

Leavitt, N. (January 2009). Is cloud computing really ready for prime time? *IEEE Computer*, 15–20.

Sun Microsystems (June 2009). *Introduction to cloud computing architecture.* White Paper, Sun Microsystems.

Voas, J., & Zhang, J. (March/April 2009). Cloud computing: New wine or just a new bottle? *IEEE ITPro*, 15–17.

Vouk, M. A. (June 2008). Cloud computing – Issues, research and implementations. *Proceedings of the ITI 30th International Conference on Information Technology Interfaces, Cavtat, Croatia*, 31–40.

Wang, L., Tao, J., & Kunze, M. (2008). Scientific cloud computing: Early definition and experience. *Proceedings of the 10th IEEE International Conference on High Performance Computing and Communications, Austin, TX*, 825–830.

Chapter 2
Cloud Computing Technologies and Applications

Jinzy Zhu

2.1 Cloud Computing: IT as a Service

In a nutshell, the existing Internet provides to us content in the forms of videos, emails and information served up in web pages. With Cloud Computing, the next generation of Internet will allow us to "buy" IT services from a web portal, drastic expanding the types of merchandise available beyond those on e-commerce sites such as eBay and Taobao. We would be able to rent from a virtual storefront the basic necessities to build a virtual data center: such as CPU, memory, storage, and add on top of that the middleware necessary: web application servers, databases, enterprise server bus, etc. as the platform(s) to support the applications we would like to either rent from an Independent Software Vendor (ISV) or develop ourselves. Together this is what we call as "IT as a Service," or ITaaS, bundled to us the end users as a virtual data center.

Within ITaaS, there are three layers starting with Infrastructure as a Service, or IaaS, comprised of the physical assets we can see and touch: servers, storage, and networking switches. At the IaaS level, what cloud computing service provider can offer is basic computing and storage capability, such as the cloud computing center founded by IBM in Wuxi Software Park and Amazon EC2. Taking computing power provision as an example, the basic unit provided is the server, including CPU, memory, storage, operating system and system monitoring software.

In order to allow users to customize their own server environment, server template technology is resorted to, which means binding certain server configuration and the operating system and software together, and providing customized functions as required at the same time.

Using virtualization technology, we could provide as little as 0.1 CPU in a virtual machine to the end user, therefore drastically increasing the utilization potential of a physical server to multiple users.

J. Zhu (✉)
IBM Cloud Computing Center, China
e-mail: jinzyzhu@cn.ibm.com

B. Furht, A. Escalante (eds.), *Handbook of Cloud Computing*,
DOI 10.1007/978-1-4419-6524-0_2, © Springer Science+Business Media, LLC 2010

With virtualization increasing the number of machines to manage, service provision becomes crucial since it directly affects service management and the IaaS maintenance and operation efficiency. Automation, the next core technology, can make resources available for users through self-service without getting the service providers involved. A stable and powerful automation management program can reduce the marginal cost to zero, which in turn can promote the scale effect of cloud computing.

On the basis of automation, dynamic orchestration of resources can be realized. Dynamic orchestration of resources aims to meet the requirements of service level. For example, IaaS platform will add new servers or storage spaces for users automatically according to the CPU utilization of the server, so as to fulfill the terms of service level made with users beforehand. The intelligence and reliability of dynamic orchestration of resources technology is a key point here. Additionally, virtualization is another key technology. It can maximize resource utilization efficiency and reduce cost of IaaS platform and user usage by promoting physical resource sharing. The dynamic migration function of virtualization technology can dramatically improve the service availability and this is attractive for many users.

The next layer within ITaaS is Platform as a Service, or PaaS. At the PaaS level, what the service providers offer is packaged IT capability, or some logical resources, such as databases, file systems, and application operating environment. Currently, actual cases in the industry include Rational Developer Cloud of IBM, Azure of Microsoft and AppEngine of Google. At this level, two core technologies are involved. The first is software development, testing and running based on cloud. PaaS service is software developer-oriented. It used to be a huge difficulty for developers to write programs via network in a distributed computing environment, and now due to the improvement of network bandwidth, two technologies can solve this problem: the first is online development tools. Developers can directly complete remote development and application through browser and remote console (development tools run in the console) technologies without local installation of development tools. Another is integration technology of local development tools and cloud computing, which means to deploy the developed application directly into cloud computing environment through local development tools. The second core technology is large-scale distributed application operating environment. It refers to scalable application middleware, database and file system built with a large amount of servers. This application operating environment enables application to make full use of abundant computing and storage resource in cloud computing center to achieve full extension, go beyond the resource limitation of single physical hardware, and meet the access requirements of millions of Internet users.

The top of the ITaaS is what most non-IT users will see and consume: Software as a Service (SaaS). At the SaaS level, service providers offer consumer or industrial applications directly to individual users and enterprise users. At this level, the following technologies are involved: Web 2.0, Mashup, SOA and multi-tenancy.

The development of AJAX technology of Web 2.0 makes Web application easier to use, and brings user experience of desktop application to Web users, which in turn make people adapt to the transfer from desktop application to Web application easily. Mashup technology provide a capability of assembling contents on Web, which can allow users to customize websites freely and aggregate contents from different websites, and enables developers to build application quickly.

Similarly, SOA provides combination and integration function as well, but it provides the function in the background of Web. Multi-tenancy is a technology that supports multi tenancies and customers in the same operating environment. It can significantly reduce resource consumptions and cost for every customer.

The following Table 2.1 shows the different technologies used in different cloud computing service types.

Table 2.1 IaaS, PaaS and SaaS

Service type	IaaS	PaaS	SaaS
Service category	VM Rental, Online Storage	Online Operating Environment, Online Database, Online Message Queue	Application and Software Rental
Service Customization	Server Template	Logic Resource Template	Application Template
Service Provisioning	Automation	Automation	Automation
Service accessing and Using	Remote Console, Web 2.0	Online Development and Debugging, Integration of Offline Development Tools and Cloud	Web 2.0
Service monitoring	Physical Resource Monitoring	Logic Resource Monitoring	Application Monitoring
Service level management	Dynamic Orchestration of Physical Resources	Dynamic Orchestration of Logic Resources	Dynamic Orchestration of Application
Service resource optimization	Network Virtualization, Server Virtualization, Storage Virtualization	Large-scale Distributed File System, Database, Middleware etc	Multi-tenancy
Service measurement	Physical Resource Metering	Logic Resource Usage Metering	Business Resource Usage Metering
Service integration and combination	Load Balance	SOA	SOA, Mashup
Service security	Storage Encryption and Isolation, VM Isolation, VLAN, SSL/SSH	Data Isolation, Operating Environment Isolation, SSL	Data Isolation, Operating Environment Isolation, SSL, Web Authentication and Authorization

Transform any IT capability into a service may be an appealing idea, but to realize it, integration of the IT stack needs to happen. To sum up, key technologies used in cloud computing are: automation, virtualization, dynamic orchestration, online development, large-scale distributed application operating environment, Web 2.0, Mashup, SOA and multi-tenancy etc. Most of these technologies have matured in recent years to enable the emergence of Cloud Computing in real applications.

2.2 Cloud Computing Security

One of the biggest user concerns about Cloud Computing is its security, as naturally with any emerging Internet technology. In the enterprise data centers and Internet Data Centers (IDC), service providers offer racks and networks only, and the remaining devices have to be prepared by users themselves, including servers, firewalls, software, storage devices etc. While a complex task for the end user, he does have a clear overview of the architecture and the system, thus placing the design of data security under his control. Some users use physical isolation (such as iron cages) to protect their servers. Under cloud computing, the backend resource and management architecture of the service is invisible for users (and thus the word "Cloud" to describe an entity far removed from our physical reach). Without physical control and access, the users would naturally question the security of the system.

A comparable analogy to data security in a Cloud is in financial institutions where a customer deposits his cash bills into an account with a bank and thus no longer have a physical asset in his possession. He will rely on the technology and financial integrity of the bank to protect his now virtual asset. Similarly we'll expect to see a progression in the acceptance of placing data in physical locations out of our reach but with a trusted provider.

To establish that trust with the end users of Cloud, the architects of Cloud computing solutions do indeed designed rationally to protect data security among end users, and between end users and service providers.

From the point of view of the technology, the security of user data can be reflected in the following rules of implementation:

1. The privacy of user storage data. User storage data cannot be viewed or changed by other people (including the operator).
2. The user data privacy at runtime. User data cannot be viewed or changed by other people at runtime (loaded to system memory).
3. The privacy when transferring user data through network. It includes the security of transferring data in cloud computing center intranet and internet. It cannot be viewed or changed by other people.
4. Authentication and authorization needed for users to access their data. Users can access their data through the right way and can authorize other users to access.

To ensure security, cloud computing services can use corresponding technologies shown in the Table 2.2 below:

Table 2.2 Recommendations to operators and users on cloud security

	To Other Users	To Operators
The privacy of user storage data	SAN network zoning, mapping Clean up disks after callback File system authentication	Bare device encryption, file system encryption
The privacy of user data at runtime	VM isolation, OS isolation	OS isolation
The privacy when transferring user data through network	SSL, VLAN, VPN	SSL, VPN
Authentication and authorization needed for users to access their data	Firewall, VPN authentication, OS authentication	VPN authentication, OS authentication

In addition to the technology solutions, business and legal guidelines can be employed to enforce data security, with terms and conditions to ensure user rights to financial compensation in case of breached security.

2.3 Cloud Computing Model Application Methodology

Cloud computing is a new model for providing business and IT services. The service delivery model is based on future development consideration while meeting current development requirements. The three levels of cloud computing service (IaaS, PaaS and SaaS) cover a huge range of services. Besides computing and the service delivery model of storage infrastructure, various models such as data, software application, programming model etc. can also be applicable to cloud computing. More importantly, the cloud computing model involves all aspects of enterprise transformation in its evolution, so technology architecture is only a part of it, and multi-aspect development such as organization, processes and different business models should also be under consideration. Based on standard architecture methodology with best practices of cloud computing, a Cloud Model Application Methodology can be used to guide industry customer analysis and solve potential problems and risks emerged during the evolution from current computing model to cloud computing model. This methodology can also be used to instruct the investment and decision making analysis of cloud computing model, determine the process, standard, interface and public service of IT assets deployment and management to promote business development. The diagram below shows the overall status of this methodology (Fig. 2.1).

2.3.1 Cloud Computing Strategy Planning Phase

Cloud strategy contains two steps to ensure a comprehensive analysis for the strategy problems that customers might face when applying cloud computing mode. Based on Cloud Computing Value Analysis, these two steps will analyze the model

IBM Cloud Computing Blueprint Model

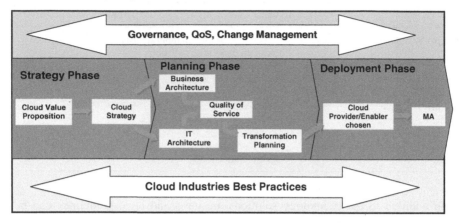

Fig. 2.1 Cloud computing methodology overview

condition needed to achieve customers' target, and then will establish a strategy to function as the guideline.

(1) Cloud Computing Value Proposition

The target of this step is to analyze the specific business value and possible combination point between cloud computing mode and specific users by leveraging the analysis of cloud computing users' requirement model and considering the best practices of cloud computing industry. Analyze the key factors that might influence customers to apply cloud computing mode and make suggestions on the best customer application methods. In this analysis, we need to identify the main target for customer to apply cloud computing mode, and the key problems they wish to solve. Take some common targets as examples: IT management simplification, operation and maintenance cost reduction; business mode innovation; low cost out-sourcing hosting; high service quality out-sourcing hosting etc.

The analysis result will be provided to support decision-making level to make condition assessments and strategy for future development and prepare for the strategy establishment and organization of the following cloud computing.

(2) Cloud Computing Strategy Planning

This step is the most important part of strategy phase. Strategy establishment is based on the analysis result of the value step, and aims to establish the strategy documentation according to the good understanding of various conditions that customers might face when applying cloud computing mode to plan for future vision and perspective. Professional analysis made by the method above typically involves broad customer business model research, organization structure analysis and operation process identification; also, there are some non-functional requirement and limitation in the plan, such as the concern for security standard, reliability requirement and rules and regulations.

2.3.2 Cloud Computing Tactics Planning Phase

At the phase of cloud planning, it is necessary to make a detailed investigation on customer position and to analyze the problems and risks in cloud application both at present and in the future. After that, concrete approaches and plans can be drawn to ensure that customers can use cloud computing successfully to reach their business goals. This phase includes some practicable planning steps in multiple orders listed as follows,

(1) Business Architecture Development
    While capturing the organizational structures of enterprises, the business models also get the information on business process support. As various business processes and relative networks in enterprise architecture are being set down one after another, gains and losses brought by relative paths in the business development process will also come into people's understanding. We categorize these to business interests and possible risks brought by cloud computing application from a business perspective.

(2) IT Architecture Development
    It is necessary to identify the major applications needed to support enterprises business processes and the key technologies needed to support enterprise applications and data systems. Besides, cloud computing maturity models should be introduced and the analysis of technological reference models should be made, so as to provide help, advices and strategy guide for the design and realization of cloud computing mode in the enterprise architecture.

(3) Requirements on Quality of Service Development
    Compared with other computing modes, the most distinguishing feature of cloud computing mode is that the requirements on quality of service (also called non-functional needs) should be rigorously defined beforehand, for example, the performance, reliability, security, disaster recovery, etc. This requirement is a key factor in deciding whether a cloud computing mode application is successful or not and whether the business goal is reached; it is also an important standard in measuring the quality of cloud computing service or the competence in establishing a cloud computing center.

(4) Transformation Plan Development
    It is necessary to formulate all kinds of plans needed in the transformation from current business systems to the cloud computing modes, including the general steps, scheduling, quality guarantee, etc. Usually, an infrastructure service cloud cover different items such as infrastructure consolidation plan report, operation and maintenance management system plan, management process plan, application system transformation plan, etc.

2.3.3 Cloud Computing Deployment Phase

The deployment phase focuses mainly on the programming of both strategy realization phase and the planning phases. Two steps are emphasized in this phase:

(1) Cloud Computing Provider or Enabler Chosen

According to the past analysis and programming, customers may have to choose a cloud computing provider or an enabler. It is most important to know that the requirement on service level agreement (SLA) is still a deciding factor for providers in winning a project.

(2) Maintenance and Technical Service

As for maintenance and technical service, different levels of standards are adopted; these standards are defined by the requirement on quality of services made beforehand. Cloud computing providers or builders have to ensure the quality of services, for example, the security of customers in service operation and the reliability of services.

2.4 Cloud Computing in Development/Test

Economic crises can bring with enterprise unprecedented business challenges and more competitions for the same markets. To address these challenges, enterprises have to optimize and update their business operations. At this critical moment, only by offering agile operating systems to end users can enterprises turn the crisis into opportunities and promote better development.

Years of IT development has closely linked IT with the business systems, and operation and maintenance systems of enterprises. To a large extent, the optimization and updating of business is indeed that of the IT system, which requires enterprises to keep innovating in business system. As a result, how to develop new IT systems quickly while doing rigorous tests to provide stable and trustworthy services for customers have become the key to enterprise development. Thus, the development testing centers have become the engines of enterprises growth and how to keep the engines operating in a quick and effective way has become a major concern for enterprise CIOs.

As the importance of development centers in companies grows, there will be more and more projects, equipments and staff in these centers. How to establish a smart development center has become many people's concern. As the latest IT breakthrough, how will cloud computing help to transform development test centers and bring competitive advantages to enterprises? We want to illustrate this problem through the following case:

Director A is the manager of an information center and he is now in charge of all development projects. Recently, he is thinking about how to best optimize his development and testing environment. After investigation, he concludes that the requirements of the new test center are as follows:

- Reducing the investment on hardware
- Providing environment quickly for new development testing projects
- Reusing equipments
- Ensuring project information security

Based on A's requirement analysis, he can use Cloud Computing solutions to establish a cloud-computing-based test development center for his company.

- Reducing the cost

In traditional test development systems, companies would set up an environment for each test and development project. Different test systems may have different functions, performances, or stabilities and thus software and hardware configurations will vary accordingly. However, in a cloud test development platform, all the servers, memories and networks needed in test development are pooling-managed; and through the technology of virtualization, each test or development project is provided with a logical hardware platform.

The virtual hardware platforms of multiple projects can share the same set of hardware resources, thus through integrating the development test project, the hardware investment will be greatly reduced.

- Providing environment for new projects

Cloud can automatically provide end users with IT resources, which include computing resources, operating system platforms and application software. All of these are realized through the automation module of Cloud.

Automation of computing resources: In the Cloud service interface, when end users input the computing resources (processor, storage and memory) needed according to the requirements of the application system, the Cloud platform will dynamically pick out the resources in the corresponding resource pool and prepare for the installation of the system platform.

Automation of system platforms: When the computing resources allocation is finished, the automation of system platforms will help you to install the system with the computing resources on the base of the chosen system platform (windows, Linux, AIX, etc.) dynamically and automatically. It can concurrently install operation system platforms for all computers in need and customize an operation system with customization parameters and system service for customers. Moreover, the users, networks and systems can all be set automatically.

Automation of application software: The software of enterprises would be controlled completely. The software distribution module can help you to deploy complex mission-critical applications from one center spot to multiple places quickly and effectively.

Through automation, Cloud can provide environment for new development test projects and accelerate the process of development tests.

- Reusing equipments

Cloud has provided a resource management process based on development life cycles test. The process covers many operations such as computing resource establishment, modification, release and reservation. When the test development projects are suspended or completed, Cloud Platform can make a back-up of the existing test environment and release the computing resources, thereby realizing the reuse of computing resources.

- Ensuring project information security

The cloud computing platform has provided perfect means to ensure the security and isolation of each project. There are two ways for users to access the system: accessing the Web management interface or accessing the project virtual machine. To access a Web interface, one needs a user ID and a password. To control a virtual machine access, the following methods can be adopted:

- User authentication is conducted through the VPN equipment in the external interface of the system.
- Each project has one and only a Vlan, and the virtual machine of each project is located inside the Vlan. The switches and the hypervisors in the hosts can guarantee the isolation of the Vlan.
- The isolation of virtual machine is guaranteed by virtual engine itself.
- Besides, user authentication of the operation systems can also protect user information.

Vlan is created dynamically along with the establishment of the project. Unicast or broadcast messages can be sent among project virtual machines or between the virtual machine and the workstation of the project members. Virtual machines of different projects are isolated from each other, thereby guaranteeing the security of project data. A user can get involved in several projects and meanwhile visit several virtual machines of different projects.

The new generation of intelligent development test platforms needs the support of intelligent IT infrastructure platforms. By establishing intelligent development test platforms through cloud computing, a new IT resource supply mode can be formed. Under this mode, the test development center can automatically manage and dynamically distribute, deploy, configure, reconfigure and recycle IT resources based on the requirements of different projects; besides, it can also install software and application systems automatically. When projects are over, the test development center can recycle the resources automatically, thereby making the best use of the computing capabilities.

2.5 Cloud-Based High Performance Computing Clusters

In the development history of information science from the last half a century, High Performance Computing (HPC) has always been a leading technology at the time. It has become a major tool for future innovations of both theoretical and research science. As new cross-disciplines combining traditional subjects and HPC emerge in the areas of computational chemistry, computational physics and bioinformatics, computing technology need to take a leap forward as well to meet the demands of these new research topics.

With the current financial crisis, how to provide higher computing performance with less resource input has become a big challenge for the HPC centers. In the construction of a new generation of computing center with high performance, we should not only pay attention to the choice of software and hardware, but also take fully account of the center operation, utilization efficiency, technological innovation cooperation and other factors. The rationality of the general framework and the effectiveness of resource management should also be fully considered. Only by doing these can the center gain long-term high-performance capacity in computing research and supply.

In other words, the new generation of high-performance computing center does not only provide traditional high-performance computing, nor it is only a high-performance equipment solution. The management of resources, users and virtualization, the dynamic resource generation and recycling should also be taken into account. In this way, the high-performance computing based on cloud computing technology is born.

The cloud computing-based high-performance computing center aims to solve the following problems:

- High-performance computing platform generated dynamically
- Virtualized computing resources
- High-performance computer management technology combined with tradition ones
- High-performance computing platform generated dynamically

In traditional high-performance computing environment, physical equipments are configured to meet the demands of customers; for example, Beowulf Linux and WCCS Architecture are chosen to satisfy customers' requirements on computing resources. All of the operation systems and parallel environment are set beforehand, and cluster management software is used to manage the computing environment. However, as high-performance computing develops, there are more and more end users and application software; thus, the requirements on the computing platform become more diverse. Different end users and application software may require different operation systems and parallel environment. High-performance computing requires a new way of resource supply, in which the platform should be dynamically generated according to the needs of every end user and application software; the platform can be open, including Linux, Windows or UNIX.

- Virtualized computing resources

Since few virtualized architecture are used in traditional high-performance computing, this kind of platform cannot manage virtualized resources. However, as high-performance computing develops, in many cases we need to attain more virtualized resources through virtualization, for example, the development and debugging of parallel software, and the support for more customer application etc.

In the cloud computing-based high-performance computing center, the virtualization of physical resources can be realized through the Cloud platform; moreover, virtualized resources can be used to establish high-performance computing platform and generate high-performance computing environment whose scale is larger than that of the actual physical resource so as to meet the requirements of customers.

• Combination with traditional management technology

The cloud computing-based high-performance computing platform can not only manage computers though the virtualization and dynamic generation technology, but also work together with traditional cluster and operation management software in enabling users to manage the virtualized high-performance computers in a traditional way, and submit their own works.

A new IT resources provision model can be attained by the adoption of cloud computing infrastructure and high-performance computing center construction. In this model, the computing center can automatically manage and dynamically distribute, deploy, configure, reconfigure and recycle the resources. The automatic installation of software and application can be realized, too. By use of the model, the high-performance computing resources can be distributed efficiently and dynamically. When the project is finished, the computing center can automatically recycle the resources to make full use of the computing power. Taking advantage of cloud computing, the high-performance computing center can not only provide high calculating power for scientific research institutions, but also expand the service content of computing center. In other words, it can serve as a data center to support other applications and promote higher utilization efficiency of entire resources.

2.6 Use Cases of Cloud Computing

2.6.1 Case Study: Cloud as Infrastructure for an Internet Data Center (IDC)

In the 1990's, Internet portals spent huge amount of investment to attract eyeballs. Rather than profits and losses, their market valuation were based on the number of unique "hits" or visitors. This strategy proved to work out well as these portals begin to offer advertisement opportunities targeting their installed user base, as well as new paid services to the end user, thereby increasing revenue per capita in a theoretically infinite growth curve.

Similarly Internet Data Centers (IDC) have become a strategic initiative for Cloud service providers to attract users. With a critical mass of users consuming computing resources and applications, an IDC would become a portal attracting more applications and more users in a positive cycle.

The development of the next generation of IDC hinges on two key factors. The first is the growth of Internet. By the end of June 2008, for example, Internet users in China totaled 253 million and the annual growth rate is as high as 56.2%.[1] As a result, the requirement on Internet storage and traffic capacity grows, which means Internet operators have to provide more storage and servers to meet users' needs. The second is the development of mobile communication. By the end of 2008, the number of mobile phone users in China has amounted to 4 billion. The development of mobile communication drives server-based computing and storage, which enables users to access to the data and computing services needed via Internet through lightweight clients.

In the time of dramatic Internet and mobile communication expansion, how can we build new IDC with core competency? Cloud computing provides an innovative business model for data centers, and thereby can help telecom operators to promote business innovation and higher service capabilities against the backdrop of the whole business integration of fixed and mobile networks.

2.6.1.1 The Bottleneck on IDC Development

Products and services offered by a traditional IDC are highly homogenized. In almost of all the

IDC's, basic collocation services account for majority of the revenue, while value-added services add only a small part of it. For example, in one of the IDC's of a telecom operator, the hosting service claims 90% of its revenue, while value-added service takes only 10%. This makes it impossible to meet customers' requirements on load balance, disaster recovery, data flow analysis, resource utilization analysis, and etc.

The energy utilization is low, but the operation costs are very high. According to CCID research statistics,[2] the energy costs of IDC enterprises make up about 50% of their operating costs and more servers will lead to an exponential increase in the corresponding power consumption.[3] With the increase of the number of Internet users and enterprise IT transformation, IDC enterprises will have to face a sharp increase in power consumption as their businesses grow. If effective solutions are not taken immediately, the high costs will undermine the sustained development of these enterprises.

Besides, as online games and Web 2.0 sites become increasingly popular, all types of content including audio, videos, images and games will need a massive storage and relevant servers to support transmission. This will result in a steady increase in enterprises' requirements for IDC services, and higher standards on the utilization efficiency of resources in data centers as well as the service level.

---

[1] www.ibm.com/cloud

[2] blog.irvingwb.com/blog/2008/07/what-is-cloud-c.html

[3] Source: CCIDConsulting, 2008–2009 China IDC Market Research Annual Report

Under the full service operation model emerged after the restructuring of telecom operators, the market competition becomes more and more fierce. The consolidation of fixed network and mobile services imposes higher requirements on telecom IDC operators as they have to introduce new services to meet market demands in time.

2.6.1.2 Cloud Computing Provides IDC with a New Infrastructure Solution

Cloud computing provides IDC with a solution that takes into consideration of both future development strategies and the current requirement for development. Cloud computing builds up a resource service management system where physical resources is on the input, and the output is the virtual resources on right time and with the right volume and right quality. Thanks to the virtualization technology, the resources of IDC centers including servers, storage and networks are put into a huge resource pool by cloud computing. With cloud computing management platform, administrators are able to dynamically monitor, schedule and deploy all the resources in the pool and provide them for the users via network. A unified resource management platform can lead to higher efficiency of IDC operation and schedule efficiency and utilization of the resources in the center and lower management complexity. The automatic resource deployment and software installation help to guarantee the timely introduction of new services and can lower the time-to-market. Customers can use the resources in data centers by renting based on their business needs. Besides, as required by business development needs, they are allowed to adjust the resources that they rent timely, and pay fees according to resource usage. This kind of flexible charging mode makes IDC more appealing. The management through a unified platform is also helpful to IDC expansion. When an IDC operator needs to add resources, new resources can be added to the existing cloud computing management platform to be managed and deployed uniformly.

Cloud computing will make it an unceasing process to upgrade software and add new functions and services, which can be done through intelligent monitoring and automatic installation program instead of manual operation.

According to the Long Tail Theory, cloud computing builds infrastructures based on the scale of market head, and provides marginal management costs that are nearly zero in market tail as well as a plug-and-play technological infrastructure. It manages to meet diversified requirements with variable costs. In this way, the effect of the Long Tail is realized to keep a small-volume production of various items and by the use of innovative IT technology, and it sets up a market economy model which is open to competition and favorable to the survival of the fittest.

2.6.1.3 The Value of Cloud Computing for IDC Service Providers

First of all, based on cloud computing technology, IDC is flexible and scalable, and can realize the effect of Long Tail at a relatively low cost. The cloud computing platform is able to develop and launch new products at a low marginal cost of management. Therefore, startup costs of new business can be reduced to nearly zero, and the resources would not be limited to a single kind of products or services. So under a specified investment scope, the operators can greatly expand products lines, and meet the needs of different services through the automatic scheduling of resources, thereby making a best use of the Long Tail.

Secondly, the cloud computing dynamic infrastructure is able to deploy resources in a flexible way to meet business needs at peak times. For example, during the Olympics, the websites related to the competitions are flooded with visitors. To address this problem, the cloud computing technology would deploy other idle resources provisionally to support the requirements on resources at the peak hours. The United States Olympic Committee has applied the cloud computing technologies provided by AT&T to support competitions viewing during Olympics. Besides, SMS and telephone calls on holidays, as well as the application and inquiry days for examinations also witness the requirements for resources at the peak.

Thirdly, cloud computing improves the return on investment for IDC service providers. By improving the utilization and management efficiency of resources, cloud computing technologies can reduce computing resources, power consumption, and human resource costs. Additionally, it can lead to shorter time-to-market for a new service, thereby helping IDC service providers to occupy the market.

Cloud computing also provides an innovative charging mode. IDC service providers charge users based on the resource renting conditions and users only have to pay for what they use. This makes the payment charging more transparent and can attract more customers (Table 2.3).

2.6.1.4 The Value Brought by Cloud Computing for IDC Users

First, initial investments and operating costs can be lowered, and risks can be reduced. There is no need for IDC users to make initial investments in hardware

Table 2.3 Value comparison on co-location, physical server renting and IaaS for providers

	Co-location	Physical server renting	IaaS with cloud computing
Profit margin	Low. Intense competition	Low. Intense competition	High. Cost saving by resource sharing
Value add service	Very few	Few	Rich, such as IT service management, software renting, etc
Operation	Manual operation. Complex	Manual operation. Complex	Automatic and integrated operation. End to end request management
Response to customer request	Manual action. Slow	Manual action. Slow	Automatic process. Fast
Power consumption	Normal	Normal	Reduce power by server consolidation and sharing. Scheduled power off

and expensive software licenses. Instead, users only have to rent necessary hardware and software resources based on their actual needs, and pay according to usage conditions. In the era of enterprise informatization, more and more subject matter experts have begun to establish their own websites and information systems. Cloud computing can help these enterprises to realize informatization with relatively less investment and fewer IT professionals.

Secondly, an automatic, streamlined and unified service management platform can rapidly meet customers' increased requirements for resources, and can enable them to acquire the resources in time. In this way customers can become more responsive to market requirements and enhance business innovation.

Thirdly, IDC users are able to access more value-added services and achieve faster requirement response. Through the IDC cloud computing unified service delivery platform, the customers are allowed to put forward personalized requirements and enjoy various kinds of value-added services. And their requirements would get a quick response, too (Table 2.4).

2.6.1.5 Cloud Computing Can Make Fixed Costs Variable

An IDC can provide 24*7 hosting services for individuals and businesses. Besides traditional hosting service, these clients also need the cloud to provide more applications and services. In so doing, enterprises are able to gain absolute control on their own computing environment. Furthermore, when necessary, they can also purchase online the applications and services that are needed quickly at any time, as well as adjust the rental scale timely.

Table 2.4 Value comparison on co-location, physical server renting and IaaS for users

	Co-location	Physical server renting	IAAS using Cloud
Performance	Depend on hardware	Depend on hardware	Guaranteed performance
Price	Server investment plus bandwidth and space fee	Bandwidth and server renting fee	CPU, memory, storage, bandwidth fee. Pay per use
Availability	Depend on single hardware	Depend on single hardware	High available by hardware failover
Scalability	Manual scale out	Manual scale out	Automated scale out
System management	Manual hardware setup and configuration. Complex	Manual hardware setup and configuration. Complex	Automated OS and software installation. Remote monitoring and control. Simple
Staff	High labor cost and skill requirement	High labor cost and skill requirement	Low labor cost and skill requirement
Usability	Need on site operation	Need on site operation	All work is done through Web UI. Quick action

2.6.1.6 An IDC Cloud Example

In one example, an IDC in Europe serves industry customers in four neighboring countries, which covers sports, government, finance, automobile and the healthcare.

This IDC attaches great importance to cloud computing technology in the hope of establishing a data center that is flexible, demand-driven and responsive. It has decided to work with cloud computing technology to establish several cross-Europe cloud centers. The first five data centers are connected by virtual SAN and the latest MPLS technology. Moreover, the center complies with the ISO27001 security standard, and other security functions that are needed by the banks and government organizations, including auditing function provided by certified partners, are also realized (Fig. 2.2).

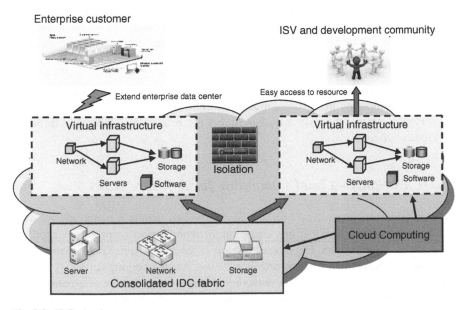

Fig. 2.2 IDC cloud

The IDC uses the main Data Center to serve customers in its sister sites. The new cloud computing center will enable this IDC to pay for fixed or usage-based changeable services according to credit card bill. In the future, the management scope of this hosting center expand to even more data centers in Europe.

2.6.1.7 The Influence of Cloud Computing in 3G Era

Ever since 3G services are launched by the major communication operators, the simple voice and information service can no longer meet the growing requirements of users. The 3G data services have become the focus of competition among operators. Many operators have introduced some specialized services. And with the growth of 3G clients and the expansion and improvement of 3G networks, operators have to provide more diversified 3G services to survive in the fierce market competition. Cloud can be used as a platform to provide such value added services.

In this 3G era, mobile TV, mobile securities and data backup will all become critical businesses. Huge amounts of videos, images, and documents are to be stored in data centers so that users can download and view them at any time, and they can promote interaction. Cloud computing can effectively support this kind of business requirements, and get maximal storage with limited resources. Besides, it can also search and provide the resources that are needed to users promptly to meet their needs.

After the restructuring of operators, the businesses of leading service providers will all cover fixed network and mobile service, and they may have to face up to fierce competition in 3G market. Cloud computing can support unified monitoring and dynamic deployment of resources. So, during the business consolidation of the operators, the cloud computing platform can deploy necessary resources in time to support business development, and respond quickly to market requirements to help operators to gain larger market share.

The 3G-enabled high bandwidth makes it easier and quicker to surf Internet through mobile phones and it has become a critical application of 3G technologies. Cloud computing makes it compatible among different equipments, software and networks, so that the customers can access the resources in the cloud through any kinds of clients.

2.6.2 Case Study – Cloud Computing for Software Parks

The traditional manufacturing industry has helped to maintain economic growth in previous generations, but it has also brought along a host of problems such as labor market deterioration, huge consumption of energy resources, environmental pollution, and ever-more drive towards lower cost. As an emerging economy begins its social transformation, software outsourcing has gained an edge compared with traditional manufacturing industry: on one hand, it can attract and develop top-level talent to enhance the technical level and competitive power of a nation; on the other hand, it can also prompt the smooth structural transformation to a sustainable and green service industry, thereby ensuring continuous prosperity and endurance even in difficult times.

As such, software outsourcing has become a main business line for many emerging economies to ramp up their service economy, based on economies of scale and affordable costs. To reach this goal, software firms in these emerging economies need to conform their products and services to international standards and absorb experiences from developed nations to enhance the quality of their outsourcing services. More importantly, good policy support from the government and necessary infrastructures are critical components in the durability of these software outsourcing firms.

The IT infrastructure is surely indispensable for software outsourcing and software businesses. To ensure the success of software outsourcing, there are two prerequisites: a certification standard of software management which is of international level (such as CMM Level 5), and an advanced software designing, programming and testing pipeline, namely the software development platform of data center. The traditional data center only put together all the hardware devices of the enterprise, leading to the monopolization of some devices by a certain project or business unit. This would create huge disparity within the system and can't guarantee the quality of applications and development. Besides, it would result in cost increase and unnecessary spending and in the long term undermine the enterprise's competitive power in the international market of software outsourcing. Furthermore, when a new project is put on the agenda, it would take a long time to prepare for and address the bottleneck of the project caused by the traditional IT equipments.

To pull the software enterprises out of this dilemma, IBM firstly developed a brand-new management mode for software developing environment: the management and development platform of "cloud computing". The platform was constructed with the aid of the accumulated experience of IBM itself in the field of software outsourcing service and data center management. The valuable experience from the long-term cooperation with other software outsourcing powers is also taken into consideration. This platform is a new generation of data center management platform. Compared with traditional data center, it has outstanding technical advantages.

Below is the schematic diagram of the relationship between Cloud Computing platform and software outsourcing ecosystems (Fig. 2.3):

Firstly, the platform can directly serve as a data service center for software outsourcing companies in the Software Park and neighboring enterprises. As soon as a software outsourcing order is accepted, the company can turn to the management and development platform of "cloud computing" to look for IT resources suitable for use, the process of which is as simple and convenient as booking hotel via Internet. Besides, by relying on IBM's advanced technology, the cloud computing platform is able to promote unified administrative standard to ensure the confidentiality, security, stability and expandability of the platform. That is to

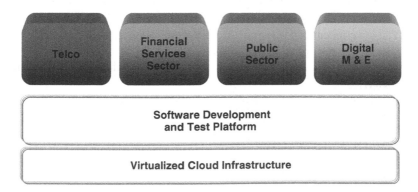

Fig. 2.3 Cloud computing platform and software outsourcing ecosystems

say, thanks to its brand effect, the platform developed by the software demonstration plot is up to international advanced level, and could thereby enhance the service level of software outsourcing in the entire park. The final aim is to measure up to international standards and to meet the needs of international and Chinese enterprises. Meanwhile, a platform of unified standard can lower IT maintenance costs and raise the response speed for requirements, making possible the sustainable development of the Software Park. Lastly, the management and development platform of cloud computing can directly support all kinds of applications and provide enterprise users with various services including outsourcing and commercial services as well as services related to academic and scientific researches.

The following are the benefits brought to the outsourcing services companies and outsourcing demonstration plot of Wuxi government by the management and development platform of cloud computing:

(1) For outsourcing service companies which apply cloud computing platform:

- An advanced platform with unified standard is provided and the quality is guaranteed;
- IT management becomes easier and the costs of developing products is greatly lowered;
- Response speed for business demand is enhanced and expandability is ensured;
- Existing applications and newly-emerged data-intensive applications are supported;
- Miscellaneous functions for expediting the speed of innovation is also provided for outsourcing service companies, colleges and universities and research institutes.

(2) Below are the advantages brought to the outsourcing demonstration plot of Wuxi government through the application of cloud computing platform:

- The government can transform from a supervision mode to a service mode which is in favor of attracting investments;
- It is conducive to environmental protection and the build-up of a harmonious society;
- It can support the development of innovative enterprises and venture companies.

Detailed information about the major functions and technical architectures of the management and development platform of Cloud Computing is introduced as below:

2.6.2.1 Cloud Computing Architecture

The management and development platform of Cloud Computing is mainly composed of two functional sub-platforms: outsourcing software research and development platform and operation management platform.

- Outsourcing software research and development platform: an end-to-end software development platform is provided for the outsourcing service companies in the park. In terms of functions, the platform generally covers the entire software developing lifecycle including requirement, designing, developing and testing of the software. It helps the outsourcing service companies in establishing a software developing procedure that is effective and operable.
- Operation management platform: according to the outsourcing service company's actual demand in building the research and development platform, as well as the practical situation of the software and hardware resources distribution in data center, the platform provides automatic provisioning services on demand of software and hardware resources. Also, management on resources distribution is based on different processes, posts and roles and resource utilization report will also be provided.

Through the cooperative effect of the two platforms mentioned above, the management and development platform of "cloud computing" could fully exert its advantage. The construction of outsourcing software research and development platform can be customized according to different project needs (e.g., games development platform, e-business development platform, etc), which can show the best practices of IBM's outsourcing software development services. And the operation management platform can provide supporting functions such as management on the prior platform, as well as operation and maintenance, and rapid configuration. It is also significant in that it can reduce the workload and costs of operation and management. Unlike the handmade software research and development platform, it is both time-saving and labor-saving, and it is not that easy to make mistakes in it.

2.6.2.2 Outsourcing Software Research and Development Platform

The outsourcing software research and development at the enterprise level have to put an emphasis on the cooperation and speed of software development. It manages to combine the software implantation with verification, so as to ensure the high quality of software and shorten the period of development. The program is targeted at and suitable for different types of outsourcing research and development companies with a demand for code development cooperation and document management. The detailed designing of the program varies according to different enterprise needs (Fig. 2.4).

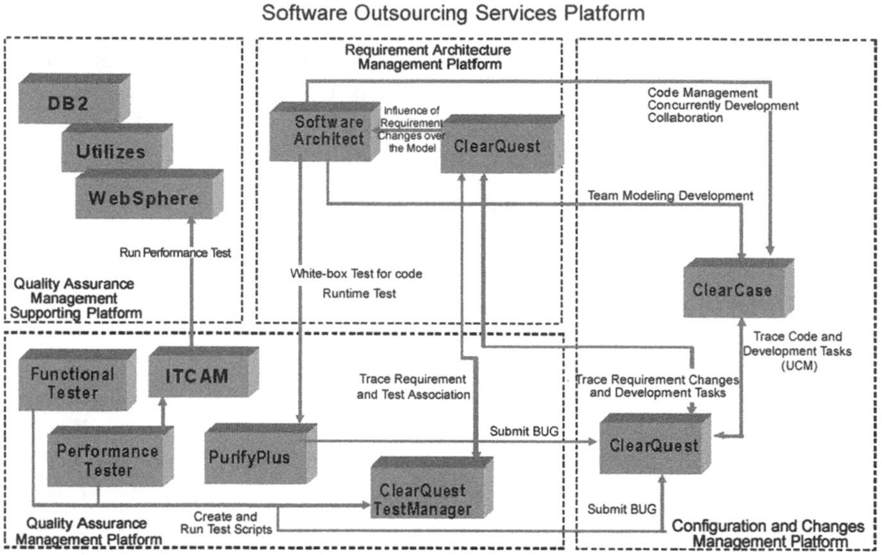

Software Outsourcing Services Platform

Fig. 2.4 Software outsourcing services platform

As can be seen in the chart, the primary construction of the outsourcing software research and development platform consists of the construction of 4 sub-platforms:

- Requirement architecture management platform
- Quality assurance management platform
- Quality assurance management supporting platform
- Configuration and changes management platform

The integrated construction and operation of these four sub-platforms cover the entire developing lifecycle of requirement, designing, developing and testing of the software. They are customer-oriented and are featured by high quality, and good awareness of quality prevention. With the help of these four sub-platforms, the outsourcing service companies can manage to establish a software development process with high efficiency and operability.

2.6.3 Case Study – an Enterprise with Multiple Data Centers

Along with China's rapid economic growth, the business of one state-owned enterprise is also gearing up for fast expansion. Correspondingly, the group has increasingly higher demand for the supporting IT environment. How can the group achieve maximum return on its IT investment? For the IT department, on one hand is repetitive and time-consuming work of system operation and management; while there is an increasingly higher demand from the managers to support the company's business and raise its competitive power and promote business transformation. Faced with this problem, this enterprise is now searching for solutions in Cloud Computing.

Enterprise Resources Plan (ERP) plays an important role as supporting the entire business in the company. The existing EAR system is not able to apply automatic technology. Repeated, manual work accounts for the majority of the system maintenance operation, which leads to lower efficiency and higher pressure on the IT system maintenance operation. Meanwhile, on the technical level, it lacks of technology platform to perform the distribution, deployment, as well as state control and recycle of system resources. As a result, the corresponding information resources management is performed through traditional manual work, which is in contradiction with the entire information strategy of the company. The specifics are listed as below:

- The contradiction between the increasing IT resources and limited human resources
- The contradiction between automatic technology and traditional manual work
- The effectiveness and persistence of resources information (including configuration information)

The company has invested a lot in information technology. It has not only constructed the ERP system for the management and control of enterprise production, but also upgraded the platform, updated host computer and improved IT management in infrastructure. In a word, the SAP system is of great significance in the IT system of Sinochem Group.

The implementation of Cloud Computing platform has helped to solve the problems faced by the IT department in this company.

2.6.3.1 Overall Design of the Cloud Computing Platform in an Enterprise

The Cloud Computing Platform mainly is related to three discrete environments of the company's data centers: the training, development/test and the disaster recovery environment. These systems involved in cloud computing are respectively located in Data Center A, Datacenter B and the disaster center in Data Center C. It shows the benefits of Cloud Computing virtualization crossing physical sites. See the following Fig. 2.5:

Combined with the technical characteristics of the Cloud Computing platform and the application characteristics of the ERP system in the company, the construction project has provided the following functions:

- The installation and deployment of the five production systems of ERP
- The automatic deployment of hardware: logical partition and distribution of hardware resources.
- The installation and recovery of the centralized AIX operating system
- Display of system resource usage: CPU/memory/disk usage.

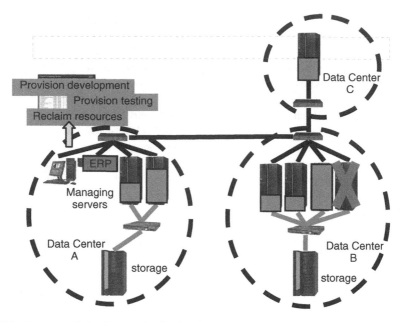

Fig. 2.5 Coverage of cloud computing in sinochem group

Fig. 2.6 SaaS cloud

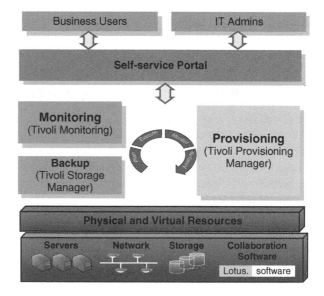

2.6.4 Case Study: Cloud Computing Supporting SaaS

By adopting cloud computing solutions, a telco can address the IT challenges faced by SMEs. Thanks to the services provided by the Blue Cloud system, VNTT has provided the customers with IBM Lotus Foundation and WebSphere Portal Express business OA service based on Redhat, CentOS and Windows platform. Besides, VNTT also provides customers with email services, file sharing and Web server that are always ready for use. For better internal and external communication, these enterprises need only one portal to rent the portal server based on IBM WebSphere Portal.

By applying Cloud Computing as the underlying infrastructure, a telecommunications company can provide its customers with a larger scale of IT services, including infrastructure hosting, collaborative platform, applications, process and information service; meanwhile, it can also ensure data security, convenience of access and the easy management of the environment. In this instance, Cloud will provide a strong technical infrastructure support as well as an effective combination with business model innovation (Fig. 2.6).

2.7 Conclusion

With Cloud Computing as a new way to consume IT services, we can be much more flexible and productive in utilizing dynamically allocated resources to create and to operate. Cloud will continue to evolve as the foundation for the future Internet where we will be interconnected in a web of content and services.

Chapter 3
Key Enabling Technologies for Virtual Private Clouds

Jeffrey M. Nick, David Cohen, and Burton S. Kaliski Jr.

Abstract The concept of a virtual private cloud (VPC) has emerged recently as a way of managing information technology resources so that they appear to be operated for a single organization from a logical point of view, but may be built from underlying physical resources that belong to the organization, an external service provider, or a combination of both. Several technologies are essential to the effective implementation of a VPC. *Virtual data centers* provide the insulation that sets one organization's virtual resources apart from those of other organizations and from the underlying physical infrastructure. *Virtual applications collect* those resources into separately manageable units. *Policy-based deployment* and *policy compliance* offer a means of control and verification of the operation of the virtual applications across the virtual data centers. Finally, *service management integration* bridges across the underlying resources to give an overall, logical and actionable view. These key technologies enable cloud providers to offer organizations the cost and efficiency benefits of cloud computing as well as the operational autonomy and flexibility to which they have been accustomed.

3.1 Introduction

It is becoming relatively commonplace for organizations to outsource some or all of their IT operations to an external "cloud" service provider that offers specialized services over the Internet at competitive prices. This model promises improved total

J.M. Nick (✉)
Office of the CTO, EMC Corporation, Hopkinton, MA, USA
e-mail: jeff.nick@emc.com

D. Cohen
Cloud Infrastructure Group, EMC Corporation, Cambridge, MA, USA
e-mail: david.cohen@emc.com

B.S. Kaliski Jr.
Office of the CTO, EMC Corporation, Hopkinton, MA, USA
e-mail: burt.kaliski@emc.com

B. Furht, A. Escalante (eds.), *Handbook of Cloud Computing*,
DOI 10.1007/978-1-4419-6524-0_3, © Springer Science+Business Media, LLC 2010

cost of ownership (TCO) through the leverage of large-scale commodity resources that are dynamically allocated and shared across many customers. The problem with this model to date is that organizations have had to give up control of the IT resources and functions being outsourced. They may gain the cost efficiencies of services offered by the external provider, but they lose the autonomy and flexibility of managing the outsourced IT infrastructure in a manner consistent with the way they manage their internal IT operations.

The concept of a *virtual private cloud (VPC)* has emerged recently (Cohen, 2008; Wood, Shenoy, Gerber, Ramakrishnan, & Van der Merwe, 2009; Extend Your IT Infrastructure with Amazon Virtual Private Cloud, http://aws.amazon.com/vpc/) as answer to this apparent dilemma of cost vs. control. In a typical approach, a VPC connects an organization's information technology (IT) resources to a dynamically allocated subset of a cloud provider's resources via a virtual private network (VPN). Organizational IT controls are then applied to the collective resources to meet required service levels. As a result, in addition to improved TCO, the model promises organizations direct control of security, reliability and other attributes they have been accustomed to with conventional, internal data centers.

The VPC concept is both fundamental and transformational. First, it proposes a distinct abstraction of public resources combined with internal resources that provides equivalent functionality and assurance to a physical collection of resources operated for a single organization, wherein the public resources may be shared with many other organizations that are also simultaneously being provided their own VPCs. Second, the concept provides an actionable path for an organization to incorporate cloud computing into its IT infrastructure. Once the organization is managing its existing resources as a private cloud (i.e., with virtualization and standard interfaces for resource management), the organization can then seamlessly extend its management domain to encompass external resources hosted by a cloud provider and connected over a VPN.

From the point of view of the organization, the path to a VPC model is relatively straightforward. In principle, it should be no more complicated, say, than the introduction of VPNs or virtual machines into the organization's IT infrastructure, because the abstraction preserves existing interfaces and service levels, and isolates the new implementation details. However, as with introduction of any type of abstraction, the provider's point of view is where the complexities arise. Indeed, the real challenge of VPCs is not whether organizations will embrace them once they meet organizational IT requirements, but how to meet those requirements – especially operational autonomy and flexibility – without sacrificing the efficiency that motivated the interest in the cloud to begin with.

With the emergence of VPCs as a means to bring cloud computing to organizations, the next question to address is: *What are the key technologies cloud providers and organizations need to realize VPCs?*

3.2 Virtual Private Clouds

A *cloud*, following NIST's definition that has become a standard reference (Mell & Grance, 2009), is a pool of configurable computing resources (servers, networks, storage, etc.). Such a pool may be deployed in several ways, as further described in Mell and Grance (2009):

- A *private cloud* operated for a single organization;
- A *community cloud* shared by a group of organizations;
- A *public cloud* available to arbitrary organizations; or
- A *hybrid cloud* that combines two or more clouds.

The full definition of a private cloud given in Mell and Grance (2009) is

Private cloud. The cloud infrastructure is operated solely for an organization. It may be managed by the organization or a third party and may exist on premise or off premise.

The definition suggests three key questions about a cloud deployment:

1. Who uses the cloud infrastructure?
2. Who runs the infrastructure?
3. Where is the infrastructure?

The distinction among private, community, public, and hybrid clouds is based primarily on the answer to the first question. The second and third questions are implementation options that may apply to more than one deployment model. In particular, a cloud provider may run and/or host the infrastructure in all four cases.

Although NIST's definition does not state so explicitly, there is an implication that the cloud infrastructure refers to physical resources. In other words, the computing resources in a private cloud are physically dedicated to the organization; they are used only (i.e., "solely") by that organization for a relatively long period of time. In contrast, the computing resources in a public or community cloud are potentially used by multiple organizations over even a short period of time.

The physical orientation of the definition motivates the concept of a *virtual private cloud*, which, following the usual paradigm, gives an appearance of physical separation, i.e., extending (Mell & Grance, 2009):

Virtual private cloud (VPC). The cloud infrastructure appears as though it is operated solely for an organization. It may be managed by the organization or a third party and may exist on premise or off premise — or some combination of these options.

In other words, a VPC offers the *function* of a private cloud though not necessarily its *form*. The VPC's underlying, physical computing resources may be operated for many organizations at the same time. Nevertheless, the virtual resources presented to a given organization – the servers, networks, storage, etc. – will satisfy the same requirements as if they were physically dedicated.

The possibility that the underlying physical resources may be run and/or hosted by a *combination* of the organization and a third party is an important aspect of the definition, as was first articulated by R. Cohen in a May 2008 blog posting (Cohen, 2008) that introduced the VPC concept:

> A VPC is a method for partitioning a public computing utility such as EC2 into quarantined virtual infrastructure. A VPC may encapsulate multiple local and remote resources to appear as a single homogeneous computing environment bridging the ability to securely utilize remote resources as part of [a] seamless global compute infrastructure.

Subsequent work has focused on a specific implementation profile where the VPC encompasses just the resources from the public cloud. Wood et al. in a June 2009 paper (Wood, Shenoy, Gerber, Ramakrishnan, & Van der Merwe, 2009) write:

> A VPC is a combination of cloud computing resources with a VPN infrastructure to give users the abstraction of a private set of cloud resources that are transparently and securely connected to their own infrastructure.

Likewise, Amazon describes its virtual private cloud in a January 2010 white paper (Extend Your IT Infrastructure with Amazon Virtual Private Cloud, http://aws.amazon.com/vpc/) as "an isolated portion of the AWS cloud," again connected to internal resources via a VPN.

In both Wood et al. and Amazon, a VPC has the appearance of a private cloud, so meets the more general definition stated above. However, the implementation profile imposes the limitation that the physical resources underlying the VPC are hosted and run by a cloud provider. In other words, the answer to the second and third questions above is "external." Although internal resources, e.g., the "enterprise site" of Wood et al., are connected to the VPC over the VPN, they are not part of the VPC proper.

This article maintains R. Cohen's broader definition because the cloud for which an organization will be responsible, ultimately, will encompass most or all of its resources, not just the external portions. The primary VPC implementation profile considered here is therefore one in which the underlying resources are drawn from a public cloud and an internal, private cloud – or, in other words, from a hybrid cloud that combines the two (see Fig. 3.1) (Note 1). How those resources are managed in order to meet organizational IT requirements is the focus of the sections that follow.

Note

1. The implementation profile is most relevant to medium to large enterprises that already have substantial internal IT investments and are likely to maintain some of those resources while incorporating IT services from an external cloud provider. For enterprises that outsource all IT to a cloud provider, the implementation profile would include only the external resources. The key enabling technologies for VPCs are relevant in either case.

Fig. 3.1 Primary virtual
private cloud (VPC)
implementation profile: VPC
is built from the hybrid of an
external public cloud and an
internal private cloud

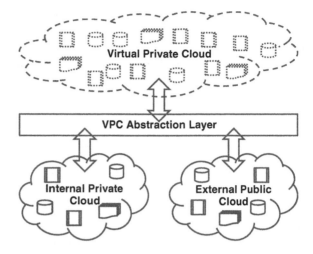

3.3 Virtual Data Centers and Applications

An organization's objective for its use of IT in general is to realize certain information-based business processes while conforming with applicable laws and regulations, and optimizing the cost/benefit tradeoff. Whether implemented with cloud computing or with conventional IT, the organization's high-level IT objectives are the same. The promise of cloud computing is that over time, organizations may be able to meet those objectives with an ever improving cost/benefit tradeoff.

3.3.1 Virtual Data Centers

In conventional IT, *data centers* provide a convenient way of organizing resources into locally connected pools. The locality provides an opportunity for common physical oversight and improved network performance among resources within the data center. In effect, a data center can be viewed as a local *container* of IT resources that can be managed together from a resource, security, and/or information perspective.

Virtualization, as a general paradigm, *insulates* resources and functionality from the underlying physical implementation, with the consequent advantage that virtual resources can be dynamically allocated to an organization without concern (by the organization) for the underlying physical implications. Moreover, the virtual resources can readily be migrated from one physical environment to another – for instance, between the organization's data centers and data centers operated by a cloud provider.

From this perspective, virtual resources "in the cloud" are, in principle, location- and container-independent. However, for the same reasons as in conventional IT, containers and location-type attributes may play an important role in practice, because organizations will continue to call for the performance advantages that locality brings, and it will be convenient to manage resources as sets. Accordingly, just as data centers are the basic, high-level unit of management in conventional IT, it is reasonable to expect that *virtual data centers* – the first key enabling technology for VPCs – will be the basic, high-level unit of resource management (Notes 1, 2):

Virtual data center (VDC). A pool of virtual resources that appear in terms of performance to be locally connected, and can be managed as a set.

For practical reasons, a VDC will typically be implemented based on a single, underlying physical data center; the apparent local connectivity would otherwise be difficult to achieve (although there are recent high-performance network technologies that do span physical data centers). The limitation is only in one direction, of course: A given physical data center can host more than one VDC. Furthermore, a data center operated by a public cloud provider may offer VDCs to multiple organizations, or henceforth, *tenants*, so the underlying computing environment is *multi-tenant*.

In addition to local connectivity, the placement of resources in a particular location may offer geographical advantages such as proximity to certain users or energy resources, or diversity of applicable laws and regulations. The placement of resources across multiple, independent locations can also help improve resilience. Such geographical aspects may be "virtualized" by policy-based management (see Section 3.4 below). The VDC (and/or its resources) would be selected so that they achieve the desired properties, with the actual location left to the implementation (although certain properties may only be achievable in a specific geography).

In addition, VDCs, like physical data centers, may vary in terms of the capabilities they offer, such as:

1. The types of virtual resources that are supported;
2. The cost, performance, security, and other attributes of those resources (and of the VDC in general), including the types of energy used; and
3. Specific resources that are already present and may be inconvenient to obtain elsewhere, such as large data sets or specialized computing functionality.

Rather than presenting the appearance of a physical data center as it actually is, the VDC abstraction can show a data center *as it ideally should be*. As a result, a VDC offers the opportunity for simplified, unified management from the point of view of the organization using it.

Given the VDC as a basic unit of management, the primary VPC implementation profile may be further refined as one in which the virtual resources are organized into the combination of

- One or more private VDCs hosted by a cloud provider; and
- One or more internal, private VDCs hosted by the organization

The cloud provider's VDCs would be based on scalable partitions of the cloud provider's public data centers; the internal VDCs could be simply the virtualization of the organization's existing data centers, or perhaps again scalable partitions. In either case, the modifier *private*, is essential: In order for the resulting VPC to appear as though it is operated solely for an organization, the component VDCs must be so as well.

Building a VPC thus decomposes into the problem of building a private VDC, or, expanding the definition, a pool of virtual resources that appears to be locally connected and to be operated for a single organization. The specific translation between a private VDC and underlying physical resources is, of course, a matter of implementation, but several technologies clearly will play a key role, including, obviously, virtualization and resource management, as well as, possibly, encryption of some form (Note 3).

With this first enabling technology in place, an organization using a VPC will have at its disposal some number of private VDCs or containers into which it may deploy resources, as well as the possibility of obtaining additional VDCs if needed. How those containers are used is the subject of the next enabling technology.

Notes

1. Cloud computing can be realized without the data center metaphor, for instance in settings where local connectivity is not important, such as highly distributed or peer-to-peer applications. The focus here is on the typical enterprise use cases, which are data-center based.
2. Virtualization, in principle, gives an appearance of privacy in the sense that if all tenants interact only through the VDC abstraction, then, by definition, they cannot access one another's resources (assuming of course that in the physical counterpart whose appearance is being presented, they cannot do so). Thus, virtualization of servers, networks, storage, etc., alone is arguably sufficient to build a VDC (as far as appearances; resource management is also needed to handle scheduling, etc.).

   There are two main problems with this position. First, there may be paths outside the abstraction by which parties may interact with the underlying resources. At the very least, the infrastructure operator will have such access, both physical and administrative. Second, there may be paths *within* the abstraction that inadvertently enable such interaction, whether due to errors or to side channels that are not completely concealed. This introduces the possibility of malevolent applications or *mal-apps* that target other tenants sharing the same public computing environment. The *cloud cartography* and *cross-VM information leakage* techniques of Ristenpart, Tromer, Shacham, and Savage (2009) are recent examples.

It is worth noting that comparable vulnerabilities are already being dealt with by conventional data centers through a range of security controls, from encryption to behavioral analysis. The difference in cloud computing is not as much the nature of the vulnerabilities, as the number of potential adversaries "in the camp." A base and adaptive set of security controls will be essential for the abstraction, robustly, to maintain its assurances, while applications implement additional controls above the abstraction just as they would if running directly on physical infrastructure. A good example of such an additional control is the VPN (which is visible to applications in the private VDC model).

Trusted computing (Mitchell, 2005) may also play a role in private VDCs by providing a *root of trust* with respect to which tenants may verify the integrity of their resources. The integration of trusted computing and virtualization is explored more fully in projects such as Terra by Garfinkel, Pfaff, Chow, Rosenblum, and Boneh (2003) and Daonity (now continued as Daoli) (Chen, Chen, Mao, & Yan, 2007).

3.3.2 Virtual Applications

Information-based processes in conventional IT are realized by various *applications* involving interactions among collections of resources. The resources supporting a given application may run in a single data center or across multiple data centers depending on application requirements.

Continuing the analogy with conventional IT, one may expect that *virtual applications* – the second key enabling technology – will be the basic, high-level unit of resource *deployment*:

> *Virtual application.* A collection of interconnected virtual resources deployed in one or more virtual data centers that implement a particular IT service.

A virtual application consists not only of the virtual machines that implement the application's software functionality, but also of the other virtual resources needed to realize the application such as virtual networks and virtual storage. In this sense, a virtual application extends the concept of a *virtual appliance* (Sapuntzakis & Lam, 2003), which includes the complete software stack (virtual machines and operating system, with network interfaces) implementing a single service, to encompass the set of services supporting the application.

Just as a VDC can show a data center in a more ideal form, a virtual application can present the appearance of an application as it ideally should be. Today, security, resource management, and information management are typically enforced by the operating system and application stack, which makes them complex and expensive to implement and maintain. With the simplified, unified management provided by the virtual application abstraction and encapsulation of application components in virtual machine containers, the virtual application container becomes a new control point for consistent application management. Instead of orchestrating each resource individually, an organization can operate on the full set in concert, achieving the equivalent of "one click" provisioning, power on, snapshot, backup, and so on.

The primary VPC implementation profile may be now refined again as one in which virtual applications consisting of virtual resources run across one or more VDCs (see Fig. 3.2) (Note 1).

Fig. 3.2 Virtual applications run across one or more private virtual data centers (VDCs), connected by virtual private networks (VPNs)

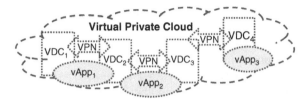

The Open Virtualization Format (OVF, 2009) recently standardized by the Data Management Task Force offers a convenient way to specify collections of virtual machines. Through metadata, the interconnections between those machines and their dependencies on other resources such as networks and storage may also be expressed, supporting full virtual applications as defined here. In addition to several commercial products, the specification is also supported in the Open-OVF open source project (open-ovf.sourceforge.net).

An organization using a VPC with the first two enabling technologies now introduced will be able to use its private VDCs to deploy virtual applications. The next enabling technologies address the contract between those applications and the VPC that enables the automatic assembly of components to meet organizational IT objectives while maintaining flexibility for optimization.

Note

1. The interplay between VDCs and virtual applications is an important aspect of meeting organizational IT requirements with a VPC, which do depend in some cases on (possibly relative) location, as noted in Section 3.3.1. Thus, in addition to the primary role of virtual applications in enabling portability *between* clouds, virtual applications may also be viewed as a way to enable effective deployment *within* a cloud by describing the desired relationships among virtual resources.

3.4 Policy-Based Management

Over time, a VPC will be populated with resources supporting virtual applications running at various VDCs. Those resources will be deployed and assembled with the ultimate intent of meeting the organizational IT requirements. This is the essence of the "contract," formal or otherwise, between an organization and the VPC.

The role of such a contract can be viewed in two parts: putting its terms into practice, and checking that the practice is correct.

3.4.1 Policy-Based Deployment

Consider an organization that wants to deploy an e-commerce application with certain objectives for security, performance, and business continuity. In a conventional data center, that application might be implemented as the combination of resources starting with a web server and a database. A firewall would be added to meet the security objectives, and a load-balancer to assign transactions to additional web servers as needed to meet the performance objectives. To address the business continuity objectives, a second instance of these components might be placed in another data center, coordinated with the first through a business continuity manager.

Suppose further that the organization also wants to deploy a customer relationship management (CRM) application with similar service objectives. That application might likewise be implemented as a combination of web servers, databases, firewalls, load-balancers, and so on, across two data centers.

Now, consider what happens when the organization decides to deploy these applications in a VPC (under some "contract," per the comments above). Following the model in Section 3.3.2, each application would be deployed as a collection of virtual resources. The VPC would thus be hosting the combination of the sets of resources for the two applications: two sets of virtual web servers, two virtual databases, two firewalls, two load-balancers, etc., and this collection would be repeated across two VDCs.

Deploying an application in a VPC as just described has some advantages, such as dynamic allocation of resources and economies of scale. However, such a process is really no more than a migration of components from one environment to another, or what could also be called a *literal virtualization*. Infrastructure sprawl in the data center translates directly into *virtual sprawl*, with as many components to manage as before, just consolidated onto fewer servers, and arranged in a more flexible way.

Cloud computing environments can improve this situation by organizing components and capabilities into searchable lists of virtual applications and resources that can readily be deployed. By selecting services from an offering catalog and inventory, rather than imposing entirely unique choices, an organization can take advantage of optimizations that the cloud provider may already have put in place. The load-balancer would be a good example. Instead of each application contributing its own load-balancer, the VPC would offer one itself for use by multiple applications.

Once an application designer knows that load-balancing will be available, he or she no longer needs to specify a virtual application as a combination of, say, two web servers and a load-balancer. Instead, the virtual application may be expressed as the combination of a single web server (and other functional components) and a *policy* that the VPC should create additional instances of the web server and balance transactions among them to maintain a specified performance level. This policy has the further benefit that the application may automatically be scaled *beyond* the two instances originally specified in a literal counterpart to the data center version.

Load-balancing is one instance of a general pattern: Applications are designed as a combination of *functionality* and *qualities* relating to *service-level agreements*

(SLAs). These qualities, sometimes also called, "*ilities*" (from the common suffix of scalability, availability, etc.), generally are implemented with quite similar components across different applications, like load-balancers, firewalls, business continuity managers, and so on. They are therefore natural candidates for services supporting multiple applications in a VPC.

A simple formula illustrates both the pattern and the problem. A typical application is constructed first by building a *base application* that meets some functional requirements, then adding "ilities" that address the non-functional ones. The resulting full application is then virtualized and deployed in the VPC. This pattern may be summarized as follows:

$$
\text{functionality} \xrightarrow{\text{build}} \textbf{base application} \xrightarrow{\text{add "ilities"}} \textbf{full application}
$$

$$
\begin{array}{c} \text{virtualize} \\ |\,@ \quad \downarrow \\ \text{full} \\ \text{virtual} \\ \text{application} \end{array}
$$

Given only the full virtual application, the VPC will likely have a problem recognizing the "ilities," and therefore, managing them or optimizing their delivery, as much as it is difficult to optimize object code without the original source. However, given some of that original source, the VPC will have much more opportunity to add value. Consequently, the preferred model for application deployment in a VPC is for the computing environment to *add "ilities" as part of deployment.*

The pattern now becomes the following:

$$
\text{functionality} \xrightarrow{\text{build}} \textbf{base application}
$$

$$
\begin{array}{c} \text{virtualize} \\ |\,@ \quad \downarrow \\ \text{base} \\ \text{virtual} \xrightarrow{\text{add "ilities"}} \textbf{full virtual} \\ \text{application} \qquad\qquad \textbf{application} \end{array}
$$

The "ilities" may be added by configuring the base virtual application or by deploying additional services. Following the VDC model in Section 3.3.2, policy may also be realized according to the placement of virtual resources into specific VDCs, e.g., where local connectivity, proximity to certain users, independence, etc. are required.

The paradigm may be summarized as the third key enabling technology, *policy-based deployment*:

Policy-based deployment. The assembly of application components in a computing environment according to predefined policy objectives.

Although policy-based deployment can also be applied in other types of clouds (as well as in data centers), the technology is particularly important for VPCs because of their primary use case: organizations that need to meet well-defined IT objectives.

Shifting the introduction of some policy features from development to deployment doesn't necessarily make deployment easier, and in fact may make it harder, as Matthews, Garfinkel, Hoff, and Wheeler (2009) observe, due to number of stakeholders and administrators potentially involved. *Automation* is essential to simplifying deployment, and a well-defined language for expressing policy is essential to automation. The separation of "ilities" from functionality is therefore necessary but not sufficient. In addition, the "ilities" must be expressed as machine-readable rules that the computing environment can implement. In Matthews et al. (2009), such rules take the form of a *Virtual Machine Contract*, defined as:

> A Virtual Machine Contract is a complex declarative specification of a simple question, should this virtual machine be allowed to operate, and if so, is it currently operating within acceptable parameters? (Matthews et al., 2009)

A specification like OVF can be employed to carry contracts and other policy information so that they travel along with the virtual machines, and, more generally, virtual applications, to which the conditions apply.

With automated policy-based deployment in place, an organization is able to specify to the VPC its expectations as far as security, performance and other SLAs, and the VPC can then, automatically, optimize its operations toward those objectives. The military expression, "You get what you inspect, not what you expect," motivates the challenge addressed by the next enabling technology: How to verify that the terms of the contract with the VPC are actually met.

3.4.2 Policy Compliance

In whatever computing environment an organization chooses to deploy an application, the organization will need some evidence that the application is running as intended. This evidence serves both the organization's own assurances and those of auditors or customers. Even if no malice is involved, the environment's optimizations may only approximate the intended result.

Policy objectives are particularly difficult to achieve in a multi-application setting because of the potential for resource contention. A physical computing resource, for instance, may reliably meet the computational performance objectives for the one application it supports, but when that resource interacts with another resource, the presence of network traffic from other applications may make the communication performance unpredictable. Network reservation schemes and similar approaches for storage play an important role in meeting SLAs for this reason. There may also be opportunities for different applications, by design, to operate in a complementary fashion that reduces the contention.

A multi-tenant computing environment such as a public cloud hosting VPCs for multiple organizations introduces further complexities. As with any Internet service, tenants are affected by one another's behavior, which can be unpredictable. Because there is no direct opportunity for negotiation among different tenants with respect to the underlying computing environment (in principle, they cannot even detect one another), any contention must be resolved by the cloud provider itself.

The objectives of different tenants are not necessarily aligned with one another, so in addition to the basic resource contention, there may also be contention among optimization strategies. The potential for interference is another strong motivation for placing the policy services within the computing environment rather than individual applications.

Finally, the tenants' objectives are not necessarily aligned with those of the public cloud provider. Although serving customers will presumably be the first priority of a successful cloud provider, staying in business is another, and there is a definite motivation for implementing further optimizations that cut costs for the provider without necessarily increasing benefit for any of the tenants (Note 1).

Given the difficulty of meeting policy requirements perfectly across multiple applications and tenants, it becomes particularly important for the VPC to provide, and the tenant to receive, some information about the extent to which those requirements are met, or not, at various points in time. This leads to the fourth key enabling technology, *policy compliance.*

Policy compliance. Verification that an application or other IT resource is operating according to predefined policy objectives.

Per the separation of "ilities" from based functionality discussed in Section 3.4.1, it is reasonable to expect that policy compliance itself will eventually be considered as just another service to be added to the application when deployed in the VPC (Note 2). Such a capability goes hand in hand with policy-based deployment: It will be much easier for a VPC to gather appropriate evidence from resources it has already assembled with policy objectives in mind, than to have to discover the objectives, the resources, and the evidence after the fact.

As far as the evidence itself, for precisely the purpose of evaluating performance, many IT resources are instrumented with activity logs that record transactions and other events. For instance, a physical network router may keep track of the source, destination, size, timestamp and other metadata of the packets it transfers (or is unable to transfer); a physical storage array may record similar information about the blocks it reads and write. With appropriate interfaces, the virtual environment can leverage these features to gather evidence of policy compliance. For example, *I/O tagging* embeds virtual application identifiers as metadata in physical requests, with the benefit that the identifiers are then automatically included in activity logs for later analysis by the virtual environment with minimal impact on performance (Note 3).

The collection of system logs from physical computing, networking, and storage resources, containing information about virtual applications and resources and their activities, provides an information set from which policy compliance evidence may be derived. This information set, keyed by the virtual application identifiers and related quantities, enables *distributed application context and correlation* – in effect, a virtual view of the activity of the virtual application, across the VPC.

Constructing such a view, especially from heterogeneous unstructured system logs that were designed only for local visibility, and management interfaces that were intended only for local control, depends on a fifth and final enabling technology, one that responds to the question: How to bring all this information together intelligently?

Notes

1. A related situation where a cloud storage provider may lose some portion of tenants' data as a result of its own performance optimizations (or actual malice) is explored in Juels and Kaliski (2007) and Bowers, Juels, and Oprea (2009), which also propose mechanisms for detecting and recovering from such loss before it reaches an irreversible stage. The detection mechanism may be viewed as an example of policy compliance for stored data.
2. If an application includes built-in policy compliance and the components involved are portable, then the compliance will continue to function in the VPC. Such verification provides a helpful checkpoint of the service levels achieved within a given computing environment. However, as more applications with built-in policy compliance are deployed, the VPC will see a sprawl of application-specific compliance components. This is another motivation for building policy compliance into the VPC.
3. I/O tagging offers the additional benefit of enabling virtualization-aware physical resource managers to enforce policies based on the virtual identifiers. This is a promising area for further exploration, particularly for methods to resolve the contention, as observed above, among policies for different applications and tenants.

3.5 Service-Management Integration

Virtual data centers, the first of the enabling technologies, may be viewed as providing *insulation* that sets one organization's virtual resources apart from those of other organizations, and from the underlying physical resources. Virtual applications, the second, collect those resources into separately manageable units. Policy-based deployment and policy compliance, the third and fourth, offer a means of *control* and verification of the operation of the virtual applications across the VDCs. All four rest on a fifth technology: a more basic foundation that bridges across underlying boundaries, one oriented toward seamless *integration*.

Recall the original implementation profile for the VPC, per Section 3.2: a hybrid of an internal, private cloud and a public cloud. Following Section 3.3, the VPC provides the appearance of some number of VDCs, some drawn from the internal cloud, some from the public cloud. Throughout Section 3.4, this VPC is essentially viewed as seamless, which it is in appearance (the exposure of multiple VDCs is an architectural feature). Thus, Section 3.4 can speak of deploying an application into the VPC, collecting evidence from the VPC, and so on, without regard to the fact that the deployment and collection ultimately involve interactions with physical resources, and more significantly, that these physical resources are in multiple data centers operated by at least two different entities.

The fundamental challenge for satisfaction of policy-based management in a VPC is *how* to enable such seamless interaction between resource, service and policy management components: across data center infrastructure boundaries, and then across federated service provider boundaries.

Such bridges are not easy to build, because the various management interfaces – like the local logs in Section 3.4.2 – were designed for separate purposes. At the physical layer, they may use different names for the same entity or function, employ incompatible authentication and access control systems, and express the same conditions in different ways. The information the organization and the VPC need is available, but is not immediately useful without some translation. Moreover, that translation is not simply a matter of converting between formats, but, in effect, virtualizing the interfaces to the management metadata across the borders of the underlying management component.

The fifth and final key enabling technology, *service-management integration*, addresses this last challenge:

Service-management integration. The translation of heterogeneous management information from separate domains into an overall, logical and actionable view.

Service-management integration is a special case of the broader technology of *information integration*, which is concerned, similarly, with translating of federating general information from multiple domains. The special case of VPCs is concerned in particular with federating three things: (1) the underlying *infrastructure* into one virtual computing environment, (2) *identities* interacting with the resources in the environment, and (3) *information* about the resources.

By its nature, service-management integration for VPCs is amenable to an *eventing* paradigm where the basic unit of information is an event published by one entity in the system, and consumed by another. This paradigm is a good match for a policy compliance manager that is interested in the content of multiple physical logs, as that evidence accumulates. It also provides a deployment manager with a current view of the underlying resources as they continually change. Further, the architectural separation between publisher and subscriber lends itself to the physical separation and distribution of participating elements across data center and cloud federation boundaries.

The intermediation between publisher and consumer can be achieved through a *messaging system*. As a dedicated communication layer for events, such a system provides a federated information delivery "backplane" that bridges multiple management domains (e.g., internal data centers, cloud provider data centers) into a single service-oriented architecture, translating back and forth among the security and management languages of the various domains. Events published in one domain can be consumed in another according to various subscription rules or filters; the policy compliance manager for a particular tenant, for instance, will only be interested in (and should only know about) events related to that tenant's virtual applications.

The messaging system can implement its translations through a set of adapters, informed by an understanding of the connections among the identities and events in the different domains. The system's learning of those connections can occur automatically, or it may require manual intervention, and in some cases it may need to be augmented with a significant amount of computation, for instance to search for correlated events in the different domains. In a cloud computing environment, the resources for such computation will not be hard to find. (How to balance between the use of resources to make the overall environment more efficient, versus allocating them directly to tenants, is another good question for further exploration.)

3.6 Conclusions

This article started with the simple premise that cloud computing is becoming more important to organizations, yet, as with any new paradigm, faces certain challenges.

One of the challenges is to define a type of cloud computing most appropriate for adoption. A virtual private cloud built with IT resources from both the organization's own internal data centers and a cloud provider's public data centers has been offered as a preferred implementation profile. To ensure privacy, i.e., the appearance that the cloud is operated solely for the organization, certain additional protections are also needed.

Another challenge is to make good use of the collective resources. A literal type of virtualization where applications are basically ported from a data center to the VPC would realize some of benefits, but the greater potential comes from enabling the VPC itself to optimize the assembly of applications. The starting point for that advance is the separation of functionality from policy within the specification of a virtual application so that policy requirements can be met in a common and therefore optimized way by the VPC. Commonality of policy management also enables the VPC to verify that policies are met.

Finally, information infrastructure rests as the foundation of realizing a VPC. Indeed, virtualization is all about turning resources into information. The better the VPC can engage with that information, rising from the shadows of the private data center past and the public cloud present, the more effectively organizations can move into the promise of the virtual private cloud future.

References

Bowers, K. D., Juels, A., & Oprea, A. (November 2009). HAIL: A high-availability and integrity layer for cloud storage. *Proceedings of the 16th ACM Conference on Computer and Communications Security (CCS), ACM, Chicago, IL, USA,* 187–198.

Cohen, R. (May 2008). *Virtual private cloud, Elastic vapor: Life in the cloud.* Retrieved January 2010, from http://www.elasticvapor.com/2008/05/virtual-private-cloud-vpc.html.

Chen, H., Chen, J., Mao, W., & Yan, F. (June 2007). Daonity – Grid security from two levels of virtualization. *Elsevier Journal of Information Security Technical Report, 12*(3), 123–138.

Garfinkel, T., Pfaff, B., Chow, J., Rosenblum, M., & Boneh, D. (December 2003). Terra: A virtual machine-based platform for trusted computing. *ACM SIGOPS Operating Systems Review, 37*(5), 193–206.

Juels, A., & Kaliski, B. S., Jr. (October 2007). PORs: Proofs of retrievability for large files. *Proceedings of the 14th ACM Conference on Computer and Communications Security (CCS), ACM, Alexandria, VA, USA,* 584–597.

Matthews, J., Garfinkel, T., Hoff, C., & Wheeler, J. (June 2009). Virtual machine contracts for datacenter and cloud computing environments. *Proceedings of the 1st Workshop on Automated Control for Datacenters and Clouds, ACM, Barcelona,* 25–30.

Mell, P., & Grance, T. (2009). *The NIST definition of cloud computing, version 15, NIST.* Retrieved January 2010, from http://csrc.nist.gov/groups/SNS/cloud-computing/.

Mitchell, C. (Ed.). (2005). *Trusted computing.* London: IET.

OVF (January 2010). *Open virtualization format specification, DMTF Document DSP0243, Version 1.0.0,* Retrieved January 2010, from http://www.dmtf.org/.

Ristenpart, T., Tromer, E., Shacham, H., & Savage, S. (November 2009). Hey, you, get off of my cloud: Exploring information leakage in third-party compute clouds. *Proceedings of the 16th ACM Conference on Computer and Communications Security (CCS), ACM, Chicago, IL,* 199–212.

Sapuntzakis, C., & Lam, M. S. (May 2003), Virtual appliances in the collective: A road to hassle-free computing. *Proceedings of HotOS IX: The 9th Workshop on Hot Topics in Operating Systems, USENIX, Lihue, Hawaii,* 55–60.

Wood, T., Shenoy, P., Gerber, A., Ramakrishnan, K. K., & Van der Merwe, J. (June 2009). The case for enterprise-ready virtual private clouds. *Proceedings of HotCloud '09 Workshop on Hot Topics in Cloud Computing, San Diego, CA, USA,* Retrieved January 2010, from http://www.usenix.org/event/hotcloud09/tech/.

Chapter 4
The Role of Networks in Cloud Computing

Geng Lin and Mac Devine

4.1 Introduction

The confluence of technology advancements and business developments in Broadband Internet, Web services, computing systems, and application software over the past decade has created a perfect storm for cloud computing. The "cloud model" of delivering and consuming IT functions as services is poised to fundamentally transform the IT industry and rebalance the inter-relationships among end users, enterprise IT, software companies, and the service providers in the IT ecosystem (Armbrust et al., 2009; Lin, Fu, Zhu, & Dasmalchi, 2009).

In the center of the cloud delivery and consumption model is the network (Gartner Report, 2008). The network serves as the linkage between the end users consuming cloud services and the provider's data centers providing the cloud services. In addition, in large-scale cloud data centers, tens of thousands of compute and storage nodes are connected by a data center network to deliver a single-purpose cloud service. How do network architectures affect cloud computing? How will network architecture evolve to better support cloud computing and cloud-based service delivery? What is the network's role in security, reliability, performance, and scalability of cloud computing? Should the network be a dumb transport pipe or an intelligent stack that is cloud workload aware?

This chapter focuses on the networking aspect in cloud computing and shall provide insights to these questions. The chapter is organized as follows. In Section 4.2, we discuss the different deployment models for cloud services – private clouds, public clouds, and hybrid clouds – and their unique architectural requirements on the network. In Sections 4.3 and 4.4, we focus on the hybrid cloud model and discuss

G. Lin (✉)
IBM Alliance, Cisco Systems, San Francisco, CA, USA
e-mail: gelin@cisco.com

M. Devine
IBM Corporation, Research Triangle Park, NC, USA
e-mail: wdevine@us.ibm.com

B. Furht, A. Escalante (eds.), *Handbook of Cloud Computing*,
DOI 10.1007/978-1-4419-6524-0_4, © Springer Science+Business Media, LLC 2010

the business opportunities associated with hybrid clouds and the network architecture that enables hybrid clouds. In many ways, the hybrid cloud network architecture encompasses the characteristics of the networks for both public and private clouds. In Section 4.5, we discuss our conclusions and highlight the directions for future work in cloud-enabling network architectures.

4.2 Cloud Deployment Models and the Network

The IT industry is attracted by the simplicity and cost effectiveness represented by the cloud computing concept that IT capabilities are delivered as services in a scalable manner over the Internet to a massive amount of remote users. While the purists are still debating the precise definition of cloud computing, the IT industry views cloud computing – an emerging business model – as a new way to solve today's business challenges. A survey conducted by Oliver Wyman (Survey, private study for IBM) in November 2008 with business executives from different enterprises identified "reduce capital cost," "reduce IT management cost," "accelerate technology deployment," and "accelerate business innovation" as the main business benefits for cloud computing. Figure 4.1 shows the detailed survey results.

Despite the benefits promised by cloud computing, the IT industry also sees that significant innovation and improvement on technologies and operations governance are needed to enable broad adoption of cloud services. Chief concerns are security and performance issues. Take security as an example, while it is acceptable for individual consumers to turn to Amazon Elastic Compute Cloud (EC2) and Simple Storage Services (S3) for on-demand compute resources and storage capacities, it is

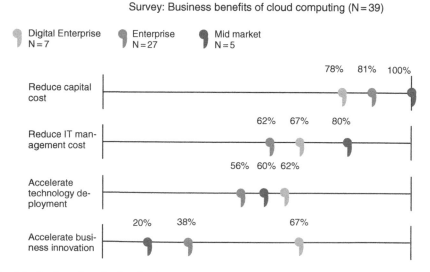

Fig. 4.1 Business benefits for cloud computing

a different matter for a bank to store its customer information in a third-party owned cloud.

Based on the differences in the deployment model, cloud services can be delivered in three principal ways: public cloud, private cloud, and hybrid cloud.

4.2.1 Public Cloud

A *public cloud* refers to a cloud service delivery model in which a service provider makes massively scalable IT resources, such as CPU and storage capacities, or software applications, available to the general public over the Internet. Public cloud services are typically offered on a usage-based model. Public cloud is the first deployment model of cloud services to enter the IT industry's vocabulary. The concept of public clouds has clearly demonstrated the long-term potential of the cloud computing model and lit up the imagination of the industry and the research community.

There are many public cloud service providers in place today, offering services ranging from infrastructure-as-as-service, to development-platform-as-a-service, to special purpose application-as-a-services. Amazon EC2, Force.com, and Google App Engine, are among some of the best known examples of public clouds, but the market now bristles with competition. See the survey by InformationWeek (Babcock, 2009a, 2009b, 2009c) on the major public cloud service providers for a detailed analysis on their services, pricing models, platforms supported, etc.

While the public cloud offers a clean, infrastructure-less model for end users to consume IT services, and intrigues the research community with its disruptive nature, migrating the majority of today's IT services, such as the various business applications in an enterprise environment (e.g. insurance applications, health care administration, bank customer account management, the list goes on and on), to a public cloud model is not feasible. Data security, corporate governance, regulatory compliance, and performance and reliability concerns prohibit such IT applications to be moved out of the "controlled domains" (i.e. within the corporate firewalls), while the public cloud infrastructure, government regulation, and public acceptation continue to improve.

4.2.2 Private Cloud

Private cloud, in contrast, represents a deployment model where enterprises (typically large corporations with multi-location presence) offer cloud services over the corporate network (can be a virtual private network) to its own internal users behind a firewall-protected environment. Recent advances in virtualization and data center consolidation have allowed corporate network and datacenter administrators to effectively become service providers that meet the needs of their customers within these corporations. Private clouds allow large corporations to benefit from the "resource pooling" concept associated with cloud computing and their very own size, yet in the mean time addressing the concerns on data security, corporate governance, government regulation, performance, and reliability issues associated with public clouds today.

Critics of private clouds point out that these corporations "still have to buy, build, and manage clouds" and as such do not benefit from lower up-front capital costs and less hands-on management, essentially "lacking the economic model that makes cloud computing such an intriguing concept." While these criticisms are true from a purist's point view, private clouds are a viable and necessary deployment model in the overall adoption of cloud computing as a new IT model. We believe that without large corporations embracing it, cloud computing will never become a main stream computing and IT paradigm (for this one can refer to the previous example of Grid Computing). Private cloud represents an enabling as well as a transitional step towards the broader adoption of IT services in public clouds. As the public cloud infrastructure, government regulation, and public acceptance continue to improve, more and more IT applications will be first offered as services in a private cloud environment and then migrated to the public cloud. The migration path of Email service in a corporation environment – from initially multiple departmental email servers, to today's single corporate-level "email cloud", to a public email cloud – offers an exemplary representation. While purists might argue in black and white terms, we believe private cloud as a viable deployment model for cloud computing will exist for a long time and deserves the attention from both business and research communities.

4.2.3 Hybrid Cloud

While public and private clouds represent the two ends of the cloud computing spectrum in terms of ownership and efficiency of shared resources – and each is finding acceptance in accordance to the services offered and customer segments targeted – a third deployment model of cloud computing, the hybrid cloud model that blends the characteristics of public and private clouds, is emerging.

A hybrid cloud is a deployment model for cloud services where an organization provides cloud services and manages some supporting resources in-house and has others provided externally. For example, an organization might store customer data within its own data center and have a public cloud service, such as Amazon's EC2, to provide the computing power in an on-demand manner when data processing is needed. Another example is the concept of "public cloud as an overflow for private clouds" where an IT manager does not need to provision its enterprise private cloud for the worst-case workload scenario (doing so will certainly defeat the economics of a private cloud), but to leverage a public cloud for overflow capacities to move less-mission-critical workloads on and off premise dynamically and transparently to accommodate business growth or seasonal peak load demands. One can find different variations of the "overflow" scenario, such as "follow-the-sun" operations in a global organization where workloads are moved around the globe based on the time zones of the working teams. Architecturally, a hybrid cloud can be considered a private cloud extending its boundary into a third party cloud environment (e.g. a public cloud) to obtain additional (or non-mission critical) resources in a secure and on-demand manner.

Adoption of cloud services is a gradual process: Enterprise IT (which represents the majority of IT industry spending and service consumption) needs a migration path to move today's on-premise IT applications to the services offered by public cloud providers through a utility model. As such, the hybrid cloud represents a prevalent deployment model. Large enterprises often have substantial investments in the IT infrastructure required to provide resources in house already. Meanwhile, organizations need to keep sensitive data under their own control to ensure security and compliance to government regulations. The tantalizing possibility offered by the hybrid cloud model – enterprise IT organizations managing an internal cloud that meshes seamlessly with a public cloud, which charges on a pay-as-you-go basis – embodies the promise of the amorphous term cloud computing. To enable hybrid clouds, virtualization, seamless workload mobility, dynamic provisioning of cloud resources, and transparent user experience, are among the critical technical challenges to be resolved.

4.2.4 An Overview of Network Architectures for Clouds

There are three principal areas in which the network architecture is of importance to cloud computing: (1) a *data center network* that interconnects the infrastructure resources (e.g. servers and storage devices) within a cloud service data center, (2) a *data center interconnect network* that connects multiple data centers in a private, public, or hybrid cloud to supporting the cloud services, (3) the *public Internet* that connect end users to the public cloud provider's data centers. The last area has mostly to do with today's telecommunications network infrastructure, and is a complex topic by itself from the architectural, regulatory, operational and regional perspectives. It is beyond the scope of this chapter. We shall focus only on the first two areas (data center network and the data center interconnect network) in this chapter.

4.2.4.1 Data Center Network

Cloud providers offer scalable cloud services via massive data centers. In such massive-scale data centers, *Data Center Network* (DCN) is constructed to connect tens, sometimes hundreds, of thousands of serves to deliver massively scalable cloud services to the public. Hierarchical network design is the most common architecture used in data center networks. Figure 4.2 show a conceptual view of a hierarchical data center network as well as an example of mapping the reference architecture to a physical data center deployment.

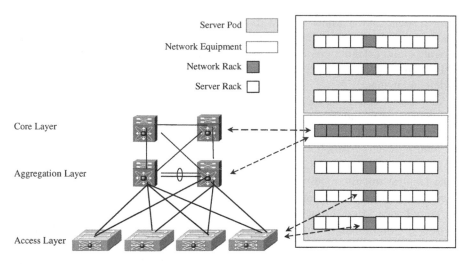

Fig. 4.2 Data center network architecture

The *access layer* of a data center network provides connectivity for server resource pool residing in the data center. Design of the access layer is heavily influenced by the decision criteria such as server density, form factor, and server virtualization that can result in higher interface count requirements. The commonly used approaches for data center access layer connectivity are *end-of-row* (EoR) switch, *top-of-rack* (ToR) switch, and *integrated switch* (typically in the form of blade switches inside a modular blade server chassis). Another form of the integrated switch is the embedded *software switch* in a server end point (see Virtual Ethernet Switch in this section). Each design approach has pros and cons, and is dictated by server hardware and application requirements.

The *aggregation layer* of the data center provides a consolidation point where access layer switches are connected providing connectivity between servers for multi-tier applications, as well as connectivity across the core of the network to the clients residing within the campus, WAN, or Internet. The aggregation layer typically provides the boundary between Layer-3 routed links and Layer-2 Ethernet broadcast domains in the data center. The access switches are connected to the aggregation layer using 802.1Q VLAN trunks to provide the capability of connecting servers belonging to different VLANs and IP subnets to the same physical switch.

The primary function of the *core layer* in a data center network is to provide highly available, high performance Layer-3 switching for IP traffic between the data center and the Telco's Internet edge and backbone. In some situations, multiple geographically distributed data centers owned by a cloud service provider may be connected via a private WAN or a Metropolitan Area Network (MAN). For such environments, expanding Layer 2 networks across multiple data centers is a better architecture design (readers can refer to Section 4.2 "Data Center

Interconnect Network" for details). In other situations, the traffic has to be carried over the public Internet. The typical network topologies for this kind of geographically distributed data centers is Layer-3 Peering Routing between the data center core switches. By configuring all links connecting to the network core as point-to-point Layer-3 connections, rapid convergence around any link failure is provided, and the control plane of the core switches is not exposed to broadcast traffic from end node devices or required to participate in STP for Layer-2 network loop prevention.

The evolution of networking technology to support large-scale data centers is most evident at the access layer due to rapid increase of number of servers in a data center. Some research work (Greenberg, Hamilton, Maltz, & Patel, 2009; Kim, Caesar, & Rexford, 2008) calls for a large Layer-2 domain with a flatter data center network architecture (2 layers vs. 3 layers). While this approach may fit a homogenous, single purpose data center environment, a more prevalent approach is based on the concept of *switch virtualization* which allows the function of the logical Layer-2 access layer to span across multiple physical devices. There are several architectural variations in implementing switch virtualization at the access layer. They include Virtual Blade Switch (VBS), Fabric Extender, and Virtual Ethernet Switch technologies. The VBS approach allows multiple physical blade switches to share a common management and control plane by appearing as a single switching node (Cisco Systems, 2009d). The Fabric Extender approach allows a high-density, high-throughput, multi-interface access switch to work in conjunction with a set of fabric extenders serving as "remote I/O modules" extending the internal fabric of the access switches to a larger number of low-throughput server access ports (Cisco Systems, 2008). The Virtual Ethernet Switch is typically software based access switch integrated inside a hypervisor at the server side. These switch virtualization technologies allow the data center to support multi-tenant cloud services and provide flexible configurations to scale up and down the deployment capacities according to the level of workloads (Cisco Systems, 2009a, 2009c).

While we have discussed the general design principles for the data center network in a massively scalable data center, some cloud service providers, especially some public cloud providers, have adopted a two-tier data center architecture to optimize data center cost and service delivery (Greenberg, Lahiri, Maltz, Patel, & Sengupta, 2008). In this architecture, the creation and delivery of the cloud service are typically accomplished by two tiers of data centers – a front end tier and a back end tier – with significant difference in their sizes. Take the Web search service as an example, the massive data analysis applications (e.g., computing the web search index) is a natural fit for the centralized mega data centers (measured by hundreds of thousands of servers) while the highly interactive user front-end applications (e.g. the query/response process) is a natural fit for geographically distributed micro data centers (measured by hundreds or thousands of servers) each placed close to major population centers to minimize network latency and delivery cost. The hierarchical data center network architecture is scalable enough to support both mega data centers and micro data centers with the same design principles discussed in this section.

4.2.4.2 Data Center Interconnect Network

Data center interconnect networks (DCIN) are used to connect multiple data centers to support a seamless customer experience of cloud services. While a conventional, statically provisioned virtual private network can interconnect multiple data centers and offer secure communications, to meet the requirements of seamless user experience for cloud services (high-availability, dynamic server migration, application mobility), the DCIN for cloud services has emerged as a special class of networks based on the design principle of *Layer 2 network extension across multiple data centers* (Cisco Systems, 2009b). For example, in the case of server migration (either in a planned data center maintenance scenario or in an unplanned dynamic application workload balancing scenario) when only part of the server pool is moved at any given time, maintaining the Layer 2 adjacency of the entire server pool across multiple data centers as opposed to renumbering IP addresses of servers is a much better solution. The Layer 2 network extension approach, on one hand, is a must from business-continuity perspective; on the other hand, is cost effective from the operations perspective because it maintains the same server configuration and operations policies.

Among the chief technical requirements and use cases for data center interconnect networks are data center disaster avoidance (including data center maintenance without downtime), dynamic virtual server migration, high-availability clusters, and dynamic workload balancing and application mobility across multiple sites. These are critical requirements for cloud computing. Take the application mobility as an example. It provides the foundation necessary to enable compute elasticity – a key characteristics of cloud computing – by providing the flexibility to move virtual machines between different data centers.

Figure 4.3 shows a high level architecture for the data center interconnect network based on the Layer 2 network extension approach.

Since the conventional design principle for Layer 2 network is to reduce its diameter to increase performance and manageability (usually limiting it to the access layer, hence advocating consolidating servers to a single mega data center and limiting the Layer 2 connectivity to intra data center communications), there are many areas of improvement and further research needed to meet the needs of data center interconnect networks. Listed below are some of the key requirements for Layer 2 network extension across multiple data centers.

End-to-End Loop Prevention

To improve the high availability of the Layer 2 VLAN when it extends between data centers, this interconnection must be duplicated. Therefore, an algorithm must be enabled to control any risk of a Layer 2 loop and to protect against any type of global disruptions that could be generated by a remote failure. An immediate option to consider is to leverage Spanning Tree Protocol (STP), but it must be isolated between the remote sites to mitigate the risk of propagating unwanted behaviors such as topology change or root bridge movement from one data center to another.

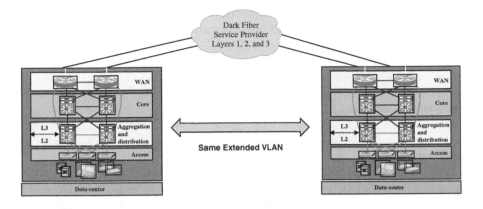

WAN Transport	Description	LNA Extension Encapsulation Options
Dark Fiber and Service Provider Layer1	Customer Owned or Service Provider Leased	Native Ethernet, IP, MPLS
Service Provider Layer 2	Service Provider Layer 2 Service, Ethernet Private Line L (EPL)	Native Ethernet,IP, MPLS
Service Provider Layer 3	IP Leased-Line Service	IP, MPLS

Fig. 4.3 Data center interconnect LAN extension encapsulation options

WAN Load Balancing

Typically, WAN links are expensive, so the uplinks need to be fully utilized, with traffic load-balanced across all available uplinks. A mechanism to dynamically balance workloads at the virtual machine level is an area of research.

Core Transparency

The LAN extension solution needs to be transparent to the existing enterprise core network, if available, to reduce any effect on operations. This is more common in the private cloud or hybrid cloud environments than in a public cloud.

Encryption

The requirement for LAN extension cryptography is increasingly prevalent, for example, to meet the needs for cloud services and for federal and regulatory requirements.

4.3 Unique Opportunities and Requirements for Hybrid Cloud Networking

IT industry is in the midst of a transformation. Globalization, explosion of business information, unprecedented levels of interconnectedness and dynamic collaboration among different business assets both within a corporation and across multiple corporations (on-demand supply chain as an example) require today's enterprise businesses to move to an IT infrastructure that is truly economical, highly integrated, agile and responsive.

As discussed in the previous section, the hybrid cloud model provides a seamless extension to an enterprise's private IT infrastructure by providing elastic compute, storage and network services in a cost-effective manner. This seamless extension could allow enterprises to streamline business processes, be more responsive to change, have flexible collaboration with business partners, leverage more rapidly the emerging technologies that address growing business challenges and increase competitiveness by delivering more services to customers.

To achieve this vision of business agility that hybrid clouds promise to enable, significant challenges lay ahead. Challenging requirements for hybrid cloud deployments in terms of deployment and operational costs, quality of service delivery, business resiliency and security must be addressed. Hybrid clouds will need to support a large number of "smart industry solution workloads" – business applications in the form of smart transportation solutions, smart energy solutions, smart supply chain solutions, etc. In these "smart industry solutions," large amount of business information and control data will be collected, analyzed and reacted upon in a time-constrained fashion across multiple tiers of cloud centers; workloads and data will be dynamically shifted within a hybrid cloud environment. This will require significant improvements to today's network. Using the metaphor of a bridge design, we can describe these requirements in three categories – the foundation, the span and the superstructure.

4.3.1 Virtualization, Automation and Standards – The Foundation

Virtualization, automation and standards are the pillars of the foundation of all good cloud computing infrastructures. Without this foundation firmly in place across the servers, storage and network layers, only minimal improvements on the adoption of cloud services can be made; conversely, with this foundation in place, dramatic improvements can be brought about by "uncoupling" applications and services from the underlying infrastructure to improve application portability, drive up resource utilization, enhance service reliability and greatly improve the underlying cost structures. However, this "uncoupling" must be done harmoniously such that the network is "application aware" and that the application is "network aware". Specifically, the networks – both the data center network and the data center interconnect network (and in the long run the public core network) – need to embrace virtualization and automation services. The network must coordinate with the upper layers of the cloud (i.e. the application workloads – both physical and virtual) to provide the needed level of operational efficiency to break the lock between IT resources in today's client-server model.

This transformation to a dynamic infrastructure which is "centered" on service delivery not only requires enterprise IT to transcend the daily IT break-and-fix routine but to create a paradigm shift within the user community toward a shared environment with repeatable, standardized processes. Centralized delivery of a standardized set of services instead of the distributed delivery of a highly customized set of services must be accompanied by new levels of flexibility via self-service mechanisms. In other words, easier and faster access to services make the standardization acceptable or even attractive to users since they sacrifice the ability to customize but gain convenience and time.

There is also a strong need for open standards to enable interoperability and federation across not only the individual layers of a private cloud behind an enterprise's firewall but also when consuming public cloud based services. This type of hybrid cloud environment allows scalable and flexible collaboration and global integration in support of evolving business model changes with clients (e.g. customer relationship management) and partners (e.g. supply chain partners).

4.3.2 Latency, Bandwidth, and Scale – The Span

The span of the network requirements for latency, bandwidth and scale which are needed to support traditional enterprise business applications and those needed to support cloud based applications can be very wide. Accurate forecasting of the quality of user experience and potential business impact for any network failure is already a major challenge for IT managers and planners even for today's traditional enterprise business applications. This challenge will become even more difficult as businesses will depend increasingly on more high performance cloud based applications which often have more variability than traditional enterprise business applications.

Meeting this challenge is essential to the "quality of experience" required to get the user community to accept a shared set of standardized services which are cloud delivered. Without this acceptance, the transformation of today's data center to a private or hybrid cloud environment with the dynamic and shared infrastructure needed for a reduced total cost of ownership will be much more difficult. Some users may choose to "get around" IT by trying to leverage services from the public cloud without the proper integration with existing IT and business processes which can have significant negative impacts on their businesses.

"Quality of experience" for access to some cloud based applications and services may require LAN-like performance to allow a portion of the user community to use real-time information to respond instantaneously to new business needs and to meet the demands of their customers. For these use cases, latency and bandwidth matter. Furthermore, there does not have to be problems at any one hop in order for end to end performance to be affected. Mild congestion at a number of hops can create problems in latency and packet loss. Therefore, content distribution, optimized routing and application acceleration services are usually required especially for hybrid cloud deployments with regional to global network connectivity.

Other users may only want the ability to simply request a new service without needing to know how or where it is built and delivered. It is not that the performance is not important to these users. In fact, communication and application delivery optimizations may still be required to increase the performance of applications and data that must traverse the cloud. It is just that their main criterion for quality has more to do with the ability to easily provision as many services as needed than it is dependent on latency and bandwidth optimizations. For these use cases, the ability to easily provision system resources including network resources (physical or virtual) is essential. It is also important to note that the two most prevailing techniques to help server-side scaling, i.e. physical density and virtualization, both drive an increased dependence on network integration.

For each user group and their corresponding use cases, the evolution to global-scale service delivery may best be accomplished via a hybrid cloud environment. The hybrid cloud can enable the visibility, control and automation needed to deliver quality services at almost any scale by leveraging not only the private network but also the public internet via managed network service providers.

Public clouds can be used for off-loading certain workloads. This off-load could be so that the private network infrastructure can be available and optimized for other latency and bandwidth sensitive workloads and/or for the provisioning of additional services due to a shortage of available infrastructure on-premise. Application platforms and tooling available on the public cloud can also be used to provide even greater flexibility for development and test environments which are often the best workloads for this type of off-loading. SaaS applications can also be consumed by the user community within a hybrid cloud environment. Under the hybrid cloud model, the consumption of public cloud services can be fully integrated with the existing on-premise IT and business processes to maximize the return of investment as well as ensure regulatory compliance.

4.3.3 Security, Resiliency, and Service Management – The Superstructure

Like the superstructure which ensures the integrity of a bridge's design, the elements of cloud computing environment – security, resiliency and service management – ensure the integrity of its design. Without these "superstructure" elements the value proposition associated with cloud computing will collapse and the economic benefits promised by cloud computing will be just illusions.

For some workloads, compliance with industry regulations like HIPAA (Health Insurance Portability and Accountability Act) and SOX (Sarbanes Oxley) require businesses to keep complete control over the security of their data. While there is much innovation happening for security within public clouds, the maturity level of these technologies may not yet be at a level where security and regulatory compliance can be guaranteed. However, even in these cases, an enterprise can still off-load non sensitive/critical workloads onto a public cloud while using a private cloud to ensure the needed SLAs for sensitive/critical workloads.

The network plays a key role in the establishment of these regulatory compliant clouds. Private WAN services must be enabled to provide the security needed for the private portion of the cloud. If a hybrid cloud environment is being used then the network must also be able to provide the federated connectivity and isolation needed and support the proper level of encryption for VPN tunnels which will be used by the public clouds to access data which remains behind a corporate firewall. Although there are other cloud deployment options available for workloads which do not have the need for the same level of compliance, networking connectivity and security functions are still central for a successful deployment of these cloud services.

Service management and automation also plays a critical role in hybrid clouds. As cloud services continue to advance, it is more likely that in the future networking services for cloud applications will be offered through an application-oriented abstraction layer APIs, rather than in specific networking technologies. Within this network architecture paradigm, modification and provisioning of network resources can be made in a more automated and optimized manner via service management or network self-adjustment. Specifically, these modifications can be made via operator-initiated provisioning through service management systems to assert direct control on network services, or via "smart" networking technologies which can also adapt services in an autonomic or self-adjusting fashion. Furthermore, it is critical that the network service management and the smart networking technologies are tightly integrated with the overall management for the cloud service delivery so that the changes required by the upper layers of the cloud "stack" in network resources can be carried through by the network service management or self-adaptations in an automated fashion.

Many of these "smart" networking technologies are focused on maximizing the resiliency of cloud deployments in terms of the availability, performance and workload mobility. For example, application delivery networking services optimize the flow of information and provide application acceleration by the classification and prioritization of application, content and user access; virtual switching technology provides an "abstraction" of the switching fabric and allows virtual machine mobility.

As these "smart" networking technologies mature, their capabilities will extend beyond the current capabilities for a single cloud to the "intra-cloud" as well as to the "intercloud." With this maturation, the hybrid cloud will provide unprecedented levels of global interconnectedness for real time or near real time information access, application-to-application integration and collaboration.

4.4 Network Architecture for Hybrid Cloud Deployments

Hybrid clouds play a key role in the adoption of cloud computing as the new generation IT paradigm. While the IT industry and the research community are still in the early stage to understand the implementation technologies for hybrid clouds, a

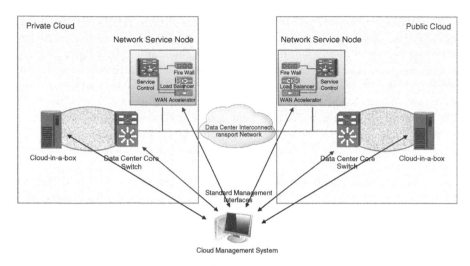

Fig. 4.4 A Functional view of network architecture for hybrid clouds

number of major functional components in the hybrid cloud network architecture have been identified. Figure 4.4 shows a functional view of the network architecture for hybrid clouds.

4.4.1 Cloud-in-a-Box

As large enterprises start to build their own private clouds and further expand them into hybrid clouds, a significant need is to simplify the design, deployment, and management of clouds. The traditional data center deployment model of having separated physical devices focusing on server units, networking units, and storage units presents a significant challenge. A new trend in the design and deployment of private and hybrid clouds is the concept of "*cloud-in-a-box.*"

A cloud-in-a-box, sometimes also called a *cloud cell,* is a pre-integrated, pre-packaged and self-contained service delivery platform that can be used easily and quickly to implement private cloud centers. Physically, it is typically delivered in a single chassis containing multiple blades; some blades are computing units, some switching units, and some storage units. They are interconnected by a combination of a common backplane (e.g. a PCI-type backplane) and high-speed converged Ethernet connections (e.g. 10G FCoE). From the networking perspective, the switches that are pre-integrated into a cloud-in-a-box are typically the access layer switches.

Software wise, a common hypervisor environment typically expands across the computing units, the networking units, and storage units in a cloud-in-a-box device. From the networking perspective, this requires a virtual Ethernet switch to be embedded in the hypervisor. In the VMware environment, the VMware's vNetwork Distributed Switch and Cisco's Nexus 1000v virtual switch are the two well known examples of hypervisor-embedded virtual Ethernet switches. On top of the common virtualization layer, a service management application is typically included to allow the management and automation of cloud services provisioning, accounting and billing, security, dynamic resource reallocation and workload mobility. Furthermore, some of today's purpose-built cloud-in-a-box platforms also include a cloud service application to offer the specific cloud service. For example, a development-and-test oriented cloud-in-a-box platform may pre-integrate and pre-package a cloud-ready Integrated Development Environment (IDE) as part of the product.

At the time of this chapter is written, there are a number of cloud-in-a-box products offered in the industry. See (VCEC, 2009; IBM Corporation, 2009) for further information.

4.4.2 Network Service Node

Layer 4 network services play an important role in the network architecture for hybrid clouds. Application firewalls ensure the secure transport of user data and application workloads between the data centers in a hybrid cloud; server load balancers ensure the workloads distributed evenly or according to operations policies both within a single data center and across multiple data centers; WAN accelerators provide WAN optimization that accelerates the targeted cloud workloads over the WAN, and ensure a transparent user experience regardless where the applications reside.

While these Layer 4 services exist in today's data center environments, the proliferation of server virtualization in the cloud delivery model has created a significant challenge to the traditional network service architecture, as the Layer 4 services now need to be virtualization aware.

Visibility into virtual machine activity and isolation of server traffic becomes more difficult when virtual machine-sourced traffic can reach other virtual machines both within the same server and across the data center network and data center interconnect network. In the traditional access model, each physical server is connected to an access port. Any communication to and from a particular server or between servers goes through a physical access switch and any associated services such as a firewall or a load balancer. But what happens when applications now reside on virtual machines and multiple virtual machines reside within the same physical server? It might not be necessary for traffic to leave the physical server and pass through a physical access switch for one virtual machine to communicate with another. On the other hand, application residing in a virtual machine can be "moved" to another data center for load balancing. How to ensure the WAN accelerator to recognize an application residing within a virtual machine and optimize the WAN treatment for a virtual machine? Enforcing network policies in this type of environment can be a significant challenge. A network service node is a logical or a physical unit that provides the layer-4 network services to support cloud service deployment. The goal remains to provide many of the same network services and features used in the traditional access layer in the new virtualization-aware access layer. We believe this will be a fertile area for future research.

4.4.3 Data Center Network and Data Center Interconnect Network

Data center network and data center interconnect network are described before. Due to the length limitation of this chapter, we shall not expand beyond what has been described in Sections 4.2.4.1 and 4.2.4.2.

4.4.4 Management of the Network Architecture

Management of the network architecture in a hybrid cloud is part of the overall cloud management system. Key topics include the "physical" system management of the network infrastructure in the hybrid cloud and the "virtualization" management aspect that spans across the entire network path, starting from the virtual Ethernet switch embedded in the Hypervisor, through the access and core switches in the data center network, and across the data center interconnect network, as well as the network service modules along the network path.

Virtualization brings a new dimension to the management architecture. Similar to traditional "physical" system management, the network virtualization management needs to dynamically provision, monitor and manage end-to-end network resources and services between virtual machines in a cloud environment. In this context, a way to express workloads, network resources and operation policies in a virtualization-aware but hypervisor independent manner is the first step. Readers interested in more details in this area can start from DMTFb (2009). Once this is achieved, algorithms and systems can be developed to derive the network configurations and resource allocation based on the requirements from the virtual machine workloads. Similar to the "physical" system management, interoperability between the systems (e.g. between management system and the network, and between management systems) is an important requirement. For this purpose, common standards, open interfaces, common data model (management information model) are key. Currently this is still a less coordinated area where a number of standards bodies, including the Distributed Management Task Force (DMTF), the Object Management Group (OMG), the Open Grid Forum (OGF), etc., are working on various "standards" for cloud management. This is an area that needs more efforts to mature. Interested readers can start from DMTFa and Cloud Standards Coordination, http://cloud-standards.org.

4.5 Conclusions and Future Directions

As the next paradigm shift for IT industry, cloud computing is still in the early stage. Just as the previous major IT paradigm shift – from centralized computing to distributed computing – has had tremendous impact on IP networking (and vice versa), we see a similar impact with regard to cloud computing and the next generation networks. In many ways, supporting cloud computing represents a natural evolution for the IP networking; we see the Layer 2 domain in the data center network becoming wider, flatter and virtualization aware; we see the data center interconnect network and Layer 4 network services becoming virtualization aware and self-adaptable to security, performance and SLA constraints; we see virtual machine mobility and cloud service elasticity not only within a single data center but also over metro networks or WAN across multiple data centers. As the IT industry creates and deploys more cloud services, more requirements will be put on to the networks and more intelligence will be implemented by the cloud-enabling network.

Our belief is that the hybrid cloud will emerge as the ideal cloud deployment model for most enterprises since it blends the best of private and public clouds. The networks play an extremely critical role in enabling hybrid cloud deployments. For core services with critical business data, the private network within the hybrid cloud can allow full control over network security, performance, management, etc. The public side of the hybrid cloud provides the ability to extend an enterprise's reach to Internet deployed applications and services which can then be integrated with its on-premise assets and business processes. We believe more cloud-enabling innovations will occur in both data center networks and data center interconnect networks. Furthermore, we believe the public Internet will embrace many of the capabilities exhibited in today's data center interconnect networks (and expand beyond). Somewhat contrary to today's loosely coupled IP networking architecture (with respect to other IT assets – servers, storage, and applications) which was the resulted from the distributed client server computing model, we believe the cloud computing model will drive a more tightly integrated network architecture with other IT assets Mell & Grance (October 2009).

References

Armbrust, M., Fox, A., Griffith, R., Joseph, A. D., Katz, R., Konwinski, A., et al. (February 2009). *Above the clouds: A Berkeley view of cloud computing* (Tech. Rep. No. UCB/EECS-2009-28).

Babcock, C. (August 2009a). *Private clouds take shape*. Information Week Article.

Babcock, C. (September 2009b). *The public cloud: Infrastructure as a service*. InformationWeek Analytics Alerts.

Babcock, C. (September 2009c). *Hybrid clouds*. Information Week Analytics Alerts.

Cisco Systems (December 2008). *Cisco Nexus 2000 series fabric extenders*. http://www. ciscopowered.info/en/US/prod/collateral/switches/ps9441/ps10110/at_a_glance_c45-511599. pdf.

Cisco Systems (May 2009a). *Security and virtualization in the data center*. http://www.cisco. com/en/US/prod/collateral/switches/ps5718/ps708/white_paper_c11_493718.pdf.

Cisco Systems (July 2009b). *Data center interconnect: Layer 2 extension between remote data.* http://www.cisco.com/en/US/prod/collateral/switches/ps5718/ps708/white_paper_c11_493718. pdf. Accessed on February 2010.

Cisco Systems and VMware Inc. (August 2009c). *Virtual machine mobility with Vmware Vmotion and cisco data center interconnect technologies.* http://www.cisco.com/en/US/ solutions/collateral/ns340/ns517/ns224/ns836/white_paper_c11-557822.pdf. Accessed on February 2010.

Cisco Systems (October 2009d). *Data center design – IP network infrastructure.* http://www. cisco.com/en/US/docs/solutions/Enterprise/Data_Center/DC_3_0/DC-3_0_IPInfra.pdf. Accessed on February 2010.

Distributed Management Task Force (DMTFa). *DMTF open cloud standards incubator.* http://www.dmtf.org/about/cloud-incubator. Accessed on February 2010.

Distributed Management Task Force (DMTFb) (February 2009). *Open virtualization format specification.* http://www.dmtf.org/standards/published_documents/DSP0243_1.0.0.pdf. Accessed on February 2010.

Greenberg, A., Lahiri, P., Maltz, D. A., Patel, P., & Sengupta, S. (August 2008). Towards a next generation data center architecture: Scalability and commoditization. *Proceedings of the ACM Workshop on Programmable Router and Extensible Services for Tomorrow (PRESTO), Seattle, WA, USA,* 55–62.

Greenberg, A., Hamilton, J., Maltz, D. A., & Patel, P. (January 2009). The cost of a cloud: Research problems in data center networks. *ACM SIGCOMM Computer Communication Review, 39*(1), 68–73.

IBM Corporation (December 2009). *A breakthrough in service delivery for data center workloads – IBM cloudburst.* ftp://ftp.software.ibm.com/software/tivoli/products/cloudburst/ IBM_CloudBurst_data_sheet-December_2009.pdf.

Skorupa, J., Fabbi, M., Leong, L., Chamberlin, T., Pultz, J. E., Willis, D. A., (June 2008). You can't do cloud computing without the right cloud (Network) (Gartner Rep. no.: G00158513).

Kim, C., Caesar, M., & Rexford, J. (August 2008). Floodless in SEATTLE: A scalable Ethernet architecture for large enterprises. *Proceedings of the ACM SIGCOMM, Seattle, WA, USA,* 3–14.

Lin, G., Fu, D., Zhu, J., & Dasmalchi, G. (March/April 2009). Cloud computing: IT as a service. *IT Professional, 11*(2), 10–13.

Mell, P., & Grance, T. (October 2009). *The NIST definition of cloud computing.* http://csrc.nist. gov/groups/SNS/cloud-computing. Accessed on February 2010.

Chapter 5
Data-Intensive Technologies for Cloud Computing

Anthony M. Middleton

5.1 Introduction

As a result of the continuing information explosion, many organizations are drowning in data and the resulting "data gap" or inability to process this information and use it effectively is increasing at an alarming rate. Data-intensive computing represents a new computing paradigm (Kouzes, Anderson, Elbert, Gorton, & Gracio, 2009) which can address the data gap using scalable parallel processing to allow government, commercial organizations, and research environments to process massive amounts of data and implement applications previously thought to be impractical or infeasible. Cloud computing provides the opportunity for organizations with limited internal resources to implement large-scale data-intensive computing applications in a cost-effective manner.

The fundamental challenges of data-intensive computing are managing and processing exponentially growing data volumes, significantly reducing associated data analysis cycles to support practical, timely applications, and developing new algorithms which can scale to search and process massive amounts of data. Researchers at LexisNexis believe that the answer to these challenges is a scalable, integrated computer systems hardware and software architecture designed for parallel processing of data-intensive computing applications. This chapter explores the challenges of data-intensive computing and offers an in-depth comparison of commercially available system architectures including the LexisNexis Data Analytics Supercomputer (DAS) also referred to as the LexisNexis High-Performance Computing Cluster (HPCC), and Hadoop, an open source implementation based on Google's MapReduce architecture.

Cloud computing emphasizes the ability to scale computing resources as needed without a large upfront investment in infrastructure and associated ongoing operational costs (Napper & Bientinesi, 2009; Reese, 2009; Velte, Velte, & Elsenpeter, 2009). Cloud computing services are typically categorized in three models:

A.M. Middleton (✉)
LexisNexis Risk Solutions, Boca Raton, FL, USA
e-mail: tony.middleton@lexisnexis.com

B. Furht, A. Escalante (eds.), *Handbook of Cloud Computing*,
DOI 10.1007/978-1-4419-6524-0_5, © Springer Science+Business Media, LLC 2010

(1) *Infrastructure as a Service (IaaS)*. Service includes provision of hardware and software for processing, data storage, networks and any required infrastructure for deployment of operating systems and applications which would normally be needed in a data center managed by the user; (2) *Platform as a Service (PaaS)*. Service includes programming languages and tools and an application delivery platform hosted by the service provider to support development and delivery of end-user applications; and (3) *Software as a Service (SaaS)*. Hosted software applications are provided and managed by the service provider for the end-user replacing locally-run applications with Web-based applications (Lenk, Klems, Nimis, Tai, & Sandholm, 2009; Levitt, 2009; Mell & Grance, 2009; Vaquero, Rodero-Merino, Caceres, & Lindner, 2009; Viega, 2009).

Data-intensive computing applications are implemented using either the IaaS model which allows the provisioning of scalable clusters of processors for data-parallel computing using various software architectures, or the PaaS model which provides a complete processing and application development environment including both infrastructure and platform components such as programming languages and applications development tools. Data-intensive computing can be implemented in a *public cloud* (cloud infrastructure and platform is publicly available from a cloud services provider) such as Amazon's Elastic Compute Cloud (EC2) and Elastic MapReduce or as a *private cloud* (cloud infrastructure and platform is operated solely for a specific organization and may exist internally or externally to the organization) (Mell & Grance, 2009). IaaS and PaaS implementations for data-intensive computing can be either dynamically provisioned in virtualized processing environments based on application scheduling and data processing requirements, or can be implemented as a persistent high-availability configuration. A persistent configuration has a performance advantage since it uses dedicated infrastructure instead of virtualized servers shared with other users.

5.1.1 Data-Intensive Computing Applications

Parallel processing approaches can be generally classified as either *compute-intensive*, or *data-intensive* (Skillicorn & Talia, 1998; Gorton, Greenfield, Szalay, & Williams, 2008; Johnston, 1998). Compute-intensive is used to describe application programs that are compute bound. Such applications devote most of their execution time to computational requirements as opposed to I/O, and typically require small volumes of data. Parallel processing of compute-intensive applications typically involves parallelizing individual algorithms within an application process, and decomposing the overall application process into separate tasks, which can then be executed in parallel on an appropriate computing platform to achieve overall higher performance than serial processing. In compute-intensive applications, multiple operations are performed simultaneously, with each operation addressing a particular part of the problem. This is often referred to as functional parallelism or control parallelism (Abbas, 2004).

Data-intensive is used to describe applications that are I/O bound or with a need to process large volumes of data (Gorton et al., 2008; Johnston, 1998; Gokhale, Cohen, Yoo, & Miller, 2008). Such applications devote most of their processing time to I/O and movement of data. Parallel processing of data-intensive applications typically involves partitioning or subdividing the data into multiple segments which can be processed independently using the same executable application program in parallel on an appropriate computing platform, then reassembling the results to produce the completed output data (Nyland, Prins, Goldberg, & Mills, 2000). The greater the aggregate distribution of the data, the more benefit there is in parallel processing of the data. Gorton et al. (2008) state that data-intensive processing requirements normally scale linearly according to the size of the data and are very amenable to straightforward parallelization. The fundamental challenges for data-intensive computing according to Gorton et al. (2008) are managing and processing exponentially growing data volumes, significantly reducing associated data analysis cycles to support practical, timely applications, and developing new algorithms which can scale to search and process massive amounts of data. Cloud computing can address these challenges with the capability to provision new computing resources or extend existing resources to provide parallel computing capabilities which scale to match growing data volumes (Grossman, 2009).

5.1.2 Data-Parallelism

Computer system architectures which can support data-parallel applications are a potential solution to terabyte and petabyte scale data processing requirements (Nyland et al., 2000; Ravichandran, Pantel, & Hovy, 2004). According to Agichtein and Ganti (2004), parallelization is considered to be an attractive alternative for processing extremely large collections of data such as the billions of documents on the Web (Agichtein, 2004). Nyland et al. (2000) define data-parallelism as a computation applied independently to each data item of a set of data which allows the degree of parallelism to be scaled with the volume of data. According to Nyland et al. (2000), the most important reason for developing data-parallel applications is the potential for scalable performance, and may result in several orders of magnitude performance improvement. The key issues with developing applications using data-parallelism are the choice of the algorithm, the strategy for data decomposition, load balancing on processing nodes, message passing communications between nodes, and the overall accuracy of the results (Nyland et al., 2000; Rencuzogullari & Dwarkadas, 2001). Nyland et al. (2000) also note that the development of a data-parallel application can involve substantial programming complexity to define the problem in the context of available programming tools, and to address limitations of the target architecture. Information extraction from and indexing of Web documents is typical of data-intensive processing which can derive significant performance benefits from data-parallel implementations since Web and other types of document collections can typically then be processed in parallel (Agichtein, 2004).

5.1.3 The "Data Gap"

The rapid growth of the Internet and World Wide Web has led to vast amounts of information available online. In addition, business and government organizations create large amounts of both structured and unstructured information which needs to be processed, analyzed, and linked. Vinton Cerf of Google has described this as an "Information Avalanche" and has stated "we must harness the Internet's energy before the information it has unleashed buries us" (Cerf, 2007). An IDC white paper sponsored by EMC estimated the amount of information currently stored in a digital form in 2007 at 281 exabytes and the overall compound growth rate at 57% with information in organizations growing at even a faster rate (Gantz et al., 2007). In another study of the so-called information explosion it was estimated that 95% of all current information exists in unstructured form with increased data processing requirements compared to structured information (Lyman & Varian, 2003). The storing, managing, accessing, and processing of this vast amount of data represents a fundamental need and an immense challenge in order to satisfy needs to search, analyze, mine, and visualize this data as information (Berman, 2008). In 2003, LexisNexis defined this issue as the "Data Gap": the ability to gather information is far outpacing organizational capacity to use it effectively.

Organizations build the applications to fill the storage they have available, and build the storage to fit the applications and data they have. But will organizations be able to do useful things with the information they have to gain full and innovative use of their untapped data resources? As organizational data grows, how will the "Data Gap" be addressed and bridged? Researchers at LexisNexis believe that the answer is a scalable computer systems hardware and software architecture designed for data-intensive computing applications which can scale to processing billions of records per second (BORPS) (Note: the term BORPS was introduced by Seisint, Inc. in 2002. Seisint was acquired by LexisNexis in 2004). What are the characteristics of data-intensive computing systems and what system architectures are available to organizations to implement data-intensive computing applications? Can these capabilities be implemented using cloud computing to reduce risk and upfront investment in infrastructure and to allow a pay-as-you-go model? This chapter will explore those issues and offer a comparison of commercially available system architectures.

5.2 Characteristics of Data-Intensive Computing Systems

The National Science Foundation believes that data-intensive computing requires a "fundamentally different set of principles" than current computing approaches (NSF, 2009). Through a funding program within the Computer and Information Science and Engineering area, the NSF is seeking to "increase understanding of the capabilities and limitations of data-intensive computing." The key areas of focus are:

- Approaches to parallel programming to address the parallel processing of data on data-intensive systems
- Programming abstractions including models, languages, and algorithms which allow a natural expression of parallel processing of data
- Design of data-intensive computing platforms to provide high levels of reliability, efficiency, availability, and scalability.
- Identifying applications that can exploit this computing paradigm and determining how it should evolve to support emerging data-intensive applications.

Pacific Northwest National Labs has defined data-intensive computing as "capturing, managing, analyzing, and understanding data at volumes and rates that push the frontiers of current technologies" (Kouzes et al., 2009; PNNL, 2008). They believe that to address the rapidly growing data volumes and complexity requires "epochal advances in software, hardware, and algorithm development" which can scale readily with size of the data and provide effective and timely analysis and processing results. The HPCC architecture developed by LexisNexis represents such an advance in capabilities.

5.2.1 Processing Approach

Current data-intensive computing platforms use a "divide and conquer" parallel processing approach combining multiple processors and disks in large computing clusters connected using high-speed communications switches and networks which allows the data to be partitioned among the available computing resources and processed independently to achieve performance and scalability based on the amount of data (Fig. 5.1). Buyya, Yeo, Venugopal, Broberg, and Brandic (2009) define a cluster as "a type of parallel and distributed system, which consists of a collection

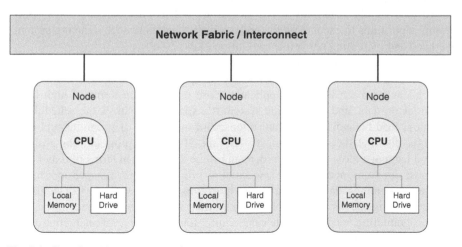

Fig. 5.1 Shared nothing computing cluster

of inter-connected stand-alone computers working together as a single integrated computing resource." This approach to parallel processing is often referred to as a "shared nothing" approach since each node consisting of processor, local memory, and disk resources shares nothing with other nodes in the cluster. In parallel computing this approach is considered suitable for data processing problems which are "embarrassingly parallel" , i.e. where it is relatively easy to separate the problem into a number of parallel tasks and there is no dependency or communication required between the tasks other than overall management of the tasks. These types of data processing problems are inherently adaptable to various forms of distributed computing including clusters and data grids and cloud computing.

5.2.2 Common Characteristics

There are several important common characteristics of data-intensive computing systems that distinguish them from other forms of computing. First is the principle of collocation of the data and programs or algorithms to perform the computation. To achieve high performance in data-intensive computing, it is important to minimize the movement of data (Gray, 2008). In direct contrast to other types of computing and supercomputing which utilize data stored in a separate repository or servers and transfer the data to the processing system for computation, data-intensive computing uses distributed data and distributed file systems in which data is located across a cluster of processing nodes, and instead of moving the data, the program or algorithm is transferred to the nodes with the data that needs to be processed. This principle – "Move the code to the data" – which was designed into the data-parallel processing architecture implemented by Seisint in 2003, is extremely effective since program size is usually small in comparison to the large datasets processed by data-intensive systems and results in much less network traffic since data can be read locally instead of across the network. This characteristic allows processing algorithms to execute on the nodes where the data resides reducing system overhead and increasing performance (Gorton et al., 2008).

A second important characteristic of data-intensive computing systems is the programming model utilized. Data-intensive computing systems utilize a machine-independent approach in which applications are expressed in terms of high-level operations on data, and the runtime system transparently controls the scheduling, execution, load balancing, communications, and movement of programs and data across the distributed computing cluster (Bryant, 2008). The programming abstraction and language tools allow the processing to be expressed in terms of data flows and transformations incorporating new dataflow programming languages and shared libraries of common data manipulation algorithms such as sorting. Conventional supercomputing and distributed computing systems typically utilize machine dependent programming models which can require low-level programmer control of processing and node communications using conventional imperative programming languages and specialized software packages which adds complexity to the parallel programming task and reduces programmer productivity. A machine dependent programming model also requires significant tuning and is more susceptible to single points of failure.

A third important characteristic of data-intensive computing systems is the focus on reliability and availability. Large-scale systems with hundreds or thousands of processing nodes are inherently more susceptible to hardware failures, communications errors, and software bugs. Data-intensive computing systems are designed to be fault resilient. This includes redundant copies of all data files on disk, storage of intermediate processing results on disk, automatic detection of node or processing failures, and selective re-computation of results. A processing cluster configured for data-intensive computing is typically able to continue operation with a reduced number of nodes following a node failure with automatic and transparent recovery of incomplete processing.

A final important characteristic of data-intensive computing systems is the inherent scalability of the underlying hardware and software architecture. Data-intensive computing systems can typically be scaled in a linear fashion to accommodate virtually any amount of data, or to meet time-critical performance requirements by simply adding additional processing nodes to a system configuration in order to achieve billions of records per second processing rates (BORPS). The number of nodes and processing tasks assigned for a specific application can be variable or fixed depending on the hardware, software, communications, and distributed file system architecture. This scalability allows computing problems once considered to be intractable due to the amount of data required or amount of processing time required to now be feasible and affords opportunities for new breakthroughs in data analysis and information processing.

5.2.3 Grid Computing

A similar computing paradigm known as grid computing has gained popularity primarily in research environments (Abbas, 2004). A computing grid is typically heterogeneous in nature (nodes can have different processor, memory, and disk resources), and consists of multiple disparate computers distributed across organizations and often geographically using wide-area networking communications usually with relatively low-bandwidth. Grids are typically used to solve complex computational problems which are compute-intensive requiring only small amounts of data for each processing node. A variation known as data grids allow shared repositories of data to be accessed by a grid and utilized in application processing, however the low-bandwidth of data grids limit their effectiveness for large-scale data-intensive applications.

In contrast, data-intensive computing systems are typically homogeneous in nature (nodes in the computing cluster have identical processor, memory, and disk resources), use high-bandwidth communications between nodes such as gigabit Ethernet switches, and are located in close proximity in a data center using high-density hardware such as rack-mounted blade servers. The logical file system typically includes all the disks available on the nodes in the cluster and data files are distributed across the nodes as opposed to a separate shared data repository such as a storage area network which would require data to be moved to nodes for processing. Geographically dispersed grid systems are more difficult to manage, less reliable, and less secure than data-intensive computing systems which are usually located in secure data center environments.

5.2.4 Applicability to Cloud Computing

Cloud computing can take many shapes. Most visualize the cloud as the Internet or Web which is often depicted in this manner, but a more general definition is that cloud computing shifts the location of the computing resources and infrastructure providing computing applications to the network (Vaquero et al., 2009). Software accessible through the cloud becomes a service, application platforms accessible through the cloud to develop and deliver new applications become a service, and hardware and software to create infrastructure and virtual data center environments accessible through the cloud becomes a service (Weiss, 2007). Other characteristics usually associated with cloud computing include a reduction in the costs associated with management of hardware and software resources (Hayes, 2008), pay-per-use or pay-as-you-go access to software applications and on-demand computing resources (Vaquero et al., 2009), dynamic provisioning of infrastructure and scalability of resources to match the size of the data and computing requirements which is directly applicable to the characteristics of data-intensive computing (Grossman & Gu, 2009). Buyya et al. (2009) provide the following comprehensive definition of a cloud: "A Cloud is a type of parallel and distributed system consisting of a collection of inter-connected and virtualized computers that are dynamically provisioned and presented as one or more unified computing resource(s) based on service-level agreements established through negotiation between the service provider and consumer."

The cloud computing models directly applicable to data-intensive computing characteristics are Infrastructure as a Service (IaaS) and Platform as a Service (PaaS). IaaS typically includes a large pool of configurable virtualized resources which can include hardware, operating systems, middleware, and development platforms or other software services which can be scaled to accommodate varying processing loads (Vaquero et al., 2009). The computing clusters typically used for data-intensive processing can be provided in this model. Processing environments such as Hadoop MapReduce and LexisNexis HPCC which include application development platform capabilities in addition to basic infrastructure implement the Platform as a Service (PaaS) model. Applications with a high degree of data-parallelism and a requirement to process very large datasets can take advantage of cloud computing and IaaS or PaaS using hundreds of computers provisioned for a short time instead of one or a small number of computers for a long time (Armbrust et al., 2009). According to Armbrust et al. in a University of California Berkeley research report (Armbrust et al., 2009), this processing model is particularly well-suited to data analysis and other applications that can benefit from parallel batch processing. However, the user cost/benefit analysis should also include the cost of moving large datasets into the cloud in addition the speedup and lower processing cost offered by the IaaS and PaaS models.

5.3 Data-Intensive System Architectures

A variety of system architectures have been implemented for data-intensive and large-scale data analysis applications including parallel and distributed relational database management systems which have been available to run on shared nothing clusters of processing nodes for more than two decades (Pavlo et al., 2009). These include database systems from Teradata, Netezza, Vertica, and Exadata/Oracle and others which provide high-performance parallel database platforms. Although these systems have the ability to run parallel applications and queries expressed in the SQL language, they are typically not general-purpose processing platforms and usually run as a back-end to a separate front-end application processing system. Although this approach offers benefits when the data utilized is primarily structured in nature and fits easily into the constraints of a relational database, and often excels for transaction processing applications, most data growth is with data in unstructured form (Gantz et al., 2007) and new processing paradigms with more flexible data models were needed. Internet companies such as Google, Yahoo, Microsoft, Facebook, and others required a new processing approach to effectively deal with the enormous amount of Web data for applications such as search engines and social networking. In addition, many government and business organizations were overwhelmed with data that could not be effectively processed, linked, and analyzed with traditional computing approaches.

Several solutions have emerged including the MapReduce architecture pioneered by Google and now available in an open-source implementation called Hadoop used by Yahoo, Facebook, and others. LexisNexis, an acknowledged industry leader in information services, also developed and implemented a scalable platform for data-intensive computing which is used by LexisNexis and other commercial and government organizations to process large volumes of structured and unstructured data. These approaches will be explained and contrasted in terms of their overall structure, programming model, file systems, and applicability to cloud computing in the following sections. Similar approaches using commodity computing clusters including Sector/Sphere (Grossman & Gu, 2008; Grossman, Gu, Sabala, & Zhang, 2009; Gu & Grossman, 2009), SCOPE/Cosmos (Chaiken et al., 2008), DryadLINQ (Yu, Gunda, & Isard, 2009), Meandre (Llor et al., 2008), and GridBatch (Liu & Orban, 2008) recently described in the literature are also suitable for data-intensive cloud computing applications and represent additional alternatives.

5.3.1 Google MapReduce

The MapReduce architecture and programming model pioneered by Google is an example of a modern systems architecture designed for processing and analyzing large datasets and is being used successfully by Google in many applications to process massive amounts of raw Web data (Dean & Ghemawat, 2004). The MapReduce architecture allows programmers to use a functional programming style to create a map function that processes a key-value pair associated with the input data to generate a set of intermediate key-value pairs, and a reduce function that merges all intermediate values associated with the same intermediate key (Dean & Ghemawat, 2004). According to Dean and Ghemawat (2004), the MapReduce programs can be used to compute derived data from documents such as inverted indexes and the processing is automatically parallelized by the system which executes on large clusters of commodity type machines, highly scalable to thousands of machines. Since the system automatically takes care of details like partitioning the input data, scheduling and executing tasks across a processing cluster, and managing the communications between nodes, programmers with no experience in parallel programming can easily use a large distributed processing environment.

The programming model for MapReduce architecture is a simple abstraction where the computation takes a set of input key-value pairs associated with the input data and produces a set of output key-value pairs. The overall model for this process is shown in Fig. 5.2. In the Map phase, the input data is partitioned into input splits

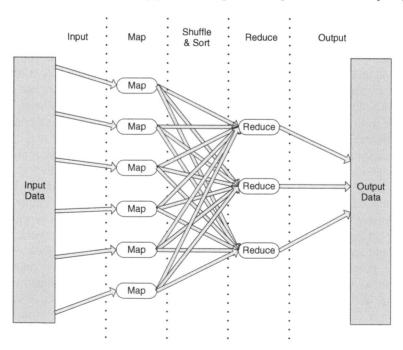

Fig. 5.2 MapReduce processing architecture (O'Malley, 2008)

and assigned to Map tasks associated with processing nodes in the cluster. The Map task typically executes on the same node containing its assigned partition of data in the cluster. These Map tasks perform user-specified computations on each input key-value pair from the partition of input data assigned to the task, and generates a set of intermediate results for each key. The shuffle and sort phase then takes the intermediate data generated by each Map task, sorts this data with intermediate data from other nodes, divides this data into regions to be processed by the reduce tasks, and distributes this data as needed to nodes where the Reduce tasks will execute. All Map tasks must complete prior to the shuffle and sort and reduce phases. The number of Reduce tasks does not need to be the same as the number of Map tasks. The Reduce tasks perform additional user-specified operations on the intermediate data possibly merging values associated with a key to a smaller set of values to produce the output data. For more complex data processing procedures, multiple MapReduce calls may be linked together in sequence.

Figure 5.3 shows the MapReduce architecture and key-value processing in more detail. The input data can consist of multiple input files. Each Map task will produce an intermediate output file for each key region assigned based on the number of Reduce tasks R assigned to the process (hash(key) modulus R). The reduce function then "pulls" the intermediate files, sorting and merging the files for a specific region from all the Map tasks. To minimize the amount of data transferred across the network, an optional Combiner function can be specified which is executed on the same node that performs a Map task. The combiner code is usually the same as

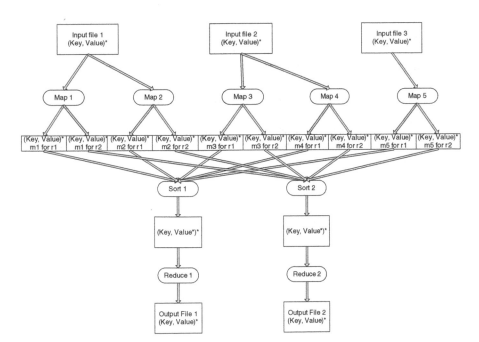

Fig. 5.3 MapReduce key-value processing (Nicosia, 2009)

the reducer function code which does partial merging and reducing of data for the local partition, then writes the intermediate files to be distributed to the Reduce tasks. The output of the Reduce function is written as the final output file. In the Google implementation of MapReduce, functions are coded in the C++ programming language.

Underlying and overlayed with the MapReduce architecture is the Google File System (GFS). GFS was designed to be a high-performance, scalable distributed file system for very large data files and data-intensive applications providing fault tolerance and running on clusters of commodity hardware (Ghemawat, Gobioff, & Leung, 2003). GFS is oriented to very large files dividing and storing them in fixed-size chunks of 64 Mb by default which are managed by nodes in the cluster called chunkservers. Each GFS consists of a single master node acting as a nameserver and multiple nodes in the cluster acting as chunkservers using a commodity Linux-based machine (node in a cluster) running a user-level server process. Chunks are stored in plain Linux files which are extended only as needed and replicated on multiple nodes to provide high-availability and improve performance. Secondary nameservers provide backup for the master node. The large chunk size reduces the need for MapReduce clients programs to interact with the master node, allows filesystem metadata to be kept in memory in the master node improving performance, and allows many operations to be performed with a single read on a chunk of data by the MapReduce client. Ideally, input splits for MapReduce operations are the size of a GFS chunk. GFS has proven to be highly effective for data-intensive computing on very large files, but is less effective for small files which can cause hot spots if many MapReduce tasks are accessing the same file.

Google has implemented additional tools using the MapReduce and GFS architecture to improve programmer productivity and to enhance data analysis and processing of structured and unstructured data. Since the GFS filesystem is primarily oriented to sequential processing of large files, Google has also implemented a scalable, high-availability distributed storage system for structured data with dynamic control over data format with keyed random access capabilities (Chang et al., 2006). Data is stored in Bigtable as a sparse, distributed, persistent multi-dimensional sorted map structured which is indexed by a row key, column key, and a timestamp. Rows in a Bigtable are maintained in order by row key, and row ranges become the unit of distribution and load balancing called a tablet. Each cell of data in a Bigtable can contain multiple instances indexed by the timestamp. Bigtable uses GFS to store both data and log files. The API for Bigtable is flexible providing data management functions like creating and deleting tables, and data manipulation functions by row key including operations to read, write, and modify data. Index information for Bigtables utilize tablet information stored in structures similar to a B+Tree. MapReduce applications can be used with Bigtable to process and transform data, and Google has implemented many large-scale applications which utilize Bigtable for storage including Google Earth.

Google has also implemented a high-level language for performing parallel data analysis and data mining using the MapReduce and GFS architecture called Sawzall and a workflow management and scheduling infrastructure for Sawzall jobs called

Workqueue (Pike, Dorward, Griesemer, & Quinlan, 2004). According to Pike et al. (2004), although C++ in standard MapReduce jobs is capable of handling data analysis tasks, it is more difficult to use and requires considerable effort by programmers. For most applications implemented using Sawzall, the code is much simpler and smaller than the equivalent C++ by a factor of 10 or more. A Sawzall program defines operations on a single record of the data, the language does not allow examining multiple input records simultaneously and one input record cannot influence the processing of another. An emit statement allows processed data to be output to an external aggregator which provides the capability for entire files of records and data to be processed using a Sawzall program. The system operates in a batch mode in which a user submits a job which executes a Sawzall program on a fixed set of files and data and collects the output at the end of a run. Sawzall jobs can be chained to support more complex procedures. Sawzall programs are compiled into an intermediate code which is interpreted during runtime execution. Several reasons are cited by Pike et al. (2004) why a new language is beneficial for data analysis and data mining applications: (1) a programming language customized for a specific problem domain makes resulting programs "clearer, more compact, and more expressive"; (2) aggregations are specified in the Sawzall language so that the programmer does not have to provide one in the Reduce task of a standard MapReduce program; (3) a programming language oriented to data analysis provides a more natural way to think about data processing problems for large distributed datasets; and (4) Sawzall programs are significantly smaller that equivalent C++ MapReduce programs and significantly easier to program.

Google does not currently make available its MapReduce architecture in a public cloud computing IaaS or PaaS environment. Google however does provide the Google Apps Engine as a public cloud computing PaaS environment (Lenk et al., 2009; Vaquero et al., 2009).

5.3.2 Hadoop

Hadoop is an open source software project sponsored by The Apache Software Foundation (http://www.apache.org). Following the publication in 2004 of the research paper describing Google MapReduce (Dean & Ghemawat, 2004), an effort was begun in conjunction with the existing Nutch project to create an open source implementation of the MapReduce architecture (White, 2009). It later became an independent subproject of Lucene, was embraced by Yahoo! after the lead developer for Hadoop became an employee, and became an official Apache top-level project in February of 2006. Hadoop now encompasses multiple subprojects in addition to the base core, MapReduce, and HDFS distributed filesystem. These additional subprojects provide enhanced application processing capabilities to the base Hadoop implementation and currently include Avro, Pig, HBase, ZooKeeper, Hive, and Chukwa. More information can be found at the Apache Web site.

The Hadoop MapReduce architecture is functionally similar to the Google implementation except that the base programming language for Hadoop is Java instead of C++. The implementation is intended to execute on clusters of commodity processors (Fig. 5.4) utilizing Linux as the operating system environment, but can also be run on a single system as a learning environment. Hadoop clusters also utilize the "shared nothing" distributed processing paradigm linking individual systems with local processor, memory, and disk resources using high-speed communications switching capabilities typically in rack-mounted configurations. The flexibility of Hadoop configurations allows small clusters to be created for testing and development using desktop systems or any system running Unix/Linux providing a JVM environment, however production clusters typically use homogeneous rack-mounted processors in a data center environment.

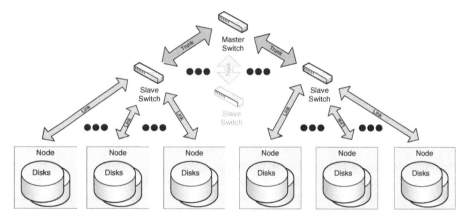

Fig. 5.4 Commodity hardware cluster (O'Malley, 2008)

The Hadoop MapReduce architecture is similar to the Google implementation creating fixed-size input splits from the input data and assigning the splits to Map tasks. The local output from the Map tasks is copied to Reduce nodes where it is sorted and merged for processing by Reduce tasks which produce the final output as shown in Fig. 5.5.

Hadoop implements a distributed data processing scheduling and execution environment and framework for MapReduce jobs. A MapReduce job is a unit of work that consists of the input data, the associated Map and Reduce programs, and user-specified configuration information (White, 2009). The Hadoop framework utilizes a master/slave architecture with a single master server called a jobtracker and slave servers called tasktrackers, one per node in the cluster. The jobtracker is the communications interface between users and the framework and coordinates the execution of MapReduce jobs. Users submit jobs to the jobtracker, which puts them in a job queue and executes them on a first-come/first-served basis. The jobtracker manages the assignment of Map and Reduce tasks to the tasktracker nodes which then execute these tasks. The tasktrackers also handle data movement between the Map and Reduce phases of job execution. The Hadoop framework assigns the Map tasks to

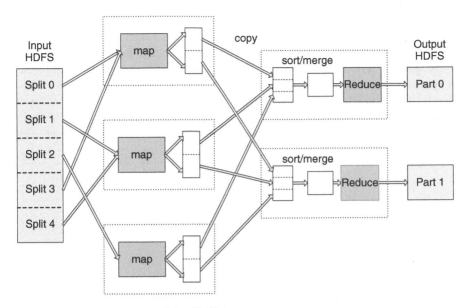

Fig. 5.5 Hadoop MapReduce (White, 2008)

every node where the input data splits are located through a process called data locality optimization. The number of Reduce tasks is determined independently and can be user-specified and can be zero if all of the work can be accomplished by the Map tasks. As with the Google MapReduce implementation, all Map tasks must complete before the shuffle and sort phase can occur and Reduce tasks initiated. The Hadoop framework also supports Combiner functions which can reduce the amount of data movement in a job.

The Hadoop framework also provides an API called Streaming to allow Map and Reduce functions to be written in languages other than Java such as Ruby and Python and provides an interface called Pipes for C++.

Hadoop includes a distributed file system called HDFS which is analogous to GFS in the Google MapReduce implementation. A block in HDFS is equivalent to a chunk in GFS and is also very large, 64 Mb by default but 128 Mb is used in some installations. The large block size is intended to reduce the number of seeks and improve data transfer times. Each block is an independent unit stored as a dynamically allocated file in the Linux local filesystem in a datanode directory. If the node has multiple disk drives, multiple datanode directories can be specified. An additional local file per block stores metadata for the block. HDFS also follows a master/slave architecture which consists of a single master server that manages the distributed filesystem namespace and regulates access to files by clients called the Namenode. In addition, there are multiple Datanodes, one per node in the cluster, which manage the disk storage attached to the nodes and assigned to Hadoop. The Namenode determines the mapping of blocks to Datanodes. The Datanodes are responsible for serving read and write requests from filesystem clients such as

MapReduce tasks, and they also perform block creation, deletion, and replication based on commands from the Namenode. An HDFS system can include additional secondary Namenodes which replicate the filesystem metadata, however there are no hot failover services. Each datanode block also has replicas on other nodes based on system configuration parameters (by default there are 3 replicas for each datanode block). In the Hadoop MapReduce execution environment it is common for a node in a physical cluster to function as both a Tasktracker and a datanode (Venner, 2009). The HDFS system architecture is shown in Fig. 5.6.

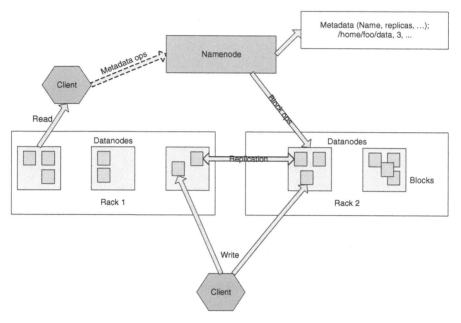

Fig. 5.6 HDFS architecture (Borthakur, 2008)

The Hadoop execution environment supports additional distributed data processing capabilities which are designed to run using the Hadoop MapReduce architecture. Several of these have become official Hadoop subprojects within the Apache Software Foundation. These include HBase, a distributed column-oriented database which provides similar random access read/write capabilities as and is modeled after Bigtable implemented by Google. HBase is not relational, and does not support SQL, but provides a Java API and a command-line shell for table management. Hive is a data warehouse system built on top of Hadoop that provides SQL-like query capabilities for data summarization, ad-hoc queries, and analysis of large datasets. Other Apache sanctioned projects for Hadoop include Avro – A data serialization system that provides dynamic integration with scripting languages, Chukwa – a data collection system for managing large distributed systems, ZooKeeper – a high-performance coordination service for distributed applications, and Pig – a high-level data-flow language and execution framework for parallel computation.

Pig is high-level dataflow-oriented language and execution environment originally developed at Yahoo! ostensibly for the same reasons that Google developed the Sawzall language for its MapReduce implementation – to provide a specific language notation for data analysis applications and to improve programmer productivity and reduce development cycles when using the Hadoop MapReduce environment. Working out how to fit many data analysis and processing applications into the MapReduce paradigm can be a challenge, and often requires multiple MapReduce jobs (White, 2009). Pig programs are automatically translated into sequences of MapReduce programs if needed in the execution environment. In addition Pig supports a much richer data model which supports multi-valued, nested data structures with tuples, bags, and maps. Pig supports a high-level of user customization including user-defined special purpose functions and provides capabilities in the language for loading, storing, filtering, grouping, de-duplication, ordering, sorting, aggregation, and joining operations on the data (Olston, Reed, Srivastava, Kumar, & Tomkins, 2008a). Pig is an imperative dataflow-oriented language (language statements define a dataflow for processing). An example program is shown in Fig. 5.7. Pig runs as a client-side application which translates Pig programs into MapReduce jobs and then runs them on an Hadoop cluster. Figure 5.8 shows how the program listed in Fig. 5.7 is translated into a sequence of MapReduce jobs. Pig compilation and execution stages include a parser, logical optimizer, MapReduce compiler, MapReduce optimizer, and the Hadoop Job Manager (Gates et al., 2009).

According to Yahoo! where more than 40% of Hadoop production jobs and 60% of ad-hoc queries are now implemented using Pig, Pig programs are 1/20th the size of the equivalent MapReduce program and take 1/16th the time to develop (Olston, 2009). Yahoo! uses 12 standard benchmarks (called the PigMix) to test Pig performance versus equivalent MapReduce performance from release to release. With the

```
visits       = load '/data/visits' as (user, url, time);
gVisits      = group visits by url;
visitCounts  = foreach gVisits generate url, count(urlVisits);

urlInfo      = load '/data/urlInfo' as (url, category, pRank);

visitCounts  = join visitCounts by url, urlInfo by url;
gCategories  = group visitCounts by category;
topUrls = foreach gCategories generate top(visitCounts,10);

store topUrls into '/data/topUrls';
```

Fig. 5.7   Sample pig latin program (Olston et al., 2008a)

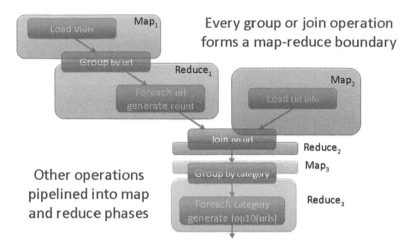

Fig. 5.8 Pig program translation to MapReduce (Olston et al., 2008a)

current release, Pig programs take approximately 1.5 times longer than the equivalent MapReduce (http://wiki.apache.org/pig/PigMix). Additional optimizations are being implemented that should reduce this performance gap further.

Hadoop is available in both public and private cloud computing environments. Amazon's EC2 cloud computing platform now includes Amazon Elastic MapReduce (http://aws.amazon.com/elasticmapreduce/) which allows users to provision as much capacity as needed for data-intensive computing applications. Data for MapReduce applications can be loaded to the HDFS directly from Amazon's S3 (Simple Storage Service).

5.3.3 LexisNexis HPCC

LexisNexis, an industry leader in data content, data aggregation, and information services independently developed and implemented a solution for data-intensive computing called the HPCC (High-Performance Computing Cluster) which is also referred to as the Data Analytics Supercomputer (DAS). The LexisNexis vision for this computing platform is depicted in Fig. 5.9. The development of this computing platform by the Seisint subsidiary of LexisNexis began in 1999 and applications were in production by late 2000. The LexisNexis approach also utilizes commodity clusters of hardware running the Linux operating system as shown in Figs. 5.1 and 5.4. Custom system software and middleware components were developed and layered on the base Linux operating system to provide the execution environment and distributed filesystem support required for data-intensive computing. Because LexisNexis recognized the need for a new computing paradigm to address its growing volumes of data, the design approach included the definition of a new high-level language for parallel data processing called ECL (Enterprise Data Control

Language). The power, flexibility, advanced capabilities, speed of development, and ease of use of the ECL programming language is the primary distinguishing factor between the LexisNexis HPCC and other data-intensive computing solutions. The following provides an overview of the HPCC systems architecture and the ECL language and a general comparison to the Hadoop MapReduce architecture and platform.

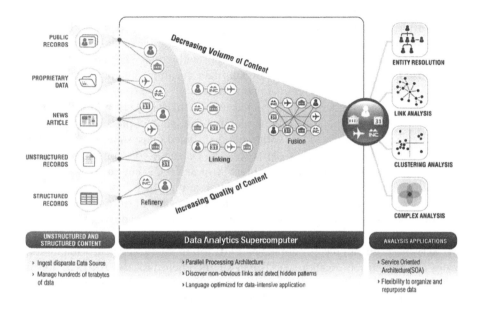

Fig. 5.9 LexisNexis vision for a data analytics supercomputer

LexisNexis developers recognized that to meet all the requirements of data-intensive computing applications in an optimum manner required the design and implementation of two distinct processing environments, each of which could be optimized independently for its parallel data processing purpose. The first of these platforms is called a Data Refinery whose overall purpose is the general processing of massive volumes of raw data of any type for any purpose but typically used for data cleansing and hygiene, ETL processing of the raw data (extract, transform, load), record linking and entity resolution, large-scale ad-hoc analysis of data, and creation of keyed data and indexes to support high-performance structured queries and data warehouse applications. The Data Refinery is also referred to as Thor, a reference to the mythical Norse god of thunder with the large hammer symbolic of crushing large amounts of raw data into useful information. A Thor system is similar in its hardware configuration, function, execution environment, filesystem, and capabilities to the Hadoop MapReduce platform, but offers significantly higher performance in equivalent configurations.

The second of the parallel data processing platforms designed and implemented by LexisNexis is called the Data Delivery Engine. This platform is designed as an online high-performance structured query and analysis platform or data warehouse delivering the parallel data access processing requirements of online applications through Web services interfaces supporting thousands of simultaneous queries and users with sub-second response times. High-profile online applications developed by LexisNexis such as Accurint utilize this platform. The Data Delivery Engine is also referred to as Roxie, which is an acronym for Rapid Online XML Inquiry Engine. Roxie uses a special distributed indexed filesystem to provide parallel processing of queries. A Roxie system is similar in its function and capabilities to Hadoop with HBase and Hive capabilities added, but provides significantly higher throughput since it uses a more optimized execution environment and filesystem for high-performance online processing. Most importantly, both Thor and Roxie systems utilize the same ECL programming language for implementing applications, increasing continuity and programmer productivity.

The Thor system cluster is implemented using a master/slave approach with a single master node and multiple slave nodes for data parallel processing. Each of the slave nodes is also a data node within the distributed file system for the cluster. This is similar to the Jobtracker, Tasktracker, and Datanode concepts in an Hadoop configuration. Multiple Thor clusters can exist in an HPCC environment, and job queues can span multiple clusters in an environment if needed. Jobs executing on a Thor cluster in a multi-cluster environment can also read files from the distributed file system on foreign clusters if needed. The middleware layer provides additional server processes to support the execution environment including ECL Agents and ECL Servers. A client process submits an ECL job to the ECL Agent which coordinates the overall job execution on behalf of the client process. An ECL Job is compiled by the ECL server which interacts with an additional server called the ECL Repository which is a source code repository and contains shared ECL code. ECL programs are compiled into optimized C++ source code, which is subsequently compiled into executable code and distributed to the slave nodes of a Thor cluster by the Thor master node. The Thor master monitors and coordinates the processing activities of the slave nodes and communicates status information monitored by the ECL Agent processes. When the job completes, the ECL Agent and client process are notified, and the output of the process is available for viewing or subsequent processing. Output can be stored in the distributed filesystem for the cluster or returned to the client process. ECL is analogous to the Pig language which can be used in the Hadoop environment.

The distributed filesystem used in a Thor cluster is record-oriented which is different from the block format used by Hadoop clusters. Records can be fixed or variable length, and support a variety of standard (fixed record size, CSV, XML) and custom formats including nested child datasets. Record I/O is buffered in large blocks to reduce latency and improve data transfer rates to and from disk Files to be loaded to a Thor cluster are typically first transferred to a landing zone from some external location, then a process called "spraying" is used to partition the file and load it to the nodes of a Thor cluster. The initial spraying process divides the file on user-specified record boundaries and distributes the data as evenly as possible in order across the available nodes in the cluster. Files can also be "desprayed" when needed to transfer output files to another system or can be directly copied between Thor clusters in the same environment.

Nameservices and storage of metadata about files including record format information in the Thor DFS are maintained in a special server called the Dali server (named for the developer's pet Chinchilla), which is analogous to the Namenode in HDFS. Thor users have complete control over distribution of data in a Thor cluster, and can re-distribute the data as needed in an ECL job by specific keys, fields, or combinations of fields to facilitate the locality characteristics of parallel processing. The Dali nameserver uses a dynamic datastore for filesystem metadata organized in a hierarchical structure corresponding to the scope of files in the system. The Thor DFS utilizes the local Linux filesystem for physical file storage, and file scopes are created using file directory structures of the local file system. Parts of a distributed file are named according to the node number in a cluster, such that a file in a 400-node cluster will always have 400 parts regardless of the file size. The Hadoop fixed block size can end up splitting logical records between nodes which means a node may need to read some data from another node during Map task processing. With the Thor DFS, logical record integrity is maintained, and processing I/O is completely localized to the processing node for local processing operations. In addition, if the file size in Hadoop is less than some multiple of the block size times the number of nodes in the cluster, Hadoop processing will be less evenly distributed and node to node disk accesses will be needed. If input splits assigned to Map tasks in Hadoop are not allocated in whole block sizes, additional node to node I/O will result. The ability to easily redistribute the data evenly to nodes based on processing requirements and the characteristics of the data during a Thor job can provide a significant performance improvement over the Hadoop approach. The Thor DFS also supports the concept of "superfiles" which are processed as a single logical file when accessed, but consist of multiple Thor DFS files. Each file which makes up a superfile must have the same record structure. New files can be added and old files deleted from a superfile dynamically facilitating update processes without the need to rewrite a new file. Thor clusters are fault resilient and a minimum of one replica of each file part in a Thor DFS file is stored on a different node within the cluster.

Roxie clusters consist of a configurable number of peer-coupled nodes functioning as a high-performance, high availability parallel processing query platform. ECL source code for structured queries is pre-compiled and deployed to the cluster. The Roxie distributed filesystem is a distributed indexed-based filesystem which uses a custom B+Tree structure for data storage. Indexes and data supporting queries are pre-built on Thor clusters and deployed to the Roxie DFS with portions of the index and data stored on each node. Typically the data associated with index logical keys is embedded in the index structure as a payload. Index keys can be multi-field and multivariate, and payloads can contain any type of structured or unstructured data supported by the ECL language. Queries can use as many indexes as required for a query and contain joins and other complex transformations on the data with the full expression and processing capabilities of the ECL language. For example, the LexisNexis Accurint comprehensive person report which produces many pages of output is generated by a single Roxie query.

A Roxie cluster uses the concept of Servers and Agents. Each node in a Roxie cluster runs Server and Agent processes which are configurable by a System Administrator depending on the processing requirements for the cluster. A Server process waits for a query request from a Web services interface then determines the nodes and associated Agent processes that have the data locally that is needed for a query, or portion of the query. Roxie query requests can be submitted from a client application as a SOAP call, HTTP or HTTPS protocol request from a Web application, or through a direct socket connection. Each Roxie query request is associated with a specific deployed ECL query program. Roxie queries can also be executed from programs running on Thor clusters. The Roxie Server process that receives the request owns the processing of the ECL program for the query until it is completed. The Server sends portions of the query job to the nodes in the cluster and Agent processes which have data needed for the query stored locally as needed, and waits for results. When a Server receives all the results needed from all nodes, it collates them, performs any additional processing, and then returns the result set to the client requestor.

The performance of query processing varies depending on factors such as machine speed, data complexity, number of nodes, and the nature of the query, but production results have shown throughput of a thousand results a second or more. Roxie clusters have flexible data storage options with indexes and data stored locally on the cluster, as well as being able to use indexes stored remotely in the same environment on a Thor cluster. Nameservices for Roxie clusters are also provided by the Dali server. Roxie clusters are fault-resilient and data redundancy is built-in using a peer system where replicas of data are stored on two or more nodes, all data including replicas are available to be used in the processing of queries by Agent processes. The Roxie cluster provides automatic failover in case of node failure, and the cluster will continue to perform even if one or more nodes are down. Additional redundancy can be provided by including multiple Roxie clusters in an environment.

Load balancing of query requests across Roxie clusters is typically implemented using external load balancing communications devices. Roxie clusters can be sized as needed to meet query processing throughput and response time requirements, but are typically smaller that Thor clusters.

The implementation of two types of parallel data processing platforms (Thor and Roxie) in the HPCC processing environment serving different data processing needs allows these platforms to be optimized and tuned for their specific purposes to provide the highest level of system performance possible to users. This is a distinct advantage when compared to the Hadoop MapReduce platform and architecture which must be overlayed with different systems such as HBase, Hive, and Pig which have different processing goals and requirements, and don't always map readily into the MapReduce paradigm. In addition, the LexisNexis HPCC approach incorporates the notion of a processing environment which can integrate Thor and Roxie clusters as needed to meet the complete processing needs of an organization. As a result, scalability can be defined not only in terms of the number of nodes in a cluster, but in terms of how many clusters and of what type are needed to meet system performance goals and user requirements. This provides a distinct advantage when compared to Hadoop clusters which tend to be independent islands of processing.

LexisNexis HPCC is commercially available to implement private cloud computing environments (http://risk.lexisnexis.com/Article.aspx?id=51). In addition, LexisNexis provides hosted persistent HPCC environments to external customers. Public cloud computing PaaS utilizing the HPCC platform is planned as a future offering.

5.3.4 ECL

The ECL programming language is a key factor in the flexibility and capabilities of the HPCC processing environment. ECL was designed to be a transparent and implicitly parallel programming language for data-intensive applications. It is a high-level, declarative, non-procedural dataflow-oriented language that allows the programmer to define what the data processing result should be and the dataflows and transformations that are necessary to achieve the result. Execution is not determined by the order of the language statements, but from the sequence of dataflows and transformations represented by the language statements. It combines data representation with algorithm implementation, and is the fusion of both a query language and a parallel data processing language. ECL uses an intuitive syntax which has taken cues from other familiar languages, supports modular code organization with a high degree of reusability and extensibility, and supports high-productivity for programmers in terms of the amount of code required for typical applications compared to traditional languages like Java and C++. Similar to the benefits Sawzall provides in the Google environment, and Pig provides to Hadoop users, a 20 times increase in programmer productivity is typical significantly reducing development cycles. ECL is compiled into optimized C++ code for execution on the HPCC system platforms, and can be used for complex data processing and analysis jobs on a Thor cluster or for comprehensive query and report processing on a Roxie cluster. ECL allows inline C++ functions to be incorporated into ECL programs, and external programs in other languages can be incorporated and parallelized through a PIPE facility. External services written in C++ and other languages which generate DLLs can also be incorporated in the ECL system library, and ECL programs can access external Web services through a standard SOAPCALL interface.

The basic unit of code for ECL is called an attribute. An attribute can contain a complete executable query or program, or a shareable and reusable code fragment such as a function, record definition, dataset definition, macro, filter definition, etc. Attributes can reference other attributes which in turn can reference other attributes so that ECL code can be nested and combined as needed in a reusable manner. Attributes are stored in ECL code repository which is subdivided into modules typically associated with a project or process. Each ECL attribute added to the repository effectively extends the ECL language like adding a new word to a dictionary, and attributes can be reused as part of multiple ECL queries and programs. With ECL a rich set of programming tools is provided including an interactive IDE similar to Visual C++, Eclipse and other code development environments.

The ECL language includes extensive capabilities for data definition, filtering, data management, and data transformation, and provides an extensive set of built-in functions to operate on records in datasets which can include user-defined transformation functions. Transform functions operate on a single record or a pair of records at a time depending on the operation. Built-in transform operations in the ECL language which process through entire datasets include PROJECT, ITERATE, ROLLUP, JOIN, COMBINE, FETCH, NORMALIZE, DENORMALIZE, and PROCESS. The transform function defined for a JOIN operation for example receives two records, one from each dataset being joined, and can perform any operations on the fields in the pair of records, and returns an output record which can be completely different from either of the input records. Example syntax for the JOIN operation from the ECL Language Reference Manual is shown in Fig. 5.10. Other important data operations included in ECL which operate across datasets and indexes include TABLE, SORT, MERGE, MERGEJOIN, DEDUP, GROUP, APPLY, ASSERT, AVE, BUILD, BUILDINDEX, CHOOSESETS, CORRELATION, COUNT, COVARIANCE, DISTRIBUTE, DISTRIBUTION, ENTH, EXISTS, GRAPH, HAVING, KEYDIFF, KEYPATCH, LIMIT, LOOP, MAX, MIN, NONEMPTY, OUTPUT, PARSE, PIPE, PRELOAD, PULL, RANGE, REGROUP, SAMPLE, SET, SOAPCALL, STEPPED, SUM, TOPN, UNGROUP, and VARIANCE.

The Thor system allows data transformation operations to be performed either locally on each node independently in the cluster, or globally across all the nodes in a cluster, which can be user-specified in the ECL language. Some operations such as PROJECT for example are inherently local operations on the part of a distributed file stored locally on a node. Others such as SORT can be performed either locally or globally if needed. This is a significant difference from the MapReduce architecture in which Map and Reduce operations are only performed locally on the input split assigned to the task. A local SORT operation in an HPCC cluster would sort the records by the specified key in the file part on the local node, resulting in the records being in sorted order on the local node, but not in full file order spanning all nodes. In contrast, a global SORT operation would result in the full distributed file being in sorted order by the specified key spanning all nodes. This requires node to node data movement during the SORT operation. Figure 5.11 shows a sample ECL program using the LOCAL mode of operation which is the equivalent of the sample PIG program for Hadoop shown in Fig. 5.7. Note the explicit programmer control over distribution of data across nodes. The colon-equals ":=" operator in an ECL program is read as "is defined as". The only action in this program is the OUTPUT statement, the other statements are definitions.

An additional important capability provided in the ECL programming language is support for natural language processing (NLP) with PATTERN statements and the built-in PARSE function. The PARSE function cam accept an unambiguous grammar defined by PATTERN, TOKEN, and RULE statements with penalties

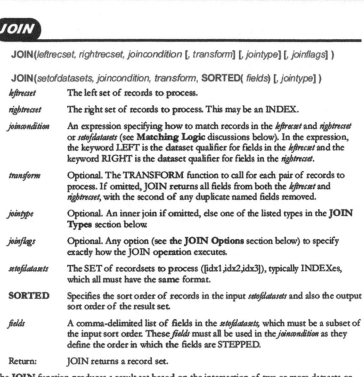

JOIN

JOIN(*leftrecset, rightrecset, joincondition* [*, transform*] [*, jointype*] [*, joinflags*])

JOIN(*setofdatasets, joincondition, transform,* **SORTED**(*fields*) [*, jointype*])

leftrecset	The left set of records to process.
rightrecset	The right set of records to process. This may be an INDEX.
joincondition	An expression specifying how to match records in the *leftrecset* and *rightrecset* or *setofdatasets* (see **Matching Logic** discussions below). In the expression, the keyword LEFT is the dataset qualifier for fields in the *leftrecset* and the keyword RIGHT is the dataset qualifier for fields in the *rightrecset*.
transform	Optional. The TRANSFORM function to call for each pair of records to process. If omitted, JOIN returns all fields from both the *leftrecset* and *rightrecset*, with the second of any duplicate named fields removed.
jointype	Optional. An inner join if omitted, else one of the listed types in the **JOIN Types** section below.
joinflags	Optional. Any option (see the **JOIN Options** section below) to specify exactly how the JOIN operation executes.
setofdatasets	The SET of recordsets to process ([idx1,idx2,idx3]), typically INDEXes, which all must have the same format.
SORTED	Specifies the sort order of records in the input *setofdatasets* and also the output sort order of the result set.
fields	A comma-delimited list of fields in the *setofdatasets*, which must be a subset of the input sort order. These *fields* must all be used in the *joincondition* as they define the order in which the fields are STEPPED.
Return:	JOIN returns a record set.

The **JOIN** function produces a result set based on the intersection of two or more datasets or indexes (as determined by the *joincondition*).

Fig. 5.10 ECL Sample syntax for JOIN operation

```
      Go                                          Queue:  dev_edserver_ ∨  Cluster:  thor400_88_de ∨    More
 1   // Sample ECL Code
 2   layout_visits := RECORD string user; string url; string time; END;
 3   visits := DATASET('~thor_data400::data::visits',layout_visits,FLAT);
 4
 5   layout_urlInfo := RECORD string url; string category; string pRank; END;
 6   urlInfo := DATASET('~thor_data400::data::urlinfo',layout_urlInfo,FLAT);
 7
 8   // Distribute Visits by URL, Count vists by URL
 9   layout_visitCounts := RECORD visits.url; visits_cnt := COUNT(GROUP); END;
10   visitCounts := TABLE(DISTRIBUTE(visits,HASH(url)),layout_visitCounts,url,LOCAL);
11
12   // Distribute Category by URL, Join category to URLs
13   visitCountsCat := JOIN(visitCounts,DISTRIBUTE(urlInfo,HASH(url)),LEFT.URL=RIGHT.URL,LOCAL);
14
15   // Distribute and Group by Category, Output top 10 URLs for each category
16   topUrls := TOPN(GROUP(DISTRIBUTE(visitCountsCat,HASH(category)),category,ALL,LOCAL),10, -visits_cnt);
17   OUTPUT(topUrls,,'~thor_data400::data::topurls',OVERWRITE);
```

Fig. 5.11 ECL code example

or preferences to provide deterministic path selection, a capability which can significantly reduce the difficulty of NLP applications. PATTERN statements allow matching patterns including regular expressions to be defined and used to parse information from unstructured data such as raw text. PATTERN statements can be combined to implement complex parsing operations or complete grammars from BNF definitions. The PARSE operation function across a dataset of records on a specific field within a record, this field could be an entire line in a text file for example. Using this capability of the ECL language it is possible to implement parallel processing for information extraction applications across document files including XML-based documents or Web pages. The key benefits of ECL can be summarized as follows:

- ECL incorporates transparent and implicit data parallelism regardless of the size of the computing cluster and reduces the complexity of parallel programming increasing the productivity of application developers.
- ECL enables implementation of data-intensive applications with huge volumes of data previously thought to be intractable or infeasible. ECL was specifically

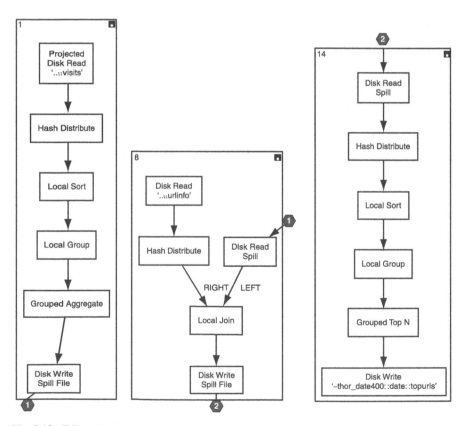

Fig. 5.12 ECL code example execution graph

designed for manipulation of data and query processing. Order of magnitude performance increases over other approaches are possible.

- ECL provides a comprehensive IDE and programming tools that provide a highly-interactive environment for rapid development and implementation of ECL applications.
- ECL is a powerful, high-level, parallel programming language ideal for implementation of ETL, Information Retrieval, Information Extraction, and other data-intensive applications.
- ECL is a mature and proven language but still evolving as new advancements in parallel processing and data-intensive computing occur.

5.4 Hadoop vs. HPCC Comparison

Hadoop and HPCC can be compared directly since it is possible for both systems to be executed on identical cluster hardware configurations. This permits head-to-head system performance benchmarking using a standard workload or set of application programs designed to test the parallel data processing capabilities of each system. A standard benchmark available for data-intensive computing platforms is the Terasort benchmark managed by an industry group led by Microsoft and HP. The Terabyte sort has evolved to be the GraySort which measures the number of terabytes per minute that can be sorted on a platform which allows clusters with any number of nodes to be utilized. However, in comparing the effectiveness and equivalent cost/performance of systems, it is useful to run benchmarks on identical system hardware configurations. A head-to-head comparison of the original Terabyte sort on a 400-node cluster will be presented here. An additional method of comparing system platforms is a feature and functionality comparison, which is a subjective evaluation based on factors determined by the evaluator. Although such a comparison contains inherent bias, it is useful in determining strengths and weaknesses of systems.

5.4.1 Terabyte Sort Benchmark

The Terabyte sort benchmark has its roots in benchmark tests sorting conducted on computer systems since the 1980s. More recently, a Web site originally sponsored by Microsoft and one of its research scientists Jim Gray has conducted formal competitions each year with the results presented at the SIGMOD (Special Interest Group for Management of Data) conference sponsored by the ACM each year (http://sortbenchmark.org). Several categories for sorting on systems exist including the Terabyte sort which was to measure how fast a file of 1 Terabyte of data formatted in 100 byte records (10,000,000 total records) could be sorted. Two categories were allowed called Daytona (a standard commercial computer system and software

with no modifications) and Indy (a custom computer system with any type of modification). No restrictions existed on the size of the system so the sorting benchmark
could be conducted on as large a system as desired. The current 2009 record holder
for the Daytona category is Yahoo! using a Hadoop configuration with 1460 nodes
with 8 GB Ram per node, 8000 Map tasks, and 2700 Reduce tasks which sorted
1 TB in 62 seconds (O'Malley & Murthy, 2009). In 2008 using 910 nodes, Yahoo!
performed the benchmark in 3 minutes 29 seconds. In 2008, LexisNexis using the
HPCC architecture on only a 400-node system performed the Terabyte sort benchmark in 3 minutes 6 seconds. In 2009, LexisNexis again using only a 400-node
configuration performed the Terabyte sort benchmark in 102 seconds.

However, a fair and more logical comparison of the capability of data-intensive
computer system and software architectures using computing clusters would be to
conduct this benchmark on the same hardware configuration. Other factors should
also be evaluated such as the amount of code required to perform the benchmark which is a strong indication of programmer productivity, which in itself is
a significant performance factor in the implementation of data-intensive computing
applications.

On August 8, 2009 a Terabyte Sort benchmark test was conducted on a development configuration located at LexisNexis Risk Solutions offices in Boca Raton,
FL in conjunction with and verified by Lawrence Livermore National Labs (LLNL).
The test cluster included 400 processing nodes each with two local 300 MB SCSI
disk drives, Dual Intel Xeon single core processors running at 3.00 GHz, 4 GB memory per node, all connected to a single Gigabit ethernet switch with 1.4 Terabytes/sec
throughput. Hadoop Release 0.19 was deployed to the cluster and the standard
Terasort benchmark written in Java included with the release was used for the benchmark. Hadoop required 6 minutes 45 seconds to create the test data, and the Terasort
benchmark required a total of 25 minutes 28 seconds to complete the sorting test
as shown in Fig. 5.13. The HPCC system software deployed to the same platform
and using standard ECL required 2 minutes and 35 seconds to create the test data,
and a total of 6 minutes and 27 seconds to complete the sorting test as shown in

---

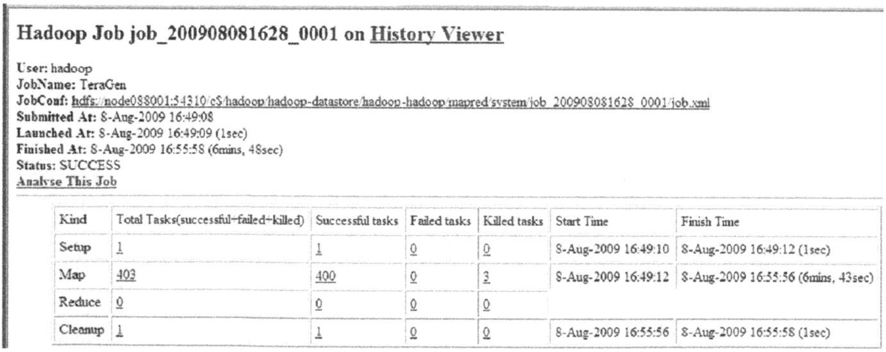

Hadoop Job job_200908081628_0001 on History Viewer

User: hadoop
JobName: TeraGen
JobConf: hdfs://node088001:54310/c$/hadoop/hadoop-datastore/hadoop-hadoop/mapred/system/job_200908081628_0001/job.xml
Submitted At: 8-Aug-2009 16:49:08
Launched At: 8-Aug-2009 16:49:09 (1sec)
Finished At: 8-Aug-2009 16:55:58 (6mins, 48sec)
Status: SUCCESS
Analyse This Job

Kind	Total Tasks(successful+failed+killed)	Successful tasks	Failed tasks	Killed tasks	Start Time	Finish Time
Setup	1	1	0	0	8-Aug-2009 16:49:10	8-Aug-2009 16:49:12 (1sec)
Map	403	400	0	3	8-Aug-2009 16:49:12	8-Aug-2009 16:55:56 (6mins, 43sec)
Reduce	0	0	0	0		
Cleanup	1	1	0	0	8-Aug-2009 16:55:56	8-Aug-2009 16:55:58 (1sec)

Fig. 5.13 Hadoop terabyte sort benchmark results

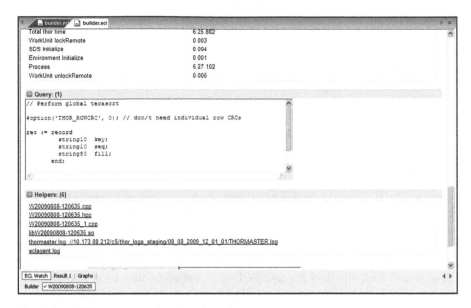

Fig. 5.14   HPCC terabyte sort benchmark results

Fig. 5.14. Thus the Hadoop implementation using Java running on the same hardware configuration took 3.95 times longer than the HPCC implementation using ECL.

The Hadoop version of the benchmark used hand-tuned Java code including custom TeraSort, TeraInputFormat and TeraOutputFormat classes with a total of 562 lines of code required for the sort. The HPCC system required only 10 lines of ECL code for the sort, a 50-times reduction in the amount of code required.

5.4.2 Pig vs. ECL

Although many Hadoop installations implement applications directly in Java, the Pig Latin language is now being used to increase programmer productivity and further simplify the programming of data-intensive applications at Yahoo! and other major users of Hadoop (Gates et al., 2009). Google also added a high-level language for similar reasons called Sawzall to its implementation of MapReduce to facilitate data analysis and data mining (Pike et al., 2004). The HPCC platform includes a high-level language discussed previously which is analogous to Pig and Sawzall called ECL. ECL is the base programming language used for applications on the HPCC platform even though it is compiled into C++ for execution. When comparing the Hadoop and HPCC platforms, it is useful to compare the features and functionality of these high-level languages.

Both Pig and ECL are intrinsically parallel, supporting transparent data-parallelism on the underlying platform. Pig and ECL are translated into programs that automatically process input data for a process in parallel with data distributed across a cluster of nodes. Programmers of both languages do not need to know the underlying cluster size or use this to accomplish data-parallel execution of jobs. Both Pig and ECL are dataflow-oriented, but Pig is an imperative programming language and ECL is a declarative programming language. A declarative language allows programmers to focus on the data transformations required to solve an application problem and hides the complexity of the underlying platform and implementation details, reduces side effects, and facilitates compiler optimization of the code and execution plan. An imperative programming language dictates the control flow of the program which may not result in an ideal execution plan in a parallel environment. Declarative programming languages allow the programmer to specify "what" a program should accomplish, instead of "how" to accomplish it. For more information, refer to the discussions of declarative (http://en.wikipedia.org/wiki/Declarative_programming) and imperative (http://en.wikipedia.org/wiki/Imperative_programming) programming languages on Wikipedia.

The source code for both Pig and ECL is compiled or translated into another language – Pig source programs are translated into Java language MapReduce jobs for execution and ECL programs are translated into C++ source code which is then compiled into a DLL for execution. Pig programs are restricted to the MapReduce architecture and HDFS of Hadoop, but ECL has no fixed framework other than the DFS (Distributed File System) used for HPCC and therefore can be more flexible in implementation of data operations. This is evident in two key areas: (1) ECL allows operations to be either global or local, where standard MapReduce is restricted to local operations only in both the Map and Reduce phases. Global operations process the records in a dataset in order across all nodes and associated file parts in sequence maintaining the records in sorted order as opposed to only the records contained in each local node which may be important to the data processing procedure; (2) ECL has the flexibility to implement operations which can process more than one record at a time such as its ITERATE operation which uses a sliding window and passes two records at a time to an associated transform function. This allows inter-record field-by-field dependencies and decisions which are not available in Pig. For example the DISTINCT operation in Pig which is used to remove duplicates does not allow this on a subset of fields. ECL provides both DEDUP and ROLLUP operations which are usually preceded by a SORT and operate on adjacent records in a sliding window mode and any condition relating to the field contents of the left and right record of adjacent records can be used to determine if the record is removed. ROLLUP allows a custom transformation to be applied to the de-duplication process.

An important consideration of any software architecture for data is the underlying data model. Pig incorporates a very flexible nested data model which allows non-atomic data types (atomic data types include numbers and strings) such as set, map, and tuple to occur as fields of a table (Olston, Reed, Srivastava, Kumar, & Tomkins, 2008b). Tuples are sequences of fields, bags are collections of tuples, and maps are a collection of data items where each data item has a key with which it can be looked up. A data record within Pig is called a relation which is an outer bag, the bag is a collection of tuples, each tuple is an ordered set of fields, and a field is a piece of data. Relations are referenced by a name assigned by a user. Types can be assigned by the user to each field, but if not assigned will default to a bytearray and conversions are applied depending on the context in which the field is used. The ECL data model also offers a nested data structure using child datasets. A user-specified RECORD definition defines the content of each record in a dataset which can contain fixed or variable length fields or child datasets which in turn contain fields or child datasets etc. With this format any type of data structure can be represented. ECL offers specific support for CSV and XML formats in addition to flat file formats. Each field in a record has a user-specified identifier and data type and an optional default value and optional field modifiers such as MAXLENGTH that enhance type and use checking during compilation. ECL will perform implicit casting and conversion depending on the context in which a field is used, and explicit user casting is also supported. ECL also allows in-line datasets allowing sample data to be easily defined and included in the code for testing rather than separately in a file.

The Pig environment offers several programmer tools for development, execution, and debugging of Pig Latin programs (Pig Latin is the formal name for the language, and the execution environment is called Pig, although both are commonly referred to as Pig). Pig provides command line execution of scripts and an interactive shell called Grunt that allows you to execute individual Pig commands or execute a Pig script. Pig programs can also be embedded in Java programs. Although Pig does not provide a specific IDE for developing and executing PIG programs, add-ins are available for several program editing environments including Eclipse, Vim, and Textmate to perform syntax checking and highlighting (White, 2009). PigPen is an Eclipse plug-in that provides program editing, an example data generator, and the capability to run a Pig script on a Hadoop cluster.

The HPCC platform provides an extensive set of tools for ECL development including a comprehensive IDE called QueryBuilder which allows program editing, execution, and interactive graph visualization for debugging and profiling ECL programs. The common code repository tree is displayed in QueryBuilder and tools are provided for source control, accessing and searching the repository. ECL jobs can be launched to an HPCC environment or specific cluster, and execution can be monitored directly from QueryBuilder. External tools are also provided including ECLWatch which provides complete access to current and historical workunits (jobs executed in the HPCC environment are packaged into workunits), queue management and monitoring, execution graph visualization, distributed filesystem utility functions, and system performance monitoring and analysis.

Although Pig Latin and the Pig execution environment provide a basic high-level language environment for data-intensive processing and analysis and increases the productivity of developers and users of the Hadoop MapReduce environment, ECL is a significantly more comprehensive and mature language that generates highly optimized code, offers more advanced capabilities in a robust, proven, integrated data-intensive processing architecture. Table 5.1 provides a feature to feature comparison between the Pig and ECL languages and their execution environments.

Table 5.1 Pig vs. ECL feature comparison

Language feature or capability	Pig	ECL
Language type	Data-flow oriented, imperative, parallel language for data-intensive computing. All Pig statements perform actions in sequentially ordered steps. Pig programs define a sequence of actions on the data.	Data-flow oriented, declarative, non-procedural, parallel language for data-intensive computing. Most ECL statements are definitions of the desired result which allows the execution plan to be highly optimized by the compiler. ECL actions such as OUTPUT cause execution of the dataflows to produce the result defined by the ECL program.
Compiler	Translated into a sequence of MapReduce Java programs for execution on a Hadoop Cluster. Runs as a client application.	Compiled and optimized into C++ source code which is compiled into DLL for execution on an HPCC cluster. Runs as a server application.
User-defined functions	Written in Java to perform custom processing and transformations as needed in Pig language statements. REGISTER is used to register a JAR file so that UDFs can be used.	Processing functions or TRANSFORM functions are written in ECL. ECL supports inline C++ in functions and external Services compiled into DLL libraries written in any language
Macros	Not supported	Extensive support for ECL macros to improve code reuse of common procedures. Additional template language for use in macros provides unique naming and conditional code generation capabilities.

Table 5.1 (continued)

Language feature or capability	Pig	ECL
Data model	Nested data model with named relations to define data records. Relations can include nested combinations of bags, tuples, and fields. Atomic data types include int, long, float, double, chararray, bytearray, tuple, bag, and map. If types not specified, default to bytearray then converted during expressions evaluation depending on the context as needed.	Nested data model using child datasets. Datasets contain fields or child datasets containing fields or additional child datasets. Record definitions describe the fields in datasets and child datasets. Indexes are special datasets supporting keyed access to data. Data types can be specified for fields in record definitions and include Boolean, integer, real, decimal, string, qstring, Unicode, data, varstring, varunicode, and related operators including set of (type), type of (expression) and record of (dataset) and ENUM (enumeration). Explicit type casting is available and implicit type casting may occur during evaluation of expressions by ECL depending on the context . Type transfer between types is also supported. All datasets can have an associated filter expression to include only records which meet the filter condition, in ECL a filtered physical dataset is called a recordset.
Distribution of data	Controlled by Hadoop MapReduce architecture and HDFS, no explicit programmer control provided. PARALLEL allows number of Reduce tasks to be specified. Local operations only are supported, global operations require custom Java MapReduce programs.	Explicit programmer control over distribution of data across cluster using DISTRIBUTE function. Helps avoid data skew. ECL supports both local (operations are performed on data local to node) and global (operations performed across nodes) modes.
Operators	Standard comparison operators; standard arithmetic operators and modulus division, Boolean operators AND, OR, NOT; null operators (is null, is not null); dereference operators for tuples and maps; explicit cast operator; minus and plus sign operators; matches operator.	Supports arithmetic operators including normal division, integer division, and modulus division; bitwise operators for AND, OR, and XOR; standard comparison operators; Boolean operators NOT, AND, OR; explicit cast operator; minus and plus sign operators; set and record set operators; string concatenation operator; sort descending and ascending operator; special operators IN, BETWEEN, WITHIN.

Table 5.1 (continued)

Language feature or capability	Pig	ECL
Conditional expression evaluation	The bincond operator is provided (condition ? true_value : false_value)	ECL includes an IF statement for single expression conditional evaluation, and MAP, CASE, CHOOSE, WHICH, and REJECTED for multiple expression evaluation. The ASSERT statement can be used to test a condition across a dataset. EXISTS can be used to determine if records meeting the specified condition exist in a dataset. ISVALID determines if a field contains a valid value.
Program loops	No capability exists other than the standard relation operations across a dataset. FOREACH … GENERATE provides nested capability to combine specific relation operations.	In addition to built-in data transform functions, ECL provides LOOP and GRAPH statements which allow looping of dataset operations or iteration of a specified process on a dataset until a loopfilter condition is met or a loopcount is satisfied.
Indexes	Not supported directly by Pig. HBase and Hive provide indexed data capability for Hadoop MapReduce which are accessible through custom user-defined functions in Pig.	Indexes can be created on datasets to support keyed access to data to improve data processing performance and for use on the Roxie data delivery engine for query applications.
Language statement types	Grouped into relational operators, diagnostic operators, UDF (user-defined function) statements, Eval functions, and load/store functions. The Grunt shell offers additional interactive file commands.	Grouped into dataset, index and record definitions, built-in functions to define processing and dataflows and workflow management, and actions which trigger execution. Functions include transform functions such as JOIN which operate on data records, and aggregation functions such as SUM. Action statements result in execution based on specified ECL definitions describing the dataflows and results for a process.
External program calls	PIG includes the STREAM statement to send data to an external script or program. The SHIP statement can be used to ship program binaries, jar files, or data to the Hadoop cluster compute nodes. The DEFINE statement, with INPUT, OUTPUT, SHIP, and CACHE clauses allow functions and commands to be associated with STREAM to access external programs.	ECL includes PIPE option on DATASET and OUTPUT and a PIPE function to execute external 3rd-party programs in parallel on nodes across the cluster. Most programs which receive an input file and parameters can adapted to run in the HPCC environment.

Table 5.1 (continued)

Language feature or capability	Pig	ECL
External web services access	Not supported directly by the Pig language. User-defined functions written in Java can provide this capability.	Built-in ECL function SOAPCALL for SOAP calls to access external Web Services. An entire dataset can be processed by a single SOAPCALL in an ECL program.
Data aggregation	Implemented in Pig using the GROUP, and FOREACH ... GENERATE statements performing EVAL functions on fields. Built-in EVAL functions include AVG, CONCAT, COUNT, DIFF, ISEMPTY, MAX, MIN, SIZE, SUM, TOKENIZE.	Implemented in ECL using the TABLE statement with group by fields specified and an output record definition that includes computed fields using expressions with aggregation functions performed across the specified group. Built-in aggregation functions which work across datasets or groups include AVE, CORRELATION, COUNT, COVARIANCE, MAX, MIN, SUM, VARIANCE.
Natural language processing	The TOKENIZE statement splits a string and outputs a bag of words. Otherwise no direct language support for parsing and other natural language processing. User-defined functions are required.	Includes PATTERN, RULE, TOKEN, and DEFINE statements for defining parsing patterns, rules, and grammars. Patterns can include regular expression definitions and user-defined validation functions. The PARSE statement provides both regular expression type parsing or Tomita parsing capability and recursive grammars. Special parsing syntax is included specifically for XML data.
Scientific function support	Not supported directly by the Pig language. Requires the definition and use of a user-defined function.	ECL provides built-in functions for ABS, ACOS, ASIN, ATAN, ATAN2, COS, COSH, EXP, LN, LOG, ROUND, ROUNDUP, SIN, SINH, SQRT, TAN, TANH.
Hashing functions for dataset distribution	No explicit programmer control for dataset distribution. PARALLEL option on relational operations allows the number of Reduce tasks to be specified.	Hashing functions available for use with the DISTRIBUTE statement include HASH, HASH32 (32-bit FNV), HASH64 (64-bit FNV), HASHCRC, HASHMD5 (128-bit MD5)

Table 5.1 (continued)

Language feature or capability	Pig	ECL
Creating sample datasets	The SAMPLE operation selects a random data sample with a specified sample size.	ECL provides ENTH which selects every nth record of a dataset, SAMPLE which provides the capability to select non-overlapping samples on a specified interval, CHOOSEN which selects the first n records of a dataset and CHOOSESETS which allows multiple conditions to be specified and the number of records that meet the condition or optionally a number of records that meet none of the conditions specified. The base dataset for each of the ENTH, SAMPLE, CHOOSEN, and CHOOSESETS can have a associated filter expression.
Workflow management	No language statements in Pig directly affect Workflow. The Hadoop cluster does allow Java MapReduce programs access to specific workflow information and scheduling options to manage execution.	Workflow Services in ECL include the CHECKPOINT and PERSIST statements allow the dataflow to be captured at specific points in the execution of an ECL program. If a program must be rerun because of a cluster failure, it will resume at last Checkpoint which is deleted after completion. The PERSIST files are stored permanently in the filesystem. If a job is repeated, persisted steps are only recalculated if the code has changed, or any underlying data has changed. Other workflow statements include FAILURE to trap expression evaluation failures, PRIORITY, RECOVERY, STORED, SUCCESS, WHEN for processing events, GLOBAL and INDEPENDENT.

Table 5.1 (continued)

Language feature or capability	Pig	ECL
PIG relation operations:		
Cogroup	The COGROUP operation is similar to the JOIN operation and groups the data in two or more relations (datasets) based on common field values. COGROUP creates a nested set of output tuples while JOIN creates a flat set of output tuples. INNER and OUTER joins are supported. Fields from each relation are specified as the join key. No support exists for conditional processing other than field equality.	In ECL, this is accomplished using the DENORMALIZE function joining to each dataset and adding all records matching the join key to a new record format with a child dataset for each child file. The DENORMALIZE function is similar to a JOIN and is used to form a combined record out of a parent and any number of children.
Cross	Creates the cross product of two or more relations (datasets).	In ECL the JOIN operation can be used to create cross products using a join condition that is always true.
Distinct	Removes duplicate tuples in a relation. All fields in the tuple must match. The tuples are sorted prior to this operation. Cannot be used on a subset of fields. A FOREACH … GENERATE statement must be used to generate the fields prior to a DISTINCT operation in this case.	The ECL DEDUP statement compares adjacent records to determine if a specified conditional expression is met, in which case the duplicate record is dropped and the remaining record is compared to the next record in a sliding window manner. This provides a much more flexible deduplication capability than the Pig DISTINCT operation. A SORT is required prior to a DEDUP unless using the ALL option. Conditions can use any expression and can reference values from the left and right adjacent records. DEDUP can use any subset of fields.
Dump	Displays the contents of a relation.	ECL provides an OUTPUT statement that can either write files to the filesystem or for display. Display files can be named and are stored in the Workunit associated with the job. Workunits are archived on a management server in the HPCC platform.

Table 5.1 (continued)

Language feature or capability	Pig	ECL
Filter	Selects tuples from a relation based on a condition. Used to select the data you want or conversely to filter out remove the data you don't want.	Filter expressions can be used any time a dataset or recordset is referenced in any ECL statement with the filter expression in parenthesis following the dataset name as dataset_name(filter_expression). The ECL compiler optimizes filtering of the data during execution based on the combination of filtering expressions.
Foreach … Generate	Generates data transformations based on columns of data. This action can be used for projection, aggregation, and transformation, and can include other operations in the generation clause such as FILTER, DISTINCT, GROUP, etc.	Each ECL transform operation such as PROJECT, JOIN, ROLLUP, etc. include a TRANSFORM function which implicitly provides the FOREACH … GENERATE operation as records are processed by the TRANSFORM function. Depending on the function, the output record of the transform can include fields from the input and computed fields selectively as needed and does not have to be identical to the input record.
Group	Groups together the tuples in a single relation that have the same group key fields.	The GROUP operation in ECL fragments a dataset into a set of sets based on the break criteria which is a list of fields or expressions based on fields in the record which function as the group by keys. This allows aggregations and transform operations such as ITERATE, SORT, DEDUP, ROLLUP and others to occur within defined subsets of the data as it executes on each subset individually.

Table 5.1 (continued)

Language feature or capability	Pig	ECL
Join	Joins two or more relations based on common field values. The JOIN operator always performs an inner join. If one relation is small and can be held in memory, the "replicated" option can be used to improve performance.	The ECL JOIN operation works on two datasets or a set of datasets. For two datasets INNER, FULL OUTER, LEFT OUTER, RIGHT OUTER, LEFT ONLY and RIGHT ONLY joins are permitted. For the set of datasets JOIN, INNER, LEFT OUTER, LEFT ONLY, and MOFN (min, max) joins are permitted. Any type of conditional expression referencing fields in the datasets to be joined can be used as a join condition. JOIN can be used in both a global and local modes also provides additional options for distribution including HASH which distributes the datasets by the specified join keys, and LOOKUP which copies one dataset if small to all nodes and is similar to the "replicated" join feature of Pig. Joins can also use keyed indexes to improve performance and self-joins (joining the same dataset to itself) is supported. Additional join-type operations provided by ECL include MERGEJOIN which joins and merges in a single operation, and smart stepping using STEPPED which provides a method of doing n-ary join/merge-join operations.
Limit	Used to limit the number of output tuples in a relation. However, there is no guarantee of which tuples will be output unless preceded by an ORDER statement.	The LIMIT function in ECL is to restrict the output of a recordset resulting from processing to a maximum number or records, or to fail the operation if the limit is exceeded. The CHOOSEN function can be use to select a specified number of records in a dataset.
Load	Loads data from the filesystem.	Since ECL is declarative, the equivalent of the Pig LOAD operation is a DATASET definition which also includes a RECORD definition. The examples shown in Figs 5.7 and Fig. 5.11 demonstrate this difference.

Table 5.1 (continued)

Language feature or capability	Pig	ECL
Order	Sorts a relation based on one or more fields. Both ascending and descending sorts are supported. Relations will be in order for a DUMP, but if the result of an ORDER is further processed by another relation operation, there is no guarantee the results will be processed in the order specified. Relations are considered to be unordered in Pig.	The ECL SORT function sorts a dataset according to a list of expressions or key fields. The SORT can be global in which the dataset will be ordered across the nodes in a cluster, or local in which the dataset will be ordered on each node in the cluster individually. For grouped datasets, the SORT applies to each group individually. Sorting operations can be performed using a quicksort, insertionsort, or heapsort, and can be stable or unstable for duplicates.
Split	Partitions a relation into two or more relations.	Since ECL is declarative, partitions are created by simply specifying filter expressions on the base dataset. Example for dataset DS1, you could define DS2 := DS1(filter_expression_1), DS3 := DS1(filter_expression_2), etc.
Store	Stores data to the file system.	The OUTPUT function in ECL is used to write a dataset to the filesystem or to store it in the workunit for display. Output files can be compressed using LZW compression. Variations of OUTPUT support flat file, CSV, and XML formats. Output can also be written to a PIPE as the standard input to the command specified for the PIPE operation. Output can write not only the filesystem on the local cluster, but to any cluster filesystem in the HPCC processing environment.
Union	The UNION operator is used to merge the contents of two or more relations into a single relation. Order of tuples is not preserved, both input and output relations are interpreted as an unordered bag of tuples. Does not eliminate duplicate tuples.	The MERGE function returns a single dataset or index containing all the datasets or indexes specified in a list of datasets. Datasets must have the same record format. A SORTED option allows the merge to be ordered according to a field list that specifies the sort order. A DEDUP option causes only records with unique keys to be included. The REGROUP function allows multiple datasets which have been grouped using the same fields to be merged into a single dataset.

Table 5.1 (continued)

Language feature or capability	Pig	ECL
Additional ECL transformation functions		ECL includes many additional functions providing important data transformations that are not available in Pig without implementing custom user-defined processing.
Combine	Not available	The COMBINE function combines two datasets into a single dataset on a record-by-record basis in the order in which they appear in each. Records from each are passed to the specified transform function, and the record format of the output dataset can contain selected fields from both input datasets and additional fields as needed.
Fetch	Not available	The FETCH function processes through all the records in an index dataset in the order specified by the index fetching the corresponding record from the base dataset and passing it through a specified transform function to create a new dataset.
Iterate	Not available	The ITERATE function processes through all records in a dataset one pair of records at a time using a sliding window method performing the transform record on each pair in turn. If the dataset is grouped, the ITERATE processes each group individually. The ITERATE function is useful in propagating information and calculating new information such as running totals since it allows inter-record dependencies to be considered.
Normalize	Use of FOREACH ... GENERATE is required	The NORMALIZE function normalizes child records out of a dataset into a separate dataset. The associated transform and output record format does not have to be the same as the input.

Table 5.1 (continued)

Language feature or capability	Pig	ECL
Process	Not available	The PROCESS function is similar to ITERATE and processes through all records in a dataset one pair of records at a time (left record, right record) using a sliding window method performing the associated transform function on each pair of records in turn. A second transform function is also specified that constructs the right record for the next comparison.
Project	Use of FOREACH . . . GENERATE is required	The PROJECT processes through all the records in a dataset performing the specified transform on each record in turn.
Rollup	Not available	The ROLLUP function is similar to the DEDUP function but includes a specified transform function to process each pair of duplicate records. This allows you to retrieve and use valuable information from the duplicate record before it is thrown away. Depending on how the ROLLUP is defined, either the left or right record passed to the transform can be retained, or any mixture of data from both.
Diagnostic operators	Pig includes diagnostic operators to aid in the visualization of data structures. The DESCRIBE operator returns the schema of a relation. The EXPLAIN operator allows you to review the logical, physical, and MapReduce execution plans that are used to compute an operation in a Pig script. The ILLUSTRATE operator displays a step-by-step execution of a sequence of statements allow you to see how data is transformed through a sequence of Pig Latin statements essentially dumping the output of each statement in the script.	The DISTRIBUTION action produces a crosstab report in XML format indicating how many records there are in a dataset for each value in each field in the dataset to aid in the analysis of data distribution in order to avoid skews. The QueryBuilder and ECLWatch program development environment tools provide a complete visualization tool for analyzing, debugging, and profiling execution of ECL jobs. During the execution of a job, the dataflows expressed by ECL can be viewed as a directed acyclic graph (DAG) which shows the execution plan, dataflows as they occur, and the results of each processing step. Users can double click on the graph to drill down for additional information. An example of the graph corresponding to the ECL code shown in Fig. 5.11 is shown in Fig. 5.12.

5.4.3 Architecture Comparison

Hadoop MapReduce and the LexisNexis HPCC platform are both scalable architectures directed towards data-intensive computing solutions. Each of these system platforms has strengths and weaknesses and their overall effectiveness for any application problem or domain is subjective in nature and can only be determined through careful evaluation of application requirements versus the capabilities of the solution. Hadoop is an open source platform which increases its flexibility and adaptability to many problem domains since new capabilities can be readily added by users adopting this technology. However, as with other open source platforms, reliability and support can become issues when many different users are contributing new code and changes to the system. Hadoop has found favor with many large Web-oriented companies including Yahoo!, Facebook, and others where data-intensive computing capabilities are critical to the success of their business. Amazon has implemented new cloud computing services using Hadoop as part of its EC2 called Amazon Elastic MapReduce. A company called Cloudera was recently formed to provide training, support and consulting services to the Hadoop user community and to provide packaged and tested releases which can be used in the Amazon environment. Although many different application tools have been built on top of the Hadoop platform like Pig, HBase, Hive, etc., these tools tend not to be well-integrated offering different command shells, languages, and operating characteristics that make it more difficult to combine capabilities in an effective manner.

However, Hadoop offers many advantages to potential users of open source software including readily available online software distributions and documentation, easy installation, flexible configurations based on commodity hardware, an execution environment based on a proven MapReduce computing paradigm, ability to schedule jobs using a configurable number of Map and Reduce tasks, availability of add-on capabilities such as Pig, HBase, and Hive to extend the capabilities of the base platform and improve programmer productivity, and a rapidly expanding user community committed to open source. This has resulted in dramatic growth and acceptance of the Hadoop platform and its implementation to support data-intensive computing applications.

The LexisNexis HPCC platform is an integrated set of systems, software, and other architectural components designed to provide data-intensive computing capabilities from raw data processing and ETL applications, to high-performance query processing and data mining. The ECL language was specifically implemented to provide a high-level dataflow parallel processing language that is consistent across all system components and has extensive capabilities developed and optimized over a period of almost 10 years. The LexisNexis HPCC is a mature, reliable, well-proven, commercially supported system platform used in government installations, research labs, and commercial enterprises. The comparison of the Pig Latin language and execution system available on the Hadoop MapReduce platform to the ECL language used on the HPCC platform presented here reveals that ECL provides significantly more advanced capabilities and functionality without the need for extensive user-defined functions written in another language or resorting to a native MapReduce application coded in Java.

The following comparison of overall features provided by the Hadoop and HPCC system architectures reveals that the HPCC architecture offers a higher level of integration of system components, an execution environment not limited by a specific computing paradigm such as MapReduce, flexible configurations and optimized processing environments which can provide data-intensive applications from data analysis to data warehousing and high-performance online query processing, and high programmer productivity utilizing the ECL programming language and tools. Table 5.2 provides a summary comparison of the key features of the hardware and software architectures of both system platforms based on the analysis of each architecture presented in this chapter.

5.5 Conclusions

As a result of the continuing information explosion, many organizations are drowning in data and the data gap or inability to process this information and use it effectively is increasing at an alarming rate. Data-intensive computing represents a new computing paradigm which can address the data gap and allow government and commercial organizations and research environments to process massive amounts of data and implement applications previously thought to be impractical or infeasible. Some organizations with foresight recognized early that new parallel-processing architectures were needed including Google who initially developed the MapReduce architecture and LexisNexis who developed the HPCC architecture. More recently the Hadoop platform has emerged as an open source alternative for the MapReduce approach. Hadoop has gained momentum quickly, and additional add-on capabilities to enhance the platform have been developed including a dataflow programming language and execution environment called Pig. These architectures, their relative strengths and weaknesses, and their applicability to cloud computing are described in this chapter, and a direct comparison of the Pig language of Hadoop to the ECL language used with the LexisNexis HPCC platform was presented. Availability of a high-level parallel dataflow-oriented programming language has proven to be a critical success factor in data-intensive computing.

The suitability of a processing platform and architecture for an organization and its application requirements can only be determined after careful evaluation of available alternatives. Many organizations have embraced open source platforms while others prefer a commercially developed and supported platform by an established industry leader. The Hadoop MapReduce platform is now being used successfully at many so-called Web companies whose data encompasses massive amounts of Web information as its data source. The LexisNexis HPCC platform is at the heart of a premier information services provider and industry leader, and has been adopted by government agencies, commercial organizations, and research laboratories because of its high-performance cost-effective implementation. Existing HPCC applications

Table 5.2 Hadoop vs. HPCC feature comparison

Architecture characteristic	Hadoop	HPCC
Hardware type	Processing clusters using commodity off-the-shelf (COTS) hardware. Typically rack-mounted blade servers with Intel or AMD processors, local memory and disk connected to a high-speed communications switch (usually Gigabit Ethernet connections) or hierarchy of communications switches depending on the total size of the cluster. Clusters are usually homogenous (all processors are configured identically), but this is not a requirement.	Same
Operating system	Unix/Linux and Windows (requires the installation of Cygwin)	Linux/Windows.
System configurations	Hadoop system software implements cluster with MapReduce processing paradigm. The cluster also functions as a distributed file system running HDFS. Other capabilities are layered on top of the Hadoop MapReduce and HDFS system software including HBase, Hive, etc.	HPCC clusters can be implemented in two configurations: Data Refinery (Thor) is analogous to the Hadoop MapReduce Cluster; Data Delivery Engine (Roxie) provides separate high-performance online query processing and data warehouse capabilities. Both configurations also function as distributed file systems but are implemented differently based on the intended use to improve performance. HPCC environments typically consist of multiple clusters of both configuration types. Although filesystems on each cluster are independent, a cluster can access files the filesystem on any other cluster in the same environment.
Licensing cost	None. Hadoop is an open source platform and can be freely downloaded and used.	License fees currently depend on size and type of system configurations. Does not preclude a future open source offering.

Table 5.2 (continued)

Architecture characteristic	Hadoop	HPCC
Core software	Core software includes the operating system and Hadoop MapReduce cluster and HDFS software Each slave node includes a Tasktracker service and Datanode service. A master node includes a Jobtracker service which can be configured as a separate hardware node or run on one of the slave hardware nodes. Likewise, for HDFS, a master Namenode service is also required to provide name services and can be run on one of the slave nodes or a separate node	For a Thor configuration, core software includes the operating system and various services installed on each node of the cluster to provide job execution and distributed file system access. A separate server called the Dali server provides filesystem name services and manages Workunits for jobs in the HPCC environment. A Thor cluster is also configured with a master node and multiple slave nodes. A Roxie cluster is a peer-coupled cluster where each node runs Server and Agent tasks for query execution and key and file processing. The filesystem on the Roxie cluster uses a distributed B+Tree to store index and data and provides keyed access to the data. Additional middleware components are required for operation of Thor and Roxie clusters.
Middleware components	None. Client software can submit jobs directly to the Jobtracker on the master node of the cluster. A Hadoop Workflow Scheduler (HWS) which will run as a server is currently under development to manage jobs which require multiple MapReduce sequences.	Middleware components include an ECL code repository implemented on a MySQL server, and ECL server for compiling of ECL programs and queries, an ECLAgent acting on behalf of a client program to manage the execution of a job on a Thor cluster, an ESPServer (Enterpise Services Platform) providing authentication, logging, security, and other services for the job execution and Web services environment, and the Dali server which functions as the system data store for job workunit information and provides naming services for the distributed filesystems. Flexibility exists for running the middleware components on one to several nodes. Multiple copies of these servers can provide redundancy and improve performance.

Table 5.2 (continued)

Architecture characteristic	Hadoop	HPCC
System tools	The dfsadmin tool provides information about the state of the filesystem; fsck is a utility for checking the health of files in HDFS; datanode block scanner periodically verifies all the blocks stored on a datanode; balancer re-distributes blocks from over-utilized datanodes to underutilized datanodes as needed. The MapReduce Web UI includes the JobTracker page which displays information about running and completed jobs, drilling down on a specific job displays detailed information about the job. There is also a Tasks page that displays info about Map and Reduce tasks.	HPCC includes a suite of client and operations tools for managing, maintaining, and monitoring HPCC configurations and environments. These include QueryBuilder the program development environment, an Attribute Migration Tool, Distributed File Utility (DFU), an Environment Configuration Utility, Roxie Configuration Utility. Command line versions are also available. ECLWatch is a Web based utility program for monitoring the HPCC environment and includes queue management, distributed file system management, job monitoring, and system performance monitoring tools. Additional tools are provided through Web services interfaces.
Ease of deployment	Assisted by online tools provided by Cloudera utilizing Wizards. Requires a manual RPM deployment.	Environment configuration tool. A Genesis server provides a central repository to distribute OS level settings, services, and binaries to all net-booted nodes in a configuration
Distributed file system	Block-oriented, uses large 64 MB or 128 MB blocks in most installations. Blocks are stored as independent units/local files in the local Unix/Linux filesystem for the node. Metadata information for blocks is stored in a separate file for each block. Master/Slave architecture with a single Namenode to provide name services and block mapping and multiple Datanodes. Files are divided into blocks and spread across nodes in the cluster. Multiple local files (1 containing the block, 1 containing metadata) for each logical block stored on a node are required to represent a distributed file.	The Thor DFS is record-oriented, uses local Linux filesystem to store file parts. Files are initially loaded (Sprayed) across nodes and each node has a single file part which can be empty for each distributed file. Files are divided on even record/document boundaries specified by the user. Master/Slave architecture with name services and file mapping information stored on a separate server. Only one local file per node required to represent a distributed file. Read/write access is supported between clusters configured in the same environment. Utilizing special adaptors allows files from external databases such as MySQL to be accessed, allowing transactional data to be integrated with DFS data and incorporated into batch jobs. The Roxie DFS utilizes distributed B+Tree index files containing key information and data stored in local files on each node.

Table 5.2 (continued)

Architecture characteristic	Hadoop	HPCC
Fault resilience	HDFS stores multiple replicas (user-specified) of data blocks on other nodes (configurable) to protect against disk and node failure with automatic recovery. MapReduce architecture includes speculative execution, when a slow or failed Map task is detected, additional Map tasks are started to recover from node failures	The DFS for Thor and Roxie stores replicas of file parts on other nodes (configurable) to protect against disk and node failure. Thor system offers either automatic or manual node swap and warm start following a node failure, jobs are restarted from last checkpoint or persist. Replicas are automatically used while copying data to the new node. Roxie system continues running following a node failure with a reduced number of nodes.
Job execution environment	Uses MapReduce processing paradigm with input data in key-value pairs. Master/Slave processing architecture. A Jobtracker runs on the master node, and a TaskTracker runs on each of the slave nodes. Each TaskTracker can be configured with a fixed number of slots for Map and Reduce tasks depending on available memory resources. Map tasks are assigned to input splits of the input file, usually 1 per block. The number of Reduce tasks is assigned by the user. Map processing is local to assigned node. A shuffle and sort operation is done following Map phase to distribute and sort key-value pairs to Reduce tasks based on key regions so that pairs with identical keys are processed by same Reduce tasks. Multiple MapReduce processing steps are typically required for most procedures and must be sequenced and chained separately by the user or language such as Pig. Job schedulers include FIFO (default), Fair, and Capacity depending on job sharing requirements, as well as Hadoop on Demand (HOD) for creating virtual clusters within a physical cluster.	Thor utilizes a Master/Slave processing architecture. Processing steps defined in an ECL job can specify local (data processed separately on each node) or global (data is processed across all nodes) operation. Multiple processing steps for a procedure are executed automatically as part of a single job based on an optimized execution graph for a compiled ECL dataflow program. A single Thor cluster can be configured to run multiple jobs concurrently reducing latency if adequate CPU and memory resources are available on each node. Middleware components including an ECLAgent, ECLServer, and DaliServer provide the client interface and manage execution of the job which is packaged as a Workunit. Roxie utilizes a multiple Server/Agent architecture to process ECL programs accessed by queries using Server tasks acting as a manager for each query and multiple Agent tasks as needed to retrieve and process data for the query.

Table 5.2 (continued)

Architecture characteristic	Hadoop	HPCC
Programming languages	Hadoop MapReduce jobs are usually written in Java. Other languages are supported through a streaming or pipe interface. Other processing environments execute on top of Hadoop MapReduce such as HBase and Hive which have their own language interface. The Pig Latin language and Pig execution environment provides a high-level dataflow language which is then mapped into multiple Java MapReduce jobs.	ECL is the primary programming language for the HPCC environment. ECL is compiled into optimized C++ which is then compiled into DLLs for execution on the Thor and Roxie platforms. ECL can include inline C++ code encapsulated in functions. External services can be written in any language and compiled into shared libraries of functions callable from ECL. A Pipe interface allows execution of external programs written in any language to be incorporated into jobs.
Integrated program development environment	Hadoop MapReduce utilizes the Java programming language and there are several excellent program development environments for Java including Netbeans and Eclipse which offer plug-ins for access to Hadoop clusters. The Pig environment does not have its own IDE, but instead uses Eclipse and other editing environments for syntax checking. A PigPen add-in for Eclipse provides access to Hadoop Clusters to run Pig programs and additional development capabilities.	The HPCC platform is provided with QueryBuilder, a comprehensive IDE specifically for the ECL language. QueryBuilder provides access to shared source code repositories and provides a complete development and testing environment for ECL dataflow programs. Access to the ECLWatch tool is built-in, allowing developers to watch job graphs as they are executing. Access to current and historical job Workunits allows developers to easily compare results from one job to the next during development cycles.
Database capabilities	The basic Hadoop MapReduce system does not provide any keyed access indexed database capabilities. An add-on system for Hadoop called HBase provides a column-oriented database capability with keyed access. A custom script language and Java interface is provided. Access to HBase is not directly supported by the Pig environment and requires user-defined functions or separate MapReduce procedures.	The HPCC platform includes the capability to build multi-key, multivariate indexes on DFS files. These indexes can be used to improve performance and provide keyed access for batch jobs on a Thor system, or be used to support development of queries deployed to Roxie systems. Keyed access to data is supported directly in ECL

Table 5.2 (continued)

Architecture characteristic	Hadoop	HPCC
Online query and data warehouse capabilities	The basic Hadoop MapReduce system does not provide any data warehouse capabilities. An add-on system for Hadoop called Hive provides data warehouse capabilities and allows HDFS data to be loaded into tables and accessed with an SQL-like language. Access to Hive is not directly supported by the Pig environment and requires user-defined functions or separate MapReduce procedures.	The Roxie system configuration in the HPCC platform is specifically designed to provide data warehouse capabilities for structured queries and data analysis applications. Roxie is a high-performance platform capable of supporting thousands of users and providing sub-second response time depending on the application.
Scalability	1 to thousands of nodes. Yahoo! has production clusters as large as 4000 nodes.	1 to several thousand nodes. In practice, HPCC configurations require significantly fewer nodes to provide the same processing performance as Hadoop. Sizing of clusters may depend however on the overall storage requirements for the distributed file system.
Performance	Currently the only available standard performance benchmarks are the sort benchmarks sponsored by http://sortbenchmark.org. Yahoo! has demonstrated sorting 1 TB on 1460 nodes in 62 seconds, 100 TB using 3452 nodes in 173 minutes, and 1 PB using 3658 nodes in 975 minutes.	The HPCC platform has demonstrated sorting 1 TB on a high-performance 400-node system in 102 seconds. In a recent head-to-head benchmark versus Hadoop on a another 400-node system conducted with LLNL, HPCC performance was 6 minutes 27 seconds and Hadoop performance was 25 minutes 28 seconds. This result on the same hardware configuration showed that HPCC was 3.95 times faster than Hadoop for this benchmark.
Training	Hadoop training is offered through Cloudera. Both beginning and advanced classes are provided. The advanced class includes Hadoop add-ons including HBase and Pig. Cloudera also provides a VMWare based learning environment which can be used on a standard laptop or PC. Online tutorials are also available.	Basic and advanced training classes on ECL programming are offered monthly in several locations or on customer premises. A system administration class is offered and scheduled as needed. A CD with a complete HPCC and ECL learning environment which can be used on a single PC or laptop is also available.

include raw data processing, ETL, and linking of enormous amounts of data to support online information services such as LexisNexis and industry-leading information search applications such as Accurint; entity extraction and entity resolution of unstructured and semi-structured data such as Web documents to support information extraction; statistical analysis of Web logs for security applications such as intrusion detection; online analytical processing to support business intelligence systems (BIS); and data analysis of massive datasets in educational and research environments and by state and federal government agencies.

There are many tradeoffs in making the right decision in choosing a new computer systems architecture, and often the best approach is to conduct a specific benchmark test with a customer application to determine the overall system effectiveness and performance. The relative cost-performance characteristics of the system in additional to suitability, flexibility, scalability, footprint, and power consumption factors which impact the total cost of ownership (TCO) must be considered. Cloud computing alternatives which reduce or eliminate up-front infrastructure investment should also be considered if internal resources are limited.

A comparison of the Hadoop MapReduce architecture to the HPCC architecture in this chapter reveals many similarities between the platforms including the use of a high-level dataflow-oriented programming language to implement transparent data-parallel processing. Both platforms are adaptable to cloud computing to provide platform as a service (PaaS). A key advantage to using the Hadoop architecture is its availability in a public cloud computing service offering. However, private cloud computing which utilizes persistent configurations with dedicated infrastructure instead of virtualized servers shared with other users common in public cloud computing can have a significant performance advantage for data-intensive computing applications. Some additional advantages of choosing the LexisNexis HPCC platform which can be utilized in private cloud computing include: (1) an architecture which implements a highly integrated system environment with capabilities from raw data processing to high-performance queries and data analysis using a common language; (2) an architecture which provides equivalent performance at a much lower system cost based on the number of processing nodes required as demonstrated with the Terabyte Sort benchmark where the HPCC platform was almost 4 times faster than Hadoop running on the same hardware resulting in significantly lower total cost of ownership (TCO); (3) an architecture which has been proven to be stable and reliable on high-performance data processing production applications for varied organizations over a 10-year period; (4) an architecture that uses a dataflow programming language (ECL) with extensive built-in capabilities for data-parallel processing which allows complex operations without the need for extensive user-defined functions and automatically optimizes execution graphs with hundreds of processing steps into single efficient workunits; (5) an architecture with a high-level of fault resilience and language capabilities which reduce the need for re-processing in case of system failures; and (6) an architecture which is available from and supported by a well-known leader in information services and risk solutions (LexisNexis) who is part of one of the world's largest publishers of information ReedElsevier.

References

Abbas, A. (2004). *Grid computing: A practical guide to technology and applications*. Hingham, MA: Charles River Media.

Agichtein, E. (2005). Scaling information extraction to large document collections. *IEEE Data Engineering Bulletin, 28*, 3–10.

Agichtein, E., & Ganti, V. (2004). Mining reference tables for automatic text segmentation. *Proceedings of the 10th ACM SIGKDD International Conference on Knowledge Discovery and Data Mining, Seattle, WA*, 20–29.

Armbrust, M., Fox, A., Griffith, R., Joseph, A. D., Katz, R., Konwinski, A., et al. (2009). *Above the clouds: A Berkeley view of cloud computing* (University of California at Berkely, Tech. Rep. UCB/EECS-2009-28).

Berman, F. (2008). Got data? A guide to data preservation in the information age. *Communications of the ACM, 51*(12), 50–56.

Borthakur, D. (2008). *Hadoop distributed file system*. Available from: http://www.opendocs.net/apache/hadoop/HDFSDescription.pdf.

Bryant, R. E. (2008). *Data intensive scalable computing*. Retrieved January 5, 2010, from: http://www.cs.cmu.edu/~bryant/presentations/DISC-concept.ppt.

Buyya, R., Yeo, C. S., Venugopal, S., Broberg, J., & Brandic, I. (2009). Cloud computing and emerging IT platforms: Vision, hype, and reality for delivering computing as the 5th utility. *Future Generation Computer Systems, 25*(6), 599–616.

Cerf, V. G. (2007). An information avalanche. *IEEE Computer, 40*(1), 104–105.

Chaiken, R., Jenkins, B., Larson, P.-A., Ramsey, B., Shakib, D., Weaver, S., et al. (2008). SCOPE: Easy and efficient parallel processing of massive data sets. *Proceedings of the VLDB Endowment, New York, NY*.

Chang, F., Dean, J., Ghemawat, S., Hsieh, W. C., Wallach, D. A., Burrows, M., et al. (2006). Bigtable: A distributed storage system for structured data. *Proceedings of the 7th Symposium on Operating Systems Design and Implementation (OSDI'06), Seattle, WA*.

Dean, J., & Ghemawat, S. (2004). MapReduce: Simplified data processing on large clusters. *Proceedings of the 6th Symposium on Operating System Design and Implementation (OSDI), Boston, MA*.

Gantz, J. F., Reinsel, D., Chute, C., Schlichting, W., McArthur, J., Minton, S., et al. (2007). *The expanding digital universe*. IDC, White Paper.

Gates, A. F., Natkovich, O., Chopra, S., Kamath, P., Narayanamurthy, S. M., Olston, C., et al. (2009). Building a high-level dataflow system on top of map-reduce: The pig experience. *Proceedings of the 35th International Conference on Very Large Databases (VLDB 2009), Lyon, France*.

Gokhale, M., Cohen, J., Yoo, A., & Miller, W. M. (2008). Hardware technologies for high-performance data-intensive computing. *IEEE Computer, 41*(4), 60–68.

Gorton, I., Greenfield, P., Szalay, A., & Williams, R. (2008). Data-intensive computing in the 21st century. *IEEE Computer, 41*(4), 30–32.

Ghemawat, S., Gobioff, H., & Leung, S.-T. (2003). The google file system. *Proceedings of the 19th ACM Symposium on Operating Systems Principles, New York, NY*.

Gray, J. (2008). Distributed computing economics. *ACM Queue, 6*(3), 63–68.

Grossman, R. L. (2009). The case for cloud computing. *IT Professional,11*(2), 23–27.

Grossman, R., & Gu, Y. (2008). Data mining using high performance data clouds: Experimental studies using sector and sphere. *Proceedings of the 14th ACM SIGKDD International Conference on Knowledge Discovery and Data Mining, New York, NY*.

Grossman, R. L., & Gu, Y. (2009). On the varieties of clouds for data intensive computing. Available from: http://sites.computer.org/debull/A09mar/grossman.pdf, 2009.

Grossman, R. L., Gu, Y., Sabala, M., & Zhang, W. (2009). Compute and storage clouds using wide area high performance networks. *Future Generation Computer Systems, 25*(2), 179–183.

Gu, Y., & Grossman, R. L. (2009). Lessons learned from a year's worth of benchmarks of large data clouds. *Proceedings of the 2nd Workshop on Many-Task Computing on Grids and Supercomputers, Portland, OR.*

Hayes, B. (2008). Cloud computing. *Communications of the ACM, 51*(7), 9–11.

Johnston,W. E. (1998). High-speed, wide area, data intensive computing: A ten year retrospective. *Proceedings of the 7th IEEE International Symposium on High-Performance Distributed Computing. Chicago, Illinois,* 280.

Kouzes, R. T., Anderson, G. A., Elbert, S. T., Gorton, I., & Gracio, D. K. (2009). The changing paradigm of data-intensive computing. *Computer, 42*(1), 26–34.

Lenk, A., Klems, M., Nimis, J., Tai, S., & Sandholm, T. (2009). What's inside the cloud? An architectural map of the cloud landscape. *Proceedings of the 2009 ICSE Workshop on Software Engineering Challenges of Cloud Computing. Vancouver, Canada,* 23–31.

Levitt, N. (2009). Is cloud computing really ready for prime time? *Computer, 42*(1), 15–20.

Liu, H., & Orban, D. (2008). GridBatch: Cloud computing for large-scale data-intensive batch applications. *Proceedings of the 8th IEEE International Symposium on Cluster Computing and the Grid, Cardiff.*

Llor, X., Acs, B., Auvil, L. S., Capitanu, B., Welge, M. E., & Goldberg, D. E. (2008). Meandre: Semantic-driven data-intensive flows in the clouds. *Proceedings of the 4th IEEE International Conference on eScience, Nottingham.*

Lyman, P., & Varian, H. R. (2003). *How much information?* (School of Information Management and Systems, University of California at Berkeley, Research Rep.).

Mell, P., & Grance, T. (2009). *The NIST definition of cloud computing.* Retrieved January 5, 2010, from: http://csrc.nist.gov/groups/SNS/cloud-computing/cloud-def-v15.doc.

Napper, J., & Bientinesi, P. (2009). Can cloud computing reach the Top500?. *Conference On Computing Frontiers. Proceedings of the combined workshops on UnConventional high performance computing workshop plus memory access workshop, Ischia, Italy.*

Nicosia, M. (2009). *Hadoop cluster management.* Retrieved January 5, 2010, from: http://wiki.apache.org/hadoop-data/attachments/HadoopPresentations/attachments/Hadoop-USENIX09.pdf.

Nyland, L. S., Prins, J. F., Goldberg, A., & Mills, P. H. (2000). A design methodology for data-parallel applications. *IEEE Transactions on Software Engineering, 26*(4), 293–314.

NSF. (2009). *Data-intensive computing.* Retrieved January 5, 2010, from: http://www.nsf.gov/funding/pgm_summ.jsp?pims_id=503324&org=IIS.

O'Malley, O. (2008). *Introduction to Hadoop.* Available from: http://wiki.apache.org/hadoop/HadoopPresentations/attachments/YahooHadoopIntro-apachecon-us-2008.pdf.

O'Malley, O., & Murthy, A. C. (2009). *Winning a 60 second dash with a yellow elephant.* Retrieved January 5, 2010, from: http://sortbenchmark.org/Yahoo2009.pdf.

Olston, C. (2009). *Pig overview presentation – Hadoop summit.* Retrieved January 5, 2010, from: http://infolab.stanford.edu/~olston/pig.pdf.

Olston, C., Reed, B., Srivastava, U., Kumar, R., & Tomkins, A. (2008a). *Pig Latin: A not-so-foreign language for data processing (Presentation at SIGMOD 2008).* Retrieved January 5, 2010, from: http://i.stanford.edu/~usriv/talks/sigmod08-pig-latin.ppt#283,18,User-Code as a First-Class Citizen.

Olston, C., Reed, B., Srivastava, U., Kumar, R., & Tomkins, A. (2008b). Pig Latin: A not-so_foreign language for data processing. *Proceedings of the 28th ACM SIGMOD/PODS International Conference on Management of Data/Principles of Database Systems, Vancouver, BC.*

Pavlo, A., Paulson, E., Rasin, A., Abadi, D. J., Dewitt, D. J., Madden, S., et al. (2009). A comparison of approaches to large-scale data analysis. *Proceedings of the 35th SIGMOD International Conference on Management of Data, New York, NY.*

PNNL. (2008). *Data intensive computing.* Retrieved January 5, 2010, from: http://www.cs.cmu.edu/~bryant/presentations/DISC-concept.ppt.

Pike, R., Dorward, S., Griesemer, R., & Quinlan, S. (2004). Interpreting the data: Parallel analysis with Sawzall. *Scientific Programming Journal, 13*(4), 227–298.

Ravichandran, D., Pantel, P., & Hovy, E. (2004). The terascale challenge. *Proceedings of the KDD Workshop on Mining for and from the Semantic Web, Boston, MA.*

Rencuzogullari, U., & Dwarkadas, S. (2001). Dynamic adaptation to available resources for parallel computing in an autonomous network of workstations. *Proceedings of the 8th ACM SIGPLAN Symposium on Principles and Practices of Parallel Programming, San Diego, CA,* 72–81.

Reese, G. (2009). *Cloud application architectures.* Sebastopol, CA: O'Reilly.

Skillicorn, D. B., & Talia, D. (1998). Models and languages for parallel computation. *ACM Computing Surveys, 30*(2), 123–169.

Vaquero, L. M., Rodero-Merino, L., Caceres, J., & Lindner, M. (2009). A break in the clouds: Towards a cloud definition. *SIGCOMM Computer Communication Review, 39*(1), 50–55.

Velte, A. T., Velte, T. J., & Elsenpeter, R. (2009). *Cloud computing: A practical approach.* New York, NY: McGraw Hill.

Venner, J. (2009). *Pro Hadoop.* New York, NY: Apress.

Viega, J. (2009). Cloud computing and the common man. *Computer, 42*(8), 106–108.

Weiss, A. (2007). Computing in the clouds. *netWorker, 11*(4), 16–25.

White, T. (2008). Understanding map reduce with Hadoop. Available from: http://wiki.apache.org/hadoop/HadoopPresentations.

White, T. (2009). Hadoop: The definitive guide. Sebastopol, CA: O'Reilly Media.

Yu, Y., Gunda, P. K., & Isard, M. (2009). Distributed aggregation for data-parallel computing: Interfaces and implementations. *Proceedings of the ACM SIGOPS 22nd Symposium on Operating Systems Principles, Big Sky, MT.*

Chapter 6
Survey of Storage and Fault Tolerance Strategies Used in Cloud Computing

Kathleen Ericson and Shrideep Pallickara

6.1 Introduction

Cloud computing has gained significant traction in recent years. Companies such as Google, Amazon and Microsoft have been building massive data centers over the past few years. Spanning geographic and administrative domains, these data centers tend to be built out of commodity desktops with the total number of computers managed by these companies being in the order of millions. Additionally, the use of virtualization allows a physical node to be presented as a set of virtual nodes resulting in a seemingly inexhaustible set of computational resources. By leveraging economies of scale, these data centers can provision cpu, networking, and storage at substantially reduced prices which in turn underpins the move by many institutions to host their services in the cloud.

In this chapter we will be surveying the most dominant storage and fault tolerant strategies that are currently being used in cloud computing settings. There are several unifying themes that underlie the systems that we survey.

6.1.1 Theme 1: Voluminous Data

The datasets managed by these systems tend to be extremely voluminous. It is not unusual for these datasets to be several terabytes. The datasets also tend to be generated by programs, services and devices as opposed to being created by a user one character at a time. In 2000, the Berkeley "How Much Information?" report (Lyman & Varian, 2000) reported that there was an estimated 25–50 TB of data on the web. In 2003 ((Lyman & Varian, 2003), the same group reported that there were approximately 167 TB of information on the web. The Large Hadron Collider (LHC) is

K. Ericson and S. Pallickara (✉)

Department of Computer Science, Colorado State University, Fort Collins, CO, USA

e-mails: {ericson; shrideep}@cs.colostate.edu

B. Furht, A. Escalante (eds.), *Handbook of Cloud Computing*,
DOI 10.1007/978-1-4419-6524-0_6, © Springer Science+Business Media, LLC 2010

expected to produce 15 PB/year (Synodinos, 2008). The amount of data being generated has been growing on an exponential scale – there are growing challenges not only in how to effectively process this data, but also with basic storage.

6.1.2 Theme 2: Commodity Hardware

The storage infrastructure for these datasets tend to rely on commodity hard drives that have rotating disks. This mechanical nature of the disk drives limits their performance. While processor speeds have grown exponentially disk access times have not kept pace. The performance disparity between processor and disk access times is in the order of 14,000,000:1 and continues to grow (Robbins & Robbins).

6.1.3 Theme 3: Distributed Data

A given dataset is seldom stored on a given node, and is typically distributed over a set of available nodes. This is done because a single commodity hard drive typically cannot hold the entire dataset. Scattering the dataset on a set of available nodes is also a precursor for subsequent concurrent processing being performed on the dataset.

6.1.4 Theme 4: Expect Failures

Since the storage infrastructure relies on commodity components, failures should be expected. The systems thus need to have a failure model in place that can ensure continued progress and acceptable response times despite any failures that might have taken place. Often these datasets are replicated, and individual slices of these datasets have checksums associated with them to detect bit-flips and the concomitant data corruptions that often taken place in commodity hardware.

6.1.5 Theme 5: Tune for Access by Applications

Though these storage frameworks are built on top of existing file systems, the stored datasets are intended to be processed by applications and not humans. Since the dataset is scattered on a large number of machines, reconstructing the dataset requires processing the *metadata* (data describing the data) to identify the precise location of specific portions of the datasets. Manually accessing any of the nodes to look for a portion of the dataset is futile since these portions have themselves been modified to include checksum information.

6.1.6 Theme 6: Optimize for Dominant Usage

Another important consideration in these storage frameworks is optimizing the most general access patterns for these datasets. In some cases, this would mean optimizing for long, sequential reads that puts a premium on conserving bandwidth while in others it would involve optimizing small, continuous updates to the managed datasets.

6.1.7 Theme 7: Tradeoff Between Consistency and Availability

Since these datasets are dispersed (and replicated) on a large number of machines accounting for these failures entails a tradeoff between consistency and availability. Most of these storage frameworks opt for availability and rely on eventual consistency. This choice has its roots in the CAP theorem. In 2000, Brewer theorized that it was impossible for a web service to provide full guarantees of Consistency, Availability, and Partition-tolerance (Brewer, 2000). In 2002, Seth Gilbert and Nancy Lynch at MIT proved this theorem (Gilbert & Lynch, 2002). While Brewer's theorem was geared towards web services, any distributed file system can be viewed as such. In some cases, such as Amazon's S3, it is easier to see this connection than others. Before delving deeper, what do we mean by Consistency, Availability, and Partition-tolerance?

Having a consistent distributed system means that no matter what node you connect to, you will always find the same exact data. Here, we take availability to mean that as long as a request is sent to a node that has not failed it will return a result. This definition has no bound on time limit, it simply states that eventually a client will get a response. Last, there is partition tolerance. A partition occurs when one part of your distributed system can no longer communicate with another part, but can still communicate with clients. The simplest example of this is in a system with 2 nodes, **A** and **B**. If **A** and **B** can no longer communicate with each other, but both can and do keep serving clients, then the system is partition tolerant. With a partition-tolerant system, nothing short of a full system failure keeps the system from working correctly.

As a quick example, let's look at a partition-tolerant system with two nodes **A** and **B**. Let's suppose there is some network error between **A** and **B**, and they can no longer communicate with each other, but both can still connect to clients. If a client were to write a change a file v hosted on both **A** and **B** while connected to **B**, the change would go through on **B**, but if the client later connects to **A** and reads v again, the client will not see their changes, so the system is no longer consistent. You could get around this by instead sacrificing availability – if you ignore writes during a network partition, you can maintain consistency.

In this chapter we will be reviewing storage frameworks from the three dominant cloud computing providers – Google, Amazon and Microsoft. We profile each storage framework along dimensions that include inter alia replication, failure model, replication and security. Our description of each framework is self-contained, and the reader can peruse these frameworks in any order. For completeness we have included a description of the xFS system (developed in the mid-90s), which explored ideas that have now found its way into several of the systems that we discuss.

6.2 xFS

Unlike the other systems mentioned here, xFS never made it to a production environment. xFS is the original "Serverless File System", and several systems in production today build upon ideas originally brought up in (Anderson et al., 1996). xFS was designed to run on commodity hardware, and expected to handle large loads and multiple users. Based on tracking usage patterns in an NFS system for several days, one assumption xFS makes is that users other than the creator of the file rarely modify files.

6.2.1 Failure Model

In xFS, when a machine fails it is not expected to come back online. Upon failure of a machine, data is automatically shuffled around to compensate for the loss. While failures are assumed to be permanent, the system was designed to be able to come back up from a full loss of functionality.

6.2.2 Replication

xFS does not support replication of files. Instead, it supports a RAID approach for storing data, as outlined in Fig. 6.1. In xFS, servers are organized into *stripe groups*. Each stripe group is a subset of the set of data servers. When a client writes to a file, it is gathered into a *write block* that is held in the client's cache. In Fig. 6.1, there are two clients, each building their own write block. Once the write block is full, the data is sent to the server group to be written to file. For a server group with N servers, the file is split into N-1 pieces, and striped in a RAID pattern across all the servers. The Nth stripe is a parity block that contains the XOR of all the other pieces, and is shown as a striped block in Fig. 6.1. This parity block will go to the parity server for the group. This way, if a server is lost, or a piece becomes corrupted it can be restored. One downside to this approach is that if multiple servers from a group go down, the data may be permanently lost, and xFS will stop working. In general, the replication level of a file can never be greater than the number of servers in the server group.

6.2.3 Data Access

In xFS a client will connect to a system manager, which will look up the appropriate server group, and have the client connect to the server group leader. In general,

Fig. 6.1 xFS RAID approach to storing data

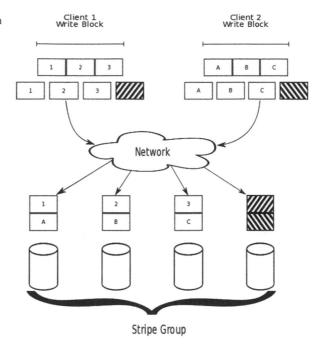

this takes about 3 hops (not including the actual transmission of data). Generally, the system will attempt to move data to be as close to the user as possible (in many cases, the design expects the client to be running on a machine that is also acting as a data server), incurring the short term penalty in network traffic of moving a file for the long term bonus of not needing further interaction with a system manager.

6.2.4 Integrity

Because of the RAID backend of xFS, data corruption can be detected and repaired using the parity block computed when data is written. xFS also uses this information to recover missing stripe blocks when a server in a stripe group fails.

6.2.5 Consistency and Guarantees

xFS guarantees *read-what-you-wrote* consistency, but it also allows users to read stale data – meaning that the best overall consistency guarantee that it can achieve is eventual. It is also not clear that the system can effectively handle concurrent writes. xFS never made it to a production environment, so there was never a strong need to establish any guarantees governing access times. Additionally, xFS was designed to handle flux in the number of available servers.

6.2.6 Metadata

The main advantage of xFS is its fully dynamic structure. The idea is to be able to move data around to handle load fluctuations and to increase locality. The system uses metadata information to help locate all files and put them back together in order.

6.2.7 Data placement

Managers in xFS try to ensure that data is being held as close to the client accessing it as possible–in some cases even shifting the location of data as a client starts writing to it. While this seems unwieldy, xFS uses a log-based storage method, so there is not too much of a network hit as data is shifted with a new write closer to the current client.

6.2.8 Security

xFS was designed to be run in a trusted environment, and it is expected that clients are running on machines that are also acting as storage servers. It is, however, possible for xFS to be mounted and accessed from an unsafe environment. Unfortunately, this is more inefficient and results in much more network traffic. It is also possible for a rogue manager to start indiscriminately overwriting data that can cause the entire system to fail.

6.3 Amazon S3

The Simple Storage Service (S3) from Amazon is used by home users, small businesses, academic institutions, and large enterprises. With S3 (Simple Storage Service), data can be spread across multiple servers around the US and Europe (S3-Europe). S3 offers low latency, infinite data durability, and 99.99% availability.

6.3.1 Data Access and Management

S3 stores data in 2 levels: a top level of *buckets* and *data objects*. Buckets are similar to folders, and can hold an unlimited number of data objects. Each Amazon Web Services (AWS) account can have up to 100 buckets. Charging for S3 is computed at the bucket level. All costs levels are tiered, but the basic costs as of January 2010 are as follows: storage costs $0.15/GB/month in the US, $0.165 in N California, and $0.15/GB/month in Europe; $0.10/GB for uploads (free until July 2010) and

$0.17/GB for downloads; and $0.01/1,000 PUT, COPY, POST, or LIST operations, $0.001/10,000 GET and all other operations. Each data object has a name, a blob of data (up to 5 GB), and metadata. S3 imposes a small set of predefined metadata entries, and allows for up to 4 KB of user generated {*name*, *value*} pairs to be added to this metadata.

While users are allowed to create, modify, and delete objects in a bucket, S3 does not support renaming data objects or moving them between buckets–these operations require the user to first download the entire object and then write the whole object back to S3. The search functions are also severely limited in the current implementation. Users are only allowed to search for objects by the name of the bucket–the metadata and data blob itself cannot be searched.

Amazon S3 supports three protocols for accessing data: SOAP, REST, and BitTorrent. While REST is most popularly used for large data transfers, BitTorrent has the potential to be very useful for the transfer of large objects.

6.3.2 Security

While clients use a Public Key Infrastructure (PKI) based scheme to authenticate when performing operations with S3, the user's public and private keys are generated by Amazon and the private key is available through the user's AWS site. This means that the effective security is down to the user's AWS password, which can be reset through email. Since S3 accounts are linked directly to a credit card, this can potentially cause the user a lot of problems.

Access control is specified using access control lists (ACL) at both the bucket and data object level. Each ACL can specify access permissions for up to 100 identities, and only a limited number of access control attributes are supported: read for buckets or data objects, write for buckets, and, finally, reading and writing the ACL itself. The user can configure a bucket to store access log records. These logs contain request type, object accessed, and the time the request was processed.

6.3.3 Integrity

The inner workings of Amazon S3 have not been published. It is hard to determine their approach to error detection and recovery. Based on the reported usage (Palankar, Iamnitchi, Ripeanu, & Garfinkel, 2008), there was no permanent data loss.

6.4 Dynamo

Dynamo is the back end for most of the services provided by Amazon. Like S3 it is a distributed storage system. Dynamo stores data in key-value pairs, and sacrifices consistency for availability. Dynamo has been designed to store relatively small files ($\sim$1 MB) and to retrieve them very quickly. A web page may have several services which each have their own Dynamo instance running in the background – this is what leads to the necessity of making sure latency is low when retrieving data.

Dynamo uses consistent hashing to make a scalable system. Every file in the system identified by a key is hashed, and this hash value is used to determine which node in the system it is assigned to. This hash space is treated as a ring, which is divided into Q equally sized partitions. Each node (server) in the system is assigned an equal number of partitions. An example of this can be seen in Fig. 6.2. In this figure, there are a total of 8 partitions. Nodes **A**, **B**, and **C** are responsible for keeping copies of all files where the hashed key falls into the striped partition that they manage.

Fig. 6.2 Dynamo hash ring

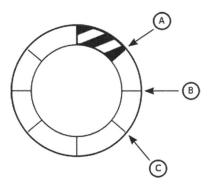

6.4.1 Checkpointing

Dynamo nodes share information via a gossip based protocol. There are no regular heartbeats sent between the nodes. All communication is pushed by client requests. If there is no request for data, the nodes do not communicate and do not care if another node is down. Periodic tests to see if a node is available occur only if a node is found to be unreachable during a client request.

6.4.2 Replication

With Dynamo, a quorum-like system is used to determine if a read or write was successful. If enough nodes reply that a write/read was successful, the whole operation is considered successful – even if not all N replicas are written to or read from. Dynamo allows the service writer to specify not only N, but R and W as well. R is the number of successful reads necessary for the whole operation to be successful, and W is the number of writes. Dynamo will report a successful write if W-1 nodes report success, so to make a system that is always up, and will never reject a write, W can be set to 1. Generally, W and R are both less than N, so that the system can make progress in the presence of failures. A suggested configuration for Dynamo is to have R + W > N. A general configuration of (N,R,W) is (3,2,2).

6.4.3 Failures

Dynamo operates under the assumption that hardware failures are expected, and trades data consistency guarantees for availability. It uses a gossip-based system to detect failures of nodes. Once a node stops responding, other nodes will eventually propagate knowledge of the failure. As a design feature, nodes are not considered removed unless an administrator issues an explicit removal command – this means the system will gracefully handle transient downtimes. If a coordinator cannot reach a node for a write, it will simply pass the data on to the next available node in the hash ring. This will contain an extra bit of metadata that marks it as belonging elsewhere. Once a node comes back online, this information can be passed back to it.

If a node is not available, the data presumed to be on that node is not immediately replicated on another node – this only happens when an administrator explicitly removes the node via a command. Dynamo is built under the expectation that there will be many transient failures, so there is no scramble to ensure replication levels are met when a node stops responding to requests. Because of this, some reads may fail if R is set equal to N. Once a node has been explicitly removed, all key ranges previously held by that node are reassigned to other nodes while ensuring that a given node is not overloaded as a result of this redistribution.

6.4.4 Accessing Data

Dynamo's gossip-based protocol for node discovery ensures that all nodes know in one step the exact node to send a read or write request to. There are two main methods of accessing data: (1) using a dedicated node to handle client requests or (2) having several dedicated nodes, or coordinators, that process client requests and forward them to the appropriate nodes. The former approach can lead to unbalanced network nodes while the latter approach results in a more balanced network and a lower latency can be assured.

6.4.5 Data Integrity

There is no specific mention of detecting corruptions in data, or how any corresponding error recovery may occur. Since data is stored as a binary object, it may be left up to the application developers to detect data corruption, and handle any sort of recovery. Reported results in live settings (DeCandia et al., 2007), do not indicate permanent data loss. Amazon requires regular archival of every system – there is a chance that this archival data is used for recovery if errors in data are found

6.4.6 Consistency and Guarantees

Dynamo guarantees eventual consistency – there is a chance that not all replications contain the same data. Due to transient network failures and concurrent writes, some changes may not be fully propagated. To solve this problem, each object also contains a context. This context contains a version vector, giving the ability to track back through changes and figure out which version of an object should carry the most precedence. There are several different schemes for handling this. Dynamo itself supports several simple schemes, including a last-write-wins method. There is also an interface that allows developers to implement more complex and data specific merging techniques. Merging of different object versions is handled on reads. If a coordinator retrieves multiple versions of an object on a read, it can attempt to merge differences before sending it to the client. Anything that cannot be resolved by the coordinator is passed onto the client. Any subsequent write from that client is assumed to have resolved any remaining conflicts. The coordinator makes sure to write back the resolved object to all nodes that responded to the object query.

The only other base guarantee provided by Dynamo is performance geared towards the 99.99th percentile of users – millisecond latencies are assured. Aside from this, service developers are allowed to tweak the system to fit the guarantees necessary for their application through the N, R and W settings.

6.4.7 Metadata

In Dynamo, the object metadata is referred to as context. Every time data is written, a context is included. The context contains system metadata and other information specific to the object such as versioning information. There may also be an extra binary field which allows developers to add any additional information needed to help their application run. The metadata is not searchable, and only seems to interact with Dynamo when resolving version conflicts as mentioned above.

6.4.8 Data Placement

According to DeCandia et al. (2007), there are guarantees in place to ensure that replicas are spread across different data centers. It is likely that Amazon has a particular scheme that allows Dynamo to efficiently determine the locations of nodes. An object key is first hashed to find its location on the network ring. Moving around the ring clockwise from that point, the first encountered node is where the first copy of the data is placed. The next N-1 nodes (still moving clockwise) will contain replicas of the data.

There are no current methods of data segregation in Dynamo – there is simply a get() and put() interface for developers, and no support for a hierarchical structure. Each service using Dynamo has its individual instance of it running. For example, your shopping cart will not be able to access the best seller's list. On the other hand, Dynamo has no guarantees that the different instances are not running on the same machine.

6.4.9 Security

Dynamo has been designed to run in a trusted environment, so there is no structure in place to handle security concerns. By design, each service that uses Dynamo has its own separate instance running. Because of this, users do have some sense of security, as there is some natural separation of data, and one application cannot access the data of another.

6.5 Google File System

The Google File System (GFS) is designed by Google to function as a backend for all of Google's systems. The basic assumption underlying its design is that components are expected to fail. A robust system is needed to detect and work around these failures without disrupting the serving of files. GFS is optimized for the most common operations – long, sequential and short, random reads, as well as large, appending and small, arbitrary writes. Additionally, a major goal in designing GFS was to efficiently allow concurrent appends to the same file. As a design goal, high sustained bandwidth was deemed more important than low latency in order to accommodate large datasets.

A GFS instance contains a *master server* and many *chunk servers*. The master server is responsible for maintaining all file system metadata and managing *chunks* (stored file pieces). There are usually also several master replicas, as well as shadow masters which can handle client reads to help reduce load on a master server. The chunk servers hold data in 64 MB-sized chunks.

6.5.1 Checkpointing

In GFS, the master server will keep logs tracking all chunk mutation. Once a log file starts to become too big, the master server will create a checkpoint. These checkpoints can be used to recover a master server, and are used by the master replicas to bring a new master process up.

6.5.2 Replication

By default, all GFS maintains a replication level of 3. This is, however, a configurable trait: "...users can designate different replication levels for different regions of the file namespace" (Ghemawat, Gobioff, & Leung, 2003). For example, a temp directory generally has a replication level of 1, and is used as a scratch space. The master server is responsible for ensuring that the replication level is met. This not only involves copying over chunks if a chunk server goes down, but also removing replicas once a server comes back up. As a general rule, the master server will try to place replicas on different racks. With Google's network setup, the master is able to deduce the network topology from IP addresses.

6.5.3 Failures

When it comes to failures, GFS always expects the worst. The master server regularly exchanges heartbeats with the chunk servers. If the master server does not receive a heartbeat from a chunk server in time, it will assume the server has died, and will immediately start to spread the chunks located on that server to other servers to restore replication levels. Should a chunk server recover, it will start to send heartbeats again and notify the master that it is back up. At this point the master server will need to delete chunks in order to drop back down to replication level and not waste space. Because of this approach, it would be possible to wreak havoc with a GFS instance by repeatedly turning on and off a chunk server. Master server failure is detected by an external management system. Once this happens, one of the master server replicas is promoted, and the master server process is started up on it. A full restart usually takes about 2 minutes – most of this time is spent polling the chunk servers to find out what chunks they contain

6.5.4 Data Access

Clients initially contact the master server to gain access to a file, after which the client interacts directly with the necessary chunk server(s). For a multi terabyte file, a client can keep track of all chunk servers in its cache. The chunk server directly interacting with clients is granted a chunk *lease* by the master server, and is now known as the *primary*. The primary is then responsible for ordering any operations on the data serially. It is then responsible for propagating these changes to the other chunk servers that hold the chunk. If a client is only looking to read data, it is possible for the client to go through the shadow master as opposed to the master server. It is possible for concurrent writes to get interleaved in unexpected ways, or for failed write attempts to show themselves as repeated data in chunks. GFS assumes that any application using it is able to handle these possible problems though redundant data may hurt the efficiency of reads.

6.5.5 Data Integrity

Each chunk in GFS keeps track of its own checksum information this informa-tion is unique for each chunk – it is not guaranteed to be the same even across replicas. Chunk servers are responsible for checking the checksums of the chunks they are holding. With this, it is possible for the system to detect corrupted files. If a corrupted chunk is detected, the chunk is deleted, and copied from another replica.

6.5.6 Consistency and Guarantees

GFS is built to handle multiple concurrent appends on a single file. It is up to a pri-mary chunk server to order incoming permutation requests from multiple clients into a sequential order, and then pass these changes on to all other replicas. Because of this, it is possible that a client will not see exactly what they wrote on a sequential read – there is a possibility that permutations from other clients have been inter-leaved with their own. Google describes this state as consistent but undefined – all clients will see the same data, regardless of which replica is primary, but mutations may be interspersed. When there is a write failure, a chunk may become inconsis-tent. This is a case where there may be redundant lines of data in some but not all replicas.

As GFS was built to maintain bandwidth, as opposed to meet a targeted latency goal there are no guarantees that pertain to latency. GFS does guarantee maintenance of the specified replication level which is achieved using system heartbeats. GFS also cannot guarantee full consistency in the face of write failures. A slightly looser definition of consistency – at least a single copy of all data is fully stored in each replica – is what GFS supplies. Any application built on top of GFS that can handle these possible inconsistencies should be able to guarantee a stronger consistency.

6.5.7 Metadata

In GFS, the master server contains metadata about all chunks contained in the sys-tem. This is how the master server keeps track of where the chunks are located. Each chunk has its own set of metadata as well. A chunk has a version number, as well as its own checksum information.

6.5.8 Data Placement

The master server attempts to place replicas on separate racks, a feat made possible by Google's network scheme. The master server also attempts to balance network load, so it will try to evenly disperse all chunks.

6.5.9 Security Scheme

GFS expects to be run in a trusted environment, and has no major security approaches. If a user could bring down a chunk server, modify the chunk versions held on it, and reconnect it to the system, GFS would slowly grind to a halt as it believes that that server has the most up-to-date chunks and begins deleting and rewriting all these chunks. This would create a lot of network traffic, and theoretically bring down not only any service that relies on GFS, but also anything else that requires network bandwidth to work.

6.6 Bigtable

As the name suggests, Bigtable stores large amounts of data in a table. While it is not a full relational model, it is essentially a multi-dimensional database. Tables are indexed by row and column keys (`strings`), as well as a timestamp (`int64`). Values inside cells are an uninterpreted array of bytes, and tables can be easily used as either inputs to or outputs of MapReduce (Dean & Ghemawat, 2004). Each table is broken up by row into tablets. Each tablet will contain a section of sequential rows, generally about 100–200 MB in size.

Bigtable has been designed by Google to handle very large files generally measuring in the petabyte range. It is in use in several products, including Google Analytics and Google Earth. Bigtable is designed to run on top of the Google File System (GFS), and inherits its features and limitations. Bottlenecks with GFS directly affect Bigtable's performance, and measures have been taken to avoid adding too much to network traffic. Additionally, Bigtable relies on Chubby for basic functionality. Chubby is a locking service which implements Lamport's Paxos theorem (Lamport, 2001) in use at Google to help clients share information about the state of their environment (Burrows, 2006). Different systems make use of Chubby to keep separate components synchronized. If Chubby goes down, then so does Bigtable. Given that, Chubby has been responsible for less than .001% of Bigtable's downtime as reported in Chang et al. (2006). Bigtable processes usually run on top of GFS servers, and have other Google processes running side-by-side. Ensuring a low latency in this environment is challenging.

There are 3 pieces to an implementation of Bigtable: First, a library is linked to every client – helping clients find the correct server when looking up data. Second, there is a single master server. This master server will generally have no interactions with clients, and as a result is usually only lightly loaded. Finally, there are many tablet servers. The tablet servers are responsible for communicating with clients, and do not necessarily serve consecutive tablets; simply what is needed. Each tablet is only served on one tablet server at a time. It is also not necessary for all tablets to be served – the master keeps a list of tablets not currently served, and will assign these tablets to a server if a client requests access to it.

Tablets are stored in GFS as in the SSTable format, and there are generally several SSTables to a tablet. An SSTable contains a set of key/value pairs, where both key

Fig. 6.3 Bigtable storage scheme

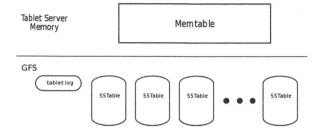

and value are arbitrary strings. Updates to tablets are kept in a commit log. Recently committed changes are stored in memory, and older tablet update records are stored as SSTable. Figure 6.3 helps to show the division of what is maintained in GFS and what is kept in tablet server memory. Both the commit logs and SSTable files are held in GFS. Storing commit files in GFS means that all commits can be recovered if a tablet server dies. These commit logs are as close as Bigtable comes to actual checkpointing – more thorough checkpointing is carried out by GFS.

6.6.1 Replication

As mentioned above, the Bigtable master server makes sure that only one server is actually modifying a tablet at a time. While this looks like Bigtable is ignoring replication entirely, every tablet's SSTables are actually being stored in GFS. Bigtable neatly bypasses the problem of replication and lets GFS handle it. Bigtable will inherit the replication level of the folders where the SSTables are stored.

6.6.2 Failures

All failure detection for Bigtable eventually comes down to Chubby. When a tablet server first starts up, it contacts Chubby and makes a server-specific file, and obtains an exclusive lock on it. This lock is kept active as long as the tablet has a connection to Chubby, and will immediately stop serving tablets if it loses that lock. If a tablet server ever contacts Chubby and finds the file gone, it will kill itself. The master server is responsible for periodically polling the tablet servers and checking to see if they are still up. If the master cannot contact a tablet server, it first checks to see if the tablet server can still communicate with Chubby. It does so by attempting to obtain an exclusive lock on the tablet server file. If the master obtains the lock, Chubby is alive and the tablet can't communicate with Chubby. The master then deletes the server file, ensuring that the server will not attempt to serve again. If the master's Chubby session expires, the master immediately kills itself without effecting tablet serving. A cluster management system running alongside Bigtable is responsible for starting up a new master server if this happens. While (Chang et al., 2006) does not explicitly state what happens if Chubby goes down, it is likely that the current master server will kill itself and the cluster manager will repeatedly try to kick start a new master until Chubby starts responding again.

6.6.3 Accessing Data

Every client is initially sent a library of tablet locations, so they should initially be able to directly contact the correct tablet server. Over time, tablet servers die, some may be added, or tablets may be deleted or split. Bigtable has a 3-tier hierarchy for tablet location. First, there is a file stored in Chubby that contains the location of the root tablet. Every Bigtable instance has its own root tablet. A root tablet specifies the location of all tablets in a METADATA table. This METADATA table holds the locations of all user tables as well as some tablet-specific information useful for debugging purposes. The root tablet is simply the first tablet of the METADATA table. The root tablet is treated specially – it is never split so that the tablet location hierarchy doesn't grow. With this scheme, 2^{34} tablet locations can be addressed. The client library caches the tablet locations from the METADATA table, and will recursively trace through the hierarchy if it doesn't have a tablet, or the tablet location is scale. With an empty cache, it will take 3 round trips but may take up to 6 with a stale cache. None of these operations need to read from GFS, so the time is negligible. The tablet servers have access to sorted SSTables, so they can usually locate required data (if not already in memory) with a single disk access.

6.6.4 Data Integrity

Bigtable is not directly involved with maintaining data integrity. All Bigtable data is stored in GFS, and that is what is responsible for actually detecting and fixing any errors that occur in data. When a tablet server goes down there is a chance that a table modification was not committed, or a tablet split was not properly propagated back to Chubby. Keeping all tablet operation logs in GFS as well solves the first problem: a new tablet server can read through the logs, and ensure all tablets are up to date. Tablet splits are even less of a problem, as a tablet server will report any tablets it has that are not referenced by Chubby.

6.6.5 Consistency and Guarantees

Bigtable guarantees eventual consistency – all replicas are eventually in sync. Tablet servers store any tablet modifications in memory, and will write permutations to a log, but will not necessarily wait for GFS to confirm that a write has succeeded before confirming it with users. This helps to improve latency, and give users a more interactive experience, such as when using Google Earth. Bigtable inherits all of the GFS guarantees pertaining to data replication, error recovery, and data placement.

6.6.6 Metadata

The METADATA table contains the metadata for all tablets held within an instance of Dynamo. This metadata includes lists of the SSTables which make up a tablet, and a set of pointers to commit logs for the tablet. When a tablet server starts serving a file, it first reads the tablet metadata to learn which SSTable files need to be loaded. After loading the SSTables into memory, it works through the commit logs, and brings the version in memory up to the point it was at when the tablet was last accessed.

6.6.7 Data Placement

All of Bigtable's data placement is handled by GFS – it has no direct concern for data placement. As far as Bigtable is concerned, there are only single copies of files – it uses GFS handles to access any files needed. While Bigtable is not directly aware of multiple versions of files, it can still take advantage of replicas through GFS.

6.6.8 Security

Bigtable is designed to run in a trusted environment, and does not really have much in the way of security measures. Theoretically, a user may be able to have encrypted row and column names, as well as the data in the fields. This would be possible since these are all arbitrary strings. While encrypting row names means you could potentially use some of the grouping abilities, there is no reason a user would not be able to gain some security with this method.

6.7 Microsoft Azure

Azure is Microsoft's cloud computing solution. It consists of three parts: storage, scalable computing, and the base fabric to hold everything together across a heterogeneous network. Figure 6.4 shows a high level overview of Azure's structure.

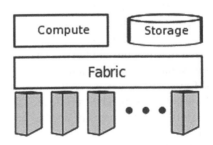

Fig. 6.4 Azure overview

Both the compute and storage levels rely on the fabric layer, which is running across many machines. Azure's scalable computing component is out of the scope of this article, but for the sake of completeness it is mentioned here. Microsoft's computing solution is designed to make sure that it worked well with the storage, but it is not necessary to use the one to use the other. Microsoft has not published very many details about Azure.

Azure's storage service allows the user to choose between three storage formats: BLOBs, tables, and queues. The BLOBs are essentially containers that can hold up to 5 GB of binary data. Azure's BLOB format is very similar to S3 – there are containers to hold the BLOBs, and there is no hierarchical support (you cannot put a container inside a container). The BLOB names have no restrictions, however, so there is nothing to keep a user from putting in "/" in a BLOB's name to help organize data. Tables in Azure are not true relational tables, but more like Bigtable – tables hold entities, and an entity is a list of named values. While you lose the ability to query Azure tables like a true relational database, it is able to scale effectively across many machines. Azures queues are primarily designed for use with the computing service. Queues are what allow different applications a user is running to communicate with each other. For example, a user may have designed a web front-end application that can communicate with several worker applications to perform back-end processing. This application suite would use queues to exchange information between the web front-end and the various workers.

6.7.1 Replication

Regardless of storage type, all data has a replication level of 3 the maintenance of which is being coordinated by the storage service itself. According to Chappell (2009a), the fabric service is not even aware of replication levels, it just sees the storage service as another application. More about how this happens is in the failure section.

6.7.2 Failure

Azure's fabric layer is made up of machines in a Microsoft Data Center. The data center is divided into *fault domains*. Microsoft defines a fault domain as a set of machines which can be brought down by the failure of a single piece of hardware. All machines dedicated to Azure are controlled by 5–7 fabric controllers. Each machine inside the fabric has a fabric controller process running which reports the status of all applications running on that machine (this includes user apps in different VMs as well as the storage service). While we are not exactly clear on how storage is handled inside the fabric, we do know that the fabric controllers see the storage service as just another application. If an application dies for any reason, the controllers are responsible for starting up another instance of the application. It stands to reason that if an instance of the storage service running on a machine dies, or if the machine itself dies, these controllers would start up another instance on a different machine. By having the fabric layer ensure that applications are spread across fault domains, it guarantees that replicas are spread out.

6.7.3 Accessing Data

If a user is using a .NET application running on Azure's compute service, ADO .NET interfaces can be used. If, on the other hand, a user is trying to access data in Azure storage through a Java application, you would use standard REST. As an example of accessing a BLOB from (Chappell, 2009b):

```
http://<StorageAccount>.blob.core.windows.net/<Container>/<BlobName>
```

Where `<StorageAccount>` is an identifier assigned when a new storage account is created, used to identify ownership of objects. `<Container>` and `<BlobName>` are the names of the container and blob that this request is accessing.

There is no specific mention of any guarantees on latency, but since it is expected to be part of a web application, it's likely low.

6.7.4 Consistency and Guarantees

Azure's storage guarantees read-what-you-write consistency – worker threads and clients will be able to immediately see changes it just wrote. Unfortunately, there is no clear picture of what this means for other threads/clients. It also guarantees a replication level of 3 for all stored data. There have also been no specific guarantees as to latency or specific mention of SLAs.

6.7.5 Data Placement

The Azure fabric layer is responsible for the placement of data. While it is not directly aware of replicas, it is able to ensure that instances of the storage service are running in different fault domains. From the whitepapers Microsoft has made available, it looks like a fabric controller only operates in one data center. There is a chance that users are able to choose which data center to use.

6.7.6 Security

All access to Azure's storage component is handled by a private key generated by Azure for a specific user. While there are no particular details about how this happens, it is likely that this is susceptible to the same problems as S3 – another person may be able to hijack this key. In Azure storage, there are no ACLs, only that single access key – developers are expected to provide their own authentication program-side.

6.8 Transactional and Analytics Debate

None of the storage systems discussed here are able to handle complex relational information. As data storage makes a shift to the cloud, where does that leave databases? Having on-site data management installations can be very difficult to maintain, requiring administrative and hardware upkeep as well as the initial hardware and software costs (Abadi, 2009). Being able to shift these applications to the cloud would allow companies to focus more on what they actually produce – possibly having the same effects that the power grid did 100 years ago (Abadi, 2009).

Transactional data management is what you generally think of first – the backbone for banks, airlines, and e-commerce sites. Transactional systems generally have a lot of writes, and files tend to be in the GB range. They usually need ACID guarantees, and thus have problems adjusting to the limitations of Brewer's CAP theorem. Transactional systems also generally contain data that needs to be secure, such as credit card numbers and other private information. Because of these reasons, it is hard to move a transactional system to the cloud. While several database companies, such as Oracle, have versions that can run in a distributed environment like Amazon's EC2 cloud, licensing can become an issue (Armbrust et al., 2009). Instead of only needing one license, the current implementation requires a separate license for each VM instance: as an application scales, this can become prohibitively expensive.

Analytical data management is slightly different. In an analytical system, there are generally more reads than writes, and writes occur in large batches. These types of systems are used to analyze large datasets, looking for patterns or trends. Files in an analytical system are also on a completely different scale – a client may need to sift through petabytes of data. For this type of system, looser eventual consistency is acceptable – making it a good fit for distributed computing. Additionally, the data analyzed usually has less need to be secure, so having a third-party such as Amazon or Google hosting the data is acceptable.

6.9 Conclusions

In this chapter we have surveyed several approaches to data storage in cloud computing settings. Data centers have, and will continue, to be built out of commodity components. The use of commodity components combined with issues related to the settings in which these components operate such as heat dissipations and scheduled downtimes imply that failures are a common occurrence and should be treated as such. In these environments, it is no longer a matter of if a system or component will fail, but simply when. Datasets are dispersed on a set of machines to cope with their voluminous nature and to enable concurrent processing on them. To cope with failures, every slice of the dataset is replicated a preset number of times; replication allows applications to sustain failures to machines that hold certain slices of the dataset and also to initiate error corrections due to data corruptions.

The European Network and Information Security Agency (ENISA) recently released a document (Catteddu & Hogben, 2009) outlining the security risks in cloud computing settings. Among the concerns raised in this document include data protection, insecure or incomplete data deletion, and the possibility of malicious insiders. Other security related concerns (Brodkin, 2008) that have been raised include data segregation, control over a data's location, and investigative support. Most of the systems that we have described here do not adequately address several of these aforementioned security concerns and also exacerbate the problem by designing systems that are presumed to operate in a trusted environment: this allows us to construct situations, in some of these systems, where a malicious entity can wreak havoc. Issues related to security and trust need to be thoroughly addressed before these settings can be used for mission critical and sensitive information.

References

Abadi, D. J. (2009). Data management in the cloud: Limitations and opportunities. *IEE Data Engineering Bulletin, 32*(1), 3–12.

Anderson, T. E., Dahlin, M. D., Neefe, J. M., Patterson, D. A., Roselli, D. S., & Wang, R. Y. (1996). Serverless network file systems. *ACM Transactions on Computer Systems (TOCS), 14*(1), 41–79.

Armbrust, M., Fox, A., Griffith, R., Joseph, A., Katz, R., Konwinski, A., et al. (2009). Above the clouds: A Berkeley view of cloud computing (University of California at Berkeley, Tech. Rep. No. UCB/EECS-2009-28).

Brewer, E. A. (2009). Towards robust distributed systems. *Principles of Distributed Computing (PODC) Keynote, Portland, OR.*

Brodkin, J. (2008). Gartner: Seven cloud-computing security risks. Retried Infoworld, July 02 2008 from: http://www.infoworld.com/d/security-central/gartner-seven-cloud-computing-security-risks-853.

Burrows, M. (2006). The chubby lock service for loosely-coupled distributed systems. *Proceedings of Operating Systems Design and Implementation (OSDI'06), Seattle, WA,* 335–350.

Catteddu, D., & Hogben, G. (Eds.). (November 2009). Cloud computing risk assessment. *European Network and Information Security Agency (ENISA).*

Chang, F., Dean, J., Ghemawat, S., Hsieh, W. C., Wallach, D. A., Burrows, M., et al. (2006). Bigtable: A distributed storage system for structured data. *Proceedings of Operating Systems Design and Implementation (OSDI'06), Seattle, WA,* 205–218.

Chappell, D. (2009a). *Introducing windows azure* (Tech. Rep., Microsoft Corporation).

Chappell, D. (2009b). *Introducing the windows azure platform: An early look at windows azure, SQL azure and NET services* (Tech. Rep., Microsoft Corporation).

Dean, J., & Ghemawat, S. (2004). MapReduce: Simplified data processing on large clusters. *Proceedings of Operating Systems Design and Implementation (OSDI'04), San Francisco, CA,* 137–149.

DeCandia, G., Hastorun, D., Jampani, M., Kakulapti, G., Lakshman, A., Pilchin, A., et al. (2007). Dynamo: Amazon's highly available key-value store. *ACM SIGOPS Operating Systems Review, 41*(6), 205–220.

Ghemawat, S., Gobioff, H., & Leung, S.-T. (2003). The google file system. *19th Symposium on Operating Systems Principles*, New York, NY, 29–43.

Gilbert, S., & Lynch, N. (2002). Brewer's conjecture and the feasibility of consistent, Available, Partition-tolerant web services. *ACM SIGACT News, 33*(2), 51–59.

Lamport, L. (2001). Paxos made simple. *ACM SIGACT News, 32*(4), 18–25.

Lyman, P., & Varian, H. R. (2000). *How Much Information?* http://www2.sims.berkeley.edu/research/projects/how-much-info/, Berkeley.

Lyman, P., & Varian, H. R. (2003). *How Much Information?* http://www2.sims.berkeley.edu/research/projects/how-much-info-2003/, Berkeley.

Palankar, M. R., Iamnitchi, A., Ripeanu, M., & Garfinkel, S. (2007). Amazon S3 for science grids: A viable solution? High performance distributed computing (HPDC). *Proceedings of the 2008 International Workshop on Data-Aware Distributed Computing (HPDC08), Boston, MA*, 55–64.

Robbins, K. A., & Robbins, S. (2003). *Unix systems programming: Communication, concurrency and threads*. Upper Saddle River, NJ: Prentice Hall.

Synodinos, D. G. (2008). *LHC Grid: Data storage and analysis for the largest scientific instrument on the planet*. Retrieved InfoQ, October 01 2008, from http://www.infoq.com/articles/lhc-grid.

Chapter 7
Scheduling Service Oriented Workflows Inside Clouds Using an Adaptive Agent Based Approach

Marc Eduard Frîncu

Abstract As more users begin to use clouds for deploying complex applications and store remote data there is an increasing need of minimizing the user costs. In addition many cloud vendors start offering specialized services and thus the need of selecting the best possible service in terms of deadline time or monetary constraints emerges. Vendors not only that provide specialized services but also prefer using their own scheduling policies and often choose their negotiation strategies. To make things even more complicated complex applications are usually comprised of smaller tasks (e.g. workflow applications) orchestrated together by a composition engine. In this highly dynamic and unpredictable environment multi-agent systems seem to provide one of the best solutions. Agents are by default independent as they act in their best interest following their own policies, but also cooperate with each other in order to achieve a common goal. In the frame of workflow scheduling the goal is represented by the minimization of the overall user cost. This paper presents some of the challenges met by multi-agent systems when trying to schedule tasks. Solutions to these issues are also given and a prototype multi agent scheduling platform is presented.

7.1 Introduction

In recent years Cloud Computing (CC) emerged as a leading solution in the field of Distributed Computing (DC). In contrast, Grid Computing lacked the open-world vision of overcoming some fundamental problems including transparent and easy access to resources, licensing or political issues, lack of virtualization support or to complicated to use architectures and end-user tools.

Clouds have emerged as a main choice for service vendors mostly due to their support for virtualization and service oriented approach. Inside clouds almost

M.E. Frîncu (✉)
Institute e-Austria, Blvd. Vasile Parvan No 4 300223, Room 045B, Timisoara, Romania
e-mail: mfrincu@info.uvt.ro

B. Furht, A. Escalante (eds.), *Handbook of Cloud Computing*,
DOI 10.1007/978-1-4419-6524-0_7, © Springer Science+Business Media, LLC 2010

everything can be offered as a service. This has led to the appearance of several paradigms including Software as a Service (SaaS), Infrastructure-as-a-Service (IaaS) or Platform-as-a-Service (PaaS).

As more users begin to use clouds for storing or executing their applications these systems become susceptible to workload related issues. The problem is even harder when considering complex tasks which require accessing services provided by different cloud vendors (see Fig. 7.1) each with their own internal policies. Selecting the optimal/fastest service for a specific task becomes in this case an important problem as sometimes users are paying for their time spent using the underlying services.

Consequently scheduling tasks on services becomes even more difficult as inside cloud environments each member uses its own policies and is not obligated to adhere to outside rules. We end up with a bundle of services from various providers that need to be orchestrated together in order to produce the desired outcome inside a given time interval. Keeping the execution inside this interval minimizes production and client costs. As service selection requires some negotiation between providers one of the simplest and straightforward solutions is to use distributed agents that play the roles of service providers and clients.

This paper presents an agent based approach to the problem of task scheduling inside clouds. Two major problems are dealt with: finding cloud resources and orchestrating services from different cloud vendors towards solving a common goal. Certain deadline and cost constrains are assumed to exist. Even though the emphasis is put on workflow tasks, independent tasks can also be handled. Towards this aim

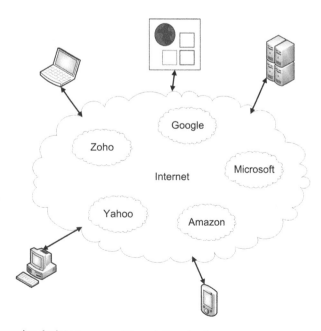

Fig. 7.1 Connecting devices to some of the existing clouds

we first present some solutions to the problem of task scheduling inside clouds. Then we present some issues regarding task scheduling inside Service Oriented Environments (SOE) together with some details on workflow scheduling. A detailed overview on a distributed agent based scheduling platform architecture capable of adapting itself to resource changes is also given. Finally a concrete experimental prototype and some conclusions are presented.

7.2 Related Work on DS Scheduling

Lot of work has been carried out in what concerns task scheduling inside Distributed Systems (DSs). This work can be divided into specialized Scheduling Algorithms (SAs) for clouds and Resource Management Systems (RMS) which discover services and allocate tasks to them. In what follows we briefly present some of the main work concerning both SAs for CC and RMS for DSs.

Concerning the development of efficient SAs for DSs, nature has proven to be a good place of inspiration. Recent papers such as Lorpunmanee, Sap, Abdullah, & Chompooinwai (2007) and Ritchie and Levine (2004) try to cope with the problem of task scheduling by offering meta-heuristics inspired from behavioral patterns observed in ant colonies. This technique also called Ant Colony Optimization (ACO) relies on the fact that ants inside a colony act as independent agents which try to find the best available resource inside their space by using global search techniques. Each time such an agent finds a resource better than the already existing one it marks the path to it by using pheromones. These attract other ants which start using the resource until a better one is found.

In Banerjee, Mukherjee, and Mahanti (2009) an ACO based approach for initiating the service load distribution inside clouds has been proposed. Simulated results on Google Application Engine (2010) and Microsoft Live Mesh (2010) have shown a slight improvement in the throughput of cloud services when using the proposed modified ACO algorithm.

The biggest disadvantage ACO has over other approaches is that it is not very effective when dynamic scheduling is considered. The reason for this is that rescheduling requires a lot of time until an optimal scenario is reached through intensive training given by multiple iterations. Because DSs are both unpredictable and heterogeneous each time a change is noticed the entire system needs to be trained again. This process which could last several hours. The large retraining time interval is not acceptable when tasks are scheduled under deadline constraints as the scheduling could take longer than the actual task execution. An improvement on this might be given by mixing the time consuming global search with local search when minor changes occur inside the DS. However defining the notion of minor changes is still an open issue.

Paper (Garg, Yeo, Anandasivam, & Buyya, 2009) deals with High Performance Computing (HPC) task scheduling inside clouds. Energy consumption is important both in what concerns the user costs and in relation to the carbon emissions.

The proposed meta-scheduler takes into consideration factors such as energy costs, carbon emission rate, CPU efficiency and resource workflows when selecting an appropriate data center belonging to a cloud provider. The designed energy based scheduling heuristics shows a significant increase in energy savings compared with other policies.

Most of the work concerning RMSs has evolved around the assumption of applying them onto grids and not clouds. This can be explained by two facts. The first one is that there are many similarities between a cloud and a grid and RMS developed for one type could also work well on the other. The second one is related with age, and as grids emerged earlier than clouds most of the solutions have been developed for the former. Nonetheless several of the grid oriented RMSs could be adapted to work for clouds too.

One example is represented by the Cloud Scheduler (2010). It allows users to set up a Virtual Machine (VM) and submit jobs to a Condor (Thain, Tannenbaum, & Livny, 2005) pool. The VM will be replicated on machines and used as container for executing the jobs.

In what follows we present some of the most known examples of RMS for DSs in general.

Notable examples include the Globus-GRAM (Foster, 2005), Nim- rod/G (Buyya, Abramson, & Giddy, 2000), Condor (Thain et al., 2005), Legion (Chapin, Katramatos, Karpovich, & Grimshaw, 1999), NetSolve (Casanova & Dongarra, 1998) and others. Many of these solutions use fixed query engines to discover and publish resources and do not rely on the advantages offered by distributed agents.

The ARMS (Cao, Jarvis, Saini, Kerbyson, & Nudd, 2002) system represents an example of agent based RMS. It uses PACE (Cao, Kerbyso, Papaefstathiou, & Nudd, 2000) for application performance predictions which are later used as inputs to the scheduling mechanism.

In paper (Sauer, Freese, & Teschke, 2000) a multi-site agent based scheduling approach consisting of two distinct decision levels one global and the other local is presented. Each of these levels has a predictive and a reactive component for dealing with workload distribution and for reacting to changes in the workloads.

Paper (Cao, Spooner, Jarvis, & Nudd, 2005) presents a grid load balancing approach by combining both intelligent agents and multi-agent approaches. Each existing agent is responsible for handling task scheduling over multiple resources within a grid. As in Sauer et al., (2000) there also exists a hierarchy of agents which cooperate with each other in a peer to peer manner towards a common goal of finding new resources for their tasks. This hierarchy is composed of a broker, several coordinators and simple agents. By using evolutionary processes the SAs are able to cope with changes in the number of tasks or resources.

Nimrod/G uses agents (Abramson, Buyya, & Giddy, 2000) to handle the setup of the running environment, the transport of the task to the site, its execution and the return of the result to the client. Agents can also record information acquired during task execution as CPU time, memory consumption etc.

Paper (Shen, Li, Genniwa, & Wang, 2002) proposes a system which can automatically select from various negotiation models, protocols or strategies the best one for the current computational needs and changes in resource environment. It does this by solving two main issues DS have to dealt with Cao et al., (2002): scalability and adaptability. The work carried in Shen et al., (2002) creates an architecture which uses several specialized agents for applications, resources, yellow pages and jobs. Job agents for example are responsible for handling a job since its submission and until its execution and their lifespan are restricted to that interval. The framework offers several negotiation models between job and resource agents including contract net protocol, auction and game theory based strategies.

AppLeS (Application-Level Scheduling) (Berman, Wolski, Casanova, Cirne, et al., 2003; Casanova, Obertelli, Berman, & Wolski, 2000) is an example of a methodology for adaptive scheduling also relying on agents. Applications using AppLeS share a common architecture and are scheduled adaptively by a customized scheduling agent. The agent follows several well established steps in order to obtain a schedule for an application: resource discovery, resource selection, schedule selection, application execution and schedule adaptation.

7.3 Scheduling Issues Inside Service Oriented Environments

Scheduling tasks inside SOE such as clouds is a particular difficult problem as there are several issues that need to be dealt with. These include: estimating task runtimes and transfer costs; service discovering and selection; negotiation between clients and different cloud vendors; and trust between involved parties. In what follows we address each of these problems separately.

7.3.1 Estimating Task Runtimes and Transfer Costs

Many SAs require some sort of user estimates in order to provide improved scheduling solutions. The estimates are either user estimated or generated by using methods involving code profiling (Maheswaran, Braun, & Siegel, 1999), statistical determination of execution times (David & Puaut, 2004), linear regression (Muniyappa, 2002) or task templating (Ali, Bunn, et al., 2004; Smith, Foster, & Taylor, 2004). When applied to SOE these methods come both with advantages and disadvantages as it is shown in the next paragraphs.

In SOE there is not much insight on the resource running behind the service and thus it is hard for users to obtain information that can help them give a correct runtime estimate.

User given estimates are dependent on the user's prior experience with executing similar tasks. Users also tend to overestimate task execution times knowing that schedulers rely on them. In this case a scheduler, depending on the scheduling heuristics, could postpone other tasks due to wrong information. To deal with these scenarios schedulers can implement penalty systems where tasks belonging to these harmful users would be intentionally delayed from execution.

Sometimes it is even difficult for users to provide runtime estimates. These situations usually occur due to the nature of the service. Considering two examples of services, one which processes satellite images and another one which solves symbolic mathematical problems we can draw the following conclusions. In the first case it is quite easy to determine runtime estimates from historical execution times as they depend on the image size and on the required operation. The second case is more complicated as mathematical problems are usually solved by services exposing a Computer Algebra System (CAS). CASs are specific applications which are focused on one or more mathematical fields and which offer several methods for solving the same problem. The choice on which method to choose depends on internal criteria which is unknown to the user. A simple example is given when considering large integer (more than 60 digits) factorizations. These operations have strong implications in the field of cryptography. In this case factorizing n does not depend on similar values as $n-1$ or $n+1$. Furthermore the factoring time is not linked to the times required to factor $n-1$ or n. It is therefore difficult for users to estimate runtimes in these situations. Refining as much as possible the notion of similarity between two tasks could be an answer to this problem but in some cases, such as the one previously presented this could require searching for identical past submissions.

Code profiling works well on CPU intensive tasks but fails to cope with data intensive applications where it is hard to predict execution time before all the input data has been received. Statistical estimations of run times face similar problems as code profiling.

Templating has also been used for assigning task estimates by placing newly arrived tasks in already existing categories. General task characteristics such as owner, solver application, machine used for submitting the task, input data size, arguments used, submission time or start time are used for creating a template. Genetic algorithms can then be used to search the global space for similarities.

Despite the difficulty in estimating runtimes there are SAs which do not require them at all. These algorithms take into consideration only resource load and move tasks only when their loads become unbalanced. This approach works well and tests have shown that scheduling heuristics such as Round-Robin (Fujimoto & Hagihara, 2004) give results comparable to other classic heuristics based on runtime estimates.

In SOE the problem of providing runtime estimates could be overcome by another important aspect related with service costs which is execution deadlines. In this case it does not matter how fast, how slow or where a task gets executed as long as it gets completed inside the specified time interval. Consequently when submitting jobs inside clouds users could attach deadline constraints instead of runtime estimates to either workflows or batch tasks and hope they will not be significantly exceeded. Deadline based scheduling heuristics are specifically useful in cases where users rent services for specific amount of times.

Related with task runtimes is the transfer costs for moving a task from a resource to another. In SOE this is a problem as usually little or nothing is known about the physical location and network route towards a particular service. When moving large amounts of data such as satellite images up to several hundreds of mega-bytes in size the transfer cost becomes an issue. In addition to the time needed to reallocate data problems including licensing and monetary cost arise. There are cases when proprietary data such as satellite images that belong to certain organizations cannot be moved outside their domain (cloud). In this case reallocation to a cloud which provides faster and/or cheaper services for image processing is not possible due to licensing issues.

Task reallocation involves more than simply moving depended data. Clouds rely heavily on virtualization and thus sometimes in orders to execute tasks VMs with certain characteristics need to be created. As a result when reallocating a task the entire VM could require relocation. This implies several other issues such as stopping and resuming preemptive tasks or restarting non-preemptive tasks once they are safely transfered. The problem of transfer costs is thus more problematic than at first glance.

7.3.2 Service Discovery and Selection

Services (SOAP-based (Pautasso et al., 2008), RESTful (Pautasso, Zimmermann, & Leymann, 2008), Grid Services (Foster, 2005)) are an important part of cloud systems. They allow for software, storage, infrastructure or entire platforms to be exposed through a unitary interface which can be used by third party clients. Each service vendor exposes its services to the general public so that the latter can use them, free or at a cost, in order to solve a particular problem.

Inside this sea of services there is also a constant need of discovering proper services for solving a particular task. Universal Description Discovery and Integration (UDDI) (UDDI, 2010) registries offer a solution to this problem. Each service provider registers its services to an UDDI which in turn is used by service consumers for searching specific services. With the occurrence of Web 2.0 these searches could be enhanced with semantic content. Once such a service is found its interface can be used for submitting tasks and for retrieving their results. Figure 7.2 shows the typical correspondence between services, UDDIs and clients.

After successfully finding a number of possible candidate services there remains the problem of selecting the best one for the task. In this direction the scheduling heuristics plays an important role as based on several criteria it will select the service which is most likely to minimize the execution costs. It should be noted that depending on whether the scheduling heuristics is adaptive or not a task could be reallocated several times before actually being executed. Task reallocation faces several problems as addressed in Section 7.3.1.

7.3.3 Negotiation Between Service Providers

Negotiation plays an important role in task scheduling when services from multiple clouds are involved in solving a given problem. Usually the negotiation is

Fig. 7.2 Finding and
invoking services using
UDDIs

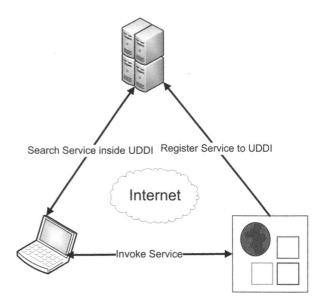

linked to the phase of service selection and involves a scheduler request for a particular service characteristic. When considering it smaller execution costs could be achieved.

Negotiation can also involve the decision on what data/tasks are allowed to be submitted to the service and whether the service provider can further use the submitted data/tasks for its own purposes or not.

As most of the times details regarding the VM or application that is exposed as a service are hidden from public the negotiation requires the introduction of negotiator entities which handle pre-selection discussions in the service/cloud name. Usually this stage is accomplished by one or more agents (Cao, Spooner, Jarvis, & Nudd, 2005; Sauer et al., 2000). Details regarding the involved agents will be given in Section 7.5. Depending on the outcome of the negotiation either access to the desired service is either granted or a new negotiation with another agent proceeds.

7.3.4 Overcoming the Internal Resource Scheduler

An important problem RMSs need to overcome in SOE is that of the internal scheduler used by the service provider. This scheduler is neither influenced nor bypassed by outside intervention. As a result it is said that scheduling between services is accomplished by a meta-scheduler (Weissman, 1998) that deals with tasks at service level, leaving the resource level scheduling to the internal Virtual Organization (VO) schedulers (see Fig. 7.3). These internal schedulers handle tasks assignments

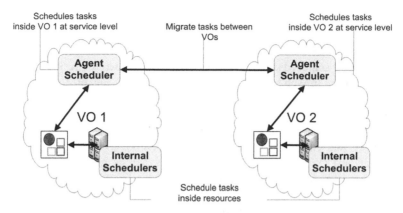

Fig. 7.3 Scheduling and meta-scheduling in multi-VOs

depending on their own policies and thus there is no guarantee that the task submitted by the meta-scheduler will be executed inside the cost constraints negotiated at the time of the submission.

As a result of the negotiation between the meta-scheduler and the service provider the latter could try to favor the task by increasing its priority. This action is in the interest of the provider as it could get penalized, with its service trust greatly diminished, for constantly exceeding the imposed deadlines. Consequently future decisions made by the meta-scheduler could ignore the service and the provider would suffer cost losses. We obtain therefore a symbiotic relationship between the meta-scheduler and the service provider that allows both of them to gain advantages: the service provider's trust will increase when executing tasks faster and thus its income will increase by receiving more tasks; and the meta-scheduler will execute tasks faster, minimizing the costs of the client that submitted them.

7.3.5 Trust in Multi-cloud Environments

When executing tasks on remote services a certain trust level between peers is needed. Trust issued occurs due to many problems including the block box approach of services and because of security issues.

Services cannot be trusted as their interfaces act as black boxes with the content changeable without notice. Thus a service requestor needs to be sure that what it accesses is the same as what was advertised by the service. If this is not the case then the VM running behind the service would not be able to solve the given task inflicting possible cost losses due to time spent for service selection and task submission.

Security issues are also important and are closely linked to the previous problem. These problems can affect both the service requestor and the service provider. The former is usually affected when the data it submits is used for other purposes than those decided during negotiation (e.g. cloning of copyrighted data). The latter can also be affected when data intended to harm the vendor is sent to it. A comprehensive insight on the security issues inside DSs is given in paper (Carstea, Macariu, Frincu, & Petcu, 2008).

Trust is usually achieved through digital certificates such as the X.509 certificates that are widely used in Web browsers, secure email services, and electronic payment systems.

When using certificates clients usually request one from service providers in order to be granted access.

Web-SOAP and Grid-SOAP services handle security issues by using the WS-Security standard (WS Security, 2010). It allows parties to add signatures and encryption headers to SOAP messages. An extension to WS-Security, WS-Trust (WS Trust, 2010), deals with issuing, renewing and validating security tokens or broker trust relationships between participants.

In addition to the WS-Security standard the Transport Layer Security (TLS) can also be used. HTTPS for example can be used to cover Web-SOAP, Grid-SOAP and RESTful services.

7.4 Workflow Scheduling

Workflows consist of several tasks bound together by data/functional dependencies that need to be executed in a specific order for achieving the goal of the problem. They are used especially in cases where the problem can be divided into smaller steps each of them being executed by a distinct solver, or in our case WS. In a cloud environment users usually submit their workflows to a service which orchestrates the execution and returns the result. Whatever happens beyond the service interface is out of reach and invisible to the client. The workflows can be created either by using graphical tools (Wolniewicz et al., 2007) or by directly writing the code in a supported format such as BPEL (WS-BPEL, 2010), YAWL (Van der Aalst & ter Hofstede, 2005), Scufl (Greenwood, 2010), etc. Once the workflow is submitted an enactment engine is responsible for executing the tasks by sending them to corresponding WSs. In our case these WSs are replaced by scheduling agents that try to schedule the tasks on the best available service through negotiation with other agents. Once a task is completed its result is sent back to the enactment engine which can proceed to the next task and so forth.

An important problem in this communication chain is the return of the result to the workflow engine. To solve this problem the address of the agent responsible for the VO in which the engine is located in is attached to each submitted task. In this way once the execution is completed the result is sent straight back to the agent that initially received the task. This task is usually achieved by messages and will be detailed in Section 7.5.2.

It can be noticed that no prior scheduling decisions are made, and that tasks are scheduled one by one as they become ready for scheduling. This is necessary due to the dynamism and unpredictability of the environment. In paper (Frincu, Macariu, & Carstea, 2009) a unified scheduling model for independent and dependent tasks has been discussed. The goal was to allow SA for independent tasks to be applied to workflows when dynamic environments and online scheduling were considered.

Although this approach is suited when global scheduling decisions are needed there are cases where the workflow engine cannot easily achieve task-to-resource mappings (Active BPEL, 2010) during runtime. Instead workflow SAs such as HEFT (Sakellariou & Zhao, 2003), Hybrid (Sakellariou & Zhao, 2004) or CPA (Radulescu & van Gemund, 2001) could be used. However they only consider the tasks in the current workflow when scheduling or rescheduling decisions are needed. These algorithms provide strategies to schedule workflow tasks on heterogeneous resources based on the analysis of the entire task graph. Every time a workflow is submitted tasks would first be assigned to resources and only then would the workflow execution begin. The negotiation for resources thus takes place prior to runtime. This static approach however is not suited for highly dynamic environments (for example clouds) where: resource availability cannot be predicted; reservations are difficult to achieve; a global perspective needs to be obtained; and deadline constraints require permanent rescheduling negotiations.

In what follows we present an agent-based solution for scheduling workflows. So called scheduling agents are used to negotiate, to schedule tasks and to send the answer back to the workflow engine. Its aim is to provide a platform for online workflow scheduling where tasks get scheduled only when they become ready for execution. This means that a task whose predecessors have not completed their execution is not considered to be submitted for execution.

7.5 Distributed Agent Based Scheduling Platform Inside Clouds

As clouds are unpredictable in what concerns resource and network load, systems need to be able to adapt to the new execution configurations so that the cost overheads are not greatly exceeded. Multi-Agent Systems (MAS) provide an answer for this problem as they rely on (semi)decentralized environments made up of several specialized agents working together towards achieving a goal through negotiation. While negotiating each agent keeps a self-centered point of view by trying to minimize its costs.

Although a good option when highly dynamic DS are involved, distributed approaches involve a great amount of transfer overhead (Weichhart, Affenzeller, Reitbauer, & Wagner, 2004) as they require permanent updated from their peers in order to maintain an up to date global view. Contrary, centralized approaches do not require a lot of communication but their efficiency peak is maximized mostly when dealing with DS that maintain a relatively stable configuration.

Decentralized agent based solutions for task scheduling also arise as suited solutions when considering a federation of multiple VOs each having its own resources and implementing its own scheduling policies. Submitting tasks in such an environment requires inter-VO cooperation in order to execute them under restrictions including execution deadlines, workloads, IO dependencies etc.

A computing agent can be defined by flexibility, agility and autonomy and as depicted in (Foster, Jennings, & Kesselman, 2004) can act as the brain for task scheduling inside the multi-cloud infrastructure. Agents allow resources to act as autonomous entities which take decisions on their own based on internal logic. Furthermore an intelligent agent (Sauer et al., 2000) can be seen as an extension to the previously given definition by adding three more characteristics: reactivity (agents react to the environment), pro-activeness (agents take initiatives driven by goals) and social ability (interaction with other agents).

In order to take scheduling decisions agents must meet all the previous requirements. They need to quickly adapt to cloud changes and to communicate with others in order to find a suitable service for tasks that need faster execution. In the context of task scheduling agent adaptiveness includes handling changes in resource workload or availability. In what follows we present a SOE oriented agent based scheduling platform.

7.5.1 The Scheduling Platform

A distributed agent scheduling platform consists of several agents working together for scheduling tasks. Inside a cloud consisting of several service providers (VOs), agents have the role of negotiating and reaching an agreement between the peers. Based on the meta-scheduling heuristics, the internal scheduler, the knowledge on the services it governs and the tasks' characteristics each agent will try to negotiate the relocation from/towards it of several tasks. In trying to achieve this goal agents will also attempt to minimize a global cost attached to each workflow.

Agent based approaches can allow each cloud provider to maintain its own internal scheduling policies (Frincu, 2009a). Furthermore they can also use their own scheduling policies at meta-scheduling level. When deciding on task relocations every agent will follow its own scheduling rules and will try to reach an agreement, through negotiation, with the rest. These aspects allow VOs to maintain autonomy and to continue functioning as independents unit inside the cloud. Autonomy is a mandatory requirement as VOs usually represent companies that want to maintain their independence while providing services to the general public.

Every VO willing to expose services will list one or more agents to a Yellow Pages online directory which can be queried by other agents wanting to negotiate for a better resource.

Agents can be designed as modular entities. In this way we can add new functionalities to agents without requiring creating new agent types. This is different from previous works (Cao, Spooner, Jarvis, & Nudd, 2005; Sauer et al., 2000) which mostly dealt with hierarchies of agents. By doing this we create a super-agent which tries to ensure that the tasks in its domain get the best resources. In addition the need of having multiple agents working together for handling the same task is eliminated. Examples of such agents include: the execution agent, the scheduling agent, the transfer agent, the interface agent, etc.

In our vision all the previously listed specialized agents become sub-modules inside every agent. Thus each agent will have: a scheduling module, a communication module, a service discovery module and an execution module. The sum of all agents forms the meta-scheduler, which is responsible for the inter-service task allocation. Figure 7.4 details this modular structure together with the interactions between agents and other cloud components.

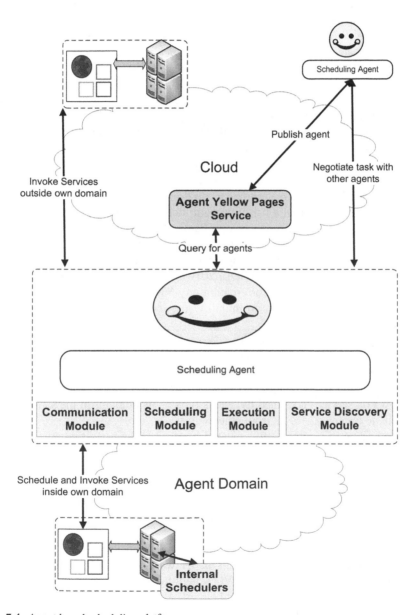

Fig. 7.4 Agent based scheduling platform

In what follows we divide the agents in two categories depending on whether they initiate the request i.e. requestor agents, or they respond to an inquiry i.e. solver agents. This division does not influence the characteristics of the agent and is only intended to depict its role.

The *communication module* handles any type of message exchange with other agents. It also facilitates the dialogue between modules such as between the scheduling module and the service discovery module, or between the scheduling module and the executor module.

The *service discovery module* allows each agent to discover services published on UDDI's located inside its own domain. Typically every resource or provider inside a VO willing to offer some functionality to the general public publishes it as services inside an UDDI. Once a service has been published it can be used by the scheduling agent when reallocating tasks. This module is not used to discover services outside the agent's domain. The reason for this behavior is simple: every service outside its domain is not controlled by the agent and thus not trusted. Trust on services is achieved through negotiation with other agents.

The *execution module* is responsible for invoking the service selected for task execution. Service invocation is usually achieved by creating a client tailored to fit the service interface. The creation has to be done dynamically during runtime as it is not feasible to maintain a list of precompiled clients on disc due to the number and diversity of the existing services. Paper (Carstea, Frincu, Macariu, Petcu, & Hammond, 2007) presents an API for accessing both SOAP-based services and Grid Services by dynamically creating clients based on the service WSDL (Web Service Description Language) (WSDL, 2010). It should be noted that the execution module is not responsible for creating any VM required by tasks. It is up to the resources behind the service to initialize any required VMs based on the task description.

The *scheduling module* deals with task-to-service or task-to-agent allocations. This module is the heart of the agent based scheduling platform and relies on scheduling heuristics for taking its decisions. Every agent has one or more tasks assigned to it. Depending on the scheduling heuristics it can choose to execute some of the tasks on services governed by agents outside its domain. In the same way it can decide to accept new tasks from other agents.

Depending on the policies implemented by the VO there are two possible scenarios that the scheduling module can face. The first occurs when the agent has no information on the resources running the applications and all it sees are the interfaces of the services. The second is the case when an agent knows all there is to know about the underlying resources i.e. workflow, characteristics, network topology, etc. Both of these scenarios are important depending on how the agent behaves when a task needs to executed on one of its services.

The requestor agent submits the job either directly to the service or to the solver agent. In what concerns the rest of this paper we deal with the latter case. The former option involves bypassing the VO scheduler represented by the agent. This happens because the task will be handled directly by the internal resource scheduler. As a consequence any further scheduling optimization at the meta-scheduling level would be hindered.

The scheduling module inside an agent implements a scheduling heuristics designed to deal with SOE. The scheduling heuristics can be seen as the strategy used by the agent to select a resource for its tasks, while the interaction with other agents represents the negotiation phase. The negotiation proceeds based on rules embedded inside the strategy. A common bargaining language and a set of predefined participation rules are used to ensure a successful negotiation.

Each agent has several services it governs (see Fig. 7.5). Attached to them there are task queues. Depending on the scheduling heuristics only a given number of tasks can be submitted (by using the execution module) to a service at any given moment. Once submitted to the service it is the job of the internal scheduler to assign

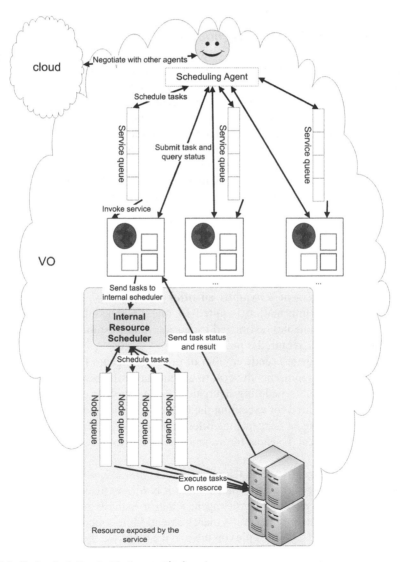

Fig. 7.5 Task scheduling inside the agent's domain

the tasks to the resources. Similarly to the service level queues there could also exist queues attached to physical resources. Reallocation between these queues is accomplished by the internal scheduler and is independent on any meta-scheduling decisions taken by agents. Each resource behind a service can implement its own scheduling policies. Usually tasks submitted to a service are not sent back to the agent for meta-scheduling. There are many ways of checking whether a task has been completed or not. One of them requires the scheduling agent to periodically query the service to which it has submitted it for the result. In Section 7.5.3 we briefly present a prototype where internal schedulers have their own scheduling heuristics and work independently from the agent meta-scheduling heuristics.

7.5.2 Scheduling Through Negotiation

The central entity of every agent based scheduling platform is the scheduling mechanism. Based on its rules agents make active/passive decisions on whether to move or to accept new tasks. Every decision is proceeded by a negotiation phase where the agent requests/receives information from other agents and decides, based on the scheduling heuristics, which offer to accept. Negotiation requires both a language and a set of participation rules (Ramchurn, 2004). Depending on the VO policy and on the adherence of other VOs to it many types of negotiation can be used. Examples include game theory models (Rosenschein & Zlotkin, 1994), heuristic approaches (Faratin, Sierra, & Jennings, 2001) and argument based (Parsons, Sierra, & Jennings, 1998) solutions.

A minimal set of locutions has been devised for the communication language used by our platform:

- *requestOffer(i,j,k)*: agent i requires an offer from agent j for a task k. Task k contains all the information required to make a scheduling decision. This may include (if available): estimated execution times, estimated transfer costs, execution deadlines, required input, etc.;
- *sendOffer(j,i,k,p)*: agent j sends an offer of price p to agent i for the execution of task k. The price p represents the cost to execute task k on resource j. Measuring costs depends on the scheduling heuristics. For example it could represent the estimated time required for executing the task on a service belonging to agent j;
- *acceptOffer(i,j,k)*: agent i accepts the offer of agent j for executing task k;
- *sendTask(i,j,k)*: agent i sends for execution task k to a service provided by agent j;
- *rejectOffer(i,j,k)*: agent i rejects the offer of agent j for executing task k;
- *requestTasks(i,j)*: agent i informs agent j that it is willing to execute more tasks;
- *requireDetails(i,j)*: agent i informs agent j that it requires more details on the services/resources under the latter's management. More specifically they refer to details (WSDL URL for example) on the service proposed by agent j;

- *sendDetails(j,i,d)*: agent *j* sends available details to agent *i*. These details contain only publicly available data as result of internal policies;
- *informTaskStatus(i,j,k,m)*: agent *i* informs by using message *m* agent *j* about the status of a task *k*. For example the message could contain the result of a task execution.

Participation rules are required in order to prohibit agents from saying something they are not allowed to say at a particular moment. Figure 7.6 shows participation rules between these locutions in the form of a finite state machine:

A negotiation starts either from a request for more tasks from an agent *j* or from a request for offers for a given task which an agent *i* decided to relocate. There is a permanent link between the workflow engine agent and the scheduling agent responsible for the VO in which the engine executes. It is to this agent where tasks

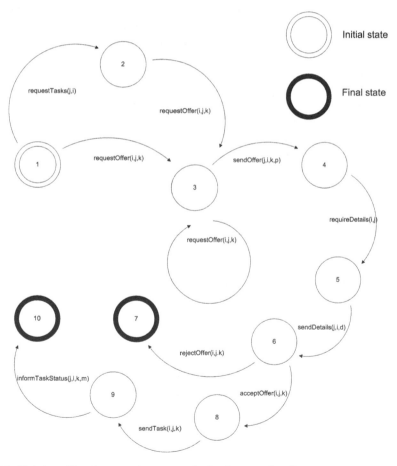

Fig. 7.6 State transitions between the communication language locutions

are placed first. Once a new task has been sent to this agent it is its responsibility to find and negotiate the execution on a resource which has the highest chance of minimizing the deadline constraint. Workflow engines agents are similar with scheduling agents and can communicate with them. However they cannot schedule tasks on resources. Their only purpose is to provide an interface between the engine and the meta-scheduling platform.

When scheduling workflows an important problem that needs to be integrated inside the negotiation phase occurs. Considering the execution of a task on a service that provides a result which can only be further used on services belonging to the same VO, any other possible solutions outside of that VO would be ignored. It is therefore the job of the requestor agent to negotiate for a solution that maximizes the search set. For that reason a balance between the best time cost at a given moment and future restrictions needs to be achieved. As an example, selecting the fastest service for executing the task could be transformed into selecting the service which executes the task faster and without restrictions on using the result.

In case an agent j has requested more tasks from another agent i the latter will ask the former for offers regarding the cost of executing some of its tasks. At this point agent j will send back to agent i an offer for the task in question.

Based on this offer agent i will ask for more details regarding the available services which will allow it to make a proper decision: it will either reject or accept the offer. In case agent i accepts the offer of agent j the task will be submitted to a service queue governed by the latter agent (see Fig. 7.5). In return it will send back a message on the task status. Once the task is completed the result will be sent back to the workflow agent which will communicate it to the engine. The engine will use the result to select consequent tasks for scheduling and execution.

In the frame of the presented negotiation protocol the key element is played by the moment a request for a relocation offer or for new tasks is made. This point in time basically marks the starting of the negotiation.

The problem of properly selecting the moment of an offer request has been addressed in our paper (Frincu, 2009a). The proposed scheduling heuristics incorporates this reallocation moment and it is shown that the schedule outcome is directly influenced by it.

In order to extend this approach to SOE, a deadline-based approach has been investigated in paper (Frincu, 2009b). The study is based on the fact that in SOE users usually want to minimize their costs with regard to usage time and thus provide an execution deadline for each task inside their workflows. The aim of the schedule is to minimize the global task lateness i.e. the difference between the actual task finish time and the user given deadline time.

The scheduling heuristics is called DMECT (Frincu, 2009a) (Dynamic Minimization of Estimated Completion Time). It periodically computes, for every task, the Time Until Deadline (TUD), the Local Waiting Time (LWT) – the time since it was assigned to the current service queue – and the Total Waiting Time (TWT) – time since the task's submission. From these values a decision on whether to move the task or not is taken by checking if the TUD/TWT – LWT is smaller than 0 or not. If the value is smaller a *requestOffer* action is taken. It must be noted that when the decision to relocate the task is taken, all the available services are taken into consideration. These include both internal (part of the current agent domain) and external (obtained from the *requestOffer* inquiry) ones. In this way every existing service gets a fair chance for competing for tasks. It can be easily seen that the relocation relation will try to relocate tasks faster as their deadline approaches.

As a response to a *requestOffer* inquiry, every agent will perform a *sendOffer* action which will inform the requestor agent on possible choices. Every reply typically contains a cost for the task's execution on its best service. If the initial inquiry also contained a lower bound for that cost a list of services offering better prices is returned. The cost for scheduling is made up of execution times possible combined with monetary costs. For example when inquired, each agent will compute the estimated execution time on every service and return only those which have values smaller than the initially provided limit. Alternatively it could return only the smallest value, ensuring that the best available offer it had was made.

In case where it is impossible to estimate the execution time due to insufficient data or internal policies the length of a service queue could be used as measure. In (Frincu, 2009b) we have shown that the smaller a queue is the likelihood that it executes tasks faster is.

After gathering all the costs the requestor agent will select the best one according to the scheduling heuristics i.e. smallest execution time in our example. All other offers will be rejected. Once selected the task will be sent to the selected solver agent which will place the task in the service queue and the LWT value for the relocated task will be set to 0. In the scenario that the task will not get executed on the newly elected service as well, i.e. TUD/TWT − LWT < 0, the solver agent will send a *requestOffer* inquiry to other agents, thus becoming the newly requestor agent for that task.

Deciding when to request for new tasks is another important case which triggers the negotiation process. In this case an agent sends a *requestTasks* message to all the other agents informing them about its willingness to accept tasks. Once this message has been sent agents will begin sending *requestOffers* to it for tasks they wish to reallocate. From this point the negotiation proceeds similarly with the previously discussed case.

Depending on the policy the request for new tasks can be done periodically or when the load of the services under an agent's supervision drops below a certain limit. Depending on the scheduling policy this approach of actively searching new tasks could be inefficient. For example in our scenario using the DMECT heuristics such a request would have no effect until at least one task exceeds its staying limit on a resource queue. Other scheduling heuristics based on simple load balancing techniques such as the one presented in Frincu et al., (2009) could be more suited for this scenario. In these cases there are no conditions preventing tasks from migrating between agents. Once an agent decides that the load on its services has dropped sufficiently new tasks can be requested.

7.5.3 Prototype Implementation Details

In this section we present some implementation aspects of the scheduling platform prototype. The platform relies on JADE (Bellifemine, Poggi, & Rimassa, 2001) as an agent platform and on the OSyRIS (OSYRIS WorkFlow Engine, 2010) engine for workflow enactment.

JADE facilitates the development of distributed applications following the agent-oriented paradigm and is in fact a FIPA (Foundation for Intelligent Physical Agents) compliant multi-agent middleware. It is implemented in the Java language and provides an Eclipse plug-in which eases the development process by integrating development, deployment and debugging graphical tools. In addition JADE can be distributed across several resources and its configuration can be controlled through a remote graphical user interface. Agents can migrate among these resources freely at any time. Also JADE provides: a standard architecture for scheduling agent activities; a standard communication protocol by using the Agent Communication Language (ACL); and allows the integration of higher functionality by allowing users to include their own Prolog modules for activity reasoning. Even though the simple model of JADE agents makes the development easier it requires a considerable amount of effort for including intelligence when complex control is required.

Paper (Poggi, Tomaiuolo, & Turci, 2004) presents an extension to JADE where the platform is augmented with two types of agents with the aim of paving the way for a more flexible agent cloud system. The two types of agents are: the *BeanShell* agent responsible for sending and executing behaviors coming from other agents; and the *Drools* agent responsible for receiving and executing rules coming from other agents. Authentication and authorization mechanisms are offered for both types of agents.

OSyRIS is a workflow enactment engine inspired by nature where rules are expressed following the Event Condition Action paradigm: tasks are executed only when some events occur and additional optional conditions are met. In OSyRIS events represent the completion of tasks and conditions are usually placed on the output values. A single instruction is used all the rest (split, join, parallel, sequence, choice, loop) deriving from it: *LHS -< RHS | condition, salience*, where *LHS* (Left Hand Side) represents the tasks that need to be completed before executing the *RHS* (Right Hand Side) tasks. The engine relies on a chemical metaphor where tasks play the role of molecules and the execution rules are the reactions.

In order to simulate VOs we have used two clusters available at the university. One consisting of 8 Pentium dual-core nodes with 4 GB of RAM each (called VO1) and the other having 42 nodes with 8 cores and 8 GB of RAM each. The latter cluster is divided into 3 blades (called VO2, VO2 and VO3) each with 14 nodes each. To each blade there is attached one scheduling agent which manages the services running on them. A single agent is used for governing the entire VO1. Nodes are paired and each pair is exposed through a service handled by the agent handling the governing VO. The agents are registered to a yellow page repository as depicted in Fig. 7.4. For inter-agent task scheduling the DMECT heuristics is used. Although different SAs could be used for local resource scheduling we have opted for a single one: the MinQL (Frincu et al., 2009) heuristics.

The scheduling scenario proceeds as follows: once a scheduling agent receives a task, it attaches it to one of its service queues (see Fig. 7.5). Tasks are received either by negotiating with other agents or directly from a workflow agent. The negotiation protocol is similar with the one in Fig. 7.6 and uses the DMECT SA's relocation condition (Frincu, 2009a) as described in Section 7.5.2. Each service can execute at most k instances simultaneously. Variable k is equal to the number of processors inside the node pair. Once sent to a service a task cannot be sent back to the agent unless explicitly specified in the scheduling heuristics. Tasks sent to services are scheduled inside the resource by using the MinQL SA which uses a simple load balancing technique. Scheduling agents periodically query the service for completed tasks. Once one is found the information inside it is used to return the result to the agent responsible for the workflow instance. This passes the information to the engine which in turn passes the consequent set of tasks to the agent for scheduling.

In order to simulate the cloud heterogeneity in terms of capabilities services offer different functionalities. In our case services offer access to both CASs and image processing methods. As each CAS offers different functions for handling mathematical problems so does the service exposing it. The same applies for the image processing services that do not implement all the available methods on every service. An insight on how CASs with different capabilities can be exposed as services is given in (Petcu, Carstea, Macariu, & Frincu, 2008).

7.6 Conclusions

In this paper we have presented some issues regarding task scheduling when services from various providers are offered. Problems such as estimating runtimes and transfer costs; service discovery and selection; trust and negotiation between providers for accessing their services; or making the independent resource scheduler cooperate with the meta-scheduler, have been discussed. As described much of the existing scheduling platforms are grid oriented and cloud schedulers are only beginning to emerge. As a consequence a MAS approach to the cloud scheduling problem has been introduced. MAS have been chosen since they provide greater flexibility and are distributed by nature. They could also represent a good choice for scheduling scenarios where negotiation between vendors is required. Negotiation is particularly important when dealing with workflows where tasks need to be orchestrated together and executed under strict deadlines in order to minimize user costs. This is due to the fact that vendors have different access and scheduling policies and therefore selecting the best service for executing a task with a provided input becomes more than just a simple reallocation problem. The prototype system uses a single type of agents which combine multiple functionalities. The resulting meta-scheduler maintains the autonomy of each VO inside the cloud.

The presented solution is under current development and future tests using various SAs and platform configurations are planned.

Acknowledgments This research is partially supported by European Union Framework 6 grant RII3-CT-2005-026133 SCIEnce: Symbolic Computing Infrastructure in Europe.

References

Abramson, D., Buyya, R., & Giddy, J. (2000). A computational economy for grid computing and its implementation in the NIMROD-G resource broker. *Future Generation Computer Systems, 18*(8), 1061–1074.

Active BPEL. (2010). Retrieved January 7, 2010, from http://www.activebpel.org/.

Ali, A., Bunn, J., Cavanaugh, R., van Lingen, F., Mehmood, M. A., Steenberg, C., & Willers, I. (2004). Predicting Resource Requirements of a Job Submission. *Proceedings of the Conference on Computing in High Energy and Nuclear Physics*.

Banerjee, S., Mukherjee, I., & Mahanti, P. K. (2009). Cloud computing initiative using modified ant colony framework. *World Academy of Science, Engineering and Technology, 56*, 221–224.

Bellifemine, F., Poggi, A., & Rimassa, G. (2001). Jade: A FIPA2000 compliant agent development environment. *Proceedings of the 5th International Conference on Autonomous Agents, Montreal, Canada*, 216–217.

Berman, F., Wolski, R., Casanova, H., Cirne, W., Dail, H., Faerman, M., et al. (2003). Adaptive Computing on the Grid using APPLES. *IEEE Transactions on Parallel and Distributed Systems, 14*(4), 369–382.

Buyya, R., Abramson, D., & Giddy, J. (2000). Nimrod/g: An architecture for a resource management and scheduling system in a global computational grid. *Proceedings of the 4th International Conference on High Performance Computing in Asia-Pacific Region, Vol. 1, Beijing, China*, 283–289.

Cao, J., Jarvis, S. A., Saini, S., Kerbyson, D. J., & Nudd, G. R. (2002). Arms: An agent-based resource management system for grid computing. *Scientific Programming, 10*(2), 135–148.

Cao, J., Kerbyso, D. J., Papaefstathiou, E., & Nudd, G. R. (2000). Performance modelling of parallel and distributed computing using pACE. *Proceedings of 19th IEEE International Performance, Computing and Communication Conference, Phoenix, AZ*, 485–492.

Cao, J., Spooner, D. P., Jarvis, S. A., & Nudd, G. R. (2005). Grid load balancing using intelligent agents. *Future Generation Computer Systems, 21*(1), 135–149.

Carstea, A., Frincu, M., Macariu, G., Petcu, D., & Hammond, K. (2007). Generic access to web an grid-based symbolic computing services. *Proceedings of the 6th International Symposium in Parallel and Distributed Computing, Miami, FL*, 143–150.

Carstea, A., Macariu, G., Frincu, M., & Petcu, D. (2008). Secure orchestration of symbolic grid services, (IeAT Tech. Rep., No. 08-08).

Casanova, H., & Dongarra, J. (1998). Applying netsolve's network-enabled server. *EEE Computational Science and Engineering, 5*(3), 57–67.

Casanova, H., Obertelli, G., Berman, F., & Wolski, R. (2000). The APPLES parameter sweep template: User-level middleware for the grid. *Proceedings of Super Computing SC'00, Dallas, TX*, 75–76.

Chapin, S. J., Katramatos, D., Karpovich, J., & Grimshaw, A. (1999). Resource management in legion. *Future Generation Computer Systems, 15*(5), 583–594.

Cloud Scheduler, (2010). Retrieved January 7, 2010, from http://cloudscheduler.org/.

David, L., & Puaut, I. (2004). Static determination of probabilistic execution times. *Proceedings of the 16th Euromicro Conference on Real-Time Systems, Catania, Sicily*, 223–230.

Faratin, P., Sierra, C., & Jennings, N. R. (2001). Using similarity criteria to make issue trade-offs in automated negotiation. *Artificial Intelligence, 142*(2), 205–237.

Foster, I. T. (2005). Globus toolkit version 4: Software for service-oriented systems. *Proceedings of International Conference on Network and Parallel Computing, Beijing, China, Vol. 3779*, 2–13.

Foster, I., Jennings, N.R., & Kesselman, C. (2004). Brain meets brawn: Why grid and agents need each other. *In Proceedings of the 3rd International Joint Conference on Autonomous Agents and Multiagent Systems, New York, NY,* 8–15.

Frincu, M. (2009). Dynamic scheduling algorithm for heterogeneous environments with regular task input from multiple requests. *In Lecture Notes in Computer Science, Vol. 5529, Geneva, Switzerland,* 199–210.

Frincu, M. (2009b). Distributed scheduling policy in service oriented environments. *Proceedings of the 11th International Symposium on Symbolic and Numeric Algorithms for Scientific Computing, Timi Soara, Romania.*

Frincu, M., Macariu, G., & Carstea, A. (2009). Dynamic and adaptive workflow execution platform for symbolic computations. *Pollack Periodica, Akademiai Kiado, 4*(1), 145–156.

Fujimoto, N., & Hagihara, K. (2004). A comparison among grid scheduling algorithms for independent coarse-grained tasks. *International Symposium on Applications and the Internet Workshops, Phoenix, AZ.*

Garg, S. K., Yeo, C. S., Anandasivam, A., & Buyya, R. (2009). *Energy-efficient scheduling of HPC applications in cloud computing environments.* (Tech. Rep.).

Google Application Engine, (2010). Retrieved January 7, 2010, from http://appengine.google.com.

Greenwood, M. (2004). *Xscufl language reference.* Retrieved January 7, 2010 from http://www.mygrid.org.uk/wiki/Mygrid/WorkFlow#XScufl.

Lorpunmanee, S., Sap, M. N., Abdullah, A. H., & Chompooinwai, C. (2007). An ant colony optimization for dynamic job scheduling in grid environment. *World Academy of Science, Engineering and Technology, 23,* 314–321.

Maheswaran, M., Braun, T. D., & Siegel, H. J. (1999). Heterogeneous distributed computing. *Encyclopedia of Electrical and Electronics Engineering* (Vol. 8, pp. 679–690). New York, NY: Wiley.

Microsoft Live Mesh. (2010). Retrieved January 7, 2010, from http://www.mesh.com.

Muniyappa, V., (2010). Inference of task execution times using linear regression techniques. (Masters Thesis, Texas Tech University, 2002).

OSyRIS Workflow Engine. (2010). Retrieved January 7, 2010, from http://gisheo.info.uvt.ro/trac/wiki/Workflow.

Parsons, S., Sierra, C., & Jennings, N. R. (1998). Agents that reason and negotiate by arguing. *Journal of Logic and Computation, 8*(3), 261–292.

Pautasso, C., Zimmermann, O., & Leymann, F. (2008). RESTful web services vs. big web services: Making the right architectural decision. *17th International World Wide Web Conference, Beijing, China.*

Petcu, D., Carstea, A., Macariu, G., & Frincu, M. (2008). Service-oriented symbolic computing with symGrid. *Scalable Computing: Practice and Experience, 9*(2), 111–124.

Poggi, A., Tomaiuolo, M., & Turci, P. (2004). Extending JADE for agent grid applications. *Proceedings of the 13th IEEE International Workshops on Enabling Technologies: Infrastructure for Collaborative Enterprise, Stanford, CA,* 352–357.

Radulescu, A., & van Gemund, A. (2001). A low-cost approach towards mixed task and data parallel scheduling. *Proceedings of the International Conference on Parallel Processing, Wakamatsu, Japan.*

Ramchurn, S. D. (2004). Multi-agent negotiation using trust and persuasion (Doctoral Thesis, University of Southampton UK, 2004).

Ritchie, G. R. S., & Levine J. (2004). A hybrid ant algorithm for scheduling independent jobs in heterogeneous computing environments. *Proceedings of the 23rd Workshop of the UK Planning and Scheduling Special Interest Group, Edinburgh.*

Rosenschein, J. S., & Zlotkin, G. (1994). *Roles of Encounter. Cambridge, MA:* MIT Press.

Sakellariou, R., & Zhao, H., (2003). Experimental investigation into the rank function of the heterogeneous earliest finish time scheduling algorithm. *In Lecture Notes in Computer Science, Klagenfurt, Austria, Vol. 2790,* 189–194.

Sakellariou, R., & Zhao, H. (2004). A hybrid heuristic for DAG scheduling on heterogeneous systems. *Proceedings of the 18th International Symposium In Parallel and Distributed Processing, Santa Fe, New Mexico.*

Sauer, J., Freese, T., & Teschke, T. (2000). Towards agent-based multi-site scheduling. *Proceedings of the ECAI 2000 Workshop on New Results in Planning, Scheduling, and Design, Berlin, Germany.*

Shen, W., Li, Y., Genniwa, H., & Wang, C. (2002). Adaptive negotiation for agent-based grid computing. *Proceedings of the Agentcities/AAMAS'02, Bologna, Italy.*

Smith, W., Foster, I., & Taylor, V. E. (2004). Predicting application run times with historical information. *Journal of Parallel and Distributed Computing, 64*(9), 1007–1016.

Thain, D., Tannenbaum, T., & Livny, M. (2005). Distributed computing in practice: The condor experience. *Concurrency and Computation: Practice and Experience, 17*(2–4) 323–356.

UDDI. (2010). Retrieved January 7, 2010, from www.uddi.org/pubs/uddi-tech-wp.pdf.

van der Aalst, W. M. P., & ter Hofstede, A. H. M. (2005). Yawl: yet another workflow language. *Information Systems, 30*(4), 245–275.

Weichhart, G., Affenzeller, M., Reitbauer, A., & Wagner, S. (2004). Modelling of an agent-based schedule optimisation system. *Proceedings of the IMS International Forum, Cernobbio, Italy,* 79–87.

Weissman, J. B. (1998). Metascheduling: A scheduling model for metacomputing systems. *The 7th International Symposium on High Performance Distributed Computing, Chicago, Illinois, USA,* 348–349.

Wolniewicz, P., Meyer, N., Stroinski, M., Stuempert, M., Kornmayer, H., Polak, M., et al. (2007). Accessing grid computing resources with G-eclipse platform. *Computational Methods in Science and Technologie, 13*(2) 131–141.

WS-BPEL 2.0. (2010). Retrieved January 7, 2010, from http://docs.oasis-open.org/wsbpel/2.0/wsbpel-v2.0.pdf.

WS Security. (2010). Retrieved January 7, 2010, from http://www.ibm.com/developerworks/library/specification/ws-secure/.

WS Trust 1.4. (2010). Retrieved January 7, 2010, from http://docs.oasis-open.org/ws-sx/ws-trust/v1.4/ws-trust.html.

WSDL. (2010). Retrieved January 7, 2010, from http://www.w3.org/TR/wsdl.

Chapter 8
The Role of Grid Computing Technologies in Cloud Computing

David Villegas, Ivan Rodero, Liana Fong, Norman Bobroff, Yanbin Liu, Manish Parashar, and S. Masoud Sadjadi

Abstract The fields of Grid, Utility and Cloud Computing have a set of common objectives in harnessing shared resources to optimally meet a great variety of demands cost-effectively and in a timely manner Since Grid Computing started its technological journey about a decade earlier than Cloud Computing, the Cloud can benefit from the technologies and experience of the Grid in building an infrastructure for distributed computing. Our comparison of Grid and Cloud starts with their basic characteristics and interaction models with clients, resource consumers and providers. Then the similarities and differences in architectural layers and key usage patterns are examined. This is followed by an in depth look at the technologies and best practices that have applicability from Grid to Cloud computing, including scheduling, service orientation, security, data management, monitoring, interoperability, simulation and autonomic support. Finally, we offer insights on how these techniques will help solve the current challenges faced by Cloud computing.

8.1 Introduction

Cloud computing exploits the advances in computing hardware and programming models that can be brought together to provide utility solutions to large-scale computing problems. At the hardware level, the last half century has seen prolific progress in computing power. This results from many improvements at the processor

D. Villegas (✉) and S. Masoud Sadjadi
CIS, Florida International University, Miami, FL, USA
e-mails: {dvill013; sadjadi}@cis.fiu.edu

I. Rodero and M. Parashar
NSF CAC, Rutgers University, Piscataway, NJ, USA
e-mails: {irodero; parashar}@rutgers.edu

L. Fong, N. Bobroff, and Y. Liu
IBM Watson Research Center, Hawthorne, NY, USA
e-mails: {llfong; bobroff; ygliu}@us.ibm.com

B. Furht, A. Escalante (eds.), *Handbook of Cloud Computing*,
DOI 10.1007/978-1-4419-6524-0_8, © Springer Science+Business Media, LLC 2010

level, and in recent years the availability of low cost multi-core circuits. Additional progress in high speed, low latency interconnects, has allowed building large-scale local clusters for distributed computing, and the extension to wide-area collaborating clusters in the Grid. Now, the recent availability of hardware support for platform virtualization on commodity machines provides a key enabler for Cloud based computing.

Software models move in lockstep to match advances in hardware. There is a considerable practical experience implementing distributed computing solutions and in supporting parallel programming models on clusters. These models now work to leverage the concurrency provided by multi-core and multi-systems. Additionally, there are two other areas of software evolution that are moving quickly to support the Cloud paradigm: one is the improving maturity and capability of software to manage virtual machines, and the other is the migration from a monolithic approach in constructing software solutions to a service approach in which complex processes are composed of loosely coupled components.

These latest steps in the evolution of hardware and software models have led to Grid and Cloud Computing as paradigms that reduce the cost of software solutions. Harnessing shared computing resources from federated organizations to execute applications is the key concept of Grid Computing, as proposed by Foster, Kesselman, and Tuecke (2001): "Grid concept is coordinated resource sharing and problem solving in dynamic, multi-institutional virtual organizations...The sharing is, necessarily, highly controlled, with resource providers and consumers defining clearly and carefully just what is shared, who is allowed to share, and the conditions under which sharing occurs."

Evolving from the technologies of Grid computing, Utility Computing is "a business model in which computing resources are packaged as metered services" (Foster, Zhao, Raicu, & Lu, 2008) to meet on demand resource requirements. The metered resource usage is similar to electric and water utility in delivery and the payment model is pay-as-you-go. The Utility Computing projects also introduced and demonstrated the ideas of dynamic provisioning of computing resources (Appleby et al., 2001).

Armbrust et al. (2009) defined Cloud Computing as providing application software delivered as services over the Internet, and the software and hardware infrastructure in the data centers that provide those services using business models that are similar to Utility Computing. Metering of services to support pay-as-you-go business models is suitable to application software (i.e., software as services, SaaS), platform (i.e., platform as services, PaaS), and infrastructure (i.e., infrastructure as services, IaaS). Moreover, Cloud Computing leverages emerging technologies such as the Web 2.0 for application services, and virtualization and dynamic provisioning support for platform services (Buyya, Yeo, Srikumar Venugopal, & Brandic, 2009).

Grid Computing, Utility Computing and Cloud Computing differ in aspects as their architectures, the types of coordinated institutions, the types of resources shared, the cost/business models, and the technologies used to achieve their objectives. However, all these computing environments have the common objectives in harnessing shared resources to optimally meet a variety of demands cost-effectively and at timely manner. Since Grid Computing started its technological journey about a decade earlier than Cloud Computing, are there lessons to learn and technologies to leverage from Grid to Cloud? In this chapter, we would like to explore the experiences learnt in Grid and the role of Grid technologies for Cloud computing.

The rest of this chapter is organized as follows:

- Introductory discussion on the basics of Grid and Cloud computing, and their respective interaction models between client, resource consumer and provider (Section 8.2)
- Comparison of key processes in Grid and Cloud computing (Section 8.3)
- Core Grid technologies and their applicability to Cloud computing (Section 8.4)
- Concluding remarks on the future directions of Grid and Cloud computing (Section 8.5).

8.2 Basics of Grid and Cloud Computing

8.2.1 Basics of Grid Computing

Grid Computing harnesses distributed resources from various institutions (resource providers), to meet the demands of clients consuming them. Resources from different providers are likely to be diverse and heterogeneous in their functions (computing, storage, software, etc.), hardware architectures (Intel x86, IBM PowerPC, etc.), and usage policies set by owning institutions. Developed under the umbrella of Grid Computing, information services, name services, and resource brokering services are important technologies responsible for the aggregation of resource information and availability, selection of resources to meet the clients' specific requirements and the quality of services criteria while adhering to the resource usage policies.

Figure 8.1 shows an exemplary relationship of resource providers and consumers for a collaborative Grid computing scenario. Clients or users submit their requests for application execution along with resource requirements from their home domains. A Resource broker selects a domain with appropriate resources to acquire from and to execute the application or route the application to domain for execution with results and status returning to the home domain.

8.2.2 Basics of Cloud Computing

IDC[1] defined two specific aspects of Clouds: Cloud Services and Cloud Computing. Cloud Services are "consumer and business products, services and solutions that are delivered and consumed in real-time over the Internet" while Cloud Computing is "an emerging IT development, deployment and delivery model, enabling real-time

---

[1] http://blogs.idc.com/ie/?p=190

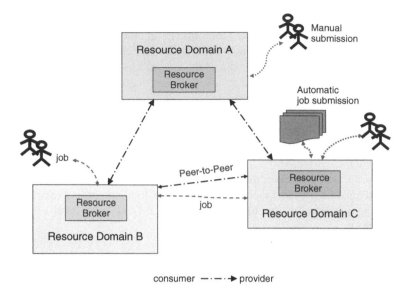

Fig. 8.1 Grid collaborating domains

delivery of products, services and solutions over the Internet (i.e., enabling Cloud services)". In this chapter, we will focus the computing infrastructure and platform aspects of the Cloud.

Amazon's Elastic Compute Cloud[2] popularized the Cloud computing model by providing an on-demand provisioning of virtualized computational resources as metered services to clients or users. While not restricted, most of the clients are individual users that acquire necessary resources for their own usage through EC2's APIs without cross organization agreements or contracts. Figure 8.2 illustrates possible usage models from clients C1 and C2 for resources/services of Cloud providers. As Cloud models evolve, many are developing the hybrid Cloud model in which enterprise resource brokers may acquire additional needed resources from external Cloud providers to meet the demands of submitted enterprise workloads (E1) and client work requests (E2). Moreover, the enterprise resource domain and Cloud providers may all belong to one corporation and thus form a private Cloud model.

8.2.3 Interaction Models of Grid and Cloud Computing

One of the most scalable interaction models of Grid domains is peer-to-peer, where most of the Grid participating organizations are both consumers and providers. In practice, there are usually agreements of resource sharing among the

---

[2]http://www.amazon.com/ec2

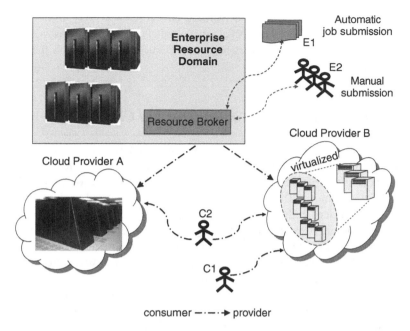

Fig. 8.2 Cloud usage models

peers. Furthermore, clients of consumer organizations in Grids use heterogeneous resources from more than one resource provider belonging to the same Virtual Organization (VO) to execute their applications. It is important for participating resource providers and consumers to have common information models, interaction protocols, application execution states, etc. The organization of Open Grid Forum (OGF)[3] has the goal of establishing relevant and necessary standards for Grid computing. Some proposed standards include Job Submission Description Language (JSDL), Basic Execution Service (BES) and others.

Currently, most of the Cloud providers offer their own proprietary service protocols and information formats. As Cloud computing becomes mature and widely adopted, clients and consumer organizations would likely interact with more than one provider for various reasons, including finding the most cost effective solutions or acquiring a variety of services from different providers (e.g., compute providers or data providers). Cloud consumers will likely demand common protocols and standardized information formats for ease of federated usage and interoperability. The Open Virtualization format (OVF) of the Distributed Management Task Force (DMTF)[4] is an exemplary proposal in this direction. Modeled after similar formations in the Grid community, OGF officially launched a workgroup, named

---

[3] http://www.ogf.org/

[4] http://www.dmtf.org/standards/

the Open Cloud Computing Interface Working Group (OCCI-WG)[5] to develop the necessary common APIs for the lifecycle management of Cloud infrastructure services. More standardization activities related to Cloud can be found in the wiki of Cloud-Standards.org.[6]

8.2.4 Distributed Computing in the Grid and Cloud

The Grid encompasses two areas of distributed system activity. One is operational with an objective of administrating and managing an interoperable collection of distributed compute resource clusters on which to execute client jobs, typically scientific/HPC applications. The procedures and protocols required to support clients from complex services built on distributed components that handle job submission, security, machine provisioning, and data staging. The Cloud has similar operational requirements for supporting complex services to provide clients with services on different levels of support such application, platform and infrastructure. The Grid also represents as a coherent entity a collection of compute resources that may be under different administrative domains, such as universities, but inter-operate transparently to form virtual organizations. Although interoperability is not a near term priority, there is a precedent for commercial Clouds to move in this direction similarly to how utilities such as power or communication contract with their competitors to provide overflow capacity.

The second aspect of distributed computing in the Grid is that job themselves are distributed, typically running on tightly coupled nodes within a cluster and leveraging middleware services such as MPICH. Jobs running in the Grid are not typically interactive, and some may be part of more complex services such as e-science workflows. Workloads in Clouds usually consist of more loosely coupled distributed jobs such as map/reduce, and HPC jobs written to minimize internode communication and leverage concurrency provided by large multi-core nodes. Service instances that form components of a larger business process workflow are likely to be deployed in the Cloud. These workload aspects of jobs running in the Cloud or Grid have implications for structuring the services that administer and manage the quality of their execution.

8.3 Layered Models and Usage patterns in Grid and Cloud

There are many similarities in Grid and Cloud computing systems. We compare the approaches by differentiating three layers of abstraction in Grid: Infrastructure, Platform and Application. Then we map these three layers to the Cloud services of IaaS, PaaS, and SaaS. An example of the relations among layers can be seen in Fig. 8.3.

---

[5]http://www.occi-wg.org/doku.php?id=start
[6]http://cloud-standards.org/

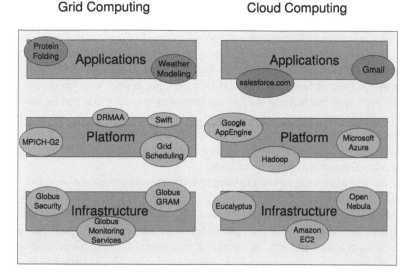

Fig. 8.3 Grid and cloud layers

8.3.1 Infrastructure

This is the layer in which Clouds share most characteristics with the original purpose of Grid middleware. Some examples are Eucalyptus (Nurmi et al., 2009), OpenNebula,[7] or Amazon EC2. In these systems users can provision execution environments in the form of virtual machines through interfaces such as APIs or command line tools. The act of defining an execution environment and sending a request to the final resource has many similarities with scheduling a job in the Grid. The main steps, shared by all of the cited Cloud environments are discussed below. We use Globus as the reference Grid technology.

- The user needs to be authorized to use the system. In Grid systems this is managed through the Community Authorization System (CAS) or by contacting a Certificate Authority that is trusted by the target institution, which issues a valid certificate. Clouds usually offer web forms to allow the registration of new users, and have additional web applications to maintain databases of customers and generate credentials, such as the case of Eucalyptus or Amazon.
- Once the user has a means of authenticating he needs to contact a gateway that can validate him and process his request. Different mechanisms are employed to carry users' requests, but Web Services are the most common of them. Users either write a custom program that consumes the WS offered by providers, or use available tools. Examples include the Amazon API tools for Amazon EC2, the euca2ools for Eucalyptus or the OpenNebula command line interface. Similarly,

---

[7]http://www.opennebula.org

Globus offers a set of console-based scripts that facilitate communication with the Grid.

- As part of the request for resource usage, users need to specify the action or task to be executed on the destination resources. Several formats are available for this purpose. Globus supports a Resource Specification Language (RSL) and a Job Submission Description Language (JSDL) that can define what process is to be run on the target machine, as well as additional constraints that can be used by a matchmaking component to restrict the class of resources to be considered, based on machine architecture, processor speed, amount of memory, etc. Alternatively, Clouds require different attributes such as the size of the execution environment or the virtual machine image to be used.

- After the job execution or the environment creation requests are received, there is a match-making and scheduling phase involved. The GRAM component from Globus is specially flexible in this regard, and multiple adapters allow different treatments for jobs: for example, the simplest job manager just performs a *fork* call to spawn a new process on the target machine. More advanced and widely used adapters transfer job execution responsibility to a local resource manager such as Condor, LoadLeveler or Sun Grid Engine. These systems are able of multiplexing jobs that are sent to a site into multiple resources. Cloud systems have simpler job management strategies, since the type of jobs are homogeneous and don't need to be adapted to a variety of resources such as in the case of the Grid. For example, Eucalyptus uses a Round Robin scheduling technique to alternate among machines. OpenNebula implements a Rank Scheduling Policy to choose the most adequate resource for a request, and supports more advance features such as advance reservations through Haizea (Sotomayor, Keahey, & Foster, 2008).

- One of the common phases involved in job submission is transferring the necessary data to and from the execution machine. The first of them, usually called *stage-in*, involves retrieving the input data for the process from a remote destination, such a GridFTP server. When the amount of data is large, a mapping service such as a Replica Location Service (RLS) can be used to translate a logical file name to a location. The second part of the process, *stage-out*, consists in either transferring the output data to the user's machine or to place it in a repository, possibly using the RLS. In the case of Cloud computing, the most important data that has to be transferred is the definition of an execution environment, usually in terms of Virtual Machine images. Users upload the data describing the operating system and packages needed to instantiate the VM and later reference it to perform operations such as booting a new machine. There is no standard method for transferring data in Cloud systems, but it is worth noting Amazon's object storage solution, the Simple Storage Service (S3), which allows users to move entities from 1 byte to 5 GB in size.

- Finally, Grid and Cloud systems need to offer users a method to monitor their jobs, as well as their resource usage. This facility can also be used by site administrators to implement usage accounting in order to track resource utilization and enforce user quotas. In the context of Globus, there are two modules that can be used for this purpose, the first is GRAM itself, which allows user to query previously submitted jobs' status. The second method of acquiring information about the Grid's resources is provided by the Monitoring and Discovery Service (MDS), which is in charge of aggregating resources' data and making it available to be queried. High-level monitoring tools have been developed on top of existing Cloud management systems such as Amazon CloudWatch.

8.3.2 Platform

This layer is built on top of the physical infrastructure and offers a higher level of abstraction to users. The interface provided by a PaaS solution allows developers to build additional services without being exposed to the underlying physical or virtual resources. These facts enable additional features to be implemented as part of the model, such as presenting seemingly infinite resources to the user or allowing elastic behavior on demand. Examples of Cloud solutions that present these features are Google App Engine,[8] Salesforce's force.com[9] or Microsoft Azure.[10]

Several solutions that can be compared to the mentioned PaaS offerings exist in the Grid, even though this exact model cannot be exactly replicated. We define Platform level solutions as those containing the following two aspects:

8.3.2.1 Abstraction from Physical Resources

The Infrastructure layer provides users with direct access to the underlying infrastructure. While this is required for the lower levels of resource interaction, in the Platform level a user should be isolated from them. This allows developers to create new software that is not susceptible to the number of provisioned machines or their network configuration, for example.

8.3.2.2 Programming API to Support New Services

The Platform layer allows developers to build new software that takes advantage of the available resources. The choice of API directly influences the programs that can be built on the Cloud, therefore each PaaS solution is usually designed with a type of application in mind.

With these characteristics Grid systems allow developers to produce new software that take advantage of the shared resources in order to compare them with PaaS solutions.

---

[8]http://code.google.com/appengine/

[9]http://www.force.com/

[10]http://www.microsoft.com/windowsazure/

- Libraries are provided by Grid middleware to access resources programmatically. The Globus Java Commodity Grid (CoG) Kit (Laszewski et al., 2001) is an example. The CoG Kit allows developers to access the Grid functionality from a higher level. However, resources have to be independently addressed, which makes programs tied to the destination sites. Additionally, it is linked to Globus and makes applications dependent on a specific middleware.
- SAGA (Goodale et al., 2006) and DRMAA[11] are higher level standards that aim to define a platform independent set of Grid operations. While the former offers a wide range of options such as job submission, security handling or data management, the later focuses on sending and monitoring jobs. These solutions provide a higher level of abstraction than the previous example, but are still tied to the Grid concept of jobs as programs that are submitted to remote resources.
- An example of an API that bypasses the underlying Grid model to offer programmers a different paradigm to develop new software is MPICH-G2 (Karonis, Toonen, & Foster, 2003). It consists of a library that can be linked to a program that uses the Message Passing Interface (MPI) to transparently enable the application to work on the Grid. The programmer can think in familiar terms even though applications are Grid enabled.
- GridSuperscalar (Badia et al., 2003) is a programming paradigm to enable applications to run on the Grid. Programmers identify the functions of their code which can be run on remote resources, then specify the data dependencies for each of those functions, and after writing the code a runtime module determines the data dependencies and places each of the tasks in the Grid, transferring data accordingly so that each task can be completed.
- Another programming paradigm aimed at building new functionality on top of the Grid is SWIFT (Zhao et al., 2007). It provides a language to define computations and data dependencies, and is specially designed to efficiently run very large numbers of jobs while easing the task of defining the order of execution or the placing of data produced from one job to be consumed by another.

Probably the main difference in Cloud PaaS paradigms compared to the described options is that Grid models need to use the lowest common denominator when implementing new services. The reason for this is that the degree of compatibility with the middleware is directly related to the number of resources available: if a user's service does not make any assumptions on the remote resources, it will be able to use all of them normally; on the other hand, services requiring additional software to be installed on the target machines would have considerably fewer candidates to for execution.

In the case of Clouds, this requirement is not as stringent for two main reasons: the first one is that PaaS solutions are deeply tied to Cloud vendors, and therefore they are designed in hand with the rest of the infrastructure, and the second is that provisioning resources with the required libraries is much easier than in Grid computing, allowing new nodes to be spawned with the required environment. In

---

[11] http://www.drmaa.org/

the case of Grids, having the required software installed in the execution resources usually involves having a human operator do it, making the process more costly.

8.3.3 Applications

There is no clear distinctions between applications developed on Grids and those that use Clouds to perform execution and storage. The choice of platform should not influence the final result, since the computations delegated to the underlying systems can take different shapes to accommodate to the available APIs and resources.

On the other hand, it is undeniable that the vast majority of Grid applications fall in the realm of scientific software, while software running in Clouds has leaned towards commercial workloads. Here we try to identify some possible causes for the different levels of adoption of these technologies for the development of applications:

- **Lack of business opportunities in Grids.** Usually Grid middleware is installed only in hardware intended for scientific usage. This phenomenon has not successfully produced business opportunities that could be exploited by industry. Conversely, Clouds are usually backed up by industry which have had better ways to monetize their investments.
- **Complexity of Grid tools.** Perhaps due to the goal of providing a standardized, one-size-fits-all solution, Grid middleware is perceived by many as complex and difficult to install and manage. On the other hand, Cloud infrastructures have usually been developed by providers to fit their organization's needs and with a concrete purpose in mind, making them easier to use and solution oriented.
- **Affinity with target software**. Most Grid software is developed with scientific applications in mind, which is not true for the majority of Cloud systems. Scientific programs need to get the most performance from execution resources and many of them cannot be run on Clouds efficiently, for example because of virtualization overhead. Clouds are more targeted to web applications. These different affinities to distinct paradigms make both solutions specially effective for their target applications.

8.4 Techniques

Here we discuss the impact of techniques used in Grid computing that can be applied in Clouds. From the time the concept of Grid was introduced, a variety of problems had to be solved in order to enable its wide adoption. Some examples of these are user interfacing (Section 8.4.1), data transfer (Section 8.4.2), resource monitoring (Section 8.4.3) or security (Section 8.4.7). These basic techniques for the enablement of Grids were designed to fulfill its main goals, namely, to allow the sharing of heterogeneous resources among individuals belonging to remote administrative domains. These goals determine the areas of application of the described techniques in Clouds, therefore we will find the most valuable set of improvements to be in the field of Cloud interoperability.

Clouds can not only benefit from the most fundamental techniques in Grid computing: additional techniques that arose on top of these building blocks to bring new functionality to Grids are also good candidates to be applied to Clouds. Among these we can find Autonomic Computing (Section 8.4.4), Grid scheduling (Section 8.4.5), interoperation (Section 8.4.6) or simulation (Section 8.4.8).

The techniques discussed in this section are therefore spread through various levels of the Grid architecture: some of them can be found in the lower layers, giving common services to other components, and others are built from the former and extend them. Following the classification discussed in Section 8.3, we find that some techniques belong to the Infrastructure layer, this is, have the main objective of resource management, and others are spread through the Infrastructure and Platform layers, such as the Metascheduling techniques described in the scheduling section.

8.4.1 Service Orientation and Web Services

The Cloud is both a provider of services (e.g. IaaS, PaaS, and SaaS) and a place to host services on behalf of clients. To implement the former operational aspects while maintaining flexibility, Cloud administrative functions should be constructed from software components. The Grid faced similar challenges in building a distributed infrastructure to support and evolve its administrative functions such as security, job submission, and creation of Virtual Organizations. The architectural principle adopted by the Grid is Service Orientation (SO) with software components connected by Web Services (WS). This section summarizes contributions of the Open Grid Forum (OGF) to SO in distributed computing and and how they apply to the Cloud. SO as an architecture, and Web Services as a mechanism of inter-component communication are explored here in the context of similarities between Grid and Cloud requirements.

Grid designers realized the advantage of the loosely-coupled client and service model being appropriately deployed in the distributed computing environments. The original Grid approach to SO was Open Grid Services Infrastructure (OGSI). OGSI was built on top of the emerging Web Services standards for expressing interfaces between components in a language neutral way based on XML schemas. While WS is an interface, OGSI attempted to make it object oriented by adding required methods. Subsequently, the Grid community worked within the WS standards to extend WS specification based on experience using a SOA. This lead to the introduction of Open Grid Services Architecture (OGSA), implemented in version 3 of the Globus toolkit. OGSA contains key extensions to the WS standard which are now described.

In Grid and Cloud the most typical service components such as provisioning an OS image, starting a virtual machine, or dispatching a job are long running. A service composition of these components requires an asynchronous programming model. A consumer service component invokes a WS provider and is immediately acknowledged so the caller does not hold his process open on a communication link. The provider component asynchronously updates the consumer as the state changes. Grid architects recognized the importance of supporting the asynchronous model and integrated this approach into Web Services through the *WS-Addressing* and *WS-Notify* extensions. WS-Addressing specifies how to reference not just service endpoints, but objects within the service endpoint. Notification is based on WS-Addressing which specifies the component to be notified on a state change.

Related to the long lived service operation and asynchronous communication model is the requirement to maintain and share state information. There are many ways to achieve statefulness, none simple, especially when multiple services can update the same state object. In principle, a WS interface is stateless although of course there are many ways to build applications on top of WS that pass state through the operation messages. The challenge is to integrate the WS specification with a standard for statefulness that does not disturb the stateless intent of WS interface model. The OGF achieved this goal, developing the *Web Service Resource Framework* (WSRF). WSRF allows factory methods in the WS implementation to create objects, which are referenced remotely using the WS-Addressing standard. Persistent resource properties are exposed to coupled services through XML. Introducing state potentially adds enormous complexity to a distributed system, and the distribution of stateful data to multiple service components has the potential for data coherence problems which would require distributed locking mechanisms. The approach introduced by the Grid passes WS endpoints to the resources so that synchronized access is provided by the service implementation.

One path to leveraging Grid technology experiences in the Cloud is to consider building operation support services with a SO. The component services interconnect using the suite of WS standards. The logic of composing the services is built with modern business process design tools which produce a workflow. The design workflow is exported in the form such as the Business Process Execution Language (BPEL) and executed by a workflow engine. This implementation path of using BPEL with WSRF to build a SOA has been demonstrated in an e-Science context (Ezenwoye & Sadjadi, 2010; Ezenwoye, Sadjadi, Carey, & Robinson, 2007).

There is already some experience using WS and WSRF in the Cloud domain. The Nimbus project[12] uses the WS and WSRF model as an interface for clients to access its Cloud workspaces.

8.4.2 Data Management

In Grid computing, data-intensive applications such as the scientific software in domains like high energy physics, bio-informatics, astronomy or earth sciences

---

[12]http://www.nimbusproject.org/

involve large amounts of data, sometimes in the scale of PetaBytes (PB) and beyond (Moore, Prince, & Ellisman, 1998).

Data management techniques to discover and access information are essential for this kind of applications. Network bandwidth, transfer latency and storage resources are as important as computational resources to determine the tasks' latency and performance. For example, a data-intensive application will preferably be run at a site that has an ample and fast network channel to its dataset so that the network overhead can be reduced, and if it generates a large amount of data, we would also prefer a site that has enough storage space close to it.

Many technologies are applied in Grid computing to address data management problems. Data Grids (Chervenak, Foster, Kesselman, Salisbury, & Tuecke, 2001) have emerged in scientific and commercial settings to specifically optimize data management. For example, one of the services provided by Data Grids is replica management (Chervenak et al., 2002; Lamehamedi, Szymanski, Shentu, & Deelman, 2002; Samar & Stockinger, 2001; Stockinger et al., 2001). In order to retrieve data efficiently and also avoid hot spots in a distributed environment, Data Grids often keep replicas, which are either complete or partial copies of the original datasets. Replica management services are responsible for creating, registering, and managing replicas. Usually, a replica catalog such as Globus Replica Catalog (Allcock et al., 2001) is created and managed to contain information of replicas that can be located by users.

Besides data replication, caching is an effective method to reduce latency at the data consumer side (Karlsson & Mahalingam, 2002). Other technologies such as streaming, pre-staging, high-speed data movement, or optimal selection of data sources and sinks are applied in Data Grids too. These data management technologies are also used in data sharing and distribution systems such as Content Delivery Networks, Peer-to-Peer Networks and Distributed Databases. In Venugopal, Buyya, and Ramamohanarao (2006), the author suggests a taxonomy of Data Grids and compares Data Grids with other related research areas.

Standards for data services have been proposed in the Grid community. The Open Grid Services Architecture (OGSA), which is adopted by the Global Grid Forum (GGF), defines OGSA Data Services (Foster, Tuecke, & Unger, 2008) which include data transfer, data access, storage resource management, data cache, data replication, data federation, and metadata catalogues services. The Database Access and Integration Services Working Group (DAIS-WG) (Antonioletti, Krause, & Paton) at GGF is also developing standards of data services with an emphasis on database management systems, which have a central role in data management such as data storage, access, organization, authorization, etc. There are other groups at GGF that work on data management in Grid computing such as Grid File System Working Group, Grid Storage Management Working Group or GridFTP Working Group. Among them, GridFTP Working Group works on improving the performance of FTP and GripFTP (Allcock, 2003). GridFTP is an extension of FTP and it supports parallel and striped data transfer and partial file transfer. FTP and GridFTP are the most widely-used transport protocols when transferring bulk data for Grid applications.

The Globus Toolkit provides multiple data management solutions including GridFTP, the Global Access to Secondary Storage(GASS), the Reliable File Transfer (RFT), the Replica Location Service (RLS) and a higher-level Data Replication Service (DRS) based on RFT and RLS. Specifically, GASS is a light-weight data access mechanism for remote storage systems. It enables pre-staging and post-staging of files and is integrated into the Globus Resource Access and Monitoring (GRAM) to stage in executables and input data and if necessary, stage out the output data and logs.

In the current state of Cloud computing, storage is usually close to computation and therefore data management is simpler than in Grids, where the pool of execution and storage resources is considerably larger and therefore efficient and scalable methods are required for placement of jobs and data location and transfer. Still, there is the need to take data access into consideration to provide better application performance. An example of this is Hadoop,[13] which schedules computation close to data to reduce transfer delays.

Same as Grid computing, Clouds need to provide scalable and efficient techniques for transferring data. For example, we may need to move virtual machine images, which are used to instantiate execution environments in Clouds, from users to a repository and from the repository to hosting machines. Techniques for improved transfer rates such as GridFTP would result in lower times for sites that have high bandwidth, since they can optimize data transfer by parallelizing the sending streams. Also, catalog services could be leveraged to improve distributed information sharing among multiple participants such that the locating of user data and data repositories is more efficient. The standards developed from Grid computing practice can be leveraged to improve interoperability of multiple Clouds. Finally, better integration of data management with the security infrastructure would enable groups of trusted users. An application of this principle could be used in systems such as Amazon EC2 where VM images are shared by individuals with no assurances about their provenance.

8.4.3 Monitoring

Although some Cloud monitoring tools have already been developed, they provide high level information and, in most cases, the monitoring functionality is embedded in the VM management system following specific mechanisms and models. The current challenge for Cloud monitoring tools is providing information from the Clouds and application/service requests with sufficient level of detail in nearly real time in order to take effective decisions rather than providing a simple and graphical representation of the Cloud status. To do this, different Grid monitoring technologies can be applied to Clouds, specially those of them that are capable to provide monitoring data in aggregate form due to the large scale and dynamic behavior of Clouds.

---

[13]http://hadoop.apache.org/

The experiences gained with the research of Grid monitoring standardization can drive the definition of unified and standard monitoring interfaces and data models to enhance interoperability among different Clouds.

Grid monitoring is a complex task, since the nature of the Grid means heterogeneous systems and resources. However, monitoring is essential in the Grid to allow resource usage to be accounted for and to let users know whether and how their jobs are running. This is also an important aspect for other tasks such as scheduling.

The OGF Performance Working Group developed a model for Grid monitoring tools called Grid Monitoring Architecture (GMA) (Tierney et al., 2002). The architecture they propose is designed to address the characteristics of Grid platforms. Performance information has a fixed, often short, lifetime of utility. Performance data is often more frequently updated than requested, whereas usual database programs are firstly designed for queries. This means that permanent storage is not always necessary, and that the tools must be able to answer quickly before the data is obsolete. A Grid performance monitoring tool also needs to handle many different types of resources and should be able to adapt when communication links or other resources go down. Thus, monitoring systems should be distributed to suit these requirements. In fact, a monitoring tool should find a good tradeoff between the following characteristics: low latency for delivering data, high data rate, scalability, security policies, and minimum intrusiveness.

The GMA is based on three types of components: producers, consumers and the directory service (see Fig. 8.4). A producer is any component that can send events to a consumer, using the producer interface (accepting subscription, queries and ability to notify). In a monitoring tool, every sensor is encapsulated in a producer; however a producer can be associated to many different sources: sensors, monitoring systems or databases, for example. A consumer is any component that can receive event data from a producer. The consumer interface contains subscription/unsubscription routines and query mechanisms. To exchange data events, producers and consumers have a direct connection, but to initiate the dialog, they need the directory service.

Several monitoring tools have been developed for Grid systems. Balaton et al. (2004) provide a description and categorization of existing performance monitoring and evaluation tools, and Serafeim et al. (Zanikolas & Sakellariou, 2005) propose a taxonomy of Grid monitoring systems, which is employed to classify a wide range of projects and frameworks. Some of these approaches are discussed below.

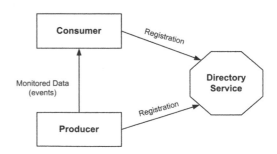

Fig. 8.4 Grid Monitoring Architecture components

Ganglia (Massie, Chun, & Culler, 2004) is a scalable distributed monitoring system for high-performance computing environments such as clusters and Grids. It is based on a hierarchical design targeted at federations of clusters, relies on a multicast-based listen/announce protocol to monitor state within clusters and uses a trace of point-to-point connections amongst representative cluster nodes to federate clusters and aggregate their state. Data is represented in XML and compressed using XDR. The Ganglia Web Frontend can be used to inspect for example CPU utilization in the last hour or last month. Ganglia has been deployed in many HPC infrastructures including supercomputing facilities and large Grid systems.

Network Weather Service (NWS) (Nolski, Spring, & Hayes, 1999) is a distributed system for producing short-term performance forecasts based on historical performance measurements. NWS provides a set of system sensors for periodically monitoring end-to-end TCP/IP performance (bandwidth and latency), available CPU percentage, and available non-paged memory. Based on collected data, NWS dynamically characterizes and forecasts the performance of network and computational resources.

Mercury (Balaton & Gombas, 2003) was designed to satisfy requirements of Grid performance monitoring: it provides monitoring data represented as metric via both pull and push access semantics and also supports steering by controls. It supports monitoring of Grid entities such as resources and applications in a generic, extensible and scalable way. Its design follows the recommendations of the OGF GMA described previously.

OCM-G (Balis et al., 2004) is an OMIS-compliant application monitor developed within the CrossGrid project. It provides configurable online monitoring via a central manager which forwards information requests to the local monitors. However, OCM-G has a distributed architecture.

The Globus Monitoring and Discovery System (MDS) (Czajkowski, Fitzgerald, Foster, & Kesselman, 2001) is another widely used monitoring tool that provides information about the available resources on the Grid and their status. It is based on the GLUE schema,[14] which is used to provide a uniform description of resources and to facilitate interoperation between Grid infrastructures. Other approaches for large-scale systems have been developed such as MonALISA (Newman et al., 2003), which is an extensible monitoring framework for hosts and networks in large-scale distributed systems, and Palantir (Guim, Rodero, Tomas, Corbalan, & Labarta, 2006) that was designed to unify the access to different monitoring and information systems for large scale resource-sharing across different administrative domains, thus providing general ways for accessing all this information. Furthermore, different Grid portal frameworks incorporate monitoring functionalities such as in the HPC-Europa Single Point of Access (Guim et al., 2007) and the P-GRADE Portal (Podhorszki & Kacsuk, 2001).

---

[14]http://forge.ogf.org/sf/projects/glue-wg.

Several data centers that provide resources to Cloud systems have adopted Ganglia as a monitoring tool. However, virtualized environments have more specific needs that have motivated Cloud computing technology providers to develop their own monitoring system. Some of them are summarized below:

Amazon CloudWatch[15] is a web service that provides monitoring for Amazon Web Services Cloud resources such as Amazon EC2. It collects raw data from Amazon Web Services and then processes the information into readable metrics that are recorded for a period of two weeks. It provides the users with visibility into resource utilization, operational performance, and overall demand patterns - including metrics such as CPU utilization, disk reads and writes, and network traffic.

Windows Azure Diagnostic Monitor[16] collects data in local storage for every diagnostic type that is enabled and can transfer the data it gathers to an Azure Storage account for permanent storage. It can be scheduled to push the collected data to storage at regular intervals or it can be requested an on-demand transfer whenever this information is required.

The **OpenNebula Information Manager** (IM) is in charge of monitoring the different nodes in a Cloud. It comes with various sensors, each one responsible for different aspects of the compute resource to be monitored (CPU, memory, hostname). Also, there are sensors prepared to gather information from different hypervisors.

The monitoring functionality of **Aneka** (Vecchiola, Chu, & Buyya, 2009) is implemented by the core middleware, which provides a wide set of services including also negotiation of the quality of service, admission control, execution management, accounting and billing. To help administrators to tune the overall performance of the Cloud, the Management Studio provides aggregated dynamic statistics.

Nimsoft Monitoring Solution[17] (NMS), built on the Nimsoft Unified Monitoring Architecture, delivers monitoring functionality to any combination of virtualized data center, on hosted or managed infrastructure, in the Cloud on IaaS or PaaS or delivered as SaaS services. Specifically, it provides unified monitoring for data centers, private Clouds and public Clouds such as Amazon WS, including service level and response time monitoring, visualization and reporting.

Hyperic CloudStatus[18] provides open source monitoring and management software for all types of web applications, whether hosted in the Cloud or on premise, including Amazon Web Services and Google App Engine. CloudStatus gives users real-time reports and weekly trends on infrastructure metrics.

---

[15]http://aws.amazon.com/cloudwatch/

[16]http://www.microsoft.com/windowsazure

[17]http://www.nimsoft.com/solutions/

[18]http://www.hyperic.com/

8.4.4 Autonomic Computing

Inspired by the the autonomic nervous system, autonomic computing aims at designing and building self-managing systems and has emerged as a promising approach for addressing the challenges due to software complexity (Jeffrey & Kephart, 2001). An autonomic system is able to make decisions to respond to changes in operating condition at runtime using high-level policies that are typically provided by an expert. Such a system constantly monitors and optimizes its operation and automatically adapts itself to changing conditions so that it continues to achieve its objectives.

There are several important and valuable milestones to reach fully autonomic computing: first, automated functions will merely collect and aggregate information to support decisions by human users. Later, they will serve as advisors, suggesting possible courses of action for humans to consider.

Self-management is the essence of autonomic computing and has been defined in terms of the following four aspects of self-management (Jeffrey & Kephart, 2001).

- *Self configuration*: Autonomic systems will configure themselves automatically in accordance with high-level policies representing business-level objectives that, for example, specify what is desired and not how it is to be accomplished. When a component is introduced, it will incorporate itself seamlessly, and the rest of the system will adapt to its presence.
- *Self optimization*: Autonomic systems will continually seek ways to improve their operation, identifying and seizing opportunities to make themselves more efficient in performance and/or cost. Autonomic systems will monitor, experiment with, and tune their own parameters and will learn to make appropriate choices about keeping functions or outsourcing them.
- *Self healing*: Autonomic computing systems will detect, diagnose, and repair localized problems resulting from bugs or failures in software and hardware.
- *Self protection*: Autonomic systems will be self-protecting in two senses. They will defend the system as a whole against large-scale, correlated problems arising from malicious attacks or cascading failures that remain uncorrected by self-healing measures. They also will anticipate problems based on early reports from sensors and take steps to avoid or mitigate them.

Figure 8.5 shows one basic structure of an autonomic element as proposed by IBM. It consists of autonomic manager which monitors, analyzes, plans and executes based on collected knowledge, and external environments including human users and managed elements. The managed element could be hardware resources such as CPU, memory and storage, software resources such as a database, a directory service or a system, or an application. The autonomic manager monitors the managed elements and its external environment including changing users requirements, and analyzes them, computes a new plan reflecting changing conditions and executes this plan.

Fig. 8.5 One basic structure
of an autonomic element.
Elements interact with other
elements and external
environments through
autonomic manager

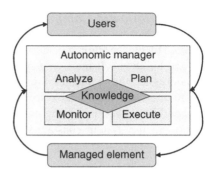

Autonomic concepts have been effectively applied to distributed computing environments such as Grids (Parashar, Li, & Chandra, 2010), and communication/networking systems (Vasilakos, Parashar, Karnouskos, & Pedrycz, 2010), to monitor resources, changing workloads or models (Quiroz, Gnana-Sambandam, Parashar, & Sharma, 2009), and then adjust resource provisioning to satisfy requirements and constraints (Quiroz, Kim, Parashar, Gnanasambandam, & Sharma, 2009). Such capabilities have been incorporated into Cloud systems. We use CometCloud (Kim, el Khamra, Jha, & Parashar, 2009) as a use case that provides autonomic capabilities at all levels. CometCloud is an autonomic computing engine for Cloud and Grid environments. It is based on decentralized coordination substrate, and supports autonomic applications on highly heterogeneous and dynamic Cloud/Grid infrastructures, as well as integration of public/private Clouds/Grids. For example, it supports autonomic cloudbursts, where the goal is to seamlessly (and securely) bridge private enterprise Clouds and data centers with public utility Clouds or Grids on-demand, to provide an abstraction of resizable computing capacity that is driven by user-defined high-level policies. It enables the dynamic deployment of application components, which typically run on internal organizational compute resources, onto a public Cloud or Grids (i.e., cloudburst) to address dynamic workloads, spikes in demands, economic/budgetary issues, and other extreme requirements. Furthermore, given the increasing application and infrastructure scales, as well as their cooling, operation and management costs, typical over-provisioning strategies are no longer feasible. Autonomic cloudbursts can leverage utility Clouds to provide on-demand scale-out and scale-in capabilities based on a range of metrics.

Other examples of Clouds technologies that are adopting autonomic computing techniques from Grid computing are Aneka (Chu, Nadiminti, Jin, Venugopal, & Buyya, 2007), VioCluster (Ruth, McGachey, & Xu, 2005) and CloudWatch.

8.4.5 Scheduling, Metascheduling, and Resource Provisioning

In the last few decades a lot of effort has been devoted to the research of job scheduling, especially in centers with High Performance Computing (HPC) facilities. The general scheduling problem consists of, given a set of jobs and requirements, a set of resources, and the system status, deciding which jobs to start executing and in which resources. In the literature there are many job scheduling policies, such as the FCFS approach and its variants (Schwiegelshohn & Yahyapour, 1998a, 1998b; Feitelson & Ruddph, 1995). Other policies use estimated application information (for example the execution time) which make no assumptions such as Smallest Job First (SJF) (Majumdar, Eager, & Bunt, 1988), Largest Job First (LJF) (Zhu & Ahuja, 1993), Smallest Cumulative Demand First (SCDF) (Leutenegger & Vernon, 1990) or Backfilling (Mu'alem & Feitelson, 2001), which is one of the most used in HPC systems.

In Grid computing, scheduling techniques have evolved to incorporate other factors, such as the heterogeneity of resources or geographical distribution. The software component responsible for scheduling tasks in Grids is usually called meta-scheduler or Grid resource broker. The main actions that are performed by a Grid resource broker are: resource discovery and monitoring, resource selection, job execution, handling and monitoring. However, it may be also responsible for other additional tasks such as security mechanisms, accounting, quality of service (QoS) ensuring, advance reservations, negotiation with other scheduling entities, policy enforcement, migration, etc. A taxonomy and survey of Grid brokering systems can be found in (Krauter, Buyya, & Maheswaran, 2002). Some of their most common characteristics are discussed as follows:

- They can involve different scheduling layers through several software components between the Grid resource broker and the resources where the application will run. Thus, the information and control available at the resource broker level is far less than that available at a cluster scheduling level.
- A Grid resource broker usually does not have ownership or control over the resources. Moreover, the cluster scheduling systems may have their own local policies that can conflict with the Grid scheduling strategy.
- There are conflicting performance goals between the users and the resource owners. While the users focus on optimizing the performance of a single application for a specified cost goal, the resource owners aim to obtain the best system throughput or minimize the response time.

While in Grid computing the most important scheduling tasks are optimizing applications response time and resource utilization, in Cloud computing other factors become crucial such as economic considerations and efficient resource provisioning in terms of QoS guarantees, utilization and energy. As virtualized data centers and Clouds provide the abstraction of nearly-unlimited computing resources through the elastic use of consolidated resources pools, the scheduling task shifts to scheduling resources (i.e. provisioning application requests with resources). The provisioning problem in question is how to dynamically allocate resources among VMs with the goal of optimizing a global utility function. Some examples are minimizing resource over-provisioning (waste of resources) and maximizing QoS (in order to prevent falling on under-provisioning that may led to providers revenue loss). Different provisioning techniques for data centers have been proposed such as those based on gang scheduling (Wiseman & Feitelson, 2003), those based on advance reservations (Sotomayor, Montero, Llorente, & Foster, 2008), those based on energy efficiency (Nathuji & Schwan, 2007; Ranganathan, Leech, Irwin, & Chase, 2006) or those based on multi-tiered resource scheduling approaches (Song, Wang, Li, Feng, & Sun, 2009).

Although Cloud computing scheduling is still challenging, several techniques developed for Grid environments can be taken into account. Existing job scheduling techniques can be also applied in virtualized environments, specially when the application requests rates are those expected in future Clouds. In fact, some approaches have started addressing this issue. Advance reservation developed for Grid scheduling is used in the Haizea lease manager for OpenNebula. Different SLA management policies for Grid computing have been extended for Clouds such as those proposed by Buyya, Yeo, Venugopal, Broberg, & Brasdic, (2009). Market-oriented allocation of resources policies developed for Grids have been realized for Clouds in Aneka (Buyya et al., 2009). Sodan (2009) proposes adaptive scheduling, which can adjust sizes of parallel jobs to consider different load situations and different resource availability through job re-shaping and VM resizing. Moreover, Cloud computing scheduling strategies can leverage Grid multi-layer architectures and strategies such the cross-layer QoS optimization policy proposed by Chunlin and Layuan (2008).

8.4.6 Interoperability in Grids and Clouds

One goal of Grid computing is to provide uniform and consistent access to resources distributed in different data centers and institutions. This is because the majority of Grids are formed based on regional as opposed to local initiatives so interoperation is a key objective. Some examples are TeraGrid in US (Catlett, Beckman, Skow, & Foster, 2006), GridX1 in Canada (Agarwal et al., 2007), Naregi in Japan (Matsuoka et al., 2005) and EGEE in Europe (Berlich, Hardt, Kunze, Atkinson, & Fergusson, 2006). Interoperation is addressed at various architectural points such as the access portal, resource brokering function, and infrastructure standardization.

Some production Grid environments, such as HPC-Europa (Oleksiak et al., 2005), DEISA (Alessandrini & Niederberger, 2004) and PRACE,[19] approach interoperability using a uniform access interface to application users. Software layers beneath the user interface then abstract the complexity of the underlying heterogeneous supercomputing infrastructures.

One tool that takes this approach for Grid interoperation is meta-brokering (Kertesz & Kacsuk, 2008), illustrated in Fig. 8.6. Meta-brokering supports the Grid interoperability from the viewpoint of the resource management and scheduling. Many projects explore this approach with varied emphases. Examples grouped loosely by primary technical foci, are reviewed below.

---

[19]http://www.prace-project.eu/

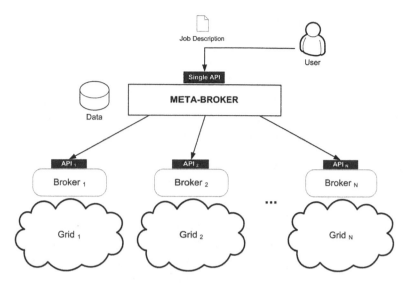

Fig. 8.6 Meta-brokering architecture

- *Infrastructure interoperability*

  GridWay (Huedo, Montero, & Llorente, 2004), which is mainly based on Globus, supports multiple Grids using Grid gateways (Huedo, Montero, & Llorente, 2004) to access resources belonging to different domains. **GridWay** forwards local user requests to another domain when the current one is overloaded.

  Latin American Grid Meta-brokering (Badia et al., 2007; Bobroff et al., 2008), proposed and implemented a common set of protocols to enable interoperability among heterogeneous meta-schedulers organized in a peer-to-peer structure. The resource domain selection is based on an aggregated resource information model (Rodero, Guim, Corbalan, Fong, & Sadjadi, 2010) and jobs from home domain can be routed to peer domains for execution.

- *Resource Optimization in interoperated Grids*

  Koala Grid Scheduler (Mohamed & Epema, 2008) is focused on data and processor co-allocation. To inter-connect different Grid domains as different Koala instances. Their policy is to use resources from a remote domain only if the local one is saturated. They use delegated matchmaking (Iosup, Epema, Tannenbaum, Farelle, & Livny, 2007) to obtain the matched resources from one of the peer Koala instances without routing the jobs to the peer domains.

  InterGrid (Assuncao, Buyya, & Venugopal, 2008) promotes interlinking different Grid systems through peering agreements based on economic approaches to enable inter-grid resource sharing. This is an economic-based approach, where business application support is a primal goal, and this also supposed to establish sustainability.

VIOLA MetaScheduling Service (Seidel, Waldrich, Zeigler, Wieder, & Yahyapour, 2007) implements Grid interoperability via SLA mechanisms (WS-Agreement) and provides co-allocation of multiple resources based on reservation.

Other projects explore the interoperability of Grid systems through the use of standard mechanisms, protocols and interfaces. For example, the Grid Interoperability Project (GRIP) (Brooke, Fellows, Garwood, & Goble, 2004), the Open Middleware Infrastructure Institute for Europe (OMII-Europe) project[20] or the work done within the P-GRADE portal (Kacsuk, Kiss, & Sipos, 2008). GRIP was one of the first proposals enabling interoperability between UNICORE and Globus Toolkit. OMII-Europe aimed to influence the adoption and development of open standards that facilitate interoperability between gLite (Lawre et al., 2006) and UNICORE such as OGSA BES (Foster et al., 2008) or JSDL (Anjomshoaa et al., 2005). The P-GRADE portal tries to bridge different Grid infrastructures by providing access to standard-based interoperable middleware. The Grid Interoperation Now Community Group (GIN-CG)[21] and the Production Grid Infrastructure Working Group (PGI-WG)[22] of the OGF also address the problem of Grid interoperability. In the former case, driving and verifying interoperation strategies and, in the latter case, oriented to production Grid infrastructures.

While significant progress on interoperation has been achieved in Grid computing, interoperability among Cloud providers has yet to be explored. While enthusiasm in establishing Cloud interoperability is limited among the for-profit Cloud providers, there are many pursuers in the academic and scientific communities.

The RESERVOIR project[23] addresses Cloud interoperability with a modular, extensible Cloud architecture based on federation of Clouds. In the RESERVOIR model, each infrastructure provider is an autonomous business with its own business goals. A provider federates with other providers based on policies aligned with the site's business goals. In the context of RESERVOIR, Grid interfaces and protocols may enable the required interoperability between the Clouds or infrastructure providers. A similar initiative is the Nuba project[24] whose main aim is the development of a federated IaaS Cloud platform to facilitate the easy and automatic deployment of Internet business services, allowing dynamic scaling based on performance and business goals criteria.

Some Cloud standardization groups have started working on defining common interfaces for interoperation. The Open Grid Forum Open Cloud Computing Interface (OCCI) working group[25] of OGF is working on defining an API

---

[20]http://www.omii-europe.org

[21]http://forge.gridforum.org/sf/projects/gin

[22]http://forge.ogf.org/sf/projects/pgi-wg

[23]http://www.reservoir-fp7.eu/

[24]http://nuba.morfeo-project.org/

[25]http://forge.ogf.org/sf/projects/occi-wg

specification for remote management of Cloud computing infrastructure, allowing for the development of interoperable tools for common tasks including deployment, autonomic scaling and monitoring. Project OpenNebula and RESERVOIR projects have provided OCCI-compiant implemenations. The Cloud Computing Interoperability Forum (CCIF)[26] is a vendor neutral, not for profit community of technology advocates, and consumers dedicated to driving the rapid adoption of global Cloud computing services.

While encouraging activities in the area of interoperable Clouds are occuring, Grid technologies are more mature. Thus, it is promising to extend these to the Cloud, particularly in the research and evaluation of scheduling and resource selection strategies. While Grid computing focuses on utilization, Cloud computing is more atuned to factors such as QoS, cost, and energy efficiency. Finally, Clouds will want to take advantage of elastic use of their resources in order to optimize both resource usage (and thus, Cloud providers revenue) and QoS given to the users.

8.4.7 Security and User Management

Clouds currently lack many of the mechanisms required for fluid intersite operation of which security is a key enabling factor. Interoperability mandates common security mechanisms that can be translated to the models chosen by local administrators. Additionally, users need to be able to submit requests regardless of the institutions involved in the process of performing the requested task or providing the necessary data. This requires introduction to the Cloud of a mechanism of privilege delegation enabling single sign-on. Finally, collaboration between multiple institutions sharing resources requires development of new methods to manage user privileges.

These challenges have already been addressed in Grid computing where a primary goal is to allow sharing of resources among Virtual Organizations. A VO defines a group of people and resources that can spawn across multiple administrative domains, and allows the definition of fine grained security policies on those resources. The Grid solutions are described in the context of Globus middleware and show how the concepts can be applied to Clouds.

Users in the Grid are granted privileges by site administrators based on their credentials, which are provided by a trusted Certificate Authority. The Grid Security Infrastructure (GSI) (Welch et al., 2003) is the component of the Globus middleware responsible for orchestrating security across different sites. GSI is used by job execution, file transfer, and resource discovery and monitoring protocols to ensure that all operations started by a user are allowed in the target resources.

GSI uses X.509 Public Key Infrastructure (PKI) and SSL/TLS protocols for transport encryption. This allows individuals belonging to organizations to trust foreign credentials issued by a CA without affecting their organization's security measures. However, two additional requirements arise from the dynamic nature of Grid systems.

---

[26]http://cloudforum.org/

Single sign-on: Users in the Grid need to access multiple resources and services with different authentication models. It would be burdensome if each time a user had to perform an action in a remote resource he had to enter a passphrase to use his private key to authenticate himself. Possible solutions such as caching the passphrase could lead to additional security problems.

Privilege delegation: Due to the dynamic nature of the Grid, users often need to delegate their privileges to other services. This occurs when requests require the orchestration of different resources, or when the user creates a new service to perform a set of capabilities. Following the principle of least privilege, a set of minimum capabilities should be transferred to these services so they can execute.

These requirements are fulfilled by an extension to X.509 certificates called proxy certificates (Welch et al., 2004). Proxy certificates are not issued by a CA, which would be burdensome given their frequency of use and dynamic nature. Instead, the issuer is identified by another public key certificate. This allows temporary certificates to be signed and used for short periods of time without the need to access the user's long time private keys. The relaxed security of proxy certificates suffices as they have a short life cycle.

Proxy certificates are also used to create new certificates with delegated subsets of privileges. The GSI architecture allows different levels of granularity when defining which of the privileges are inherited by the created proxy. Finer levels of granularity can be implemented by using policy languages to express delegation policies. This opportunity is effectively exploited by more advance security services built on the GSI such as the CAS service described below.

An example of the use of proxy certificates would be a computational job that requires access to a storage server to pull datasets to be processed. In this case, a new proxy would be created at the first site by delegating the user's privileges over the network, and in turn, the resource receiving the request would transfer the credentials to the storage server which would perform the operation based on its authorization policies.

One problem with x509 based proxy certificates is the need for users to initiate requests from a machine that has their private keys stored, in addition to the required software to generate a proxy and start a request to the Grid. Often, users access the Grid through web portals, making it difficult to generate their proxy certificates. The MyProxy credential repository was created to solve this issue and permit any user to access Grid resources through a Grid portal using a web browser (Novotny, Tuecke, & Welch, 2001). The MyProxy model adds a repository service where users delegate their credentials and associate them to a user name and password. Subsequently, users can log in to a MyProxy enabled web portal and retrieve and use a previously stored Grid certificate. Certificates delegated to MyProxy repositories have longer lifetimes than usual proxies so users just need to generate them occasionally.

The GSI infrastructure allows resource owners to define access policies in an ad-hoc fashion: usually, site administrators are in charge of defining a mapping from Distinguished Names (DNs) to the local security method. This poses a number of problems, specially when dealing with large VOs that are distributed across different institutions: the first problem is the burden added to administrators to include access policies for all users, specially if there is a need of defining finer grained ones that vary from one resource to another. Second, systems administrators in charge of assigning access policies don't have a big picture of the project's needs in terms of authorization structure.

The Community Authorization Service (CAS) (Pearlman, Welch, Foster, & Kesselman, 2002) is an extension build on the GSI that provides additional mechanisms to address the deficiencies mentioned above. The CAS abstracts the complexity of access policies for a project into a central server that acts as a repository of policies and users, freeing local resource administrators from the task of identifying authorization requirements. The immediate benefit of this separation of concerns is that project administrators can define users and access rules in the CAS server, and even create groups to define fine grained policies. Once users are added to the CAS server, they contact it when access to a resource is needed, and the CAS server confers them a capability that is equivalent to a proxy certificate. Site administrators need only to validate that the intended operation is allowed for the community the user belongs to and that the operation is allowed by the offered capability. This method scales independently from the number of users and resources. It is directly built on the GSI, which allows its deployment with minimal changes to existing technologies.

In the case of Cloud computing, the lack of standardization among vendors results in multiple security models: for example, both Amazon EC2 and Eucalyptus employ pairs of X.509 certificates and private keys for authentication. Google App Engine, an example of PaaS solution, requires users to first log-in via Google Accounts. The variety of methods makes it difficult to create new opportunities for interoperation, and the fragmentation of security models hinders the reuse of newly developed features.

The OGF Open Cloud Computing Interface Working Group (OCCI) has made a step towards proposing a standardized set of operations, and in its specification it suggests that implementations may require authentication using standard HTTP mechanisms and/or encryption via SSL/TLS. The latest versions of OpenNebula support this specification for communicating with their Cloud controller. This definition represents a possibility to create common grounds for IaaS implementations, providing uniform security paradigms among different vendors.

However, there is still much work in order to achieve a good security infrastructure in Clouds. Methods to specify trust among certificate issuers and resource owners have yet to be implemented, especially for scenarios in which different organizations participate in sharing them. Models such as the GSI infrastructure, where different providers trust various Certifying Authorities without compromising the rest of institutions, would allow the scaling of Clouds outside of single institution boundaries. In those cases, additional techniques to manage users and their associated privileges would be necessary to avoid centralization, and new distributed methods for accounting would be required. Clouds can learn from these solutions in order to define new, standardized interfaces that allow secure, inter-organizational communication.

8.4.8 Modeling and Simulation of Clouds and Grids

Since it is difficult or even not feasible to evaluate different usages on real Grid testbeds, different simulators have been developed in order to study complex scenarios. Simulations allow us to research policies for large and complex configurations with numerous jobs and high demand of resources and to easily include modifications and refinements in the policies. There are many rich simulation models developed by the Grid community.

The GridSim (Sulistio, Cibej, Venugopal, Robic, & Buyya, 2008) simulator has been widely used by many researchers to evaluate Grid scheduling strategies. As described by the GridSim project team it provides a comprehensive facility to create different classes of heterogeneous resources that can be aggregated using resource brokers. GangSim (Dumitrescu & Foster, 2005) allows the simulation of complex workloads and system characteristics. It is also capable of supporting studies for controlled resource sharing based on SLAs. The SimGrid toolkit (Legrand, Marchal, & Casanova, 2003) is a non workload based simulator that allows the evaluation of distributed applications in heterogeneous distributed environments. In these last models, almost all of them model how the jobs are scheduled at the multi-site level (by a given broker or meta-scheduler) but not how the jobs are scheduled and allocated once sent to the final computing resources. In a different approach, the Alvio simulator (Guim, Corbalan, & Labarta, 2007) and Teikoku (Grimme et al., 2007) model all the scheduling layers that are involved in Grid architectures, from meta-brokering policies (see Section 8.4.6) to local job scheduling strategies. DGSim (Iosup, Sonmez, & Epema, 2008) is another relevant simulation framework, which also allows Grid simulation with meta-brokering approaches but, as the former approaches, it does not model local scenarios.

Simulation tools are specially important for Cloud computing research due to the fact that many Clouds are also still in development. CloudSim (Buyya, Ranjan, & Calheiros, 2009) models and simulates Cloud computing environments supporting multiple VMs within a data center node. In fact, VM management is the main novelty of this simulator. It also allows simulation of multiple federated data centers to enable studies of VM migration policies for reliability and automatic scaling of applications. However, several aspects of Cloud computing have not been addressed yet such as the simulation of multiple layers simultaneously. Therefore, the lack of simulators for Cloud computing motivates the extension of existing simulators that were developed for Grid systems and have similar requirements. Some of the existing Grid simulators are described below. While many Cloud simulation models are yet to be developed, leveraging some of the simulation models and experiences would likely accelerate the development for Clouds.

Workloads are crucial to evaluate policies using simulation tools. Although different workload models have been proposed, traces from logs of production systems capture better the behavior of realistic scenarios. There are different publicly available workload traces from production system such as those provided by the Grid

Observatory,[27] which collects, publishes, and analyzes data on the behavior of
the EGEE Grid.[28] This is currently one of the most complex public Grid traces
with higher frequency of application request arrivals than other large Grids such
as Grid5000. However, any of them captures the heterogeneous nature of virtualized Cloud infrastructures with multiple geographically distributed entry points and
potential high job arrival rates.

Furthermore, since the traces from different systems are in different formats,
using standard formats is very important. Within the Parallel Workload Archive,[29] as
well as providing detailed workload logs collected from large scale parallel systems
in production use such as San Diego Supercomputer Center or Los Alamos National
Lab, Feitelson et al. proposes the Standard Workload Format (SWF) (Chapin et al.,
1999) that was defined to ease the use of workload logs and models. Iosup et al.
extended this idea for Grids with the Grid Workload Archive (Iosup et al., 2008) and
with the Failure Trace Archive (Kondo, Javadi, Iosup, & Epema, 2010) to facilitate
the design, validation, and comparison of fault-tolerant models and algorithms.

There is a lack of workload traces and standard models for Clouds. This is an
important obstacle to model and simulate realistic Cloud computing scenarios due
to Cloud workloads may be composed of different application types, including service requests that have different behavior than the modeled in the current public
traces. These existing approaches for parallel systems and Grid systems can be
extended to Cloud computing with a definition of a standard Cloud workload format.
Workload logs collected from production or research Cloud systems should be also
made publicly available to facilitate the research of Cloud computing techniques
through simulation.

8.5 Concluding Remarks

Grids and Clouds have many similarities in their architectures, technologies and
techniques. Nowadays, it seems Cloud computing is taking more significance as
a means to offer an elastic platform to access remote processing resources: this is
backed up by the blooming market interest on new platforms, the number of new
businesses that use and provide Cloud services and the interest of academia in this
new paradigm. However, there are still multiple facets of Cloud computing that
need to be addressed, such as vendor lock-in, security concerns, better monitoring
systems, etc. We believe that the technologies developed in Grid computing can be
leverage to accelerate the maturity of the Cloud, and the new opportunities presented
by the latter will in term address some of the shortcomings of the Grid.

As this chapter tries to convey, perhaps the area in which Clouds can gain the
most from Grid technologies is in multi-site interoperability. This comes naturally

---

[27] http://www.grid-observatory.org/

[28] http://www.eu-egee.org/

[29] http://www.cs.huji.ac.il/labs/parallel/workload/

from the fact that the main purpose of Grid systems is to enable remote sites under different administration policies to establish efficient and orchestrated collaboration. This is arguably one of the weakest points in Clouds, which usually are services offered by single organizations that enforce their -often proprietary- protocols, leading for examples to the already identified problem of vendor lock-in. On the other hand, Grid computing, through the use of well defined standards, has achieved site interoperability as it can be seen by the multiple computing and data Grids used by projects in fields as particle physics, earth sciences, genetics and economic sciences.

Another path worth diving into is the one exploring how the new paradigm of Cloud computing can benefit existing technologies and solutions proposed by the Grid community: the realization of utility computing, elastic provisioning of resources, or the homogenization of heterogeneous resources (in terms of hardware, operating systems and software libraries) through virtualization bring a new realm of possible uses for vast, underutilized computing resources. New consolidation techniques allow for studies on lower energy usage for data centers and diminished costs for users of computing resources. There is effectively a new range of applications that can be run on Clouds because of the improved isolation provided by virtualization techniques. Thus, existing software that was difficult to run on Grids due to hard dependencies on libraries and/or operating systems can now be executed on many more resources that have been provisioned complying with the required environment.

Finally, there are some outstanding problems that need to be considered which prevent some users from switching to new Cloud technologies. These problems need to be tackled before we can fully take advantage of all the mentioned opportunities. Other authors, such as (Armbrust et al., 2009), have already listed several of such problems. Some examples are:

1. In certain cases, when processes require intense use of I/O, virtualized environments offer lower performance than native resources. There is a range of scientific applications that have a high communication demand, such as those that rely on synchronous message passing models. Those applications do not offer good performance on Cloud systems.
2. Even though Clouds offer the promise of elasticity of computing resources that would appear to users as endless supply, there are scenarios in the scientific world for which the resources offered by a single Cloud would not be enough. Once the demand for processing power reaches the maximum capacity for a provider, there are no additional means to acquire new resources for the users, if need be. Attempting to use different providers as a back up would mean different protocols, security schemas and new APIs to be employed. For example, the Large Hadron Collider (LHC) project requires processing power not available by any single organization and if deployed to the Cloud, there is a need for interoperability among different Cloud vendors.

We hope that the efforts being taken by numerous researchers in this area identify and address these shortcomings and lead to better and more mature technologies that will improve the current Cloud computing practices. In these efforts, we believe that a good knowledge of existing technologies, techniques and architectures such as those developed in the field of Grid computing will greatly help accelerating the pace of research and development of the Cloud, and will ensure a better transition to this new computing paradigms.

Acknowledgments We would like to thank Hyunjoo Kim, from the NFS Center for Autonomic Computing at Rutgers University, for her useful insights in Section 8.4.4. This work was partially supported by the National Science Foundation under Grant No. OISE-0730065 and by IBM. Any opinions, findings, and conclusions or recommendations expressed in this chapter are those of the authors and do not necessarily reflect the views of the NSF or IBM.

References

Agarwal, A., Ahmed, M., Berman, A., Caron, B. L., Charbonneau, A., Deatrich, D., et al. (2007). GridX1: A Canadian computational grid. *Future Generation Computer Systems, 23*, 680–687.

Alessandrini, V., & Niederberger, R. (2004). The deisa project: Motivations, strategies, technologies. *19th International Supercomputer Conference, Heidelberg, Germany.*

Allcock, B., Bester, J., Bresnahan, J., Chervenak, A. L., Foster, I., Kesselman, C., et al. (2001). Secure, efficient data transport and replica management for high-performance data-intensive computing. *In Processings of IEEE Mass Storage Conference, IEEE Press, San Diego, CA, USA.*

Allcock, W., (2003). Gridftp protocol specification. Global grid forum recommendation GFD.20 (Tech. Rep., Open Grid Forum (OGF)).

Anjomshoaa, A., Drescher, M., et al. (2005). Job submission description language (JSDL) specification version 1.0, GFD-R.056 (Tech. Rep., Open Grid Forum (OGF)).

Antonioletti, M., Krause, A., & Paton, N. W. (2005). An outline of the global grid forum data access and integration service specifications. *Data Management in Grids LNCS, 3836*, 71–84.

Appleby, K., Fakhouri, S., Fong, L., Goldszmidt, G., Kalandar, M., Krishnakumar, S., et al. (2001). Oceano-sla based management of a computing utility. *Proceeding of the 7th IFIP/IEEE International Symposium on Integrated Network Management (IM 2001), Seattle, WA.*

Armbrust, M., Fox, A., & Griffith, R., et al. (2009). Above the clouds: A berkeley view of cloud computing (CoRR UCB/EECS-2009-28, EECS Department, University of California, Berkeley).

Assuncao, M. D., Buyya, R., & Venugopal, S. (2008). InterGrid: A case for internetworking Islands of grids. *Concurrency and Computation: Practice and Experience, 20*, 997–1024.

Badia, R., et al. (2007). High performance computing and grids in action, Chap. *Innovative Grid Technologies Applied to Bioinformatics and Hurricane Mitigation, IOS Press, Amsterdam*, 436–462.

Badia, R. M., Labarta, J., Sirvent, R., Pérez, J. M., Cela, J. M., & Grima, R. (2003). Programming grid applications with grid superscalar. *Journal of Grid Computing, 1*, 2003.

Balaton, Z., & Gombas, G. (2003). Resource and job monitoring in the grid. *Euro-Par 2003 Parallel Processing, Volume LNCS 2790, Klagenfurt, Austria*, 404–411.

Balis, B., Bubak, M., Funika, W., Wismüller, R., Radecki, M., Szepieniec, T., et al. (2004). Performance evaluation and monitoring of interactive grid applications. *Recent Advances in Parallel Virtual Machine and Message Passing Interface*, Volume LNCS 3241 345–352.

Berlich, R., Hardt, M., Kunze, M., Atkinson, M., & Fergusson, D. (2006). Egee: building a pan-european grid training organisation. *ACSW Frontiers '06: Proceedings of the 2006 Australasian Workshops on Grid Computing and e-Research, Darlinghurst, Australia*, 105–111.

Bobroff, N., Fong, L., Liu, Y., Martinez, J., Rodero, I., Sadjadi, S., et al. (2008). Enabling inter-operability among meta-schedulers. *IEEE International Symposium on Cluster Computing and the Grid (CCGrid), Lyon, France*, 306–315.

Brooke, J., Fellows, D., Garwood, K., & Goble, C., et al. (2004). Semantic matching of grid resource descriptions. European Acrossgrids Conference, Volume LNCS 3165 Nicosia, Greece, 240–249.

Buyya, R., Ranjan, R., & Calheiros, R. N. (2009). Modeling and simulation of scalable cloud computing environments and the cloudsim toolkit: Challenges and opportunities. *7th High Performance Computing and Simulation Conference (HPCS 2009), Leipzig, Germany.*

Buyya, R., Yeo, C. S., Srikumar Venugopal, J. B., & Brandic, I. (2009). Cloud computing and emerging it platforms: Vision, hype and reality for delivery computing as the 5th utility. *Future Generation Computer Systems, 25*, 599–616.

Buyya, R., Yeo, C. S., Venugopal, S., Broberg, J., & Brandic, I. (2009). Cloud computing and emerging it platforms: Vision, hype, and reality for delivering computing as the 5th utility. *Future Generation Computer Systems, 25*(6), 599–616.

Catlett, C., Beckman, P., Skow, D., & Foster, I. (2006). Creating and operating national-scale cyberinfrastructure services. *Cyberinfrastructure Technology Watch Quarterly, 2*, 2–10.

Chapin, S. J., Cirne, W., Feitelson, D. G., Jones, J. P., Leutenegger, S. T., Schwiegelshohn, U., et al. (1999). Benchmarks and standards for the evaluation of parallel job schedulers. *Job Scheduling Strategies for Parallel Processing (JSSPP), Volume LNCS 1659, Klagenfurt, Austria*, 66–89.

Chervenak, A., Deelman, E., Foster, I., Guy, L., Hoschek, W., Iamnitchi, A., et al. (2002). Giggle: A framework for constructing scalable replica location services. *Conference on High Performance Networking and Computing, IEEE Computer Society Press, San Jose, CA* 1–17.

Chervenak, A., Foster, I., Kesselman, C., Salisbury, C., & Tuecke, S. (2001). The data grid: Towards an architecture for the distributed management and analysis of large scientific datasets. *Journal of Network and Computer Applications, 23*, 187–200.

Chu, X., Nadiminti, K., Jin, C., Venugopal, S., & Buyya, R. (2007). Aneka: Next-generation enter-prise grid platform for e-science and e-business applications. *e-Science and Grid Computing, IEEE International Conference*, 151–159. DOI 10.1109/E-SCIENCE.2007.12.

Chunlin, L., & Layuan, L. (2008). Cross-layer optimization policy for qos scheduling in computa-tional grid. *Journal of Network and Computer Applications, 31*(3), 1084–8055.

Czajkowski, K., Fitzgerald, S., Foster, I., & Kesselman, C. (2001). Grid information services for distributed resource sharing. *IEEE International Symposium on High-Performance Distributed Computing (HPDC), Paris, France* 181.

Dumitrescu, C. L., & Foster, I. (2005). Gangsim: a simulator for grid scheduling studies. *Fifth IEEE International Symposium on Cluster Computing and the Grid (CCGrid'05), Washington, DC, USA*, 1151–1158.

Ezenwoye, O., & Sadjadi, S. M. (2010). *Developing effective service oriented architectures: con-cepts and applications in service level agreements, quality of service and reliability*, chap Applying concept reuse for adaptive service composition. IGI Global (2010).

Ezenwoye, O., Sadjadi, S. M., Carey, A., & Robinson, M. (2007). Grid service composition in BPEL for scientific applications. *In Proceedings of the International Conference on Grid computing, high-performAnce and Distributed Applications (GADA'07), Vilamoura, Algarve, Portugal*, 1304–1312.

Feitelson, D., & Rudolph, L. (1995). Parallel job scheduling: Issues and approaches. *Job Scheduling Strategies for Parallel Processing (JSSPP), Vol. LNCS 949, Santa Barbara, CA, USA*, 1–18.

Foster, I., Kesselman, C., & Tuecke, S. (2001). The anatomy of the grid – enabling scalable virtual organizations. *International Journal of Supercomputer Applications, 15*, 200–222.

Foster, I., Tuecke, S., & Unger, J. (2003). Osga data services. *Global Grid Forum, 9.*

Foster, I., et al. (2008). *OGSA basic execution service version 1.0, GFD-R.108* (Tech. Rep., Open Grid Forum (OGF)).

Foster, I., Zhao, Y., Raicu, I., Lu, S. (2008). Cloud computing and grid computing 360-degree compared. *IEEE Grid Computing Environments Workshop*, 1–10.

Gerndl, M., Wismuller, R., Balaton, Z., Gombas, G., Kacsuk, P., Nemeth, Z., et al. (2004). Performance tools for the grid: State of the art and future. White paper, Shaker Verlag.

Goodale, T., Jha, S., Kaiser, H., Kielmann, T., Kleijer, P., Laszewski, G. V., Lee, C., Merzky, A., Rajic, H., & Shalf, J. (2006). SAGA: A Simple API for Grid Applications. High-level application programming on the Grid. *Computational Methods in Science and Technology*.

Grimme, C., Lepping, J., Papaspyrou, A., Wieder, P., Yahyapour, R., Oleksiak, A., et al. (2007). Towards a standards-based grid scheduling architecture (Tech. Rep. TR-0123, CoreGRID Institute on Resource Management and Scheduling).

Guim, F., Corbalan, J., & Labarta, J. (2007). Modeling the impact of resource sharing in backfilling policies using the alvio simulator. *Annual Meeting of the IEEE International Symposium on Modeling, Analysis, and Simulation of Computer and Telecommunication Systems (MASCOTS), Istambul*.

Guim, F., Rodero, I., Corbalan, J., Labarta, J., Oleksiak, A., Kuczynski, T., et al. (2007). Uniform job monitoring in the hpc-europa project: Data model, api and services. *International Journal of Web and Grid Services, 3*(3), 333–353.

Guim, F., Rodero, I., Tomas, M., Corbalan, J., & Labarta, J. (2006). The palantir grid meta-information system. *7th IEEE/ACM International Conference on Grid Computing (Grid), Washington, DC, USA* 329–330.

Huedo, E., Montero, R., & Llorente, I. (2004). A framework for adaptive execution in grids. *Software—Practice & Experience*, 34, 631–651.

Huedo, E., Montero, R., & Llorente, I. (2009). A recursive architecture for hierarchical grid resource management. *Future Generation Computer Systems, 25*, 401–405.

Iosup, A., Epema, D., Tannenbaum, T., Farrelle, M., & Livny, M. (2007). Inter-operable grids through delegated matchMaking. *International Conference for High Performance Computing, Networking, Storage and Analysis (SC07), Reno, Nevada*.

Iosup, A., Li, H., Jan, M., Anoep, S., Dumitrescu, C., Wolters, L., et al. (2008). The grid workloads archive. *Future Generation Computer Systems, 24*, 672–686.

Iosup, A., Sonmez, O., & Epema, D. (2008) Dgsim: Comparing grid resource management architectures through trace-based simulation. *Euro-Par '08: Proceedings of the 14th international Euro-Par conference on Parallel Processing, Las Palmas de Gran Canaria, Spain*, 13–25.

Kephart, J. O., & Chess, D. M. (2001). *The vision of autonomic computing*. http://www.research.ibm.com/autonomic/manifesto/. Accessed on July 22, 2010.

Kacsuk, P., Kiss, T., & Sipos, G. (2008). Solving the grid interoperability problem by P-GRADE portal at workflow level. *Future Generation Computer Systems, 24*, 744–751.

Karlsson, M., & Mahalingam, M. (2002). Do we need replica placement algorithms in content delivery networks. *Proceedings of the International Workshop on Web Content Caching and Distribution (WCW), Boulder, CA, USA*, 117–128.

Karonis, N. T., Toonen, B., & Foster, I. (2003). Mpich-g2: a grid-enabled implementation of the message passing interface. *Journal of Parallel and Distributed Computing, 63*(5), 551–563. DOI http://dx.doi.org/10.1016/S0743-7315(03)00002-9.

Kertesz, A., & Kacsuk, P. (2008). Grid meta-broker architecture: Towards an interoperable grid resource brokering service. *CoreGRID Workshop on Grid Middleware in Conjunction with Euro-Par, LNCS 4375, Desden, Germany*, 112–116.

Kim, H., el Khamra, Y., Jha, S., & Parashar, M. (2009). An autonomic approach to integrated hpc grid and cloud usage. *e-Science, 2009. e-Science '09. Fifth IEEE International Conference*, 366–373. DOI 10.1109/e-Science.2009.58.

Kondo, D., Javadi, B., Iosup, A., & Epema, D. (2010). The failure trace archive: Enabling comparative analysis of failures in diverse distributed systems. *10th IEEE/ACM International Symposium on Cluster, Cloud and Grid Computing (CCGrid), Melbourne, Australia*.

Krauter, K., Buyya, R., & Maheswaran, M. (2002). A taxonomy and survey of grid resource management systems for distributed computing. *Software: Practice and Experience (SPE), 32,* 135–164.

Lamehamedi, H., Szymanski, B., Shentu, Z., & Deelman, E. (2002). Data replication strategies in grid environments. *Proceedings of the 5th International Conference on Algorithms and Architectures for Parallel Processing (ICA3PP'02), IEEE Press, Beijing, China,* 378–383.

Laszewski, G. V., Foster, I. T., Gawor, J., & Lane, P. (2001). A Java commodity grid kit. *Concurrency and Computation: Practice and Experience, 13*(8–9), 645–662.

Laure, E., Fisher, S. M., Frohner, A., Grandi, C., Kunszt, P., Krenek, A., et al. (2006). Programming the grid with glite. *Computational Methods in Science and Technology, 12,* 33–45.

Legrand, A., Marchal, L., & Casanova, H. (2003). Scheduling distributed applications: the simgrid simulation framework. 3rd *International Symposium on Cluster Computing and the Grid (CCGrid'03), Washington, DC, USA,* 138.

Leutenegger, S., & Vernon, M. (1990). The performance of multiprogrammed multiprocessor scheduling algorithms. *ACM SIGMETRICS Performance Evaluation Review, 18,* 226–236.

Majumdar, S., Eager, D., & Bunt, R. (1988). Scheduling in multiprogrammed parallel systems. *ACM SIGMETRICS Performance Evaluation Review, 16,* 104–113.

Massie, M. L., Chun, B. N., & Culler, D. E. (2004). The ganglia distributed monitoring system: Design, implementation and experience. *Parallel Computing, 30*(5–6), 817–840.

Matsuoka, S., Shinjo, S., Aoyagi, M., Sekiguchi, S., Usami, H., & Miura, K. (2005). Japanese computational grid research project: Naregi. *Proceedings of the IEEE, 93*(3), 522–533.

Mohamed, H., & Epema, D. (2008). KOALA: a Co-allocating grid scheduler. *Concurrency and Computation: Practice & Experience, 20,* 1851–1876.

Moore, R., Prince, T. A., & Ellisman, M. (1998). Data-intensive computing and digital libraries. *Communications of the ACM, 41,* 56–62.

Mu'alem, A., & Feitelson, D. (2001). Utilization, predictability, workloads, and user runtime estimates in scheduling the ibm sp2 with backfilling. *IEEE Transactions on Parallel and Distributed Systems, 12, 529–543.*

Nathuji, R., & Schwan, K. (2007). Virtualpower: Coordinated power management in virtualized enterprise systems. *ACM SIGOPS Symposium on Operating Systems Principles, Stevenson, Washington, USA.*

Newman, H., Legrand, I., Galvez, P., Voicu, R., & Cirstoiu, C. (2003). Monalisa: a distributed monitoring service architecture. *Computing in High Energy and Nuclear Physics (CHEP03), La Jolla, CA.*

Novotny, J., Tuecke, S., & Welch, V. (2001). An online credential repository for the grid: Myproxy. *Proceedings of the Tenth International Symposium on High Performance Distributed Computing (HPDC-10), IEEE, San Francisco, CA, USA,* 104–111.

Nurmi, D., Wolski, R., Grzegorczyk, C., Obertelli, G., Soman, S., Youseff, L., et al. (2009). The eucalyptus open-source cloud-computing system. *Cluster Computing and the Grid, IEEE International Symposium, 0,* 124–131. DOI http://doi.ieeecomputersociety.org/ 10.1109/CCGRID.2009.93.

Oleksiak, A., Tullo, A., Graham, P., Kuczynski, T., Nabrzyski, J., Szejnfeld, D., et al. (2005). HPC-Europa: Towards uniform access to European HPC infrastructures. *IEEE/ACM International Workshop on Grid Computing, Seattle, WA, USA,* 308–311.

Parashar, M., Li, X., & Chandra, S. (2010). *Advanced computational infrastructures for parallel and distributed applications.* New York, NY: Wiley.

Pearlman, L., Welch, V., Foster, I., & Kesselman, C. (2002). A community authorization service for group collaboration. *IEEE 3rd International Workshop on Policies for Distributed Systems and Networks, George Mason University, USA,* 50–59.

Podhorszki, N., & Kacsuk, P. (2001). Semi-on-line monitoring of p-grade applications. *Parallel and Distributed Computing Practices, 4*(4), 43–60.

Quiroz, A., Gnanasambandam, N., Parashar, M., & Sharma, N. (2009). Robust clustering analysis for the management of self-monitoring distributed systems. *Cluster Computing: The Journal of Networks, Software Tools, and Applications, 12*(1), 73–85.

Quiroz, A., Kim, H., Parashar, M., Gnanasambandam, N., & Sharma, N. (2009). Towards autonomic workload provisioning for enterprise grids and clouds. *10th IEEE/ACM International Conference on Grid Computing (Grid 2009), Banff, Alberta, Canada*, 50–57.

Ranganathan, P., Leech, P., Irwin, D., & Chase, J. (2006). Ensemble-level power management for dense blade servers. *Annual International Symposium on Computer Architecture, New York, NY, USA*, 66–77.

Rodero, I., Guim, F., Corbalan, J., Fong, L., & Sadjadi, S. (2010). Broker selection strategies in interoperable grid systems. *Future Generation Computer Systems, 26*(1), 72–86.

Ruth, P., McGachey, P., & Xu, D. (2005). Viocluster: Virtualization for dynamic computational domains. *Cluster Computing, 2005, IEEE International,* DOI 1–10. 10.1109/CLUSTR.2005.347064.

Samar, A., & Stockinger, H. (2001). Grid data management pilot (gdmp): A tool for wide area replication. *IASTED International Conference on Applied Informatics (AI2001), ACTA Press, Calgary, Canada.*

Schwiegelshohn, U., & Yahyapour, R. (1998). Analysis of first-come-first-serve parallel job scheduling. *9th annual ACM-SIAM Symposium on Discrete Algorithms, Vol. 38, San Francisco, CA,* 629–638.

Schwiegelshohn, U., & Yahyapour, R. (1998). Improving first-come-first-serve job scheduling by gang scheduling. *Job Scheduling Strategies for Parallel Processing (JSSPP), Vol. LNCS 1459, Orlando, FL,* 180–198.

Seidel, J., Waldrich, O., Ziegler, W., Wieder, P., & Yahyapour, R. (2007). Using SLA for resource management and scheduling – a Survey, TR-0096 (Tech. Rep., CoreGRID Institute on Resource Management and Scheduling).

Sodan, A. (2009). Adaptive scheduling for qos virtual machines under different resource availability first experiences. *14th Workshop on Job Scheduling Strategies for Parallel Processing, Vol. LNCS 5798, Rome, Italy,* 259–279.

Song, Y., Wang, H., Li, Y., Feng, B., & Sun, Y. (2009). Multi-tiered on-demand resource scheduling for vm-based data center. *IEEE International Symposium on Cluster Computing and the Grid (CCGrid), Los Alamitos, CA, USA,* 148–155.

Sotomayor, B., Keahey, K., & Foster, I. (2008). Combining batch execution and leasing using virtual machines. *HPDC '08: Proceedings of the 17th International Symposium on High Performance Distributed Computing, ACM, New York, NY, USA,* 87–96. DOI http://doi.acm.org/10.1145/1383422.1383434.

Sotomayor, B., Montero, R., Llorente, I., & Foster, I. (2008). Capacity leasing in cloud systems using the opennebula engine. *Workshop on Cloud Computing and its Applications (CCA08), Chicago, IL, USA.*

Stockinger, H., Samar, A., Allcock, B., Foster, I., Holtman, K., & Tierney, B. (2001). File and object replication in data grids.

Sulistio, A., Cibej, U., Venugopal, S., Robic, B., & Buyya, R. (2008). A toolkit for modelling and simulating data grids: An extension to gridsim. *Concurrency and Computation: Practice and Experience (CCPE), 20*(13), 1591–1609.

Tierney, B., Aydt, R., Gunter, D., Smith, W., Swany, M., Taylor, V., et al. (2002). A grid monitoring architecture.

Vasilakos, A., Parashar, M., Karnouskos, S., & Pedrycz, W. (2010). *Autonomic communication* (XVIII, pp. 374). ISBN: 978-0-387-09752-7. New York, NY: Springer.

Vecchiola, C., Chu, X., Buyya, R. (2009). Aneka: A Software Platform for .NET-based Cloud Computing, Technical Report, GRIDS-TR-2009-4, *Grid Computing and Distributed Systems Laboratory, The University of Melbourne, Australia.*

Venugopal, S., Buyya, R., & Ramamohanarao, K. (2006). A taxonomy of data grids for distributed data sharing, management and processing. *ACM Computing Surveys (CSUR), 38*(1), 2006.

Welch, V., Foster, I., Kesselman, C., Mulmo, O., Pearlman, L., Gawor, J., et al. (2004). X.509 proxy certificates for dynamic delegation. *Proceedings of the 3rd Annual PKI R&D Workshop, Gaithersburg, MD, USA.*

Welch, V., Siebenlist, F., Foster, I., Bresnahan, J., Czajkowski, K., Gawor, J., et al. (2003). Security for grid services. *High-Performance Distributed Computing, International Symposium, 0,* 48. DOI http://doi.ieeecomputersociety.org/10.1109/HPDC.2003.1210015.

Wiseman, Y., & Feitelson, D. (2003). Paired gang scheduling. *IEEE Transactions on Parallel and Distributed Systems, 14*(6), 581–592.

Wolski, R., Spring, N. T., Hayes, J. (1999). The network weather service: A distributed resource performance forecasting service for metacomputing. *Future Generation Computer Systems, 15*(5–6), 757–768.

Zanikolas, S., & Sakellariou, R. (2005). A taxonomy of grid monitoring systems. *Future Generation Computer Systems, 21*(1), 163–188.

Zhao, Y., Hategan, M., Clifford, B., Foster, I., von Laszewski, G., Nefedova, V., et al. (2007). Swift: Fast, reliable, loosely coupled parallel computation. *Services, IEEE Congress on 0,* 199–206. DOI http://doi.ieeecomputersociety.org/10.1109/SERVICES.2007.63.

Zhu, Y., & Ahuja, M. (1993). On job scheduling on a hypercube. *IEEE Transactions on Parallel and Distributed Systems, 4,* 62–69.

Chapter 9
Cloudweaver: Adaptive and Data-Driven Workload Manager for Generic Clouds

Rui Li, Lei Chen, and Wen-Syan Li

Abstract Cloud computing denotes the latest trend in application development for parallel computing on massive data volumes. It relies on clouds of servers to handle tasks that used to be managed by an individual server. With cloud computing, software vendors can provide business intelligence and data analytic services for internet scale data sets. Many open source projects, such as Hadoop, offer various software components that are essential for building a cloud infrastructure. Current Hadoop (and many others) requires users to configure cloud infrastructures via programs and APIs and such configuration is fixed during the runtime. In this chapter, we propose a workload manager (WLM), called *CloudWeaver*, which provides automated configuration of a cloud infrastructure for runtime execution. The workload management is data-driven and can adapt to dynamic nature of operator throughput during different execution phases. *CloudWeaver* works for a single job and a workload consisting of multiple jobs running concurrently, which aims at maximum throughput using a minimum set of processors.

9.1 Introduction

Cloud Computing denotes the latest trend in application development for parallel computing on massive data volumes. It relies on clouds of servers to handle tasks that used to be managed by an individual server. With Cloud Computing, software vendors can provide business intelligence and data analytic services for Internet scale data sets.

R. Li (✉) and L. Chen
Hong Kong University of Science and Technology, Clear Water Bay, Hong Kong
e-mails: {cslr; leichen}@cse.ust.hk

W.-S. Li
SAP Technology Lab, Shanghai, China
e-mail: wen-syan.li@sap.com

B. Furht, A. Escalante (eds.), *Handbook of Cloud Computing*,
DOI 10.1007/978-1-4419-6524-0_9, © Springer Science+Business Media, LLC 2010

Due to the size of these data sets, traditional parallel database solutions can be prohibitively expensive. To be able to perform this type of web-scale analysis in a cost-effective manner, several companies have developed distributed data storage and processing systems on large clusters of shared-nothing commodity servers, including Google's App Engine[1] Antoshenkov (1996), Amazon's EC2/S3/SimpleDB[2] Acker, Roth, and Bayer (2008), Microsoft's SQL Server data services[3] DeWitt, Naughton, Schneider, and Seshadri (1992) and IBM's "Blue Cloud" service. At the same time, open source projects such as Hadoop offer various software components that are essential for building a cloud infrastructure. Current Hadoop (and many others) provides virtualization of data location, concurrent execution coordination, and load balance across servers; however, it requires users to configure cloud infrastructures via programs and APIs, moreover, such a configuration is fixed during the runtime. For example, an Hadoop programmer needs to manually set the ratio of the number of servers/tasks used by reducer functions to the number of servers used by map functions/tasks[4] Davlid, DeWitt, Shanker (2008). In addition, Hadoop's programming model is limited to SQL augmented user-defined functions and stored procedures, which is less user-friendly and flexible compared to SQL language supported by parallel databases. So far, Hadoop supports parallelism to only simply tasks, instead of general purpose computation required by ETL and DBMS.

In this chapter, motivated by the limitations of Hadoop, we deploy a workload manager (WLM), *CloudWeaver*, to provide automated configuration of a cloud infrastructure for runtime execution. Thus, programmers do not need to set the ratio between mappers and reducers while it is managed by *CloudWeaver*. Furthermore, it is not feasible for the programmers to set such ratios when pipelining is enable since there would be multiple tiers of servers/tasks running concurrently.

The load at each tier could change dynamically depending data distribution and characteristics of operators. Thus, cloud computing is actually data-driven and throughput of each operator/task changes during the phase of computation. *CloudWeaver* can adapt to dynamic nature of operator throughput during different execution phases, moreover, it works for both single job and a workload of multiple jobs running concurrently aiming at maximum throughput with a minimum set of processors.

The rest of the chapter is organized as follows: In Section 9.2, we provide an overview of CloudWeaver system architecture and define the terminologies used in this chapter. In Section 9.3, we describe the algorithm of core component of CloudWeaver. We compare our approach with related work in Section 9.4 and conclude and discuss some future work in Section 9.5.

---

[1] http://code.google.com/appengine/.

[2] http://aws.amazon.com/.

[3] http://www.microsoft.com/sql/dataservices/.

[4] http://hadoop.apache.org/core/docs/current/mapred_tutorial.htm#reducer.

9.2 System Overview

In this section, we describe the system architecture of CloudWeaver. We first briefly review Hadoop and its components. Hadoop is an integrated system for Map/Reduce jobs. It runs on a large cluster with HDFS (Hadoop Distributed File System). HDFS has a single *Namenode* which manages the file system namespace and regulates access to files by clients. Each machine has a *Datanode* which manages storage attached to the machine. Each data file in HDFS is stored as many small data blocks, typically of a fixed size. Each block has 2 or 3 replicas located in different Datanodes. Using multiple copies of small data blocks provides better availability and accessability. The Map/Reduce execution is built on top of HDFS. The user submits a Map/Reduce job configuration through *job client*. A *master node* will maintain a *job tracker* and fork many *slave nodes* to execute map/reduce tasks. Each slave node has a *task tracker* which manages map or reduce tasks instances on that node.

Compared to Hadoop, our new proposed system for generic clouds, called CloudWeaver, has the following extensions:

- *Cloud Monitor* is added to monitor the resource utilization of a processor and consumption status of processor output (i.e. results). It is also used to add new servers to the cloud or shut down some computing utilities.
- The Hadoop cloud is extended to be more generic for general purpose computing as a *generic cloud*. In order to enable a generic cloud, the jobs for CloudWeaver is also extended from MapReduce job description to DAG with general purpose operators.
- *Workload Manager (WLM)* is added to automate the assignment of processors to tasks and map jobs to processors. More detail will be given in Section 9.3.

Figure 9.1 shows the system architecture of CloudWeaver, which consists of a job client, a central workload manager (WLM), servers (or called workers or slave nodes), a name node and a storage system (called data node).

The generic cloud is provided as hardware computing facility. The user may or may not know the detail of its configurations, and the configuration can be changed. The user will submit a query job to the generic cloud from a client computer. The query job can be considered as a marked operator tree, so that we know the work flow or the data flow. The name of input files of some of the nodes are also included. We assume that these input files reside in the storage system of the generic cloud. In other words, the cloud can read the files from their names.

We also relax the assumption of HDFS by supporting both shared file systems and non-HDFS shared-nothing file systems. In shared-file systems, each processor can access storage directly through some common interface. Other distributed file systems may have different storage policy. We assume that the name node and data node in CloudWeaver provide interface to access data as small blocks similar to HDFS. Our scheduling algorithm is designed to run many tasks on small data blocks to improve performance and to achieve load balance.

Fig. 9.1 Architecture of cloudweaver

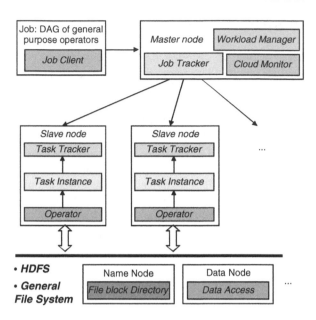

9.2.1 Components

In this section, we briefly discuss the extension components of Hadoop in CloudWeaver, which are the workload manager, cloud monitor, and generic cloud.

9.2.1.1 Workload Manager

The workload manager will accept the query job and is responsible for processing the query. It knows the status of whole system: where's the name node, where are the computing servers and where 's the storage system. Any change to the cloud environment will be noticed by workload manager. WLM looks at the operator tree of a query job and process the job in a data driven model. That is, WLM will schedule small tasks to run in servers. Each task will take a small block of input and generates some output files. WLM will schedule the intermediate result file to feed other operators until the output of the query job is generated. In this process, WLM takes care of the generic cloud change and the working progress, so it can dynamically utilize all available resources.

The name node maintain a directory of all the files. It could be considered as a file system interface. The source input files reside in the storage system. When the server process job, it will ask the name node about the accessing address of the files and then read them or write new files as result.

The storage system can be either shared storage or a share-nothing structure. We assume that there's central node to maintain all the related files in the generic cloud.

9.2.1.2 Cloud Monitor

Since the system is based on producer-consumer model, the output of lower tier tasks is used as input for the tasks of upper tiers. Each task stores its output (i.e. intermediate results) in the local disk by pre-defined block size. The intermediate result blocks are then read by upper tier tasks.

If the number of intermediate result blocks is increasing, *Cloud Monitor* can notice *WLM* to increase the number of upper tier tasks to consume increasing number of blocks.

9.2.1.3 Generic Cloud

A generic cloud has a cluster of servers with computing power. The large data set is either stored in the cloud or can be passed into the cloud to fulfill a data processing job. Each data processing job is called a job for short in the rest of this chapter. We mainly study the queries in this chapter. A task can be parallelized into small jobs. Jobs are executed on different servers. The performance is improved by using parallelism.

The servers can have different computing power and storage size. Scheduling jobs in generic cloud to achieve best response time is a hard optimization problem. A predefined scheduling algorithm is hard to deal with the changing environment. In this chapter we solves the scheduling problem in run time. We check the data processing requirement and cloud status in run time, determine the number of jobs and the assignment of jobs to servers. Because each partition and scheduling step is based on available data need to be processed, we believe that this data driven method can best balance the workload of servers in the generic cloud and achieve best performance.

We consider SQL like query processing for large data set with MapReduce support.

Query Job

A user query job can be described by an operator graph. The operators include Extract, Join, Aggregate functions and Map Reduce. The map and reduce function is provided by user.

In the parallel environment, the behavior of an operator can be partitioned into several small jobs and run in parallel on different servers. Each job can be run by an executable file. The command takes some input files and generate output files. In this way, we can direct data files and executable file to different servers, the the executable file consume the input files and produce outputs.

Input

We assume that the input data are very large. The large input files may be considered as tables in an RDMBS. We can consider each file as a big table.

In order for the WLM to make use of parallelism, WLM needs to know how each operator can be parallelized. For example, a simple table extract can be arbitrarily partitioned into many small files. Each file can be processed by an extract task on any server. The sort operator can also be parallelized by many small sorting tasks with a merger task, but the output can not be fixed unless all the input data has been processed. This kind of operator is called a blocking operator.

The join operator is another complex example. We can partition the input and use many servers to process join. When WLM wants to schedule more servers to process join, the status of original join servers need to be migrated or changed.

The differences of our work compared with others is that we do not assume SMP or cluster with identical servers. Instead, we deal with machines with different power. We deal with dynamic data throughput nature. For example, when we have a join, the processing speed is a constant, but the result output rate varies during the whole processing period.

Our framework also has a big difference compared with Map/reduce or Hadoop. The Hadoop system aims to provide abstraction (virtualization) for the underlying hardware/file location/load balancing so that the application programs can focus on writing maps and reduce functions. In our work, CloudWeaver provides similar functionality but focuses on coordinating execution of complex jobs (virtualization of execution flow management and optimization from the programmers). The execution of Map/Reduce or Hadoop is much simpler since the execution only have two phases and does not involve complex data flow. This is similar to SQL: users need to specify what they want using SQL and no need to specify how to execute the queries and how to optimize it. The optimization and execution is done automatically by the database system. Our system is a powerful implementation for data processing under cloud environment.

A whole data processing system can have three important components on top of each other. The first is user's input to describe the job. The second is the parallelization and execution. The third, like the storage is provided by the infrastructure. In our system, the workload manager focuses on the second phase of execution by conducting the processor/task mapping. We assume that how to select a set of right servers is handled by the infrastructure. Hadoop and Map/Reduce provide all three phases. This makes it not generic. Our system can deal with a user's input job and has good performance over arbitrary infrastructure.

Dryad is similar to our system in the sense that it parallelize a sequential data flow, but it only has local optimization for operators that runs slower while our algorithm schedule the whole job DAG in a data driven fashion, which is more flexible and extensible. Besides, Dryad has not been extended to schedule multiple jobs.

9.3 Workload Manager

In this section, we present the details of the proposed workload manager (WLM) of the CloudWeaver. Different from the traditional parallel scheduling algorithms, where the operator trees are extracted from the query plan and the basic execution units are identified for each node, in cloud, the scheduling algorithms are data driven. In other words, the scheduling algorithm does not know the execution units for each operator. Only when data arrive, the scheduler maps the processors to the "ready" nodes (i.e. the node with correct input data) and starts preparing processors to accept the output from each node. Moreover, the processor nodes are heterogenous (different processing power) and scanned data for each operator often change. In order to address these issues, we propose a dynamic scheduler to achieve minimum intermediate result size and balanced workload among processors. In the rest of this section,we first present the terminologies used to describe job and then, formalize the scheduling methods for single and multiple jobs.

9.3.1 Terminology

In this chapter, we describe the user's specified job with a DAG, where each node N_i is an operator that provides an abstraction of the sub-jobs and each edge $E_{i,j}$ between two operators N_i and N_j indicates the data flow from N_i to N_j. An example is shown in the left of Fig. 9.2. It has six operators named as A to F and two branches $A \rightarrow B \rightarrow C$ and $E \rightarrow F$, with the top most operator D. A and F are the leaf nodes (I/O nodes) and B,C,D, E are non-leaf nodes (computing nodes). This input job could be mapped to a query which performs join over two tables after some selections and projections are done. The input job is a high level abstraction and does not tell how to parallelize the data processing. In addition to this graph, the cloud should know the input file name as well as the function of each operator.

We use intra-operator parallelism to speed up the execution of each operator. As shown in the right part of Fig. 9.2, each operator can be executed by many small tasks. Tasks are OS level processing units and can be viewed as threads/processes.

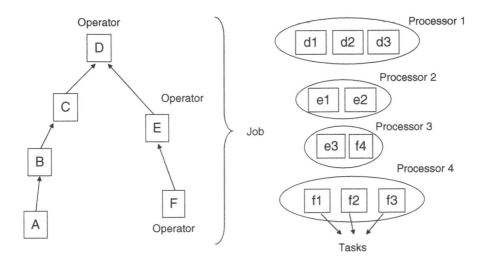

Fig. 9.2 Job, operator, processor and task

Tasks are independent and shared nothing. A task can be assigned the whole processor or part of the processor shared with other tasks. We use the term processor for being more general. Depending on the environment which the job runs, the processor could be a server (for cluster), a node (for grid/cloud), or a CPU/core for SMP/CMP setting. We allocate tasks to processors in the cloud. This step is called processor/tasks mapping. Each processor can run one or more tasks from same operator or different operators.

An example of our data driven parallelization algorithm for the example user job is illustrated in Fig. 9.3. In this running example, we assume a share-nothing architecture. Each processor has its local disks. The input is stored in some of the disks. A node can process files stored in other node by file transferring through networks. Our target is to schedule the progress of the all operators so that the intermediate result files can be minimized and the processors in the cloud can be maximally utilized.

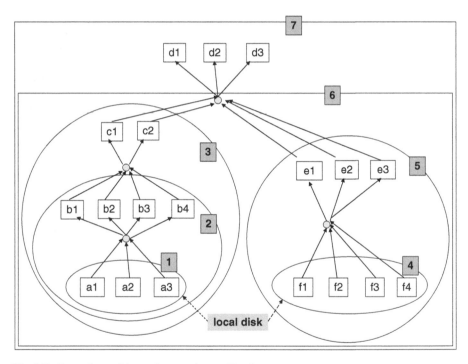

Fig. 9.3 Example machine assignment by workload management

9.3.2 Operator Parallelization Status

Workload manager will maintain status for each operator. Specifically in our data driven approach, it manages two sets of an operator O: an input queue of input data file names $Q.input$ and output queue of output data file names $Q.output$. User specification will describe how the input data could be processed (for example join or aggregation), specifically by a particular kind of operator. Input files (data) will be broken down into many data blocks as that have been done in Hadoop. Many data blocks from an operator can be executed concurrently by many tasks in different processors. A snapshot of all the input/output data and running tasks related to an operator is treated as the *operator status*. Specifically, the operator status could be: waiting for data (not started), in processing or finished. In addition to the operator status, we should also record how the operator is parallelized, how many tasks is assigned for each operator as well as the parameters. For example, we have assign 5 processors to execute a scanning operator at the beginning. The workload manager will record how much has been done by each processor. Later, when more resources (e.g. processors) are available, the workload manager may decide to distribute the remaining to-be-scanned data to a total of 10 processors. Decision should be made according to the operator status.

By monitoring an operator's status, the workload manager will know how to parallelize and schedule the execution of each operator. The input queue to each operator will be monitored as an input buffer. When the input buffer grows rapidly, it indicates that more computing power is needed at the operator and if possible, workload manager should assign more processors to consume the input. The running statistics can be collected from running tasks on different processors, which can be used by the workload manager to map appropriate processors to operator/task in order to achieve better processor usage.

9.3.3 Job Execution Algorithm

Our data driven algorithm is to determine the processors and the schedule of tasks for each operator. Given a data processing job, we have two types of operators: blocking operators and non-blocking operators. If an operator can not generate any output before all its input data are ready, we call it a blocking operator. The successor operators of a blocking operator can not be started until the blocking operator is finished, for example a sorting operator. On the contrary, a non-blocking operator and its success operators can be execute in a pipelined fashion, for example a scan operator followed by a join operator.

Given an operator tree corresponding to a user job description, we consider three scenarios that our algorithm will target. The first two are discussed for non-blocking operators and the last one is presented for blocking operators. Workload manager aims at dynamically assigning a set of processors to each operator, called *candidate processor set* of an operator. The tasks of the operator is scheduled within the candidate processor set. By controlling the size of candidate processor set, workload manager actually controls the power of execution for a specific operator. An operator with no task to execute will have an empty candidate processor set.

1. Assume that we have sufficient resources and operators are non-blocking ones. In this case, whenever an operator has a certain amount of input, the workload manager will give a non-empty candidate processor set to the operator and schedule

tasks. Workload manager will aims at full usage of each processor. When the data arrival rate change and the current candidate processor set does not match to input data queue, the workload manager can increase/reduce candidate processor set size.

2. Assume that we have limited resources and operators are non-blocking ones. In this case, when an operator has a certain amount of input, the workload manager may not always be able to find available processors for the operator. Assume that all the operators are non-blocking operators and data driven execution will start from the leaf operator. When there are data for other operators, workload manager will start the operator if possible candidate processor set can be found. The workload manager will try to schedule more operators at the same time in a pipeline fashion, not one by one. If resources are limited, we can reduce the size of candidate set for each leaf node operator and assign processors to operators in the data tree. After an operator finished all its work, its candidate processor set can be released and used by other operators.

3. Assume that some of the operators are blocking operators. In this case, we could not extend pipeline execution further to parent operators of blocking operators. For an operator N_i, all the operators under its node in the operator tree have been assigned candidate processor set. We refer to the all the processors assigned to this branch as branch processor set. If N_i is a blocking operator, the workload manager can start to assign processors to new branches in the operator tree which have input. If there are no other branch, the remaining available processors can be added to the branch of N_i. We add processor from the leaf node until N_i, which means that we assign more CPU power to the source operators and push data in a bottom up style. When a leaf operator is complete, limited resource can be moved to upper level operators.

There may exist many possible approaches to determine which operators to schedule first. Many optimization solutions may be used. In our work, we first focus on a simple bottom up manner and choose operators in a particular order (for example post order), as long as the operator can be started. We aim to achieve full usage of CPU power and better progress management of pipeline along the operator tree.

9.3.4 Dynamic Parallelization for Job Execution

Given a job DAG with n operators, $N_1, \ldots, N_n$, and a set of m processors with processing power, $P_1, \ldots, P_m$, respectively, we assume that the edge $E_{i,j}$ between N_i and N_j is a pipeline edge if N_j is not a blocking operator. A pipeline edge indicates that operator N_i's partial output can be passed to N_j for the next level execution. If N_j is a blocking operator, we call the edge $E_{i,j}$ as a blocking edge, which indicates that operator N_j has to wait all the data from its input operations until any result can be given. The dynamic scheduler aims to maximize the parallelism and balance the workloads among processors. Since the required processors for each operator(except for the leaf nodes) are unknown before the data is ready, the dynamic scheduler has to perform an "exploration" phase to estimate the input and output rate of each operator. Specifically, the scheduler first assigns one processor P_i to each *scan* operator (a.k.a *leaf* operator) N_i. For this step, we can randomly pick a processor and assign to a leaf level operator's tasks. Assume the output rate of operator N_i running on processor P_j is $R_{i,k}$, and the output results are fed into another operator N_k as its input. If the edge between N_i and N_k is a pipeline edge, in order to make the two operators N_i and N_k run concurrently, the scheduler needs to assign processors to consume the output from N_i with an approximately the same rate $R_{i,k}$. Similarly, after processing operator N_k, the scheduler will know the output rate of N_k and do the processor assignment accordingly. The same procedure is repeated until a blocking edge is encountered. In this case, the blocking operator, such as sort/aggregation, will wait until all the source data from its child operators arrive. Once all the sources are ready for processing, we schedule appropriate tasks to match the input of blocking operator.

The processor assignment discussed so far only assumes that there are enough processors to be assigned. Once the processors are not enough, the scheduler should reduce the processors assigned to the leaf nodes and adjust the processor assignment to non-leaf nodes accordingly.

Now we briefly describe how the workload manager schedule the tasks of the example job in Fig. 9.2 in a cloud with X machines. Each operator can be parallelized and executed by multiple tasks. $A \rightarrow B \rightarrow C$ or $E \rightarrow F$ can be executed as a pipeline and D can run as long as long as the corresponding data from C and E is available.

The job is scheduled as Fig. 9.3 in real time by the workload manager. We explain how data driven approach will deal with potential I/O bound by utilizing pipeline and overlapping I/O and CPU tasks. It is executed in following steps.

Step 1: identify the tasks for the operator A.

Step 2: select appropriate server nodes $a1, a2, a3$ to execute tasks of A

Step 3: the nodes $a1, a2, a3$ execute operator A's tasks and can be viewed as a single node "1".

Step 4: find a set of servers to handle output from the node "1". Say, we allocate $b1, b2, b3, b4$.

Step 5: those 7 nodes can be viewed together as a node "2".

Step 6: find a set of servers enough to handle the output from the node "2".

Step 7: all the server in the left branch can be viewed as the node "3" and in this node 3, all servers are busy since data inflow and processing power are matched, thus, the resource utilization is high.

Step 8: construct node "4", "5" similar to the way we construct node "3".

Step 9: then the nodes 3 and 5 can be combined to node "6"

Step 10: find a set of servers to handle the output of node "6", which are $d1, d2, d3$.

Note that we try to find a set of right "ratios" for servers between adjacent tiers illustrated by the numerical nodes. Once we can find such set of ratios, for any number of machines, we know how to assign servers (or computing powers by sharing servers) for each operator.

This approach ideally produces the best response time using a minimal number of servers (most of time during the execution) since all machines are fully utilized. Alternatively, we can use a lot of machines for operator A and F and then finish B, C, E, D one by one. This naive approach is apparently inefficient if we can not utilize all the machines for one of the operators.

To determine the best ratio between two tiers, we are aiming at balancing the input/output, or the size of the middle result. In other words, we want to keep the volume of waiting data blocks under a threshold T. When there are more data blocks than the threshold, we try to add more resources, so that more data blocks will be processed. When number of data blocks is smaller than threshold, we could reduce the number of processors. So finally the ratio can be reached around the threshold. Apparently, for small value of T, the whole job will be pipelined faster.

9.3.5 Balancing Pipelined Operators

In a general work flow, we could eliminate intermediate results if we can extend the execution of operators of a pipeline. If the computing resources are unlimited, we can always assign new tasks to processors. In general, computing resources for one job is limited. In such case, WLM has to assign tasks intelligently among limited number of resources.

Figure 9.4 give a real example. Three operators has different processing speeds for the same size of data. We cannot know exactly how fast data will be generated before the job start so the resource allocation is best to be done in real time. We focus

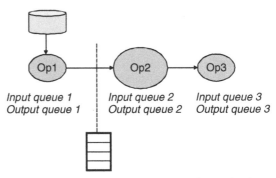

Fig. 9.4 Pipeline

on the first two operators. Op1 and Op2 have a producer/consumer relationship, the output data of Op1 will be passed to Op2. After initial execution, WLM finds that the computing power should be configured into 4:3 for two operators to match the input/output. If we have 7 equivalent servers, this can be fulfilled easily. Suppose we have only 6 equivalent servers, we propose following shared-worker scheme: we assign 3 workers dedicated for Op1's task, 2 workers for Op2's task, the remaining for tasks from both workers (interleaving). The balancing target here is to reduce the size of the intermediate result and shared-worker can execute the unbalanced tasks in real time.

To summarize, our data driven model will monitor input/output queues between neighboring operators and adjust the task assignment. If resources are limited we can request new workers or slow down the source input. The ratio in pipeline can be changed because of variation in the selectivity, our scheduling can quickly response to such change.

9.3.6 Balancing Tiers

Next, we show how to select servers to balance the workload between adjacent tiers. Because it is data driven, the easiest way is – assign one server or minimum number of servers for I/O related sub-jobs (say, operators A and B) – then we can figure out how many servers should be assigned to B,C,E,D respectively handle output from A and F respectively. The idea is something like: assign one I/O node; traverse from the leaf node to top until reach a join node at the join node, multiple branches need to be "synchronized" across all branches; to do so, the servers need to be added to traverse down the branches; after all branches at the join node are in sync (balanced), then continue to traverse upward; after we reach to the top, we have determine the ratios between servers between adjacent tiers. We can multiple a number (i.e. 2, 3, 4, ...) to the numbers of servers in each node and expand the execution graph.

We mention "ratio" for numerical nodes in the example. The actual ratio will change during run time and change from one run to another. The ratio may not mean anything unless all computation units are "normalized".

The ratio gives us a reference as how to allocation cpu power to different sub-job so that the workload can be balanced.

9.3.7 Scheduling Multiple Jobs

To schedule multiple jobs, we first map a certain number of processors for each job, which can be considered as its individual processor pool. For each job, the workload manager will only assign its task to the machines in its own pool. This processor/job mapping is done by WLM when a job is initialized. If resource is limited, we allow a processor to be shared by multiple jobs since some of processor may be powerful. Figure 9.5 shows an example of two jobs. Processor 2 is shared by both jobs.

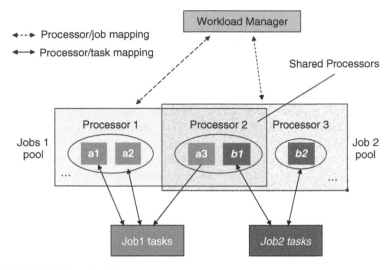

Fig. 9.5 Multiple jobs scheduling

The approach of initial worker pool is good for large scale cloud since we can then easily control the resources used by each job. The pool can be dynamically changed during run time to achieve a lot of scheduling purposes. For example, we can simply suspend a job by cleaning its worker pool. We can speed up/slow down the processing by adding/reducing processors in a job's worker pool. The initial pool size may not be suitable for all jobs. In real time, WLM may assign more/less workers to the job which has more/less tasks to be scheduled. For example, if a job is first executed by 10 machines to do I/O scan and the final step is done by a single task on one machine, the unused processors can be released early to the process other jobs with limited resources.

We list the internal design of WLM for scheduling a DAG job in Algorithm 1. The WLM is run as a central scheduler. When a new job request is submitted, WLM will first initialize the job and request an initial worker set of the job, then, WLM will schedule the job. When a scheduled task finished, WLM will schedule other tasks of its corresponding job.

9.4 Related Work

9.4.1 Parallel Databases

Parallelized processing on multi processors or shared-nothing architecture has been shown to have a high degree of scale up and speed up (DeWitt & Gray, 1992). Some of oldest systems include: GAMMA, Bubba, PRISMA/DB.

Parallelism can be done by using inter-operator parallelism (use pipeline or parallelize bushy-tree) and intra-operator parallelism (partition and parallelize inside a

---

Algorithm 1 Workload Manager (WLM) for a single job

Require: Input: A job b of DAG with n operators and a set of m processors.

1: Initialize status of all operators in DAG of job b.
2: Open the source input of leaf-level operators.
3: Request the worker set for b.
4: **while** The root operator of DAG is not complete **do**
5:     Get an operator N which has tasks to be scheduled, and the status of tasks.
6:     Get the current worker set P of b and their working status.
7:     Determine the processor/tasks mapping;
8:     Assign available tasks to workers and request/release worker if necessary.
9: **end while**
10: Job b is finished, release the work set and return.

---

relational operator). Great efforts has been put in parallelizing relational operators, such as join algorithm (Schneider & DeWitt, 1989).

The optimization of parallelized query plan and scheduling has been widely studied (Hasan, 1995), such as scheduling pipelined query operators (Hasan & Motwani, 1994; Liu & Rundensteiner, 2005), static task scheduling and resource allocation (Kwok & Ahmad, 1999; Garofalakis & Ioannidis, 1996; Lo, Chen, Ravishankar, & Yu, 1993), load balancing (Bouganim, Florescu, & Valduriez, 1996) with skew handling (DeWitt, Naughton, Schneider, & Seshadri, 1992), managing intra-operator parallelism with multi-users (Mehta & DeWitt, 1995), dynamic query plan optimization and migration according to running statistics (Antoshenkov, 1996). For the dynamic query plan optimization, we can also prepare many different query plans and select one in runtime (Hsiao, Chen, & Yu, 1994) or use parameters (Ioannidis, Ng, Shim, & Sellis, 1992).

There also exist efforts on improving database performance on other computer architectures, such as parallelizing the query processing in a multi-core environment (Acker, Roth, & Bayer, 2008), database processing designed for simultaneous multithreading processors (Zhou, Cieslewicz, Ross, & Shah, 2005), etc.

In nowadays' real internet applications, large data analysis is needed over streaming of data. A lot of effort has been put to fulfil such query requirement (Shah, Hellerstein, Chandrasekaran, & Franklin, 2003; Madden, Shah, Hellerstein, & Raman, 2002; Zhu, Rundensteiner, & Heineman, 2004; Liu, Zhu, Jbantova, Momberger, & Rundensteiner, 2005).

9.4.2 Data Processing in Cluster

Google's MapReduce (Dean & Ghemawat, 2004) is deployed in a large cluster running Google Files System. In this system, a large data set is stored as multiple copies of small standard blocks in different locations in the cluster. The MapReduce is a standard programming model, executing two user provided functions for Map tasks and Reduce tasks. Many data processing problems can be transformed into this programming model, and the scheduling of tasks over the specifically designed system improve the performance very well. But because the programming model limitation, it's not suitable for all kinds of jobs such as a relation join which takes two files as input. Because of its usefulness and simplicity, many pioneering database systems are beginning to integrate its functionality, including Aster Data[5] and GreenPlum.[6] An open source implementation of MapReduce is provided by Hadoop.[7]

Dyrad (Isard, Budiu, Yu, Birrell, & Fetterly, 2007) implements a general-purpose data parallel execution engine over a cluster. It uses low-level programming language to represent a DAG of data flow. The static plan will be given to a runtime executor to schedule over a cluster. However it has some drawbacks. First, the user need to master a complex graph description language. It also requires the user to know the detail of cluster to specify degree of parallelism (at least as a suggestion). The optimization is through changing the shape of predetermined graph plan during runtime. Currently it can only deal with one user task or query.

Recently, data driven workflow planning in cluster is proposed in Robinson and DeWitt (2007) and Shankar and DeWitt (2007). Clustera (Davlid, DeWitt, & Shankar, 2008) is a recently developed prototype of an integrated computation and data management system. It has good extensibility and can execute computationally intensive tasks, MapReduce task, as well as SQL queries in a parallel environment. It will first compile any kind of task description into a DAG of concrete jobs, then the job scheduler will optimize the execution in the cloud. It also adopts the idea of using database for the cluster management problem. It has comparable performance compared to Hadoop on the performance of MapReduce task and good scalability for SQL like queries.

Inspired by MapReduce, it has been realized that a descriptive language for SQL like data processing is needed for the parallel processing in a share-nothing environment. Industries have proposed their idea, like Yahoo's Pig Latin (Olston, Reed, Srivastava, Kumar, & Tomkins, 2008) and Microsoft's SCOPE (Ronnie Chaiken, Jenkins, & Zhou, 2008).

9.5 Conclusion

Cloud computing denotes the latest trend in application development for parallel computing on massive data volumes. It relies on clouds of servers to handle tasks that used to be managed by individual server. We observe that current Hadoop (and many others) require users to configure cloud infrastructures via programs and APIs

---

[5]http://www.asterdata.com/index.php.

[6]http://www.greenplum.com/.

[7]http://hadoop.apache.org/core/.

and such configuration is fixed during the runtime. We argue that such ad hoc configurations may result in less utilization and efficiency of the whole system. In this chapter, we provide automated configuration of a cloud infrastructure adaptive to runtime execution as well as extend supported set of operators. We evaluate the proposed framework on both synthetic computational jobs. The future work includes automated translation of SQL to Hadoop APIs (or extended APIs if required) in a prototype system for further evaluation.

References

Acker, R., Roth, C., & Bayer, R. (2008). Parallel query processing in databases on multicore architectures. *ICA3PP*, 2–13.

Antoshenkov, G. (1996). Dynamic optimization of index scans restricted by booleans. *ICDE*, 430–440.

Bouganim, L., Florescu, D., & Valduriez, P. (1996). Dynamic load balancing in hierarchical parallel database systems. *VLDB*, 436–447.

Davlid, E. R., DeWitt, J., & Shankar, S. (2008). Clustera: An integrated computation and data management system. *VLDB*.

Dean, J., & Ghemawat, S. (2004). Mapreduce: Simplified data processing on large clusters. *OSDI*, 137–150.

DeWitt, D. J., & Gray, J. (1992). Parallel database systems: The future of high performance database systems. *Communications of the ACM, 35*(6), 85–98.

DeWitt, D. J., Naughton, J. F., Schneider, D. A., & Seshadri, S. (1992). Practical skew handling in parallel joins. *VLDB*, 27–40.

Garofalakis, M. N., & Ioannidis, Y. E. (1996). Multi-dimensional resource scheduling for parallel queries. *SIGMOD Conference*, 365–376.

Hasan, W. (1995). Optimization of sql queries for parallel machines. (*Doctoral Thesis, Standford University*, 1995).

Hasan, W., & Motwani, R. (1994). Optimization algorithms for exploiting the parallelism-communication tradeoff in pipelined parallelism. *VLDB*, 36–47.

Hsiao, H.-I., Chen, M.-S., & Yu, P. S., (1994). On parallel execution of multiple pipelined hash joins. *SIGMOD Conference*, 185–196.

Ioannidis, Y. E., Ng, R. T., Shim, K., & Sellis, T. K. (1992). Parametric query optimization. *VLDB*, 103–114.

Isard, M., Budiu, M., Yu, Y., Birrell, A., & Fetterly, D. (2007). Dryad: distributed data-parallel programs from sequential building blocks. *EuroSys*, 59–72.

Kwok, Y.-K., & Ahmad, I. (1999). Static scheduling algorithms for allocating directed task graphs to multiprocessors. *ACM Computing Surveys, 31*(4), 406–471.

Liu, B., & Rundensteiner, E. A., (2005). Revisiting pipelined parallelism in multi-join query processing. *VLDB*, 829–840.

Liu, B., Zhu, Y., Jbantova, M., Momberger, B., & Rundensteiner, E. A. (2005). A dynamically adaptive distributed system for processing complex continuous queries. *VLDB*, 1338–1341.

Lo, M.-L., Chen, M.-S., Ravishankar, C. V., & Yu, P. S. (1993). On optimal processor allocation to support pipelined hash joins. *SIGMOD Conference*, 69–78.

Madden, S., Shah, M. A., Hellerstein, J. M., & Raman, V. (2002). Continuously adaptive continuous queries over streams. *SIGMOD Conference*, 49–60.

Mehta, M., & DeWitt, D. J. (1995). Managing intra-operator parallelism in parallel database systems. *VLDB*, 382–394.

Olston, C., Reed, B., Srivastava, U., Kumar, R., & Tomkins, A. (2008). Pig latin: A not-so-foreign language for data processing. *SIGMOD Conference*, 1099–1110.

Robinson, E., & DeWitt, D. J. (2007). Turning cluster management into data management; A system overview. *CIDR*, 120–131.

Ronnie Chaiken, P.-A. L. B. R. D. S. S. W., Jenkins, B., & Zhou, J. (2008). Scope: Easy and efficient parallel processing of massive data sets. *VLDB*.

Schneider, D. A., & DeWitt, D. J. (1989). A performance evaluation of four parallel join algorithms in a shared-nothing multiprocessor environment. *SIGMOD Conference*, 110–121.

Shah, M. A., Hellerstein, J. M., Chandrasekaran, S., & Franklin, M. J. (2003). Flux: An adaptive partitioning operator for continuous query systems. *ICDE*, 25–36.

Shankar, S., & DeWitt, D. J. (2007). Data driven workflow planning in cluster management systems. *HPDC*, 127–136.

Zhou, J., Cieslewicz, J., Ross, K. A., & Shah, M. (2005). Improving database performance on simultaneous multithreading processors. *VLDB*, 49–60.

Zhu, Y., Rundensteiner, E. A., & Heineman, G. T. (2004). Dynamic plan migration for continuous queries over data streams. *SIGMOD Conference*, 431–442.

Part II
Architectures

Chapter 10
Enterprise Knowledge Clouds:
Architecture and Technologies

Kemal A. Delic and Jeff A. Riley

10.1 Introduction

This chapter outlines the architectural foundations of *Enterprise Knowledge Clouds* (EKC) (Delic & Riley, 2009), describing the underlying technological fabrics and then pointing at the key capabilities of the (hypothetical) intelligent enterprise operating in constantly evolving, dynamic market conditions. Our aim is to give readers of this chapter a better understanding of knowledge cloud architectural aims and practical insights into EKC technological components. Thanks to knowledge, the enterprise will know more, will act better and react sooner in changing environment conditions, ultimately improving its performance and enabling it to show better behaviour and measurable improvements.

The *Enterprise* is an organisational structure which may take varying forms in different domains and circumstances. For our purposes here, we consider that the enterprise is an operating business employing 5000 or more people, operating globally with revenues in excess of $1 Billion, and supported by appropriate IT capabilities and facilities. There is a long, ongoing debate over the value and impact of IT use in business operations, but we can easily imagine what would happen if a business enterprise suddenly finds itself without any IT systems.

Knowledge gives distinctive capabilities to living creatures, with humans being at the top of the hierarchical tree of life. Tacit knowledge enables perception, reflection and action as the basic features of any intelligent behaviour. Technology, on the other hand, enables capturing and reuse of tacit knowledge in explicit form. Much of what we know as 'knowledge management' is about transforming tacit knowledge into explicit and vice-versa. Intuitively is clear that knowledge plays the key role in each and every part of the business enterprise. Knowledge takes various forms, has variable value and makes varying impact, and requires different technologies to

K.A. Delic and J.A. Riley (✉)
Hewlett-Packard Co., New York, NY, USA
e-mails: {kemal.delic; jeff.riley}@hp.com

B. Furht, A. Escalante (eds.), *Handbook of Cloud Computing*,
DOI 10.1007/978-1-4419-6524-0_10, © Springer Science+Business Media, LLC 2010

deal with its entire, continues life cycle. It is 'enterprise knowledge' which makes the difference in operational tasks (automation) and strategic situations (decision making).

Cloud computing is an emerging architectural paradigm driven by the sharp drop of technology costs followed by radically improved performance (commoditisation) (ITU, 2010). Social changes and economic advances have created a huge number of consumers and producers of various content artefacts (text, photo, music, video, etc.) representing huge user clouds and large communities (in the order of 100s of millions people). Thus we see cloud computing developing on an unprecedented scale and dynamics on a global basis (CLOUDSCAPE, 2009).

Highly abstracted, company operations are described as the interplay between people, machines and processes, providing either tangible goods or consumable services (Delic, 2003). Depending upon the business context, one component might be dominant over others, while each will contain something which we could label as 'knowledge'. It is important to observe is that approximately 75% of the economic activity in most advanced countries is created by service industries where knowledge is the primary resource or ingredient, thus we have been hearing about 'knowledge-based economies' for many years now. We may conclude that the services economy is driven by the power of knowledge.

We postulate that *Knowledge Clouds* (KC) will enable the global spread of economic growth, efficient delivery of services, smoother exchange and profitable trade of goods and services.

10.2 Business Enterprise Organisation

The typical business enterprise is an hierarchical organisation which has certain characteristics of military command-control layout with executives at the top of the hierarchy (numbered in tens) senior managers and managers at the next level (numbered in hundreds) and employees at the base (numbered in thousands). Depending upon the industry branch and regional specifics, it might be that certain functions are global and others regional. This usually leads to a characteristics matrix organisation which is marked by high complexity.

At the conceptual level, we can talk about key entities as Clients and Customers, Partners and Suppliers, interconnected to the business enterprise via distribution channels and supply chains (Fig. 10.1). Internally, the enterprise will have shared functions such as Human Resources, Finance, R&D Labs, IT, Sales and Marketing synchronised with Production and Services. Specialised enterprise software (e.g. CRM, ERP, SCM etc.) enables smooth operations of the enterprise and represent typical enterprise software applications today. For large enterprises, those are key systems requiring many years to perfect and require a large effort to operate. Each system encompasses knowledge embodied either as human expertise, business processes, software algorithms or analytical models. Enterprise IT plays a special, technological role for which KM will have distinctive value and lasting importance.

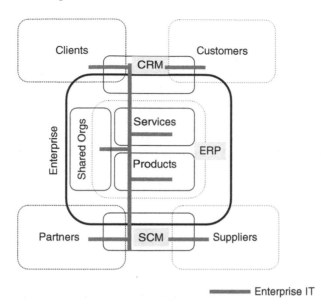

Fig. 10.1 Conceptual business enterprise organisation

From the perspective of technology components, Enterprise IT can be abstracted with the key components having their own operational indicators, such as dollars-per-call for the help-desk or cents-per-event for processing, enabling management and administrators to grasp inefficiencies and estimate the overall cost (Fig. 10.2). The ultimate objective is to minimise the cost while maximising the efficiency of the each IT unit, considering that data centres are machine intensives, help-desks are

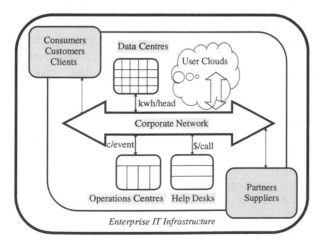

Fig. 10.2 Conceptual enterprise IT organisation

labour intensive, and operational centres and corporate networks are event intensive units.

To enable synchronised and orchestrated changes, enterprise architecture captures the overall state of the enterprise business and IT infrastructure and provides guidance expressed with a set of architecting principles.

Enterprise infrastructure used for business operations represents interconnecting, mediation fabrics which improve operational behaviour captured and indicated via key performance indicators. Removal of the IT fabrics will cripple business operations, demonstrating that today's business operations are not imaginable without the deployment of IT. In fact, the majority of businesses consider IT as a business-critical component which must be cleverly architected and well designed as a highly dependent part of the business.

10.3 Enterprise Architecture

Enterprise architecture is a strategic framework that captures the current state of the enterprise business and supporting IT, and outlines an evolutionary path towards the future state of the business and IT (Delic, 2002b). It is a very hard challenge to provide synchronised development of business and IT in dynamic and unpredictable market conditions. Thus, having a sound enterprise architecture that charts evolutionary change over 3–5 years is an important, competitive advantage. In simplistic terms, enterprise architecture is a model depicting the evolution of business, infrastructure, applications and data landscapes over 3–5 years. Each artefact of the enterprise architecture has a very high monetary value, a strong proprietary nature, and a vital importance for the future of the enterprise.

In a simplistic fashion, these artefacts can be represented in an hierarchical manner, implying the type of models appropriate for each layer, characteristic entities and key metrics (Fig. 10.3). Enterprise architecture can be viewed as a global strategic plan which will synchronise business evolution with IT development and ensure that future needs are properly addressed. As such, enterprise architecture represents the most valuable strategic planning item for enterprise executives and management: business can plan underlying technology changes after observing forthcoming technology shifts (Zachman, 1999). Such technology changes will critically improve or cripple business performance, and mature businesses should keep those plans private, current and sound.

The expanded landscape shown in Fig. 10.4 indicates some characteristic operational figures for the very large, global business enterprise, and illustrates the scope, scale and complexity of operations (Delic, 2002a). It is intuitively clear that such a complex environment contains several points of inefficiency and structural weakness which could be best dealt with via deployment of KM techniques. A more developed enterprise landscape (e.g. Delic, Greene, & Kirsch, 2010) will also contain data points, indicate the dynamics and spread of data flows, and show the key technological, business and market indicators.

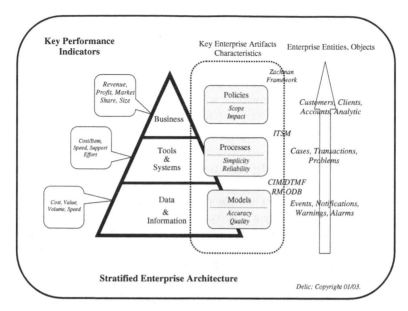

Fig. 10.3 Simplified enterprise/business/IT artefacts

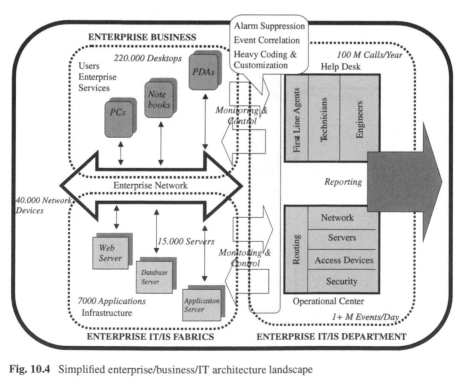

Fig. 10.4 Simplified enterprise/business/IT architecture landscape

As an example, for an enterprise business to efficiently handle in the order of 100 million calls per year, it should capture and deploy knowledge about callers, encountered problems and solution procedures. Similarly, to configure, manage and maintain in the order of 40,000 network devices, an enterprise business needs deep and reliable knowledge about network topology, fault behaviours and overall traffic flows. For each domain, knowledge will have different capture paradigms and technology, and will have a different impact on internal IT performances and cost.

10.4 Enterprise Knowledge Management

Enterprises as large, distributed and complex entities have several points of inter-operations with their environment which could be improved via deployment of KM applications. Knowledge about clients and customers will improve financial results and customer satisfaction. Knowing partners and suppliers better will help improve cooperation. Internal systems may help harvest employees ideas which then might be transformed into valuable intellectual property. As previously indicated, the IT domain is especially suitable for the deployment of KM systems, and this is an area in which the authors both have many years experience. We describe in detail three examples of KM deployment for internal IT operations, decision support, and knowledge harvesting.

An example of KM deployment in the IT domain is in the use of various knowledge repositories and systems to resolve a range of IT problems (Delic et al., 2010). Following a problem event from the IT infrastructure we see from Fig. 10.5 that

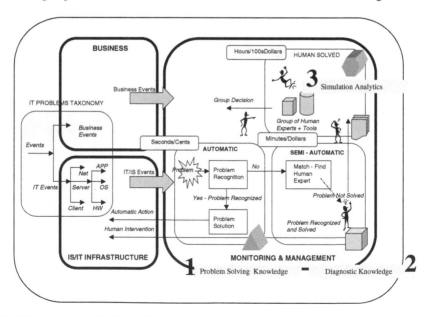

Fig. 10.5 Knowledge for IT problem solving

(1) problem recognition software will search a knowledge base containing problem solving knowledge, and if it recognises and identifies the problem it will deploy the solution found in the knowledge base; otherwise, (2) a knowledgeable human expert will be identified and, after deploying the diagnostic procedure, knowledge will be forwarded into automatic problem solving layer. For more complex, intricate or inter-dependent problems, (3) a group of human experts will be engaged to use knowledge captured in simulation analytics to resolve the problem via group decision making.

As millions of problems are solved daily, it is clear that the cost and speed of problem resolution are important parameters that illustrate the value of KM deployment for IT operations. At the very high level of abstraction, we see transformation of the raw data into information and then into knowledge and problem-solving acts, having measurable business impact and monetary value. It is important to note that KM techniques serve important roles in support and services, and that technologies deployed mainly originated from the field of Artificial Intelligence.

Another example of knowledge deployment is in decision support systems for enterprise operations, based on Enterprise Management Analytics (Casati et al., 2004; Cerri et al., 2008). We depict a layered IT architecture serving the business to orchestrate operations with clients and customers while being supported by suppliers and partners (Fig. 10.6). Those layers have distinctive architectures dictated by the general intent, so that all events from the instrumentation layer are served in

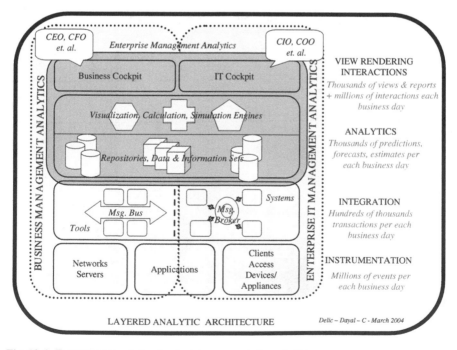

Fig. 10.6 Enterprise knowledge for decision support via analytics

timely manner and never missed; transactions in the integration layer are captured and never lost; and analytics in the interaction layer are always delivered and never inaccurate.

Decision support is provided via portals embodied as business and IT cockpits for executives, an operational workbench for managers, and working spaces for employees. Knowledge is captured in enterprise management analytics. This is yet another enterprise architecture landscape which combines the three-layers stratification principle with analytics technologies to illustrate the current state of enterprise KM systems for dependable and effective decision support.

Harvesting of employees' ideas represents an important activity as it may spawn the seeds of valuable new processes, inventive technologies, or innovative solutions. After initial triage and assessment, ideas could be suitable for transformation into valuable intellectual property – as patents for example. Figure 10.7 depicts a hypothetical example, illustrating that a large brainstorming exercise, or grand challenge, can create big idea clouds which could be harvested, transformed and potentially monetised. It is an illustration of KM deployed for innovation on a mass scale, where emerging cloud computing facilities may enable rescaling of these processes by orders of magnitude (Delic & Fulgham, 2004).

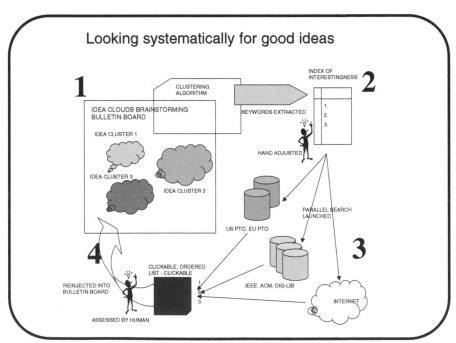

Delic & Fulgham – Dec 2004

Fig. 10.7 Knowledge harvesting and transformation: conceptual IP architecture

Figure 10.7 depicts, in stylised form, that (1) ideas are created, spawned and enriched; (2) the ideas are then organized and ordered by potential value or interestingness; then (3) the ideas are compared and measured against similar ideas in patent repositories, document libraries or internet documents; and finally, (4) the ideas are either refined and formalised or re-injected into yet another round of brainstorming.

Some large companies have arranged intensive sessions or grand challenges creating more than 100,000 ideas in a very short period of time – so the next possible challenge will be in the automation of processes related to triage, evaluation, valuation and formalization of the assessments of ideas. Due to this automation, the amount of innovation knowledge captured will be extremely large, and the power of scope and scale of such a system and its potential monetary value can only be imagined.

10.5 Enterprise Knowledge Architecture

Contemporary enterprise applications usually reside in data centres and have a typical stacked architecture (Fig. 10.8). Web servers manage interactions, deliver content and capture traces (front-end system) for enterprise applications residing in application servers (middleware). It is common for the databases capture events, transactions and analytics in the back-end system. To deal with high load and transient peaks, load-balancers are installed on the front-end and SAN (Storage Area Networks) for archiving in the back-end.

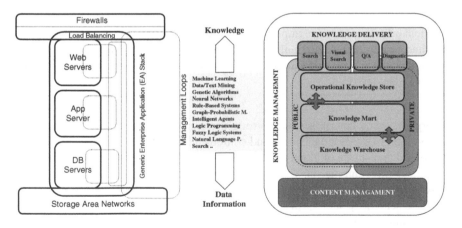

Fig. 10.8 Enterprise knowledge management stack

The abstracted enterprise knowledge management system can also be shown with a stacked architecture (Fig. 10.8), and here we recognise three characteristic layers: front-end, middle/application layer and back-end. We indicate a whole slew of technologies (middle/application layer) originating from Artificial Intelligence research which represent the essence of many KM applications (Cannataro & Talia, 2003;

Delic, Riley, Bartolini, & Salihbegovic, 2007). Knowledge is delivered via various types of portals, either to registered internal or external users or anonymous web consumers. It is typically the case that knowledge users can become knowledge producers via various discussion forums. Another channel of delivery and exchange of knowledge is via machine-to-machine exchanges.

The knowledge (idea) harvesting system described previously can be implemented as an enterprise application with three layers in which content (as externalised knowledge) is being processed and stored within three logical knowledge layers. The Operational Knowledge Store provides rapid access; the Knowledge Mart is an intermediate knowledge repository; and archived knowledge is stored in the Knowledge Warehouse.

We should stress that all conceptual drawings show an architecture which can be materialised with different logical and physical architectures, depending on the deployment domain and choice of key technologies – KM technologies such as content management, enterprise search, delivery portals, discussion forums as key enterprise components glued together via Service Oriented Architecture (SOA) into service delivery fabrics.

10.6 Enterprise Computing Clouds

Cloud computing is the next evolutionary step in the distributed computing field enabled by:

- radical price/performance improvement leading to commoditization
- technology advances with multi-core and energy-aware chip designs
- architectural interplay of warehouse-scale computing and huge number of intelligent edge devices

Large business enterprises have strong incentives to consider their architecture plans in light of developments in cloud computing (Sun Microsystems, 2009).

A possible instance of cloud computing serving billions of users can be depicted as the next wave internet in which the number of devices, gadgets and things can easily surpass 10 billion items (Delic, 2005; Delic & Walker, 2008) (Fig. 10.9). It will be served by strategically placed data centres federated into grids via efficient communication fabrics. In data centres clusters of various sizes (hundred to thousands of machines) will be dynamically allocated to handle varying enterprise workloads. At the chip level, programming of multi-core will become the principal preoccupation of designers aiming at energy efficient designs.

The entire chain from the chips, via racks, clusters and data centres should be designed with cloud computing in mind. The same should apply for the software design. At the level of large aggregation of grids, entirely new economics and legal concerns can govern traffic flows, data storage and choice of application execution

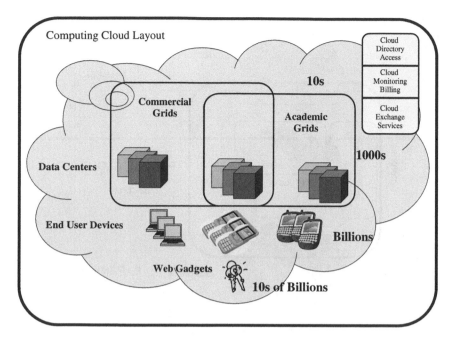

Fig. 10.9 Computing cloud layout

location. This represents a wide number of unresolved issues and particularly hard challenges.

The business reality of the enterprise and constantly changing market conditions will dictate specific choices of the Enterprise Knowledge Clouds, which we describe in the next section.

10.7 Enterprise Knowledge Clouds

Taking into account the current organisational layout of the typical enterprise, we can outline a generic architecture for enterprise clouds. This generic architecture has three principal architectural layers: *private*, *partner* and *public* cloud. We postulate that knowledge management techniques will be appropriately spread over the each and every enterprise cloud. This natural separation is dictated by the required capability of each cloud: security and privacy is a must for the private cloud; availability and reliability is a precondition for the partner cloud; and rescaling and coverage is important for the public cloud (Fig. 10.10). These requirements will be not only guiding principles but also design criteria for enterprise clouds.

We can easily imagine that Finance, Human Resources and R&D Labs will be the prime candidates for the *private* enterprise KM cloud. The Supply Chain and Delivery organisations will naturally fall into the *partner* cloud; while Sales, Marketing, Public Relations and Publicity would be natural fits for the *public* cloud.

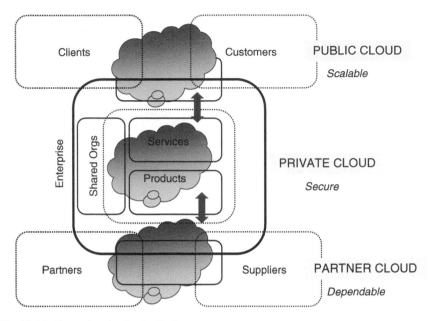

Fig. 10.10 Enterprise clouds stratified

The private, partner and public enterprise clouds should have the facility to inter-operate and exchange data, information and knowledge on a regular and intensive basis. The choice of technologies for the enterprise clouds will be critical, and the emergence of suitable standards is keenly anticipated.

It is not expected that the large enterprises will switch overnight onto cloud computing fabrics, but we expect they will start to gradually deploy cloud-based applications for a few, carefully selected domains. Each previous wave of enterprise technologies has gone through the *prototype-test-deploy* cycle, and cloud technol-ogy will not be different. It is also during this time that choices of key cloud technologies respecting the ultimate capability for each cloud type will be made.

Monitoring, measurement and calculation of key performance parameters for each enterprise cloud should be undertaken in order to measure the impact of the new cloud architecture on enterprise performance, and justify investment in the new technologies.

10.8 Enterprise Knowledge Cloud Technologies

Figure 10.11 depicts an abstracted cloud architecture and shows three principal groups of technologies that provide *virtualisation*, *automation* and *scheduling*. *Virtualisation* (of hardware and software) will provide better use of resources; *Automation* will lower support costs and improve dependability of the clouds; and *Scheduling* will enable economics-based reasoning about the use of resources and

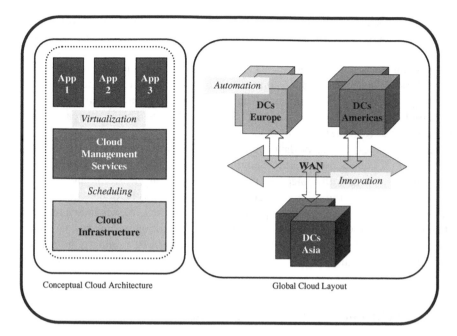

Fig. 10.11 Cloud computing technologies

dispatching of the enterprise workloads. Finally, we assume that the innovation in communication part of the clouds should lead to breakthrough improvements.

All these technologies could be categorised as open-source, proprietary or hybrid. It is beyond the scope of this discussion to delve deeper into detail, but we point out that these choices are critical for real-world deployment. Intuitively we would suggest open-source technology for the public cloud, proprietary for the private cloud, and hybrid for the partner cloud, but we also recognise that this is a difficult problem and once put into a real-world application context, the choices might not be right nor will alternatives be obvious.

The choice of technologies for enterprise clouds will be the difference between success and failure. We expect that the private, public and partner clouds will inter-operate, so the choice of technology should take into account existing or emerging standards which will enable future flows among clouds. All interested parties in this domain have reasons to participate in establishing standards: some self-serving, some altruistic. We expect to see some robust negotiations between the "proprietary" and "open" camps.

Looking back over the short history of cloud computing, we can identify some early platforms on which a very high number of users have developed and deployed a large number of applications. Thus we expect that the leading vendors will try to create very large scale platforms that will attract millions of developers. Platform development usually means the choice of programming language and associated framework. This has advantages for developers and consumers, but also for the vendors in that it tends to lock developers into a single platform. Interoperability of platforms will pose some of the greatest challenges for cloud computing in the future.

Ongoing technology developments are especially noticeable, and sometimes targeted, by small companies which aim to exit the business via a sale to an established, global vendor. We have seen this happen already with some small companies having been sold for (up to) several hundreds of million of dollars. This is an area for future and exciting developments.

Automation, especially of data centres, will represent the most intricate part of the cloud, as it must address multiple engineering issues and big challenges in which the ultimate goal is multi-objective: the maximisation of utilisation and monetary benefits, and the minimisation of energy cost. All this is to be done whilst guaranteeing dependability and achieving performance objectives. We see here a long road of future research, engineering development and technology innovation (Armbrust et al., 2009). All this will be a critical path to see enterprises running the majority of their business in the cloud.

10.9 Conclusion: Future Intelligent Enterprise

If we take a longer perspective look into past technology developments and business evolution, we observe some distinct phases characterised by a single word to describe an entire technology epoch (Delic & Dayal, 2002). For the automotive industry *automation* of production was the key technological advance; *integration* for the aviation industry and aircraft production; *optimisation* for e-commerce; and for the forthcoming service industry, it is *adaptation* (Delic & Greene, 2006) (Fig. 10.12).

Adaptive behaviour is a characteristic of living systems, while businesses are hybrid systems combining people, technology and processes into orchestrated whole. We believe that the injection of technologies will improve interconnectivity, reduce latencies and increase speeds while improving the problem solving capabilities based on a higher 'knowledge density'. We call such an enterprise an *Intelligent Enterprise* to denote improved behaviour and the ability to adapt and survive in changing circumstances. While we do not (yet) compare this artefact to intelligent living creatures, the analogy is clear.

We postulate that the synergy among big data, big mobile crowds and large infrastructures will lead to unprecedented improvements in the key indicators above (Fig. 10.12).

The emerging cloud computing paradigm embodied in useful applications for the enterprise knowledge management offers:

- radical cost reduction
- a great ability to scale
- much improved agility

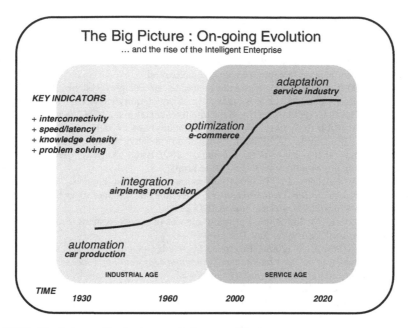

Fig. 10.12   Ninety years of technology evolution – market view

As such, it might be a good showcase for cloud computing success. In practical terms, enterprise architecture will evolve towards cloud architecture, and all architectural components and layers will be impacted and adapted accordingly.

New technologies will enable the next wave of business models, impact market developments, and see the rise of much changed business enterprise.

References

Armbrust, M., Fox, A., Griffith, R., Joseph, A. D., Katz, R. H., Konwinski, A., Lee, G., Patterson, D. A., Rabkin, A., Stoica, I., & Zaharia, M. (2009). *Above the clouds: A Berkeley view of cloud computing* (University of California, Berkeley, Tech. Rep. UCB/EECS-2009-28).

Cannataro, M., & Talia, D. (2003). The knowledge grids. *Communications of the ACM, 46*(1), 89–93.

Casati, F., Castellanos, M., Chan, P. K., Dayal, U., Delic, K. A., Greene, W. M., et al. (2004). Enterprise management analytics. *Proceedings of the Eleventh OpenView University Association Workshop (OVUA04), Paris, France.*

Cerri, D., Valle, E. D., Marcos, D. D. F., Giunchiglia, F., Naor, D., Nixon, L., Teymourian, K., Obermeier, P., Rebholz-Schuhmann, D., Krummenacher, R., & Simperl, E. (August 2008). *Towards knowledge in the cloud* (Tech. Rep. DISI-08-30, University of Trento, Italy).

CLOUDSCAPE (October 2009). The 451 group and Tier1 research. *The cloud codex,* CLOUDSCAPE. Available from: http://www.the451group.com/cloudscape/451_cloudscape.php.

Delic, K. A. (2002a). Enterprise IT complexity. *ACM Ubiquity, 2002*(Issue January). Available from: http://www.acm.org/ubiquity/views/k_delic_1.html.

Delic, K. A. (2002b). Enterprise models, strategic transformations and possible solutions. *ACM Ubiquity, 2002*(Issue July). Available from: http://www.acm.org/ubiquity/views/k_delic_2.html.

Delic, K. A. (2003). IT Services = People + Tools + Processes. *ACM Ubiquity, 4*(37). Available from: http://www.acm.org/ubiquity/views/v4i37_delic.html.

Delic, K. A. (2005). Science and engineering of large-scale complex systems. *ACM Ubiquity, 2005*(Issue March). Available from: http://www.acm.org/ubiquity/views/v6i8_delic.html.

Delic, K. A., & Dayal, U. (2002). The rise of the intelligent enterprise. *ACM Ubiquity, 2002*(Issue December). Available from: http://www.acm.org/ubiquity/views/k_delic_4.pdf.

Delic, K. A., & Dayal, U. (2004). Adaptation in large-scale enterprise systems: Research and challenges. *ACM Ubiquity, 2004*(Issue August). Available from: http://www.acm.org/ubiquity/views/v5i23_delic.html.

Delic, K. A., & Fulgham, M. T. (2004). Corporate innovation engines: Tools and processes. *Proceedings of the 5th International Conference on Practical Aspects of Knowledge Management (PAKM2004), Vienna, Austria*, 220–226 .

Delic, K. A., & Greene, W. M. (2006). Emerging sciences of service systems. *Proceedings of the R&D Management Conference, New Delhi*, 140–144.

Delic, K. A., & Riley, J. A. (2009). Enterprise knowledge clouds: Next generation KM systems?. *Proceedings of the 2009 International Conference Information, Process, and Knowledge Management (eKnow 2009), Cancun, Mexico*, 49–53.

Delic, K. A., & Walker, M. A. (2008). Emergence of the academic computing clouds. *ACM Ubiquity, 2008*(Issue August). Available from: http://www.acm.org/ubiquity/volume_9/v9i31_delic.html.

Delic, K. A., Greene, W. M., & Kirsch, H.-J. (2004). Intelligent routing of enterprise IT Events. *InfoManagement direct*, February 2004, Retrieved March 10, 2010, from http://www.information-management.com/infodirect/20040227/8166-1.html.

Delic, K. A., Riley, J. A., Bartolini, C., & Salihbegovic, A. (2007). Knowledge-based self-management of apache web servers. *Proceedings of the 21st International Symposium on Information, Communication and Automation Technologies (ICAT'07), Sarajevo, Bosnia, and Herzegovena.*

International Communication Union ITU (2009). *Distributed computing: utilities, grids & clouds* (ITU-T Tech. Watch Rep. 9, 2009). Retrieved 10 March 2010, from http://www.itu.int/dms_pub/itu-t/oth/23/01/T23010000090001PDFE.pdf.

Sun Microsystems (2009). *Introduction to cloud computing architecture.* White Paper.

Zachman, J. (December 1999). Enterprise architecture: The past and the future. *Information Management Magazine.* Available from: http://www.information-management.com/issues/19991201/1702-1.html.

Chapter 11
Integration of High-Performance Computing into Cloud Computing Services

Mladen A. Vouk, Eric Sills, and Patrick Dreher

Abstract High-Performance Computing (HPC) projects span a spectrum of computer hardware implementations ranging from peta-flop supercomputers, high-end tera-flop facilities running a variety of operating systems and applications, to mid-range and smaller computational clusters used for HPC application development, pilot runs and prototype staging clusters. What they all have in common is that they operate as a stand-alone system rather than a scalable and shared user reconfigurable resource. The advent of cloud computing has changed the traditional HPC implementation. In this article, we will discuss a very successful production-level architecture and policy framework for supporting HPC services within a more general cloud computing infrastructure. This integrated environment, called Virtual Computing Lab (VCL), has been operating at NC State since fall 2004. Nearly 8,500,000 HPC CPU-Hrs were delivered by this environment to NC State faculty and students during 2009. In addition, we present and discuss operational data that show that integration of HPC and non-HPC (or general VCL) services in a cloud can substantially reduce the cost of delivering cloud services (down to cents per CPU hour).

11.1 Introduction

The concept of cloud computing is changing the way information technology (IT) infrastructure is being considered for deployment in businesses, education, research

M.A. Vouk (✉)
Department of Computer Science, Box 8206, North Carolina State University, Raleigh, NC 27695, USA
e-mail: vouk@ncsu.edu

E. Sills
North Carolina State University, Raleigh, NC 27695, USA
e-mail: edsills@ncsu.edu

P. Dreher
Renaissance Computing Institute, Chapel Hill, NC 27517, USA
e-mail: dreher@renci.org

B. Furht, A. Escalante (eds.), *Handbook of Cloud Computing*,
DOI 10.1007/978-1-4419-6524-0_11, © Springer Science+Business Media, LLC 2010

and government. General interest in this technology has been increasing over the past several years as can be seen in a Google trends comparison graph of the number of search citations and references among grid computing, high performance computing and cloud computing (Fig. 11.1). This trend dramatically increased after about October 2007 when Google and IBM announced "cloud computing" research directions (Lohr, Google, & IBM, 2007) and IBM announced its cloud computing initiative (IBM, 2007).

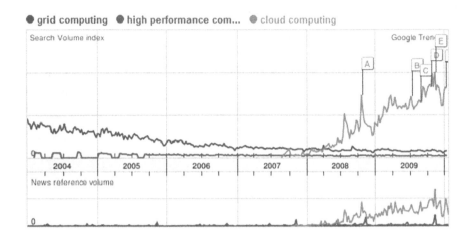

Fig. 11.1 The rise of cloud computing

This level of interest has made "cloud computing" a popular term that generally describes a flexible system providing users with access to hardware, software, applications and services. However, because there is no one generic user and the hardware, software, and services may be grouped in various combinations, this cloud computing concept has quickly fractured into many individualized descriptions and perspectives. As a result, it is very difficult to agree on one common definition of cloud computing (Dreher, Vouk, Sills, & Averitt, 2009; see also Armbrust et al., 2009; Vouk et al., 2009) that describes the number and types of cloud computing systems that are being deployed.

In the context of this chapter, we consider "cloud computing" as referring to a seamless component-based architecture that can deliver an integrated, orchestrated and rich suite of both loosely and tightly coupled on-demand information technology functions and services, significantly reduce overhead and total cost of ownership and services, and at the same time empower the end-user in terms of control. Some more obvious advantages of cloud computing are server consolidation, hardware abstraction via virtualization, better resource management and utilization, service reliability and availability, improved security and cost effectiveness.

While some organizations are still debating (e.g., Armbrust et al., 2009; Golden, 2009) if or how this technology might be applied, early adopters, such as NC State

University, have been providing cloud-based services to their students, faculty and staff with great success since 2004 (Dreher et al., 2009; Averitt et al., 2007; Seay & Tucker, 2010; Vouk et al., 2008, 2009; Vouk, 2008). Unlike some cloud computing deployments that tend to be equated to delivery of a single type or category of desktop service, specific server functionalities, applications, or application environments, the NC State cloud computing deployment (Virtual Computing Laboratory or VCL, http://vcl.ncsu.edu) offers capabilities that are very flexible and diverse ranging from Hardware-as-a-Service all the way to highly complex Cloud-as-a-Service. These capabilities can be combined and offered as individual and group IT services, including true High-Performance Computing (HPC) services. The VCL HPC implementation very successfully integrates HPC into the cloud computing paradigm by managing not only resource capabilities and capacity, but also resource topology, i.e., appropriate level of network/communication coupling among the resources.

In the remainder of the chapter we describe NC State University (NC State or NCSU) cloud computing environment (Section 11.2), the VCL cloud architecture (Section 11.3), integration of HPC into the VCL cloud architecture (Section 11.4), and the economic performance of the solution (Section 11.5). A summary is provided in Section 11.6.

11.2 NC State University Cloud Computing Implementation

A cloud computing system should be designed based on a service-oriented architecture (SOA) that can allocate resources on-demand in a location and device independent way. The system should incorporate technical efficiency and scalability through appropriate level of centralization, including sharing of cloud resource and control functions, and through explicit or implicit self-provisioning of resources and services by users to reduce administration overhead (Dreher et al., 2009; Vouk et al., 2009). One of the principal differences between "traditional" and cloud computing configurations is in the level of control delegated to the user. For example, in a traditional environment, control of resource use and management lies primarily with the service delivery site and provider, whereas in a cloud environment this control is for most part transferred to users through self-provisioning options and appropriate user delegated privileges and access controls. Similarly, other traditional functions such as operating system and environment specifications and mode of access and prioritizations now become explicit user choices. While this can increase management efficiency and reduce provisioning costs, the initial base-line configuration of a cloud computing system is more complex.

At NC State University VCL is a *high performance open-source award-winning*[1] *cloud computing technology* initially conceived and prototyped in 2004 by NC State's College of Engineering, Office of Information Technology, and Department

---

[1]2007 Computerworld Honors Program Laureate Medal (CHPLM) for Virtual Computing Laboratory (VCL), 2009 CHPLM for NC State University Cloud Computing Services

of Computer Science. Since then, VCL development has rapidly progressed in collaboration with industry, higher education, and K-12 partners to the point that today it is a large scale, production-proven system which is *emerging as a dominant force in the nascent and potentially huge open-source private-cloud market* (Seay et al., 2010; Schaffer et al., 2009; Vouk et al., 2009).

The experience at NC State University has shown that for education and research cloud computing implementations a flexible and versatile environment is needed to provide a range of differential services from Hardware-as-a-Service all the way to Cloud-as-a-Service and Security-as-a-Service. In the context of NC State's VCL we distinguish we define each of these services as follows

- Hardware as a Service (HaaS) – On-demand access to a specific computational, storage and networking product(s) and/or equipment configuration possibly at a particular site
- Infrastructure as a service (IaaS) – On-demand access to user specified hardware capabilities, performance and services which may run on a variety of hardware products
- Platform as a service (PaaS) – On-demand access to user specified combination of hypervisors (virtualizations), operating system and middleware that enables user required applications and services running on either HaaS and/or IaaS
- Application as a Service (AaaS) – On-demand access to user specified application(s) and content. Software as a Service (SaaS) may encompass anything from PaaS through AaaS
- Higher level services – A range of capabilities of a cloud to offer a composition of HaaS, IaaS, PaaS and AaaS within an envelope of particular policies, such as security policies – for example Security-as-a-Service. Another example are composites and aggregates of lower-level service such as a "Cloud-as-a-Service" – a service that allows a user to define sub-clouds (clusters of resources) that the user controls in full.

At some level all of the above services are available to the NC State VCL users, commensurate with the level and type of privileges granted to the user (Vouk et al., 2008; Vouk et al., 2009). If one wishes to construct high-performance computing (HPC) services within VCL with a particular topology, or to have the ability to deliver specific end-to-end quality of service, including application performance, it is essential to grant users both HaaS and IaaS access. We find that a carefully constructed cloud computing implementation offering the basic services listed above can result in good technical performance and increased productivity regardless of whether the cloud computing system is serving commercial customers, educational institutions, or research collaboration functions.

Campus use of VCL has expanded exponentially over the last five years (Fig. 11.2). We now have over 30,000+ users and deliver over 100,000 reservations per semester through over 800 service environments (amounting about 500,000 CPU hours annually). In addition, we deliver over 8.500,000 HPC CPU hours annually. In-state initiatives include individual UNC-System universities (e.g., ECU,

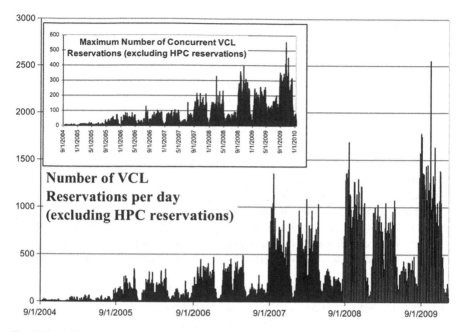

Fig. 11.2 VCL usage

NCCU, UNC-CH, UNCG, WCU – technically all UNC System campuses which implement Shibboleth authentication have access to VCL), the NC Community College System (production and pilots in 15 colleges: Beaufort County, Brunswick, Cape Fear, Catawba Valley, Central Piedmont, Cleveland, Edgecombe, Fayetteville Tech, Forsyth Tech, Guilford Tech, Nash, Sandhills, Surry, Wake Tech), and several K-12 pilots and STEM initiatives.

In addition to multiple VCL deployments in the State of North Carolina, regional, national and international interest in VCL has increased dramatically after VCL was accepted as an incubator technology and posted as an open source implementation through the Apache Software Foundation (Apache VCL, 2010). Educational institutions such as George Mason University (GMU) have become a VCL leader and innovator for the Virginia VCL Consortium, recently winning the 2009 Virginia Governor's award for technology innovation, schools such as Southern University Baton Rouge and California State University East Bay are in the process of implementing VCL-based clouds. In addition to numerous deployments within the United States, VCL cloud computing implementations, including HPC configurations, have now been deployed world-wide and are providing a rich mix of experiences working with HPC in a cloud computing environment.

VCL's typical base infrastructure (preferred but not required) is an HPC blade system. This type of system architecture provides capabilities for HPC delivery either as a whole or "sliced and diced" dynamically into smaller units/clusters. This allows the VCL cloud to be appropriately "packaged" to partition research clusters and sub-clouds and true high-performance computing from other sets of highly individualized requirements such as single desktop services, groups of "seats" in classrooms, servers and server farms. These flexible configurations permit hardware infrastructure and platforms as services to be simultaneously delivered to users within the overall VCL cloud computing cyberinfrastructure, with each of these services being capable of customized provisioning of software and application as a service.

Figure 11.2 shows the number of VCL reservations made per day by users over last five years. This includes reservations made by individual students for Linux or Windows XP desktops along with some specific applications, but also reservations that researchers may make to construct their own computational sub-clouds, or specialized resource aggregations – including high-performing ones. Figure 11.2 inset is the number of such concurrent reservations per day. What is not included in the data shown in these figures are the reservations that deliver standard queue-based VCL HPC services. We therefore label these as non-HPC (or general-VCL) services, although self-constructed high-performance sub-clouds are still in this category of services.

NC State's VCL currently has about 2000 blades distributed over three production data centers and two research and evaluation data centers. About one third of the VCL blades are in this non-HPC service delivery mode, some of the remaining blades are in our experimental test-beds and in maintenance mode, and the rest (about 600–800) operate as part of the VCL-HPC (http://hpc.ncsu.edu) and are controlled through a number of LSF[2] queues.

There three things to note with reference to Fig. 11.2:

1. VCL usage, and by implication that of the NC State Cloud Computing environment, has been growing steadily
2. Resource capacity (virtual or bare-machine loaded) kept on the non-HPC side at any one time is proportional to the needed concurrency
3. Non-HPC usage has clear gaps

The number of VCL reservations tends to decrease during the night, student vacations and holidays. On the other hand, if one looks at the NC State demand for HPC cycles we see that it has been growing (Fig. 11.3), but that it is much less subject to seasonal variations (Fig. 11.4).

An in depth analysis of the economics of cloud computing (Dreher et al., 2009) shows one of the key factors is average utilization level of the resources. Any of our resources that are in VCL-HPC mode are fully utilized (with job backlog queues as long as two to three times the number of jobs running on the clusters). In 2009 (including maintenance down time) VCL-HPC recorded nearly 8.5 million HPC CPU-hours which is greater than a 95+% utilization level, while desktop augmentation and similar non-HPC usage recorded 550,000 CPU-hours (over 220,000 individual reservations) yielding about a 10–15% utilization.

---

[2]http://www.platform.com/Products

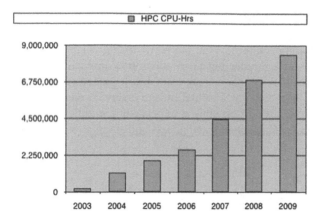

Fig. 11.3 NC State HPC usage over years

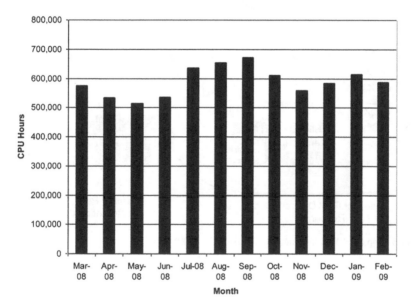

Fig. 11.4 NC State HPC usage over March 2008 – February 2009

To satisfy high demand for HPC, VCL is designed to shift hardware resources from the non-HPC desktop to HPC. To balance workloads during the times when non-HPC use is lower, such as during summer holidays, VCL can automatically move the resources into its HPC mode where they are readily used. When there is again need for non-HPC use, these resources are again automatically moved back in that use pool. As a result, the combined HPC/non-HPC resource usage runs at about 70% level. This is both a desirable and cost-effective strategy requiring an active collaboration of the underlying (cloud) components.

11.3 The VCL Cloud Architecture

The first level architecture of VCL is shown in Fig. 11.5. A user accesses VCL through a web interface, selecting a combination of applications, operating systems and services. These combinations are available in image repositories. The images can be either a bare-metal or a virtual machine service environment consisting of the operating system (with possibly a hypervisor), middleware, application stack along with security management, and access and storage tools and agents.

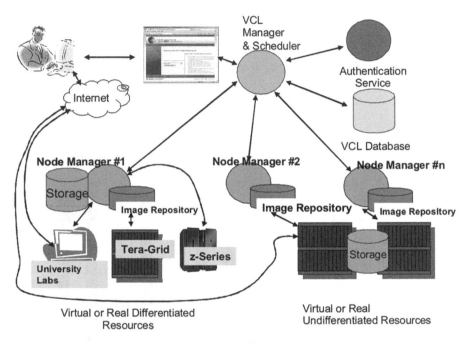

Fig. 11.5 Top level VCL architecture (Vouk et al., 2009)

If a specific combination of services and applications are not available as an "image" a user with appropriate authorizations can construct a customized image from the VCL library components. The VCL manager software then maps the request from the user onto available software application images and available hardware resources. The manager then schedules the user request for either immediate use (on demand) or for later use. The VCL manager software was developed by NCSU using a combination of off-the-shelf products (such as IBM xCAT[3]) and in-house developed open-source "glue" software (about 50,000+ lines of code – now available through Apache (2010)).

---

[3] http://xcat.sourceforge.net/

All components of the VCL can be (and are) distributed. A site installation will typically have one resource management node (Node Manager in Fig. 11.5) for about 100 physical blades. This ensures adequate image load times as well as resource fail-over redundancy. In the context of our architecture we distinguish between dirrerentiated and undifferentiated resources. Undifferentiated resources are completely malleable and can be loaded with any service environment designed for that platform. Differentiated resources can only be used "as is" without any modifications beyond the configured user access permissions. For example, during the night VCL users are allowed to use NC State computing lab Linux machines remotely, but they are not allowed to modify their images since lab machines are considered a differentiated resource. In contrast, when they make a "standard" VCL Linux or Windows reservation, students get full root/administrator privileges and can modify the image as much as they wish. However, once they are finished (typically our student reservations last 1–2 h) that resource is "wiped" clean and a fresh image is reloaded. Students save their data either onto our network-attached corporate storage or on their own laptops.

In this fashion VCL provides the ability to deliver scheduled and on-demand, virtualized and "bare-metal" infrastructures, and differentiated and undifferentiated services consistent with established NC State policy and security requirements which may vary by user and/or application. VCL dynamically constructs and deconstructs computing environments thereby enabling near continuous use of resources. These environments, consisting of intelligent software-stack image(s) and metadata specifications for building the environment, can be created, modified, and transferred as policy and authorization permit. VCL security capabilities enable wide latitude in the assignment of these permissions to faculty, staff, and students.

VCL code version 2.x has been available for about a year as part of the Apache offering (Apache VCL, 2010). Version 2.x of VCL represents a major rewrite of the base code and moves VCL to a modular software framework in which functional elements can be modified, replaced or optioned without attendant code changes elsewhere. It greatly empowers both community contribution and non disruptive customization. In addition, its already excellent security profile is being constantly enhanced through a federally funded Secure Open Source Initiative (http://sosi.ncsu.edu).

VCL environments are stored in on-line repositories, providing a low-cost high-volume retention capability that not only supports extreme scaling of access and reuse but also enables the breath of scale and scope required for intelligent real-time sequencing of multi-stage workflows. The benefit of this advance varies depending on the limitations of use imposed by software licensing agreements. Absent these limitations VCL empowers a new paradigm of *build once well and pervasively reuse*. In fact, one of the special characteristics of VCL is that its provenance and meta-data collection is sufficiently fine-grained and thorough that it allows very detailed metering of the software usage – by user, by department, by duration, by location, etc. That in itself offers an opportunity to implement a metering-based license management model that, given appropriate vendor agreements, allows porting and exchange of service environments across/among clouds.

11.3.1 Internal Structure

Figure 11.6 shows more of the internals of the VCL architecture. At the heart of the solution are (a) a user interface (including a GUI and a remote service API), (b) an authorization, provenance and service tracking data-base, (c) a service environment "image" library, and (d) a service environment management and provisioning engine.

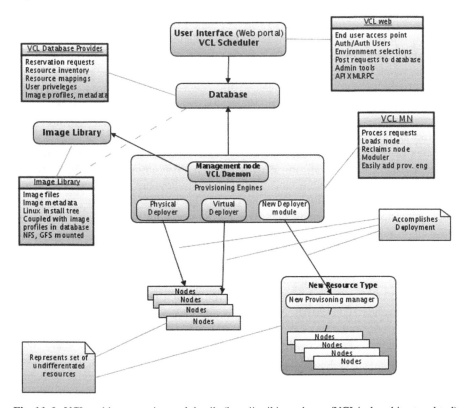

Fig. 11.6 VCL architecture – internal details (http://cwiki.apache.org/VCL/vcl-architecture.html)

Provisioning engine deploys service environments on demand either at the physical layer (bare-machine loads are typically done using xCAT, but other loaders can be used, including IBM Director), or at a virtual layer (e.g., VMWare, Xen, KVM "images"), or through a specially defined mechanism, e.g., a new service interface, a remote service, a service API to another cloud. Deployed service environments can consist of a single bare-metal image or a virtual software stack "image," such as a Windows or Linux desktop loaded onto undifferentiated resources, or it can consist of a collection of images (including their interconnection topology) loaded onto a set of (possibly) hybrid resources, or it can consist of set of HPC images loaded onto VCL resources being moved into differentiated HPC mode, and so on. In the NC

State implementation, physical server manager loads images to local disk via kick-start (only Linux environments), copies images to local disk (Linux and Windows), or loads images to local memory (stateless). For loading of virtual machine images, VCL leverages command-line management tools that come with hypervisors.

11.3.1.1 Storage

Where information is stored in the cloud is very important. Secure image storage is part of the core VCL architecture, however end-user storage is more flexible. It can range from storage on the physical VCL resource, to secure NAS or SAN accessed via the storage access utilities on the image itself, to storage on the end-user access station (e.g., laptop storage or memory key storage), etc. At NC State most of our images are equipped with agents that can connect in a secure way to our corporate storage (AFS based) and thus access backed-up storage space that students and faculty are assigned, and tools (such as visual sftp) that can access other on-line storage. Our HPC images are constructed so that they all have access to HPC scratch and permanent storage via NFS. Figure 11.7 illustrates the current extent of that storage. It is interesting to comment on the part of the storage marked as "Partners".

Fig. 11.7 NC State (baseline VCL-HPC services) and Partner storage

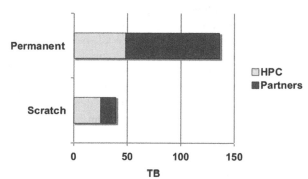

11.3.1.2 Partner's Program

NC State researchers have the option of purchasing VCL-HPC compatible hardware and any specialized or discipline-specific software licenses. NC State Office of Information Technology (OIT) provides space in an appropriate and secure operating environment, all necessary infrastructure (rack, chassis, power, cooling, networking, etc.), and the system administration and server support.

In return for infrastructure and services provided by OIT, when partner compute resources are not being used by the partner they are available to the general NC State HPC user community. This program has been working very well for us as well as for our researchers. As can be seen from Figs. 11.7 and 11.8, a large fraction of our HPC resources are partner resources.

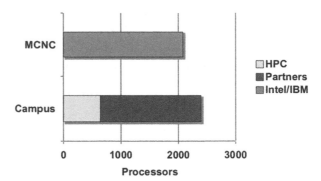

Fig. 11.8 NC State VCL-HPC distributed memory computing resources (HPC), Partner computational resources and resources acquired through gifts and joint projects with IBM and Intel

11.3.2 Access

As part of our cloud HPC services we have both distributed memory (typically IBM BladeCenter clusters) and shared memory computing resources (typically 4-socket quad core Opteron servers with at least 2 GB of memory per core). We also provide items such as resource manager/scheduler, compilers, debuggers, application software, user training and support, consulting, code porting and optimization help, algorithm development support, and general collaboration. There are two ways of reaching those resources – through VCL-based reservation of one's own login node, or through the use of a communal login node. Personal login nodes make sense if end-users wish to monitor their runs in real time. One submits jobs in the usual fashion using job queues, in our case controlled via LSF. Queue priority depends on the resources requested and partnership privileges. Partners get absolute and immediate priority on the resources they own (or an equivalent set of resources), and they get additional priority towards adding common resources beyond what they own.

11.3.2.1 Standard

All VCL-HPC resources run the same HPC service environment (typically RedHat-based), and have access to a common library of applications and middleware. However, users can add their own applications to the computational resources they are given access to. All our standard VCL-HPC nodes are bare-metal loaded for sole use on VCL blades. They are managed as differentiated resources, i.e., users have full control over them, but they cannot re-load them or change them (except for the software in the user's home directories), and they must be used with the NC State maintained scheduler and file system. Most of our HPC nodes operate in this mode and as such they are very similar to any other HPC cluster. Nodes are tightly coupled with 1 Gbps or better interconnects. A user with sufficient privileges can select appropriate run queues that then map the jobs onto the same BladeCenter chassis or same rack if that is desired, or onto low latency interconnects (e.g., Infiniband interconnected nodes).

11.3.2.2 Special needs

If a user does not wish to conform to the "standard" NC State HPC environment, a user has the option of requesting the VCL cloud to give him/her access to a customized cluster. In order to do that, the user needs to have "image creation" privileges (Vouk et al., 2008; Vouk et al., 2009), and the user needs to take ownership of that sub-cluster or sub-cloud service environment. First the user creates a "child image" – a data node or computational node image – running operating system of their choice as well as tools that allow communications with the cluster controller and access to the data exchange bus of their choice, e.g., NFS-based delivery of directories and files. The user saves that image.

Then the user creates a "parent image" in the VCL-cloud aggregation mode. Again the user picks the base-line operating system, and adds to it a cluster controller, such as an HPC scheduler of choice, e.g., PBS, or a cloud controller such as Hadoop's controller, or similar. Now the user attaches to this any number of "child images". Typically child images are of the same type, e.g., a computational HPC Linux image, if the user wishes to operate a homogenous cluster. But, the user can also attach different child images, say 20 computational Linux images, one Linux-based data-base image, one Windows web-services image, and so on. Then the user saves the "parent image". From now on, when the user loads the "parent or control image" all the children are also loaded into virtual or bare-machine resource slots, depending on how the child-images were defined. All Linux-based child images that are part of such a VCL aggregate know about each others IP numbers through a VCL placed/etc file. "Parent image" control software needs to know how to access that information and communicate with the children.

Default custom topology is random and loosely coupled, i.e., VCL maps the "parent" or anchor image, and its children onto resources on which the images can run, but it does not pay attention to inter-node latency or topology. If a specific topology is desired, such as tight low latency inter-image communication coupling suitable for HPC, and the image owner has appropriate privileges, mapping of the images onto nodes that conform to a particular topology or interconnect latency is possible.

11.3.3 Computational/Data Node Network

There are some important differences between the "standard" queue-based batch-mode VCL-HPC offering and a user-constructed user-owned cloud or HPC cluster. Following (Vouk et al., 2009):

> One of the key features of the undifferentiated VCL resources is their networking set-up. It allows for secure dynamic reconfiguration, loading of images, and for isolation of individual images and groups of images. Every undifferentiated resource is required to have at least two networking interfaces. One on a private network, and the other one on either public

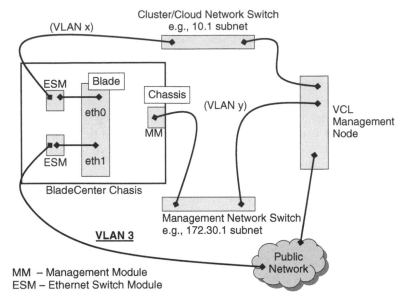

Fig. 11.9 Undifferentiated general resource node network configuration

or private network depending on the mode in which the resource operates. Also, for full functionality, undifferentiated resources need to have a way of managing the hardware state through an external channel –for example through the BladeCenter^{TM} chassis Management Module (MM).

Figure 11.9 illustrates the configuration where seats/services are assigned individually or in synchronized groups, or when we want to assign/construct an end-user image aggregate or environment where every node in the environment can be accessed from a public network (e.g., an image of a web server, plus an image of a data-base, plus an image of an application, or a cluster of relatively independent nodes). Typically eth0 interface of a blade is connected to a private network (10.1 subnet in the example) which is used to load images. Out-of-band management of the blades (e.g., power recycling) is effected through the management network (172.30.1 in the example) connected to the MM interface. The public network is typically connected to eth1 interface. The VCL node manager (which could be one of the blades in the cluster, or an external computer) at the VCL site has access to all three links, that is, it needs to have three network interfaces. If it is a stand-alone server, this means three network interface cards. If management node is a blade, the third interface is virtual and possibly on a separate VLAN.

It is worth noting that the external (public) interface is on a VLAN to provide isolation (e.g., VLAN 3 for the public interface in Fig. 11.9). This isolation can take several levels. One is just to separate resources, another one is to individually isolate each end-user within the group by giving each individual resource or group of resources a separate VLAN – and in fact end-to-end isolation through addition of VPN channels. This isolation can be effected for both real and virtual hardware resources, but the isolation of physical hardware may require extra external switching and routing equipment. In highly secure installations it is also recommended that both the private network (eth0) and the MM link be on separate VLANs. Currently, one node manager can effectively manage about 100 blades operating in the non-HPC mode.

11.4 Integrating High-Performance Computing into the VCL Cloud Architecture

"Figure 11.10 illustrates the VCL configuration when the blades are assigned to a tightly coupled VCL-HPC cluster environment, or to a large overlay (virtual) "cloud" that has relatively centralized public access and computational management. In this case the node manager is still connected to all three networks (i.e., public, management, and image-loading and preparation private network) but now eth1 is connected to what has now become a Message Passing Interface (MPI) network switch (VLAN 5 in Fig. 11.10). This network now carries intra-cluster communications needed to effect tightly coupled computing tasks usually given to an HPC cloud. Switching between non-HPC mode and this HPC mode takes place electronically, through VLAN manipulation and table entries; the actual physical set-up does not change. We use different VLANs to eth1 to separate Public Network (external) access to individual blades, when those are in the Individual-Seat mode (VLAN 3 in Fig. 11.9), from the MPI communications network to the same blade, when it is in the HPC mode (VLAN 5 in Fig. 11.10)."

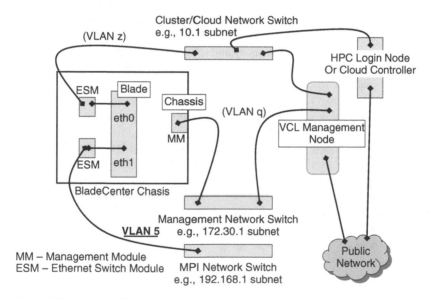

Fig. 11.10 Differentiated HPC node network configuration

The VCL code to construct this HPC configuration in a cloud computing environment is available from Apache.org (2010). While the current version of VCL can operate on any X86 hardware, we have been using primarily IBM's BladeCenters because of their reliability, power savings, ease of maintenance, and compact footprint.

When building a "starter" non-HPC version of VCL one could limit the installation to virtual environments only, e.g., all resources operate as VMWare servers,

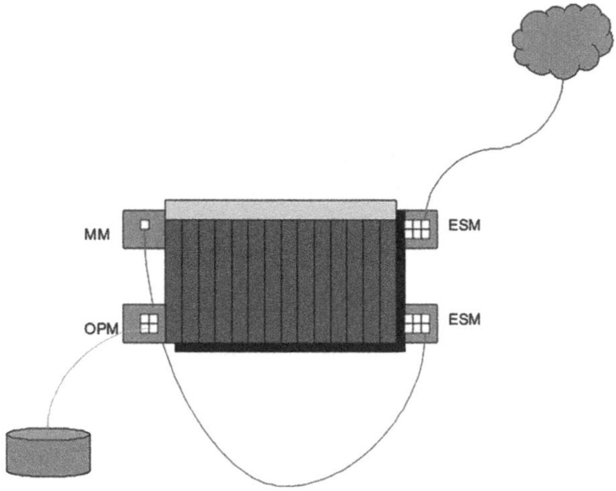

Fig. 11.11 A small "starter"
VCL cloud installation

and VCL controls and provisions only virtual images using the virtual version of the VCL non-HPC configuration in Fig. 11.9. This is quick and works well, but is probably less appealing for those who wish to have the HPC option. The HPC community is still wary of having true HPC computations run on virtualized resources. Furthermore, some of the large memory and CPU footprint engineering applications also may not behave best in a virtualized environment. In those cases, installation of VCL's bare-machine load capabilities (via XCat) is recommended.

Figures 11.12 and 11.13 show a rack of BladeCenter chasses with additional racks interconnected in a similar way. The differences between the two images are logical, i.e., switching from one mode to another is done electronically in softwareby VCL based on the image characteristics. One can mix non-HPC and HPC configurations at the granularity of chasses. An important thing to note is that in order to maintain good performance characteristics, we do not want to daisy-chain internal chassis switches. Instead, we provide an external switch, such as Cisco 6509e (or equivalent from any vendor) that is used to interconnect different chasses on three separate networks and VLANs. In non-HPC mode, one network provides public access, one network is used for managing image loads and for accessing back-end image storage, and the third one is for direct management of the hardware (e.g., power on/off, reboot). In HPC mode, the public network becomes MPI network, and special login nodes are used to access the cluster from outside. While we can use one VCL web-server and data-base for thousands of blades, with references to Figs. 11.5 and 11.6, in a scalable environment we need one resource management node for every 100 or so computational blades to insure good image reservation response times – especially when image clusters are being loaded. We also need physical connection(s) to a storage array – we typically run a shared file system (such as GFS[4] or GPFS[5]) for multiple management nodes at one site.

---

[4]http://sources.redhat.com/cluster/gfs/

[5]http://www-03.ibm.com/systems/software/gpfs/index.html

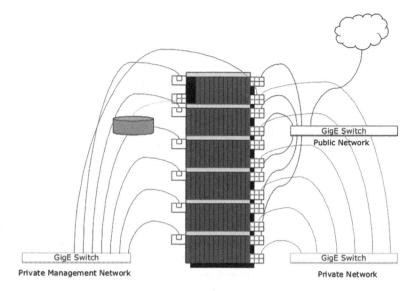

Fig. 11.12 Scaling VCL cloud

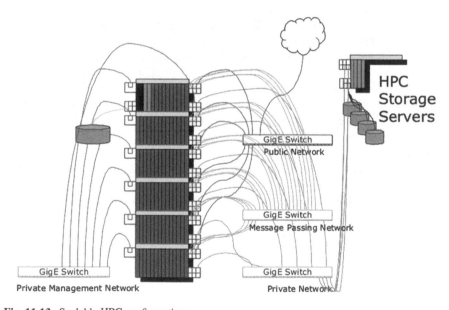

Fig. 11.13 Scalable HPC configuration

When VCL provides distributed and shared memory compute service for HPC, this is done through tightly coupled aggregation of computational resources with appropriate CPU, memory and interconnect capacity. In our case, distributed memory compute services take the form of a logical Linux cluster of different sizes with Gigabit or 10 Gigabit Ethernet interconnects. A subset of our nodes have additional

Myrinet or InfiniBand low-latency interconnects. Nodes which can be allocated for shared memory computations have large number of cores and plenty of memory.

To operate VCL in HPC mode, we dedicate one private network to message passing (Fig. 11.10) – for that we use the blade network interface that would have been used for public user access in VCL standard mode (Fig. 11.9). Also on a HPC BladeCenter chassis we configure two VLANs in one switch module, one for public Internent and for message passing interface. VCL management node makes those changes automatically based on image metadata. An HPC service environment image "knows" through its meta-data that it requires VCL-HPC network configuration (Fig. 11.10) and those actions are initiated before it is loaded. VCL-HPC environment consists of one or more login nodes, and any number of compute nodes. LSF[6] resource manager is part of a login node.

Both login nodes and compute nodes are given permanent reservations (until canceled) – as opposed to time-limited resource reservations that typically occur on the non-HPC side. An HPC compute node image consists of a minimal Linux with LSF client that, when it becomes available, automatically registers with the LSF manager. All HPC compute images also automatically connect to user home directory and to shared scratch storage for use during computations. An HPC login node image contains full Linux and LSF server. There are usually two to three login nodes through which all users access HPC facility to launch their jobs. However it is also possible to reserve, using VCL web page, a "personal" login node on a temporary basis. On these "personal" nodes users can run heavy duty visualization and analytics without impacting other users. All login nodes have access to both HPC scratch storage and user HPC home directories (with appropriate permissions), as well as long-term HPC data storage. While compute nodes conform to configuration in Fig. 11.10 – two private networks, one for MPI traffic, the other for image load and management, login nodes conform to Fig. 11.9 topology, and have a public interface to allow access from the outside, and a private side to access and control compute nodes.

If we wish to add low latency interconnects for HPC workloads, we need to make additional changes in chasses and servers that will be used for that. Chassis network modules for low-latency interconnects (Myrinet, InfiniBand) need an optical pass-through and an appropriate external switch is needed (e.g., InfiniBand). Blade servers need to be equipped with a low-latency interconnect daughtercards.

11.5 Performance and Cost

VCL delivers classroom, lab and research IT services for faculty and students. On the one hand, if users are to rely on VCL, then the VCL system must have sufficient available resources to satisfy the peak demand loads. On the other hand, if VCL is to operate cost effectively it is essential that it is not over provisioned to the point

---

[6]http://www.platform.com/Products

where the system is uneconomical to deploy. In order to provide VCL capabilities to users across widely varying demand loads, NC State decided to make a capital investment to assure that its "on-demand" level of service is available when needed. The user demand for these computing services is governed by the academic calendar of the university. The present VCL implementation delivers service availability that exceeds 99%.

One way of assuring this high level of user availability, i.e., servicing of peak loads (see Fig. 11.2), is for the university to maintain a pool of equipment in standby or idle mode for long periods of time. The consequence of this policy however would be an overall low average utilization of the resources. This is an expensive and uneconomical total cost of ownership option for the university. Therefore, one of the key VCL non-HPC – HPC design considerations is resource sharing. In a research university, such as NC State, HPC is a very useful and needed workload. Because HPC jobs are primarily batch jobs, HPC can act as an excellent "filler" load for idle computational cycles thereby providing an option to markedly decrease the total cost of ownership for both systems.

An analysis of the NC State HPC usage pattern shows that researchers who actively use HPC computational systems, do so year round and do not have their computational workloads strongly fluctuating with the academic calendar. Because HPC jobs usually have large requirements for computational cycles, they are an excellent resource utilization backfill, provided that the cloud can dynamically transfer resources between single-"seat" and HPC use modes. Furthermore, it turns out that the demand for HPC cycles increases during holidays, and when classes are not in session, since then both faculty and graduate students seem have more time to pursue computational work. In NC State's case, co-location of complementary computational workloads, i.e. on-demand desktop augmentation and HCP computations, results in a higher and more consistent utilization and in an overall savings.

In this context, one has to understand that although for most part HPC operates in "batch" mode, HPC requests are as demanding as those of the "on-demand" desktop users. HPC users expect their clusters to have very low interconnect latency, sufficient memory, and the latest computational equipment. The ability to utilize bare-machine loads for HPC images is essential, as is the ability to map onto an appropriate interconnect topology. Therefore, running HPC on virtualized resources is not really an option.. This need has recently been confirmed through a detailed comparison of NAS Parallel Benchmarks run on Amazon's EC2 cloud and NCSA clusters. Results showed wide variations in levels of degraded performance when HPC codes were loaded onto Amazon EC2 resources (Walker, 2008). While the specific results depended on the particular application and the cloud computing configuration used, the general lesson learned was that one needs to have both control over what one is running on (virtual or real machines) and low interconnect latency times among the cluster nodes. VCL was designed to allow sharing of resources while retaining full performance capability for HPC..

The VCL operational statistics over the past several years strongly support this design choice and suggest that by building a coherent integrated campus IT layer for faculty and student academic and research computational needs allows the institution flexibility in servicing both of these university functions. It also allows the educational institution itself to maximize the return on their capital investment in the IT equipment and facilities and decrease the total cost of ownership.

IT staff supporting VCL see advantages as well because the systems is scalable and serviceable with fewer customization requests and less personnel (NC State uses 2 FTEs to maintain 2000 blade system). Because the VCL hardware may be remotely located outside the classroom or laboratory, there is also better physical security of the hardware and a more organized program for computer security of the systems.

Commercial cloud computing firms are beginning to venture into the educational and research space using the pay-per-use model. Examples are Amazon EC2 services, and more recently Microsoft's Azure.[7] However, none of them (in our opinion), currently offer a viable integration of high-end low latency HPC services into their clouds (Dreher, Vouk, Averitt, & Sills, 2010).

"In the case of Amazon Web Services, they have positioned their EC2[8] cloud offering in a way that is strongly focused on marketing rental of physical hardware, storage and network components. Although Amazon's EC2 enables users to acquire on-demand compute resources, usually in the form of virtual machines, the user must configure this hardware into a working cluster at deployment time, including loading and linking the appropriate applications. Using the Amazon web service users can create and store an image containing their applications, libraries, data and associated configuration settings or load pre-configured template images from an image library. Amazon implements "availability zones" to allow users some degree of control over instance placement in the cloud. Specifically, EC2 users can choose to host images in different availability zones if they wish to try and ensure independent execution of codes and protection from a global failure in case of difficulties with their loaded image. They also have choices when to run their images, the quantity of servers to select and how to store their data. Amazon bills customers on a pay-as-you-go basis for the time rented on each component of their cloud infrastructure."

There are a number of possible open-source solutions. Very few, if any, except for VCL, offer the full HaaS to CaaS suite of cloud computing services. One example is Eucalyptus.[9] Eucalyptus has an interface that is compatible with Amazon Web Services cloud computing but treats availability zones somewhat differently. With Eucalyptus, each availability zone corresponds to a separate cluster within the Eucalyptus cloud. Under Eucalyptus, each availability zone is restricted to a single "machine" (e.g., cluster) where at Amazon, the zones are much broader.

---

[7] http://www.microsoft.com/azure/windowsazure.mspx

[8] http://aws.amazon.com/ec2/

[9] http://www.eucalyptus.com/

11.6 Summary

The Virtual Computing Laboratory open source cloud implementation was initially developed and implemented at NC State University in 2004. Since then, it has been documented to show that this combined non-HPC and HPC cloud computing architecture can provide both of these types of cloud services in a cost effective manner. This successful cloud computing architecture is now growing rapidly and has since been adopted regionally, nationally and internationally and is now delivering state-of-the-art cloud computing services to users world-wide.

Acknowledgments This work is supported in part by IBM Corp., Intel Corp., SAS Institute, NetApp, EMC, NC State University, State of North Carolina, UNC General Administration, and DOE (DE-FC02-07)ER25809. The authors would like thank the NC State VCL team for their advice, support and input.

References

Apache VCL (2010). Retrieved March 2010, from http://cwiki.apache.org/VCL/.

Armbrust, M., Fox, A., Griffith, R., Joseph, A., Katz, R., Konwinski, A., Lee, Gunho., Patterson, D., Rabkin, A., Stoica, I., & Zaharia, M. (February 2009). *Above the clouds: A Berkeley view of cloud computing* (Tech. Rep. No. UCB/EECS-2009-28), at http://www.berkeley.edu/Pubs/TechRpts.EECS-2009-28.html.

Averitt, S., Bugaev, M., Peeler, A., Schaffer, H., Sills, E., Stein, S., et al. (May 2007). The virtual computing laboratory. *Proceedings of the International Conference on Virtual Computing Initiative, IBM Corporation, Research Triangle Park, NC,* http://vcl.ncsu.edu/news/papers-publications/virtual-computing-laboratory-vcl-whitepaper, 1–16.

Dreher, P., Vouk, M. A., Sills, E., & Averitt, S. (2009). Evidence for a cost effective cloud computing implementation based upon the NC state virtual computing laboratory model. In W. Gentzsch, L. Grandinetti, & G. Joubert (Eds.), *Advances in parallel computing, high speed and large scale scientific computing* (Vol. 18, pp. 236–250), ISBN 978-1-60750-073-5.

Dreher, P., Vouk, M., Averitt, S., & Sills, E. "An Open Source Option for Cloud Computing in Education and Research", In: S. Murugesan (ed.) Cloud Computing: Technologies, Business Models, Opportunities and Challenges, CRC Press/Chapman and Hall, 2010.

Golden, B. (2009). The case against cloud computing, *CIO Magazine.*

IBM (2007). IBM introduces ready-to-use cloud computing. Retrieved November 15, 2007, from http://www-03.ibm.com/press/us/en/pressrelease/22613.wss.

Lohr, S., Google, & I.B.M. (October 2007). Join in 'Cloud Computing' research, *The New York Times.*

Schaffer, H. E., Averitt, S. F., Hoit, M. I., Peeler, A., Sills, E. D., & Vouk, M. A. (July 2009). NCSUs virtual computing lab: A cloud computing solution. *IEEE Computer, 42*(7), 94–97.

Seay, C., & Tucker, G. (March 2010). Virtual computing initiative at a small public university. *Communications of the ACM, 53*(3), 75–83.

Vouk, M., Averitt, S., Bugaev, M., Kurth, A., Peeler, A., Rindos, A., et al. (May 2008). 'Powered by VCL' – Using virtual computing laboratory (VCL) technology to power cloud computing. *Proceedings of the 2nd International Conference on Virtual Computing (ICVCI),* RTP, NC, May 15–16, 1–10. http://vcl.ncsu.edu/news/papers-publications/powered-vcl-using-virtual-computing-laboratory-vcl.

Vouk, M. (2008). Cloud computing – Issues, research and implementations. *Journal of Computing and Information Technology, 16*(4), 235–246.

Vouk, M., Rindos, A., Averitt, S., Bass, J., Bugaev, M., Peeler, A., et al. (2009). Using VCL technology to implement distributed reconfigurable data centers and computational services for educational institutions. *IBM Journal of Research and Development, 53*(4), 1–18.

Walker, E. (2008). Benchmarking Amazon EC2 for high-performance scientific computing. *Usenix Magazine,* *33*(5). http://www.usenix.org/publications/login/2008-10/openpdfs/walker.pdf. Accessed Sep 3, 2010.

Chapter 12
Vertical Load Distribution for Cloud Computing via Multiple Implementation Options

Thomas Phan and Wen-Syan Li

Abstract Cloud computing looks to deliver software as a provisioned service to end users, but the underlying infrastructure must be sufficiently scalable and robust. In our work, we focus on large-scale enterprise cloud systems and examine how enterprises may use a service-oriented architecture (SOA) to provide a streamlined interface to their business processes. To scale up the business processes, each SOA tier usually deploys multiple servers for load distribution and fault tolerance, a scenario which we term horizontal load distribution. One limitation of this approach is that load cannot be distributed further when all servers in the same tier are loaded. In complex multi-tiered SOA systems, a single business process may actually be implemented by multiple different computation pathways among the tiers, each with different components, in order to provide resilience and scalability. Such multiple implementation options gives opportunities for vertical load distribution across tiers. In this chapter, we look at a novel request routing framework for SOA-based enterprise computing with multiple implementation options that takes into account the options of both horizontal and vertical load distribution.

12.1 Introduction

Cloud computing looks to have computation and data storage moved away from the end user and onto servers located in data centers, thereby relieving users of the burdens of application provisioning and management (Dikaiakos, Pallis, Katsaros, Mehra, & Vakali, 2009; Cloud Computing, 2009). Software can then be thought of as purely a service that is delivered and consumed over the Internet, offering users the flexibility to choose applications on-demand and allowing providers to scale out their capacity accordingly.

T. Phan (✉)
Microsoft Corporation, Washington, DC, USA
e-mail: thomas.phan@acm.org

W.-S. Li
SAP Technology Lab, Shanghai, China
e-mail: wen-syan.li@sap.com

B. Furht, A. Escalante (eds.), *Handbook of Cloud Computing*,
DOI 10.1007/978-1-4419-6524-0_12, © Springer Science+Business Media, LLC 2010

As rosy as this picture seems, the underlying server-side infrastructure must be sufficiently robust, feature-rich, and scalable to facilitate cloud computing. In this chapter we focus on large-scale enterprise cloud systems and examine how issues of scalable provisioning can be met using a novel load distribution system.

In enterprise cloud systems, a service-oriented architecture (SOA) can be used to provide a streamlined interface to the underlying business processes being offered through the cloud. Such an SOA may act as a programmatic front-end to a variety of building-block components distinguished as individual services and their supporting servers (e.g. (DeCandia et al., 2007)). Incoming requests to the service provided by this composite SOA must be routed to the correct components and their respective servers, and such routing must be scalable to support a large number of requests.

In order to scale up the business processes, each tier in the system usually deploys multiple servers for load distribution and fault tolerance. Such load distribution across multiple servers within the same tier can be viewed as *horizontal* load distribution, as shown in Fig. 12.1. One limitation of horizontal load distribution is that

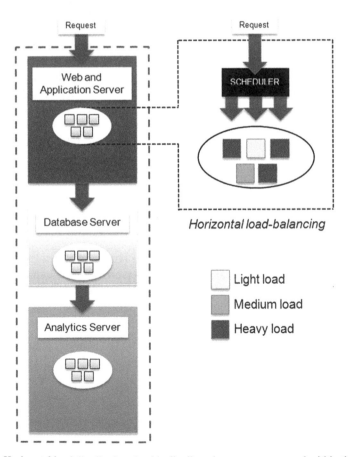

Fig. 12.1 Horizontal load distribution: load is distributed across a server pool within the same tier

load cannot be further distributed when all servers in the given tier are loaded as a result of mis-configured infrastructures – where too many servers are deployed at one tier while too few servers are deployed at another tier.

An important observation is that in complex multi-tiered SOA systems, a single business process can actually be implemented by multiple different computation pathways through the tiers (where each pathway may have different components) in order to provide resiliency and scalability. Such SOA-based enterprise computing with multiple implementation options gives opportunities for *vertical* load distribution across tiers.

Although there exists a large body of research and industry work focused on request provisioning by balancing load across the servers of one service (Cisco; F5 Networks), there has been little work on balancing load across *multiple implementations of a composite service*, where each service can be implemented via pathways through different service types.

A composite service can be represented as multiple tiers of component invocations in an SOA-based IT infrastructure. In such a system, we differentiate *horizontal* load distribution, where load can be spread across multiple servers for one service component, from *vertical* load distribution, where load can be spread across multiple implementations of a given service. The example in Fig. 12.2 illustrates these terms. Here a composite online analytic task can be represented as a call to a Web and Application Server (WAS) to perform certain pre-processing, followed by a call from the WAS to a database server (DB) to fetch required data set, after which the WAS forwards the data set to a dedicated analytic server (AS) for computationally-expensive data mining tasks.

This composite task can have multiple implementations in a modern IT data center. An alternative implementation may invoke a stored procedure on the database to perform data mining instead of having the dedicated analytic server perform this task. This alternative implementation provides *vertical* load distribution by allowing the job scheduler to select the WAS-and-DB implementation when the analytic server is not available or heavily loaded. Multiple implementations are desirable for the purpose of fault tolerance and high flexibility for load balancing. Furthermore, it is also desirable for a server to be capable of carrying out multiple instances of the same task for the same reasons.

Reusability is one of the key goals of the SOA approach. Due to the high reusability of application components, it is possible to define a complex workflow in multiple ways. However, it is hard to judge in advance which one is the best implementation, since in reality the results depend on the runtime environment (e.g. what other service requests are being processed at the same time). We believe that having multiple implementations provides fault tolerance and scalability, in particular when dealing with diverse runtime conditions and missed configured infrastructures. In this respect, an SOA plays an important role in enabling the feasibility and applicability of multiple implementations.

In this chapter we propose a framework for request-routing and load balancing *horizontally and vertically* in SOA-based enterprise cloud computing

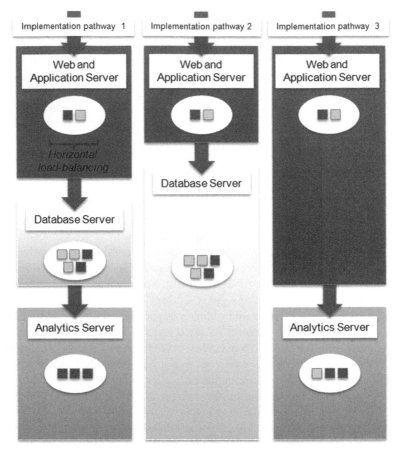

Fig. 12.2 Vertical load distribution: load can be spread across multiple implementations of the same composite service. This figure illustrates three different implementations of the same service that was shown in Fig. 12.1

infrastructures. We show that a stochastic search algorithm is appropriate to explore a very large solution space.

In our experiments, we show that our algorithm and methodology scale well up to a large scale system configuration comprising up to 1000 workflow requests to a complex composite web services with multiple implementations. We also show that our approach that considers both *horizontal and vertical* load distribution is effective in dealing with a misconfigured infrastructure (i.e. where there are too many servers in one tier and too few servers in another tier).

The key contributions of this paper are the following:

- We identify the need for QoS-aware scheduling in workloads that consist of composite web services. Our problem space lies in the relationship between consumers, service types, implementation options, and service providers.

- We provide a framework for handling both *horizontal and vertical* load distribution.
- We provide a reference implementation of a search algorithm that is able to produce optimal (or near-optimal) schedules based on a genetic search heuristic (Holland, 1992).

The rest of this chapter is organized as follows. In Section 12.2, we describe the system architecture and terminology used in this paper. In Section 12.3, we describe how we model the problem and our algorithms for scheduling load distribution for composite web services. In Section 12.4 we show experimental results, and in Section 12.5 we discuss related work. We conclude the paper in Section 12.6.

12.2 Overview

In this section we give a system architecture overview and discuss the terms that will be used in this paper. Consider a simplified cloud computing example (shown in Fig. 12.3) in which an analytic process runs on a Web and Application Server (WAS), a Database Server (DB), and a specialized Analytic Server (AS). The

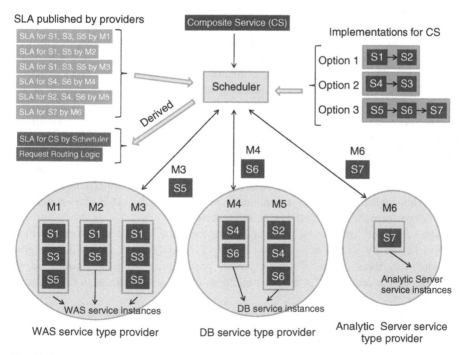

Fig. 12.3 Request routing for SOA-based enterprise computing with multiple implementation options

analytic process can be implemented by one of three options (as shown in the upper-right of the figure):

- Executing some lightweight pre-processing at WAS (S_1) and then having the DB to complete most of expensive analytic calculation (S_2); or
- Fetching data from the DB (S_4) to the WAS and then completing most of the expensive analytic calculation at the WAS (S_3); or
- Executing some lightweight pre-processing at the WAS (S_5), then having the DB fetch necessary data (S_6), and finally having the AS perform the remaining expensive analytic calculation (S_7).

The analytic process requires three different *service types*; namely, the WAS service type, the DB service type, and the AS service type. S_1, S_3, and S_5 are *instances* of the WAS service type since they are the services provided by the WAS. Similarly, S_2, S_4, and S_6 are instances of the DB service type, and S_7 is an instance of the AS service type.

Furthermore, there are three kinds of servers: WAS servers (M_1, M_2, and M_3); DB servers (M_4 and M_5); and AS servers (M_6). Although a server can typically support any instance of its assigned service type, in general this is not always the case. Our example reflects this notion: each server is able to support all instances of its service type, except M_2 and M_4 are less powerful servers so that they cannot support computationally expensive service instances, S_3 and S_2.

Each server has a service level agreement (SLA) for each service instance it supports, and these SLAs are published and available for the scheduler. The SLA includes information such as a profile of the load versus response time and an upper bound on the request load size for which a server can provide a guarantee of its response time.

The scheduler is responsible for routing and coordinating execution of composite services comprising one or more implementations. A derived SLA can only be deployed with its corresponding routing logic. Note that the scheduler can derive SLA and routing logic as well as handle the task of routing the requests. Alternatively, the scheduler can be used solely for the purpose of deriving SLA and routing logic while configuring a content aware routers, such as (Cisco System Inc), for high performance and hardware-based routing.

The scheduler can also be enhanced to perform the task of monitoring actual QoS achieved by workflow execution and by individual service providers. If the scheduler observes failure of certain service providers to their QoS published, it can re-compute feasible SLA and routing logic on demand to adapt to the runtime environment.

In this paper, we focus on the problem of automatically deriving the routing logic of a composite service with consideration of both *horizontal* and *vertical* load distribution options. The scheduler is required to find an optimal combination of a set of variables illustrated in Fig. 12.3 for a number of concurrent requests. We discuss our scheduling approach next.

12.3 Scheduling Composite Services

12.3.1 Solution Space

In this section, we formally define the problem and describe how we model its complexity. We assume the following scenario elements:

- *Requests* for a workflow execution are submitted to a scheduling agent.
- The workflow can be embodied by one of several *implementations*, so each request is assigned to one of these implementations by the scheduling agent.
- Each implementation invokes several *service types*, such as a web application server, a DBMS, or a computational analytics server.
- Each service type can be embodied by one of several *instances* of the service type, where each instance can have different computing requirements. For example, one implementation may require heavy DBMS computation (such as through a stored procedure) and light computational analytics, whereas another implementation may require light DBMS querying and heavy computational analytics. We assume that these implementations are set up by administrators or engineers.
- Each service type is executed on a *server* within a pool of servers dedicated to that service type.

Each service type can be served by a pool of servers. We assume that the servers make agreements to guarantee a level of performance defined by the completion time for completing a web service invocation. Although these SLAs can be complex, in this paper we assume for simplicity that the guarantees can take the form of a linear performance degradation under load, an approach similar to other published work on service SLAs (e.g. (DeCandia et al., 2007)). This guarantee is defined by several parameters: α is the expected completion time (for example, on the order of seconds) if the assigned workload of web service requests is less than or equal to β, the maximum concurrency, and if the workload is higher than β, the expected completion for a workload of size ω is $\alpha + \gamma(\omega - \beta)$ where γ is a fractional coefficient. In our experiments we vary α, β, and γ with different distributions.

We would like to ideally perform optimal scheduling to simultaneously distribute the load both vertically (across different implementation options) and horizontally (across different servers supporting a particular service type). There are thus two stages of scheduling, as shown in Fig. 12.4.

In the first stage, the requests are assigned to the implementations. In the second stage each implementation has a known set of instances of a service type, and each instance is assigned to servers within the pool of servers for the instance's service type. The solution space of possible scheduling assignments can be found by looking at the possible combinations of these assignments. Suppose there are R requests and M possible implementations. There are then M^R possible assignments in the first stage. Suppose further there are on average T service type invocations per

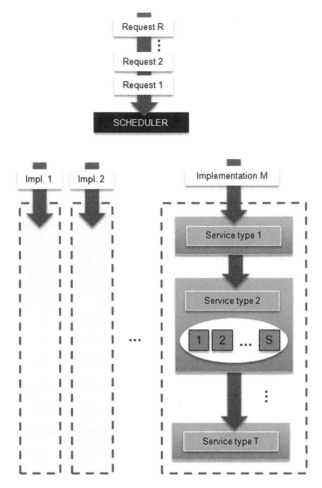

Fig. 12.4 The scheduling and assignment spans two stages. In the first stage, requests are assigned to implementations, and in the second stage, service type instances are assigned to servers

implementation, and each of these service types can be handled by one of S on average possible servers. Across all the implementations, there are then S^T combinations of assignments in the second stage. It total, there are $M^R \cdot S^T$ combinations.

Clearly, an exhaustive search through this solution space is prohibitively costly for all but the smallest configurations. In the next section we describe how we use a genetic search algorithm to look for the optimal scheduling assignments.

12.3.2 Genetic algorithm

Given the solution space of $M^R \cdot S^T$, the goal is to find the best assignments of requests to implementations and service type instances to servers in order to minimize the running time of the workload, thereby providing our desired vertical and horizontal balancing. To search through the solution space, we use a genetic algorithm (GA) global search heuristic that allows us to explore portions of the space in a guided manner that converges towards the optimal solutions (Holland, 1992; Goldberg, 1989). We note that a GA is only one of many possible approaches for a search heuristic; others include tabu search, simulated annealing, and steepest-ascent hill climbing. We use a GA only as a tool.

A GA is a computer simulation of Darwinian natural selection that iterates through various generations to converge toward the best solution in the problem space. A potential solution to the problem exists as a chromosome, and in our case, a chromosome is a specific mapping of requests-to-implementations and instances-to-servers along with its associated workload execution time. Genetic algorithms are commonly used to find optimal exact solutions or near-optimal approximations in combinatorial search problems such as the one we address. It is known that a GA provides a very good tradeoff between exploration of the solution space and exploitation of discovered maxima (Goldberg, 1989). Furthermore, a genetic algorithm does have an advantage of progressive optimization such that a solution is available at any time, and the result continues to improve as more time is given for optimization.

Note that the GA is not guaranteed to find the optimal solution since the recombination and mutation steps are stochastic.

Our choice of a genetic algorithm stemmed from our belief that other search heuristics (for example, simulated annealing) are already along the same lines as a GA. These are randomized global search heuristics, and genetic algorithms are a good representative of these approaches. Prior research has shown there is no clear winner among these heuristics, with each heuristic providing better performance and more accurate results under different scenarios (Lima, Francois, Srinivasan, & Salcedo, 2004; Costa & Oliveira, 2001; Oliveira & Salcedo, 2005). Furthermore, from our own prior work, we are familiar with its operations and the factors that affect its performance and optimality convergence. Additionally, the mappings in our problem context are ideally suited to array and matrix representations, allowing us to use prior GA research that aid in chromosome recombination (Davis, 1985). There are other algorithms that we could have considered, but scheduling and assignment algorithms are a research topic unto themselves, and there is a very wide of range of approaches that we would have been forced to omit.

Pseudo-code for a genetic algorithm is shown in Algorithm 1. The GA executes as follows. The GA produces an initial random population of chromosomes. The chromosomes then recombine (simulating sexual reproduction) to produce children using portions of both parents. Mutations in the children are produced with small probability to introduce traits that were not in either parent. The children with the best scores (in our case, the lowest workload execution times) are chosen for the next generation. The steps repeat for a fixed number of iterations, allowing the GA to converge toward the best chromosome. In the end it is hoped that the GA explores a large portion of the solution space. With each recombination, the most beneficial portion of a parent chromosome is ideally retained and passed from parent to child,

Algorithm 1 Genetic Search Algorithm

```
 1: FUNCTION Genetic algorithm
 2: BEGIN
 3: Time t
 4: Population P(t) := new random Population
 5:
 6: while ! done do
 7:     recombine and/or mutate P(t)
 8:     evaluate(P(t))
 9:     select the best P(t + 1) from P(t)
10:     t := t + 1
11: end while
12: END
```

so the best child in the last generation has the best mappings. To improve the GA's convergence, we implemented elitism, where the best chromosome found so far is guaranteed to exist in each generation.

12.3.2.1 Chromosome Representation of a Solution

We used two data structures in a chromosome to represent each of the two scheduling stages. In the first stage, R requests are assigned to M implementations, so its representative structure is simply an array of size R, where each element of the array is in the range of $[1, M]$, as shown in Fig. 12.5.

Fig. 12.5 An example chromosome representing the assignment of R requests to M implementations

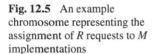

The second stage where instances are assigned to servers is more complex. In Fig. 12.6 we show an example chromosome that encodes one scheduling assignment. The representation is a 2-dimensional matrix that maps {implementation,

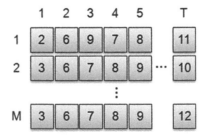

Fig. 12.6 An example chromosome representing a scheduling assignment of (implementation,service type instance) → service provider. Each row represents an implementation, and each column represents a service type instance. Here there are M workflows and T service types instances. In workflow 1, any request for service type 3 goes to server 9

service type instance} to a service provider. For an implementation i utilizing service type instance j, the $(i, j)^{th}$ entry in the table is the identifier for the server to which the business process is assigned.

12.3.2.2 Chromosome Recombination

Two parent chromosomes recombine to produce a new child chromosome. The hope is that the child contains the best contiguous chromosome regions from its parents.

Recombining the chromosome from the first scheduling stage is simple since the chromsomes are simple 1-dimensional arrays. Two cut points are chosen randomly and applied to both the parents. The array elements between the cut points in the first parent are given to the child, and the array elements outside the cut points from the second parent are appended to the array elements in the child. This is known as a 2-point crossover and is shown in Fig. 12.7.

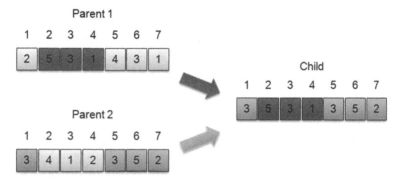

Fig. 12.7 An example recombination between two parents to produce a child for the first stage assignments. This recombination uses a 2-point crossover recombination of two one-dimensional arrays. Contiguous subsections of both parents are used to create the new child

For the 2-dimensional matrix, chromosome recombination was implemented by performing a one-point crossover scheme twice (once along each dimension). The crossover is best explained by analogy to Cartesian space as follows. A random location is chosen in the matrix to be coordinate $(0, 0)$. Matrix elements from quadrants II and IV from the first parent and elements from quadrants I and III from the second parent are used to create the new child. This approach follows GA best practices by keeping contiguous chromosome segments together as they are transmitted from parent to child, as shown in Fig. 12.8.

The uni-chromosome mutation scheme randomly changes one of the service provider assignments to another provider within the available range. Other recombination and mutation schemes are an area of research in the GA community, and we look to explore new operators in future work.

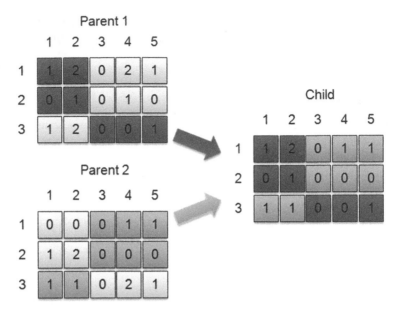

Fig. 12.8 An example recombination between two parents to produce a child for the second stage assignments. Elements from quadrants II and IV from the first parent and elements from quadrants I and III from the second parent are used to create the new child

12.3.2.3 GA Evaluation Function

The evaluation function returns the resulting workload execution time given a chromosome. Note the function can be implemented to evaluate the workload in any way so long as it is consistently applied to all chromosomes across all generations.

Our evaluation function is shown in Algorithm 2. In lines 6–8, it initialises the execution times for all the servers in the chromosome. In lines 11–17, it assigns requests to implementations and service type instances to servers using the two mappings in the chromosome. The end result of this phase is that the instances are accordingly enqueued the servers. In lines 19–21 the running times of the servers are calculated. In lines 24–26, the results of the servers are used to compute the results of the implementations. The function returns the maximum execution time among the implementations.

12.3.3 Handling Online Arriving Requests

As mentioned earlier, the problem domain we consider is that of batch-arrival request routing. We take full advantage of such a scenario through the use of the GA, which has knowledge of the request population. We can further extend this approach to online arriving requests, a lengthy discussion which we omit here due to space limits. A typical approach is to aggregate the incoming requests into a

---

Algorithm 2 GA evaluation function

1: **FUNCTION** evaluate
2: **IN:** *CHROMOSOME*, a representation of the assignments of requests to implementation and service type instances to servers
3: **OUT:** *runningtime*, the running time of this workload
4: **BEGIN**
5:
6: **for** (each *server* ∈ *CHROMOSOME*) **do**
7:     set *server*'s running time to 0
8: **end for**
9:
10: {Loop over each request and its implementations}
11: **for** (each *request* ∈ *CHROMOSOME*) **do**
12:     *implementation* := *requests*'s implementation
13:     **for** (each *instance* ∈ *implementation*) **do**
14:         *server* := *implementation*'s server
15:         Enqueue this job at *server*
16:     **end for**
17: **end for**
18:
19: **for** (each *server*) **do**
20:     Compute the running time of *server*
21: **end for**
22:
23: {Now compute the running time of the implementations}
24: **for** (each *implementation* ∈ *CHROMOSOME*) **do**
25:     Aggregate the running time of this *implementation* across its instances
26: **end for**
27:
28: *runningtime* := maximum running time of each implementation
29: **return** *runningtime*
30: **END**

---

queue, and when a designated timer expires, all requests in the queue at that time are scheduled. There may still be uncompleted requests from the previous execution, so the requests may be mingled together to produce a larger schedule. An alternative approach is to use online stochastic optimization techniques commonly found in online decision-making systems (Van Hentenryck & Bent, 2006).

First, we can continue to use the GA, but instead of having the complete collection of requests available to us, we can allow requests to aggregate into a queue first. When a periodic timer expires, we can run the GA on those requests while aggregating any more incoming requests into another queue. Once the GA is finished with the first queue, it will process the next queue when the periodic timer expires again. If the request arrival rate is faster than the GA's processing rate, we can take advantage of the fact that the GA can be run as an incomplete, near-optimal search heuristic: we can go ahead and let the timer interrupt the GA, and the GA will have *some* solutions that, although sub-optimal, is probabilistically better than a greedy solution. This typical methodology is also shown in (Dewri, Ray, Ray, & Whitley, 2008), where requests for broadcast messages are queued, and the messages are optimally distributed through the use of an evolutionary strategies algorithm (a close cousin of a genetic algorithm).

Second (and unrelated to genetic algorithms), we can use online stochastic optimization techniques to serve online arrivals. This approach approximates the offline problem by sampling historical arrival data in order to make the best online decision. A good overview is provided in (Bent & Van Hentenryck, 2004). In this technique, the online optimizer receives an incoming sequence of requests, gets historical data over some period of time from a sampling function that creates a statistical distribution model, and then calculates and returns an optimized allocation of requests to available resources. This optimization can be done on a periodic or continuous basis.

12.4 Experiments and Results

We ran experiments to show how our system compared to other well-known algorithms with respect to our goal of providing request routing with horizontal and vertical distribution. Since one of our intentions was to demonstrate how our system scales well up to 1000 requests, we used a synthetic workload that allowed us to precisely control experimental parameters, including the number of available implementations, the number of published service types, the number of service type instances per implementation, and the number of servers per service type instance. The scheduling and execution of this workload was simulated using a program we implemented in standard C++. The simulation ran on an off-the-shelf Red Hat Linux desktop with a 3.0 GHz Pentium IV and 2 GB of RAM.

In these experiments we compared our algorithm against the following alternatives:

- A *round-robin* algorithm that assigns requests to an implementation and service type instances to a server in circular fashion. This well-known approach provides a fast and simple scheme for load-balancing.
- A *random-proportional* algorithm that proportionally assigns instances to the servers. For a given service type, the servers are ranked by their guaranteed completion time, and instances are assigned proportionally to the servers based on the servers' completion time. (We also tried a proportionality scheme based on both the completion times and maximum concurrency but attained the same results, so only the former scheme's results are shown here.) To isolate the behavior of this proportionality scheme in the second phase of the scheduling, we always assigned the requests to the implementations in the first phase using a round-robin scheme.
- A *purely random* algorithm that randomly assigns requests to an implementation and service type instances to a server in random fashion. Each choice was made with a uniform random distribution.

- A *greedy* algorithm that always assigns business processes to the service provider that has the fastest guaranteed completion time. This algorithm represents a naive approach based on greedy, local observations of each workflow without taking into consideration all workflows.

In the experiments that follow, all results were averaged across 20 trials, and to help normalize the effects of any randomization used during any of the algorithms, each trial started by reading in pre-initialized data from disk. In Table 12.1 we list our experimental parameters for our baseline experiments. We vary these parameters in other experiments, as we discuss later.

Table 21.1 Experiment parameters

Experimental parameter	Comment
Requests	1 to 1000
Implementations	5, 10, 20
Service types used per implementation	uniform random: 1 – 10
Instances per service type	uniform random: 1 – 10
Servers per service type	uniform random: 1 – 10
Server completion time (α)	uniform random: 1 – 12 seconds
Server maximum concurrency (β)	uniform random: 1 – 12
Server degradation coefficient (γ)	uniform random: 0.1 – 0.9
GA: population size	100
GA: number of generations	200

12.4.1 Baseline Configuration Results

In Figs. 12.9, 12.10, and 12.11 we show the behavior of the algorithms as they schedule requests against 5, 10, and 20 implementations, respectively. In each graph, the x-axis shows the number of requests (up to 1000), and the y-axis is average response time upon completing the workload. This response time is the *makespan*, the metric commonly used in the scheduling community and calculated as the maximum completion time across all requests in the workload. As the total number of implementations increases across the three graphs, the total number of service types, instances, and servers scaled as well in accordance to the distributions of these variables from Table 12.1. In each of the figures, it can be see that the GA is able to produce a better assignment of requests to implementations and service type instances to servers than the other algorithms. The GA shows a 45% improvement over its nearest competitor (typically the round-robin algorithm) with a configuration of 5 implementations and 1000 requests and a 36% improvement in the largest configuration with 20 implementations and 1000 requests.

The relative behavior of the other algorithms was consistent. The greedy algorithm performed the worst while the random-proportional and random algorithms were close together. The round-robin came the closest to the GA.

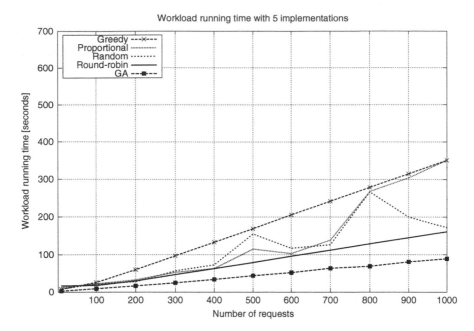

Fig. 12.9 Response time with 5 implementations

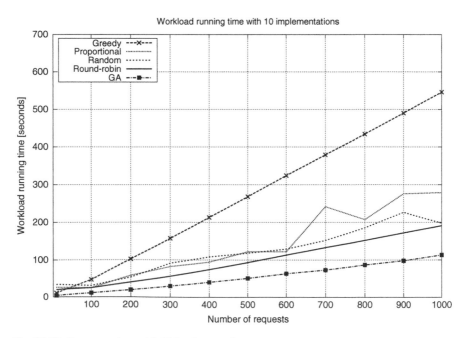

Fig. 12.10 Response time with 10 implementations

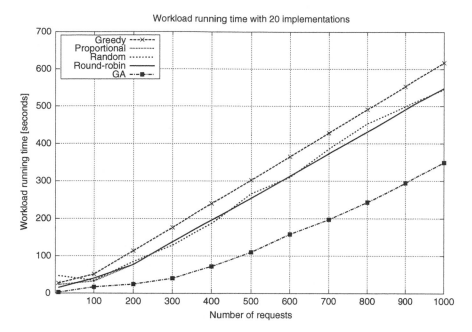

Fig. 12.11 Response time with 20 implementations

To better understand these results, we looked at the individual behavior of the servers after the instance requests were assigned to them. In Fig. 12.12 we show the percentage of servers that were saturated among the servers that were actually assigned instance requests. These results were from the same 10-implementation experiment from Fig. 12.10. For clarity, we focus on a region with up to 300 requests.

We consider a server to be saturated if it was given more requests than its maximum concurrency parameter. From this graph we see the key behavior that the GA is able to find assignments well enough to delay the onset of saturation until 300 requests. The greedy algorithm, as can be expected, always targets the best server from the pool available for a given service type and quickly causes these chosen servers to saturate. The round robin is known to be a quick and easy way to spread load and indeed provides the lowest saturation up through 60 requests. The random-proportional and random algorithms reach saturation points between that of the greedy and GA algorithms.

12.4.2 Effect of Service Types

We then varied the number of service types per implementation, modeling a scenario where there is a heavily skewed number of different web services available to each of the alternative implementations. Intuitively, in a deployment where there is a large

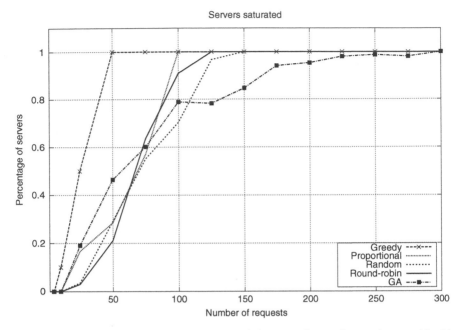

Fig. 12.12 Percentage of servers that were saturated. A saturated server is one whose workload is greater than its maximum concurrency

number of services types to be invoked, the running time of the overall workload will increase.

In Fig. 12.13 we show the results where we chose the numbers of service types per implementation from a Gaussian distribution with a mean of 2.0 service types; this distribution is in contrast to the previous experiments where the number was selected from a uniform distribution in the inclusive range of 1–10. As can be seen, the algorithms show the same relative performance from prior results in that the GA is able to find the scheduling assignments resulting in the lowest response times. The worst performer in this case is the random algorithm. In Fig. 12.14 we skew the number of service types in the other direction with a Gaussian distribution with a mean of 8.0. In this case the overall response time increases for all algorithms, as can be expected. The GA still provides the best response time.

12.4.3 Effect of Service Type Instances

In these experiments we varied the number of instances per service type. We implemented a scheme where each instance incurs a different running time on each server; that is, a unique combination of instance and server provides a different response time, which we put into effect by a Gaussian random number generator. This approach models our target scenario where a given implementation may run

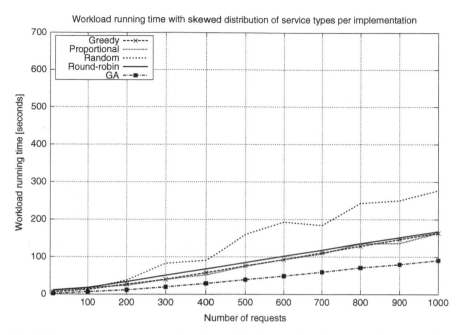

Fig. 12.13 Average response time with a skewed distribution of service types per implementation. The distribution was Gaussian ($\lambda = 2.0$, $\sigma = 2.0$ service types)

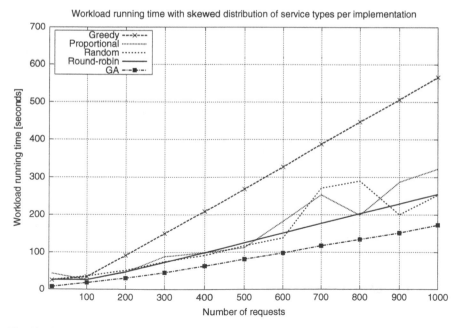

Fig. 12.14 Average response time with a skewed distribution of service types per implementation. The distribution was Gaussian ($\lambda = 8.0$, $\sigma = 2.0$ service types)

an instances that performs more or less of the work associated with the instance's service type. For example, although two implementations may require the use of a DBMS, one implementation's instance of this DBMS task may require less computation than the other implementation due to the offload of a stored procedure in the DBMS to a separate analytics server. Our expectation is that having more instances per service type allows a greater variability in performance per service type.

Figure 12.15 shows the algorithm results when we skewed the number of instances per service type with a Gaussian distribution with a mean of 2.0 instances. Again, the relative ordering shows that the GA is able to provide the lowest workload response among the algorithms throughout. When we weight the number of instances with a mean of 8.0 instances per service type, as shown in Fig. 12.16, we can see that the the GA again provides the lowest response time results. In this larger configuration, the separation between all the algorithms is more evident with the greedy algorithm typically performing the worst; its behavior is again due the fact that it assigns jobs only to the best server among the pool of servers for a service type.

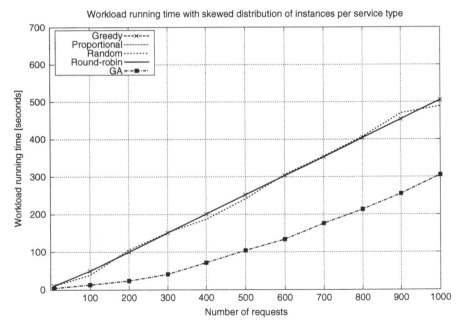

Fig. 12.15 Average response time with a skewed distribution of instances per service type. The distribution was Gaussian ($\lambda = 2.0$, $\sigma = 2.0$ instances)

12.4.4 Effect of Servers (Horizontal Balancing)

Here we explored the impact of having more servers available in the pool of servers for the service types. This experiment isolates the effect of horizontal balancing.

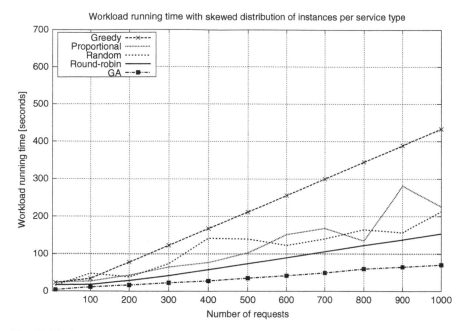

Fig. 12.16 Average response time with a skewed distribution of instances per service type. The distribution was Gaussian ($\lambda = 8.0$, $\sigma = 2.0$ instances)

Increasing the size of this pool will allow assigned requests to be spread out and thus reduce the number of requests per server, resulting in lower response times for the workload. In Figs. 12.17 and 12.18 we show the results with Gaussian distributions with means of 2.0 and 8.0, respectively. In both graphs the GA appears to provide the lowest response times. Furthermore, it is interesting to note that in the random, random-proportional, and round-robin algorithms, the results did not change substantially between the two experiments even though the latter experiment contains four times the average number of servers. We believe this result may be due to the fact that the first-stage scheduling of requests to implementations is not taking sufficient advantage of the second-stage scheduling of service type instances to the increased number of servers. Since the GA is able to better explore all combinations across both scheduling stages, it is able to produces its better results. We will explore this aspect in more detail in the future.

12.4.5 Effect of Server Performance

In this section we look at the impact on the servers' individual performance on the overall workload running time. In previous sections we described how we modeled each server with variables for the response time (α) and the concurrency (β). Here we skewed these variables to show how the algorithms performed as a result.

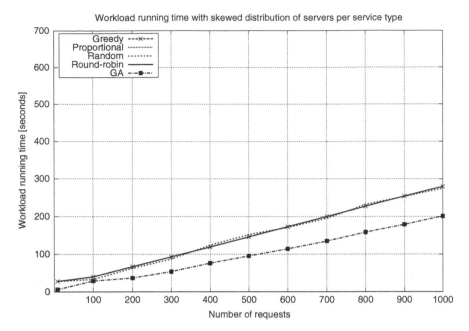

Fig. 12.17 Average response time with a skewed distribution of servers per service type. The distribution was Gaussian ($\lambda = 2.0$, $\sigma = 2.0$ instances)

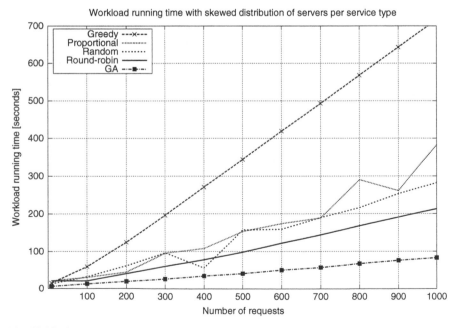

Fig. 12.18 Average response time with a skewed distribution of servers per service type. The distribution was Gaussian ($\lambda = 8.0$, $\sigma = 2.0$ instances)

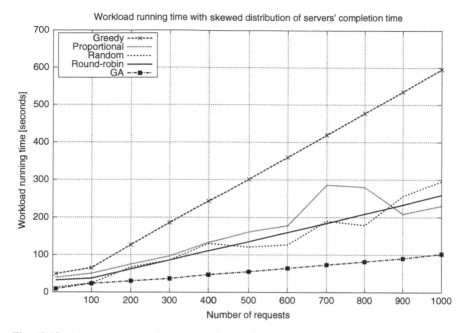

Fig. 12.19 Average response time with a skewed distribution of servers' completion time. The distribution was Gaussian ($\lambda = 2.0$, $\sigma = 2.0$ s)

In Figs. 12.19 and 12.20 we skewed the completion times with Gaussian distributions with means of 2.0 and 9.0, respectively. It can be seen that the relative orderings of the algorithms are roughly the same in each, with the GA providing best performance, the greedy algorithm giving the worst, and the other algorithms running in between. Surprisingly, the difference in response time between the two experiments was much less than we expected, although there is a slight increase in all the algorithms except for the GA. We believe that the lack of a dramatic rise in overall response time is due to whatever load balancing is being performed by the algorithms (except the greedy algorithm).

We then varied the maximum concurrency variable for the servers using Gaussian distributions with means of 2.0 and 9.0, as shown in Figs. 12.21 and 12.22. From these results it can be observed that the algorithms react well with an increasing degree of maximum concurrency. As more requests are being assigned to the servers, the servers respond with faster response times when they are given more headroom to run with these higher concurrency limits.

12.4.6 Effect of Response Variation Control

We additonally evaluated the effect of having the GA minimize the variation in the requests' completion time. As mentioned earlier, we have been been calculating the

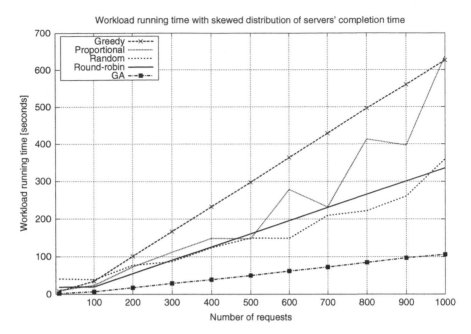

Fig. 12.20 Average response time with a skewed distribution of servers' completion time. The distribution was Gaussian ($\lambda = 9.0$, $\sigma = 2.0$ s)

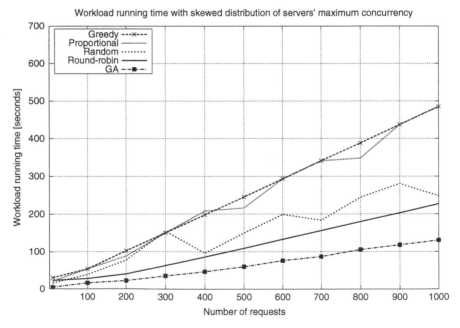

Fig. 12.21 Average response time with a skewed distribution of servers' maximum concurrency. The distribution was Gaussian ($\lambda = 4.0$, $\sigma = 2.0$ jobs)

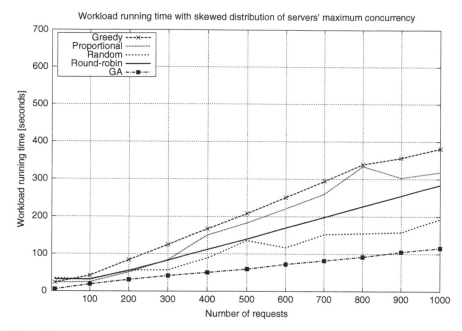

Fig. 12.22 Average response time with a skewed distribution of servers' maximum concurrency. The distribution was Gaussian ($\lambda = 11.0$, $\sigma = 2.0$ jobs)

workload completion as the maximum completion time of the requests in that workload. While this approach has been effective, it produces wide variation between the requests' completion times due to the stochastic packing of requests by the GA. This variation in response time, known as *jitter* in the computer networking community, may not be desirable, so we further provided an alternative objective function that minimizes the jitter (rather than minimizing the workload completion time). In Fig. 12.23 we show the average standard deviations resulting from these different objective functions (using the same parameters as in Fig. 12.10). With variation minimization on, the average standard deviation is always close to 0, and with variation minimization off, we observe an increasing degree of variation. The results in Fig. 12.24 show that the reduced variation comes at the cost of longer response times.

12.4.7 Effect of Routing Against Conservative SLA

We looked at the GA behavior when its input parameters were not the servers' actual parameters but rather the parameters provided by a conservative SLA. In some systems, SLAs may be defined with a safety margin in mind so that clients of the service do not approach the actual physical limits of the underlying service. In that vein, we ran an experiment similar to that shown in Fig. 12.10, but in this configuration we

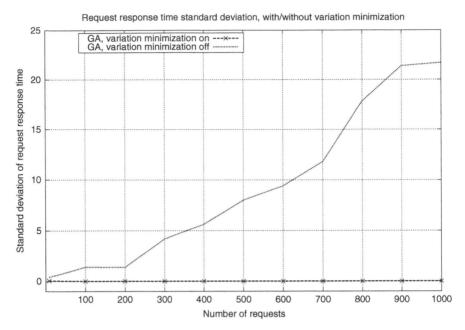

Fig. 12.23 Average standard deviation from the mean response for two different objective functions

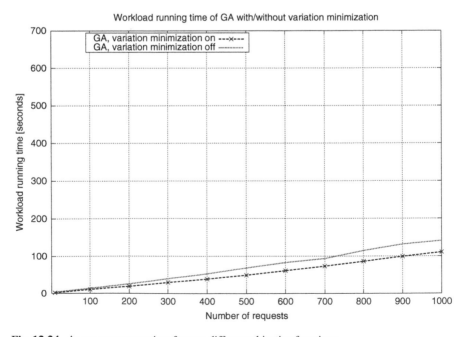

Fig. 12.24 Average response time for two different objective functions

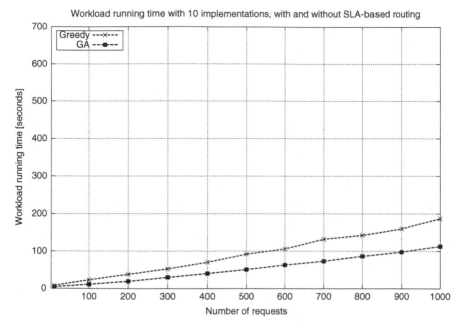

Fig. 12.25 Response time with 10 implementations with configurations for SLA and without SLA

used parameters for the underlying servers with twice the expected response time and half the available parallelism, mirroring a possible conservative SLA. As can be seen in Fig. 12.25, the GA converges towards a scheduling where the extra slack given by the conservative SLA results in a slower response time.

12.4.8 Summary of Experiments

In this section we evaluated our GA reference implementation of a scheduler that performs request-routing for horizontal and vertical load distribution. We showed that the GA consistently produces lower workload response time than its competitors. Furthermore, as can be expected, the scheduler is sensitive to a number of parameters, including the number of service types in each implementation, the number of service type instances, the number of servers, the per-server performance, the desired degree of variation, and the tightness of the SLA parameters.

12.5 Related Work

Shankar, De Miguel, and Liu (2004) described a distributed quality of service (QoS) management architecture and middleware that accommodates and manages different dimensions and measures of QoS. The middleware supports the specification, maintenance and adaptation of end-to-end QoS (including temporal requirements) provided by the individual components in complex real time application systems. Using QoS negotiation, the middleware determines the quality levels and resource allocations of the application components. This work focused on analysis tradeoff between QoS and cost instead of ensuring QoS requirements in our paper.

Yu and Lin (2005) presented two algorithms for finding replacement services in autonomic distributed business processes when web service providers fail to response or meet the QoS requirement: following alternative predefined routes or finding alternative routes on demand. The algorithms give the QoS brokerage service fault tolerance capability and is complementary to our work.

Yu et al. developed a set of algorithms for Web services selection with end-to-end QoS constraints (Yu & Lin, 2005b, 2006; Yu, Zhang, & Lin, 2007). A key difference between our work and theirs is that they simplify and reduce the complexity space considerably, something which we do not do. They take all incoming workflows, aggregate them into one singe workflow, and then schedule that one workflow onto the underlying service providers. We do not do this aggregation, and therefore our approach provides a higher degree of scheduling flexibility.

Consider the two workflows shown on the left of Fig. 12.26 where each task in the workflow invokes a particular service type. In their work, they aggregate the workflows into a single function graph, resulting in a simplified form shown on the right of Fig. 12.26.

Fig. 12.26 Aggregation of service workflows

Each service type is then mapped onto a service provider chosen from the pool of service providers for that type. It is important to note that each service type is assigned to the same chosen provider, even though the instances of that service type are different. For example, because both *workflow 1* and *workflow 2* use *S3*, both instances are mapped to the same provider.

In our work, we do not do this aggregation to reduce the complexity space. We consider unique combinations of {workflow, service type} and map these to a service provider. Thus, in our work, *S3* in *workflow 1* may map to a different provider than *S3* in *workflow 2*. This distinction allows for more flexible scheduling and potentially better turnaround time than their work.

In the work (Phan & Li, 2008b), the GA algorithm was used for load distribution for database cluster. In this work, the analytic workloads are distributed across a database cluster. The load distribution algorithm needs to consider collocation of MQTs (i.e. materialized views) with queries which can utilize them to improve performance, collocation of MQTs and the base tables which are needed to construct the MQTs, and minimizing the execution time of the whole workload on the database cluster. This work is a kind of *horizontal* load distribution. Similarly, the

GA algorithm is also used in (Phan & Li, 2008a) to schedule query execution and view materialization sequence for minimal overall execution time.

Our work is related to prior efforts in web service composition, web service scheduling, and job scheduling. A web service's interface is expressed in WSDL, and given a set of web services, a workflow can be specified in a flow language such as BPEL4WS (2005) or WSCI (Josephraj, 2007). Several research projects have looked to provide automated web services composition using high-level rules (e.g. eFlow (Casati, Ilnicki, & Jin, 2000), SWORD (Ponnekanti & Fox, 2004)). Our work is complementary to this area, as we schedule business processes within multiple, already-defined workflows to the underlying service providers.

In the context of service assignment and scheduling, (Zeng, Benatallah, Dumas, Kalagnanam, & Sheng, 2003) maps web service calls to potential servers but their work is concerned with mapping only single workflows; our principal focus is on scalably scheduling multiple workflows (up to one thousand). Urgaonkar, Shenoy, Chandra, and Goyal (2005) presents a dynamic provisioning approach that uses both predictive and reactive techniques for multi-tiered Internet application delivery. However, the provisioning techniques do not consider the challenges faced when there are alternative query execution plans and replicated data sources. Soundararajan, Manassiev, Chen, Goel, and Amza (2005) presents a feedback-based scheduling mechanism for multi-tiered systems with back-end databases, but unlike our work, it assumes a tighter coupling between the system components.

The work in (Jin & Nahrstedt, 2004) creates end-to-end paths for services (such as transcoding) and assigns servers on a hop-by-hop basis by minimising network latency between hops. Our work is complementary in that service assignment is based on business value metrics defined by agreed-upon service level agreements.

An SLA can be complex, requiring IT staff to translate from the legal document level description to system-specific requirement for deployment and enforcement. Ward et al. (2005) proposed a framework for configuring extensible SLA management systems. In this work, an SLA is represented in XML format. In Buco et al. (2003), an SLA execution manager (SAM) is proposed to manage cross-SLA execution that may involve an SLA with different terms. The work provides metadata management functionality for SLA aware scheduling presented in this paper. Thus, it is complementary to our work.

Tang, Chang, and So (2006) and Gu, Nahrstedt, Chang, and Ward (2003) applied peer-to-peer technology for support real time services, such as data dissemination across internet with QoS assurance. In their context, they create an application-layer network route across multiple service nodes in order to provide some end-to-end service. This routing occurs in two steps: the user's high-level request is mapped to a service template, and then the template is mapped to a route of servers. This approach is similar to ours in that our business processes request service from the service types, and the service types must instantiated by assigning the business processes to an underlying server. The key differences are that: (1) their work is constrained by the topology of the application-layer network. Their work looks at pipelines of service nodes in a line. The problem is finding routes through a network by adapting Dijkstra's algorithm for finding shortest path whereas our problem is assigning business processes to servers; Their work looks at pipelines of service nodes in a line; whereas our work looks at a more flexible workflow condition that may involve branches, including AND and OR; (3) their primary metrics are availability and latency, whereas we use a more flexible and generalizable business value to evaluate assignments. Furthermore, our work supports an infrastructure where a server can support multiple service types (c.f. our scenario is that business processes within a workflow must be scheduled onto web service providers). The salient differences are that the machines can process only one job at a time (we assume servers can multi-task but with degraded performance and a maximum concurrency level), tasks within a job cannot simultaneously run on different machines (we assume business processes can be assigned to any available server), and the principal metric of performance is the *makespan*, which is the time for the last task among all the jobs to complete. As we showed, optimizing on the makespan is insufficient for scheduling the business processes, necessitating different metrics.

12.6 Conclusion

Cloud computing aims to do the dirty work for the user: by moving issues of mangement and provisioning away from the end consumer and into the server-side data centers, users are given more freedom to pick and choose the applications that suit their needs. However, computing in the clouds depends heavily on the scalablity and robustness of the underlying cloud architecture.

We discussed enterprise cloud computing where enterprises may use a service-oriented architecture to publish a streamlined interface to their business processes. In order to scale up the number of business processes, each tier in the provider's architecture usually deploys multiple servers for load distribution and fault tolerance. Such load distribution across multiple servers within the same tier can be viewed as *horizontal* load distribution. One limitation of this approach is that load cannot be distributed further when all servers in the same tier are fully loaded. Another approach for providing resiliency and scalabilty is to have *multiple implementation options* that give opportunities for *vertical* load distribution across tiers.

We described in detail a request routing framework for SOA-based enterprise cloud computing that takes into account both these options for *horizontal* and *vertical* load distribution. Experiments showed that our algorithm and methodology can scale well up to a large-scale system configuration comprising up to 1000 workflow requests directed to a complex composite service with multiple implementation options available. The experimental results also demonstrate that our framework is more agile in the sense that it is effective in dealing with mis-configured infrastructures in which there are too many or too few servers in one tier. As a result, our framework can effectively utilize available multiple implementations to distribute loads across tiers.

References

Bent, R., & Van Hentenryck, P. (2004). Regrets Only! Online stochastic optimization under time constraints. *Nineteenth National Conference on Artificial Intelligence, San Jose, CA.*

Buco, M. J., Chang, R. N., Luan, L. Z., Ward, C., Wolf, J. L., Yu, P. S., et al. (2003). Managing ebusiness on demand sla contracts in business terms using the cross-sla execution manager sam. *ISADS, Washington, DC,* 157–164.

Business Process Execution Language for Web Services (Version 1.1), (2005). www-128.ibm.com/developerworks/library/ws-bpel/.

Casati, F., Ilnicki, S., Jin, L., Krishnamoorthy, V., & Shan, M.-C. (2000). Adaptive and Dynamic Service Composition in eFlow. *Proceedings of CAISE, Stockholm, Sweden,* 13–31.

Cisco. Ace application-level load balancer. http://www.cisco.com/en/US/products/ps6906/. Accessed July 29, 2010.

Cisco. Scalable Content Switching. http://www.cisco.com/en/US/products/hw/contnetw/ps792/products_white_paper09186a0080136856.shtml. Accessed July 29, 2010.

Cloud Computing (2009). Clash of the clouds. *The Economist.*

Costa, L., & Oliveira, P. (2001). Evolutionary algorithms approach to the solution of mixed integer nonlinear programming problems. *Computers and Chemical Engineering, 25*(2–3), 257–266.

Davis, L. (1985). Job shop scheduling with genetic algorithms. *Proceedings of the International Conference on Genetic Algorithms, Pittsburgh, PA.*

DeCandia, G., Hastorun, D., Jampani, M., Kakulapati, G., Lakshman, A., Pilchin, A., Sivasubramanian, S., Vosshall, P., & Vogels, W. (2007). Dynamo: Amazon's highly available key-value store. *Proceedings of SOSP, Washington D.C.,* 205–220.

Dewri, R., Ray, I., Ray, I., & Whitley, D. (2008). Optimizing on-demand data broadcast scheduling in pervasive environments. *Proceedings of EDBT, Nantes, France,* 559–569.

Dikaiakos, M., Pallis, G., Katsaros, D., Mehra, P., & Vakali, A. (2009). Cloud computing: Distributed internet computing for it and scientific research. *IEEE Internet Computing, 13*(5), 10–13.

F5 Networks. Big-ip application-level load balancer. http://www.f5.com/products/big-ip/. Accessed July 29, 2010.

Goldberg, D. (1989). *Genetic algorithms in searth, optimization, and machine learning.* Dordrecht: Kluwer.

Gu, X., Nahrstedt, K., Chang, R. N., & Ward, C. (2003). Qos-assured service composition in managed service overlay networks. *Proceedings of ICDCS, Providence, Rhode Island, USA,* 194–203.

Holland, J. (1992). *Adaptation in natural and artificial systems.* Cambridge, MA: MIT Press.

Jin, J., & Nahrstedt, K. (2004). On exploring performance optimisations in web service composition. *Proceedings of Middleware, Toronto, Canada.*

Josephraj, J. (2007). *Web services choreography in practice.* www-128.ibm.com/developerworks/library/ws-choreography. Accessed July 29, 2010.

Lima, R., Francois, G., Srinivasan, B., & Salcedo, R. (2004). Dynamic optimization of batch emulsion polymerization using MSIMPSA, a simulated-annealing-based algorithm. *Industrial and Engineering Chemistry Research, 43*(24), 7796–7806.

Oliveira, R., & Salcedo, R. (2005). Benchmark testing of simulated annealing, adaptive random search and genetic algorithms for the global optimization of bioprocesses. *International Conference on Adaptive and Natural Computing Algorithms, Coimbra, Portugal.*

Phan, T., & Li, W.-S. (2008a). Dynamic materialization of query views for data warehouse workloads. *Proceedings of the International Conference on Data Engineering, Long Beach, CA.*

Phan, T., & Li, W.-S. (2008b). Load distribution of analytical query workloads for database cluster architectures. *Proceedings of EDBT, Nantes, France,* 169–180.

Ponnekanti, S., & Fox, A. (2004). Interoperability among Independently Evolving Web Services. *Proceedings of Middleware, Toronto, Canada.*

Shankar, M., De Miguel, M., & Liu, J. W.-S. (2004). An end-to-end qos management architecture. *Proceedings of the Fifth IEEE Real Time Technology and Applications Symposium, Vancouver, British Columbia, Canada*, p. 176.

Soundararajan, G., Manassiev, K., Chen, J., Goel, A., & Amza, C. (2005). Back-end databases in shared dynamic content server clusters. *Proceedings of ICAC, Dublin, Ireland*.

Tang, C., Chang, R. N., & So, E. (2006). A distributed service management infrastructure for enterprise data centers based on peer-to-peer technology. *IEEE SCC, Chicago, IL*, 52–59.

Urgaonkar, B., Shenoy, P., Chandra, A., & Goyal, P. (2005). Dynamic provisioning of multi-tier internet applications. *Proceedings of ICAC, Seattle, WA*.

Van Hentenryck, P., & Bent, R. (2006). *Online stochastic combinatorial optimization*. Cambridge, MA: MIT Press.

Ward, C., Buco, M. J., Chang, R. N., Luan, L. Z., So, E., & Tang, C. (2005). Fresco: A web services based framework for configuring extensible sla management systems. *ICWS, Sunshine Coast, Australia* 237–245.

Yu, T., & Lin, K.-J. (2005a). Adaptive algorithms for finding replacement services in autonomic distributed business processes. *Proceedings of the 7th International Symposium on Autonomous Decentralized Systems, Chengdu, China*.

Yu, T., & Lin, K.-J. (2005b). Service selection algorithms for web services with end-to-end qos constraints. *Information Systems and E-Business Management, 3*(2), 103–126.

Yu, T., & Lin, K.-J. (2006). Qcws: An implementation of qos-capable multimedia web services. *Multimedia Tools and Applications, 30*(2), 165–187.

Yu, T., Zhang, Y., & Lin, K.-J. (2007). Efficient algorithms for web services selection with end-to-end qos constraints. *ACM Transactions on the Web (TWEB), 1*(1). http://portal.acm.org/citation.cfm?id=1232722.1232728. Accessed July 29, 2010.

Zeng, L., Benatallah, B., Dumas, M., Kalagnanam, J., & Sheng, Q. (2003). quality driven web services composition. *Proceedings of WWW, Helsinki, Finland*.

Chapter 13
SwinDeW-C: A Peer-to-Peer Based Cloud Workflow System

Xiao Liu, Dong Yuan, Gaofeng Zhang, Jinjun Chen, and Yun Yang

13.1 Introduction

Workflow systems are designed to support the process automation of large scale business and scientific applications. In recent years, many workflow systems have been deployed on high performance computing infrastructures such as cluster, peer-to-peer (p2p), and grid computing (Moore, 2004; Wang, Jie, & Chen, 2009; Yang, Liu, Chen, Lignier, & Jin, 2007). One of the driving forces is the increasing demand of large scale instance and data/computation intensive workflow applications (large scale workflow applications for short) which are common in both eBusiness and eScience application areas. Typical examples (will be detailed in Section 13.2.1) include such as the transaction intensive nation-wide insurance claim application process; the data and computation intensive pulsar searching process in Astrophysics. Generally speaking, instance intensive applications are those processes which need to be executed for a large number of times sequentially within a very short period or concurrently with a large number of instances (Liu, Chen, Yang, & Jin, 2008; Liu et al., 2010; Yang et al., 2008). Therefore, large scale workflow applications normally require the support of high performance computing infrastructures (e.g. advanced CPU units, large memory space and high speed network), especially when workflow activities are of data and computation intensive themselves. In the real world, to accommodate such a request, expensive computing infrastructures including such as supercomputers and data servers are bought, installed, integrated and maintained with huge cost by system users. However, since most of these resources are self-contained and organised in a heterogeneous way, resource scalability, i.e. how easily a system can expand and contract its resource pool to accommodate heavier or lighter work loads, is very poor. Due to such a problem, on one hand, it requires great cost, if not impossible, to recruit external resources to address "resource insufficiency" during peak periods; on the other hand,

X. Liu (✉), D. Yuan, G. Zhang, J. Chen, and Y. Yang
Faculty of Information and Communication Technologies, Swinburne University of Technology, Hawthorn, Melbourne, Australia 3122
e-mails: {xliu@groupwise.swin.edu.au; dyuan@groupwise.swin.edu.au; gzhang@groupwise.swin.edu.au; jchen@swin.edu.au; yyang@groupwise.swin.edu.au}

B. Furht, A. Escalante (eds.), *Handbook of Cloud Computing*,
DOI 10.1007/978-1-4419-6524-0_13, © Springer Science+Business Media, LLC 2010

it cannot provide services to others during off-peak periods to make full advantage of the investment. In current computing paradigms, workflow systems have to maintain their own high performance computing infrastructures rather than employ them as services from a third party according to their real needs. Meanwhile, most of the resources are idled except for bursting resource requirements of large scale workflow applications at peak periods. In fact, many workflow systems also need to deal with a large number of conventional less demanding workflow applications for large proportion of the time. Therefore, resource scalability is becoming a critical problem for current workflow systems. However, such an issue has not been well addressed by current computing paradigms such as cluster and grid computing.

In recent years, cloud computing is emerging as the latest distributed computing paradigm and attracts increasing interests of researchers in the area of Distributed and Parallel Computing (Raghavan, Ramabhadran, Yocum, & Snoeren, 2007), Service Oriented Computing (Ardagna & Pernici, 2007) and Software Engineering (SECES, 2008). As proposed by Ian Foster in (Foster, Zhao, Raicu, & Lu, 2008) and shared by many researchers and practitioners, compared with conventional computing paradigms, cloud computing can provide "a pool of abstracted, virtualised, dynamically-scalable, managed computing power, storage, platforms, and services are delivered on demand to external customers over the Internet". Therefore, cloud computing can provide scalable resources on demand to system requirement. Meanwhile, cloud computing adopts market-oriented business model where users are charged according to the usage of cloud services such as computing, storage and network services like conventional utilities in everyday life (e.g. water, electricity, gas and telephony) (Buyya, Yeo, Venugopal, Broberg, & Brandic, 2009). Evidently, it is possible to utilise cloud computing to address the problem of resource scalability for managing large scale workflow applications. Therefore, the investigation of workflow systems based on cloud computing, i.e. cloud workflow systems, is a timely issue and worthwhile for increasing efforts.

Besides scalable resources, another principal issue for large scale workflow applications is decentralised management. In order to achieve successful execution, effective coordination of system participants (e.g. service providers, service consumers and service brokers) is required for many management tasks such as resource management (load management, workflow scheduling), QoS (Quality of Service) management, data management, security management and others. One of the conventional ways to solve the coordination problem is centralised management where coordination services are set up on a centralised machine. All the communications such as data and control messages are transmitted only between the central node and other resource nodes but not among them. However, centralised management depends heavily on the central node and thus can easily result in the performance bottleneck. Some others common disadvantages also include: single point of failure, lack of scalability and the advanced computation power required for the coordination services. To overcome the problems of centralised management, decentralised management where the centralised data repository and control engine are abandoned, and both data and control messages are transmitted between all the nodes through general broadcast or limited broadcast communication mechanisms. Thus the performance bottlenecks are likely eliminated and the system scalability can be greatly enhanced. Peer to Peer (p2p) is a typical decentralised architecture. However, without any centralised coordination, pure p2p (unstructured decentralised) where all the peer nodes are communicating with each other through complete broadcasting suffers from low efficiency and high network load. Evidently, neither centralised nor unstructured decentralised management is suitable for managing large scale workflow applications since massive communication and coordination services are required. Therefore, in practice, structured p2p architecture is often applied where a super node acts as the coordinator peers for a group of peers. Through those super nodes which maintain all the necessary information about the neighbouring nodes, workflow management tasks can be effectively executed where data and control messages are transmitted in a limited broadcasting manner. Therefore, structured decentralised management is more effectively than other for managing workflow applications.

Based on the above analysis, it is evident that cloud computing is a promising solution to address the requirement of scalable resource, and structured decentralised architecture such as structured p2p is an effective solution to address the requirement of decentralised management. Therefore, in this chapter, we present SwinDeW-C (**Swin**burne **De**centralised **W**orkflow for **C**loud), a peer to peer based Cloud workflow system for managing large scale workflow applications. SwinDeW-C is not built from the scratch but based on our existing SwinDeW-G (Yang et al., 2007) (a peer-to-peer based grid workflow system) which will be introduced later in Section 13.3. As agreed among many researchers and practitioners, the general cloud architecture includes four basic layers from top to bottom: application layer (user applications), platform layer (middleware cloud services to facilitate the development/deployment of user applications), unified resource layer (abstracted/encapsulated resources by virtualisation) and fabric layer (physical hardware resources) (Foster et al., 2008). In order to support large scale workflow applications, a novel SwinDeW-C architecture is presented where the original fabric layer of SwinDeW-G is inherited with the extension of external commercial cloud service providers. Meanwhile, significant modifications are made to the other three layers: the underlying resources are virtualised at the unified resource layer; functional components are added or enhanced at the platform layer to support the management of large scale workflow applications; the user interface is modified to support Internet (Web browser) based access.

This chapter describes the novel system architecture and the new features of SwinDeW-C. Specifically, based on a brief introduction about SwinDeW-G, the architecture of SwinDeW-C as well as the architecture of SwinDeW-C peers (including both ordinary peers and coordinator peers) is proposed. Meanwhile, besides common features for cloud computing and workflow systems, additional new functional components are enhanced or designed in SwinDeW-C to facilitate large scale workflow applications. In this chapter, three new functional components including QoS Management, Data Management and Security Management are presented as the key components for managing large scale workflow applications. The SwinDeW-C prototype system is demonstrated to verify the effectiveness of

SwinDeW-C architecture and the feasibility of building cloud workflow system based on existing grid computing platform.

The remainder of the paper is organised as follows. Section 13.2 presents the motivation and system requirements. Section 13.3 introduces our SwinDeW-G grid computing environment. Section 13.4 proposes the architecture for SwinDeW-C as well as SwinDeW-C peers. Section 13.5 presents the new components in SwinDeW-C for managing large scale workflow applications. Section 13.6 presents SwinDeW-C system prototype. Section 13.7 introduces the related work. Finally, Section 13.8 addresses the conclusion and feature work.

13.2 Motivation and System Requirement

In this section, we first present some examples to illustrate the motivation for utilising cloud computing to facilitate large scale workflow applications. Afterwards, based on the introduction of our existing SwinDeW-G grid computing environment, system requirements for cloud workflow systems are presented.

13.2.1 Large Scale Workflow Applications

Here, we present two examples, one is from the business application area (insurance claim) and the other one is from the scientific application area (pulsar searching).

Insurance claim: Insurance claim process is a common business workflow which provides services for processes of such as insurance under employee benefits including, for example, medical expenses, pension, and unemployment benefits. Due to the distributed geographic locations of a large number of applicants, the insurance offices are usually deployed at many locations across a wide area serving for a vast population. Despite the differences among specific applications, the following requirements are often seen in large/medium sized insurance companies: (1) supporting a large number of processes invoked from anywhere securely on the Internet, the privacy of applicants and confidential data must be protected; (2) avoiding management of the system at many different locations due to the high cost for the setting and ongoing maintenance; (3) being able to serve for a vast population involving processes at the minimum scale of tens of thousands per day, i.e. instance intensive; and (4) for better quality of service, needing to handle workflow exceptions appropriately, particularly in the case of instance-intensive processes.

Pulsar searching: The pulsar searching process is a typical scientific workflow which involves a large number of data intensive and computation intensive activities. For a single searching process, the average data volume (not including the raw stream data from the telescope) can be over terabytes and the average execution time can be about one day. In a single searching process, many parallel execution paths need to be executed for the data collected from different beams of the telescope, and each execution path includes four major segments: data collection, data pre-processing, candidate searching and decision making. Take the operation of de-dispersion in the data pre-processing segment as an example. De-dispersion is to generate a large number of de-dispersion files to correct the pulsar signals which are dispersed by the interstellar medium. A large number of de-dispersion files need to be generated according to different choices of trial dispersion and normally take many hours on high performance computing resources. After the generation of large volume of de-dispersion files, different pulsar candidate seeking algorithms such as FFT seek, FFA seek, and single pulse seek will be further implemented. Finally, the results will be visualised to support the decision of human experts on whether a pulsar has been found or not. Generally speaking, the pulsar searching process often has the following requirements: (1) easy to scale up and down the employed computing resources for data processing at different stages; (2) for better QoS, especially efficient scheduling of parallel computing tasks so that every pulsar searching process can be finished on time; (3) decreasing the cost on data storage and data transfer, specific strategies are required to determine the allocation of generated data along workflow execution; (4) selecting trustworthy service nodes, ensuring the security of data storage, especially for those need to be stored for long term.

13.2.2 System Requirements

Based on the introduction about the above two examples, here, we present the system requirements for managing large scale workflow applications. Besides the two fundamental requirements, namely scalable resource and decentralised management, which have been discussed in the introduction section, there are three important system requirements, including: QoS Management, Data Management, and Security Management.

13.2.2.1 QoS Management

It is critical to deliver services with user satisfied quality of service (QoS) in cloud computing environment, otherwise, the reputation of the service providers will be deteriorated and finally eliminated from the cloud market. Generally speaking, there are 5 major dimensions of QoS constraints including time, cost, fidelity, reliability and security (Yu & Buyya, 2005). Among them, time, as the basic measurement for system performance, is probably the most general QoS constraint in all application systems. Especially for large scale workflow applications, temporal QoS is very important since any large delays may result in poor efficiency or even system bottlenecks. Therefore, in this paper, we mainly focus on temporal constraints as the example for QoS management.

For a general cloud workflow application, both global and local temporal constraints are assigned at workflow build time through service level agreement (SLA) (Erl, 2008). Then, at workflow run time, due to the highly dynamic system performance (activity durations with large deviations (Liu, Chen, Liu, & Yang, 2008)), workflow execution is under constant monitoring against the violations of these temporal constraints (Chen & Yang, 2008a, 2010, 2008b). If a temporal violation is detected (i.e. the workflow execution time exceeds the temporal constraint), exception handling strategies will be triggered to compensate the occurring time delays. Therefore, to deliver satisfactory temporal QoS (as well as other QoS constraints), a set of strategies should be adopted or designed to facilitate at least the following three tasks: the setting of QoS constraints, the monitoring of workflow execution against QoS constraint violations, and the handling of QoS constraint violations.

13.2.2.2 Data Management

Large scale workflow applications often come along with data intensive computing (Deelman, Gannon, Shields, & Taylor, 2008), where workflow tasks will access large datasets for query or retrieving data, and during the workflow execution similar amounts or even larger datasets will be generated as intermediate or final products (Deelman & Chervenak, 2008). Data management in cloud workflow systems has some new requirements, which becomes an important issue. Firstly, new data storage strategy is required in cloud workflow systems (Yuan, Yang, Liu, & Chen, in press). In a cloud computing, theoretically, the system can offer unlimited storage resources. All the application data can be stored, including the intermediate data, if we are willing to pay for the required resources. Hence, we need a strategy to cost-effectively store the large application data. Secondly, new data placement strategy is also required (Yuan, Yang, Liu, & Chen, 2010). Cloud computing platform contains different cloud service providers with different pricing models, where data transfers between service providers also carry a cost. The cloud workflows are usually distributed, and the data placement strategy will decide where to store the application data, in order to reduce the total system cost. Last but not least, new data replication strategy should also be designed for cloud workflow systems (Chervenak et al., 2007). A good replication strategy can not only guarantee the security of application data, but also further reduce the system cost by replicating frequently used data in different locations. Replication strategy in the cloud should be dynamic based on the application data's usage rate.

13.2.2.3 Security Management

Security always plays an important role in distributed computing systems (Lin, Varadharajan, Wang, & Pruthi, 2004). To ensure high QoS of these systems, we focus on the security problems brought by different types of components, large volume of heterogeneous data, and unpredictable execution processes. Since some general aspects of system security such as service quality and data security are partially included in the previous QoS and data management components, this chapter emphasises the trust management which plays an important role in the management of SwinDeW-C peers. In the large scale workflow applications, to meet the high requirements of quality and scalability, an efficient and adaptive trust management is an indispensable part of the SwinDeW-C platform (Bhargav-spantzel, Squicciarini, & Bertino, 2007; Winsborough & Li, 2006). Besides, User management is essential to guarantee system security and avoid illegal usage. Facing the complex network structures in the cloud environment, we also need encryption technology to protect privacy, integrity, authenticity and undeniableness.

Given these basic system requirements, cloud computing is a suitable solution to address the issue of resource scalability and p2p is an effective candidate for decentralised management. Meanwhile, to adapt the system requirements of large scale workflow applications, functional components for Data Management, QoS Management and Security Management are required to be designed or enhanced to guarantee satisfactory system performance.

13.3 Overview of SwinDeW-G Environment

Before we present SwinDeW-C, some background knowledge about SwinDeW-G needs to be introduced. SwinDeW-G (**Swin**burne **De**centralised **W**orkflow for **G**rid) is a peer-to-peer based scientific grid workflow system running on the SwinGrid (Swinburne service Grid) platform (Yang et al., 2007).

An overall picture of SwinGrid is depicted in Fig. 13.1 (bottom plane). SwinGrid contains many grid nodes distributed in different places. Each grid node contains many computers including high performance PCs and/or supercomputers composed of significant numbers of computing units. The primary hosting nodes include the

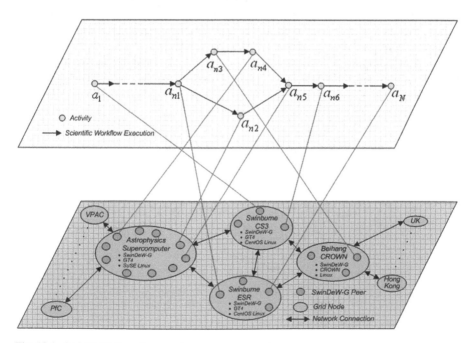

Fig. 13.1 SwinDeW-G environment

Swinburne CS3 (Centre for Complex Software Systems and Services) Node, the Swinburne ESR (Enterprise Systems Research laboratory) Node, the Swinburne Astrophysics Supercomputer Node, and the Beihang CROWN (China R&D environment Over Wide-area Network) Node in China. They are running either Linux, GT4 (Globus Toolkit) or CROWN grid toolkit 2.5 where CROWN is an extension of GT4 with more middleware, and thus is compatible with GT4. The CROWN Node is also connected to some other nodes such as those at the Hong Kong University of Science and Technology, and at the University of Leeds in the UK. The Swinburne Astrophysics Supercomputer Node is cooperating with the Australian PfC (Platforms for Collaboration) and VPAC (Victorian Partnership for Advanced Computing). Currently, SwinDeW-G is deployed at all primary hosting nodes as exemplified in the top plane of Fig. 13.1. In SwinDeW-G, a scientific workflow is executed by different peers that may be distributed at different grid nodes. As shown in Fig. 13.1, each grid node can have a number of peers, and each peer can be simply viewed as a grid service. In the top plane of Fig. 13.1, we show a sample of how a scientific workflow can be executed in the grid computing environment.

The basic service unit in SwinDeW-G is a SwinDeW-G peer which runs as a grid service along with other grid services. However, it communicates with other peers via JXTA (http://www.sun.com/software/jxta/), a platform for p2p communication. As Fig. 13.2 shows, a SwinDeW-G peer consists of the following components:

The Task Component manages the workflow tasks. It has two main functions. First, it provides necessary information to the Flow Component for scheduling and stores received tasks to Task Repository. Second, it determines the appropriate time to start, execute and terminate a particular task. The resources that a workflow task instance may require are stored in the Resource Repository.

The Flow Component interacts with all other modules. First, it receives the workflows definition and then creates the instance definition. Second, it receives tasks from other peers or redistributes them. Third, it decides whether to pass a task to

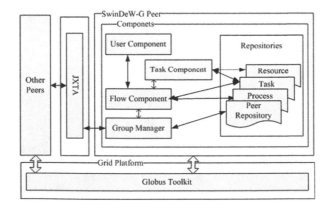

Fig. 13.2 Architecture of a SwinDeW-G Peer

the Task Component to execute locally or distribute it to other peers. The decision is made according to the capabilities and load of itself and other neighbours. And finally, it makes sure that all executions conform to the data dependency and control dependency of the process definitions which are stored in the Process Repository and the Task Repository.

The Group Manager is the interface between the peer and JXTA. In JXTA, all communications are conducted in terms of peer group, and the Group Manager maintains the peer groups the peer has joined. The information of the peer groups and the peers in them is stored in the Peer Repository. While a SwinDeW-G peer is implemented as a grid service, all direct communications between peers are conducted via p2p. Peers communicate to distribute information of their current state and messages for process control such as heartbeat, process distribution, process enactment etc.

The User component is the interface between the corresponding workflow users and the workflow environment. In SwinDeW-G, its primary function is to allow users to interfere with the workflow instances when exceptions occur.

Globus Toolkit serves as the grid service container of SwinDeW-G. Not only a SwinDeW-G peer itself is a grid service located inside Globus Toolkit, the capabilities which are needed to execute certain tasks are also in forms of grid services that the system can access. That means when a task is assigned to a peer, Globus Toolkit will be used to provide the required capability as grid service for that task.

13.4 SwinDeW-C System Architecture

In this section, the system architecture of SwinDeW-C is introduced. SwinDeW-C (**Swin**burne **De**centralised **W**orkflow for **C**loud) is built on SwinCloud cloud computing infrastructure. SwinDeW-C inherits many features of its ancestor SwinDeW-G but with significant modifications to accommodate the novel cloud computing paradigm for managing large scale workflow applications.

13.4.1 SwinCloud Infrastructure

SwinCloud is a cloud computing simulation environment, on which SwinDeW-C is currently running. It is built on the computing facilities in Swinburne University of Technology and takes advantage of the existing SwinGrid systems. We install VMWare (VMware, 2009) on SwinGrid, so that it can offer unified computing and storage resources. Utilising the unified resources, we set up data centres that can host applications. In the data centres, Hadoop (2009) is installed that can facilitate Map-Reduce computing paradigm and distributed data management. The architecture of SwinCloud is depicted in Fig. 13.3.

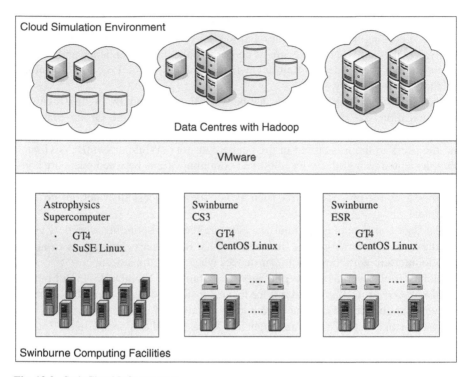

Fig. 13.3 SwinCloud Infrastructure

13.4.2 Architecture of SwinDeW-C

The architecture of SwinDeW-C is depicted in Fig. 13.4. As discussed earlier, the general cloud architecture includes four basic layers from top to bottom: application layer (user applications), platform layer (middleware cloud services to facilitate the development/deployment of user applications), unified resource layer (abstracted/encapsulated resources by virtualisation) and fabric layer (physical hardware resources). Accordingly, the architecture of SwinDeW-C can also be mapped to the four basic layers. Here, we present the lifecycle of an abstract workflow application to illustrate the system architecture. Note that here we focus on the system architecture, the introduction on the cloud management services (e.g. brokering, pricing, accounting, and virtual machine management) and other functional components are omitted here and will be introduced in the subsequent sections.

Users can easily get access to SwinDeW-C Web portal (as demonstrated in Section 13.6) via any electronic devices such as PC, laptop, PDA and mobile phone as long as they are connected to the Internet. Compared with SwinDeW-G which can only be accessed through a SwinDeW-G peer with pre-installed programs, the SwinDeW-C Web portal has greatly improved its usability. At workflow build-time stage, given the cloud workflow modelling tool provided by the Web portal on the application layer, workflow applications are modelled by users as

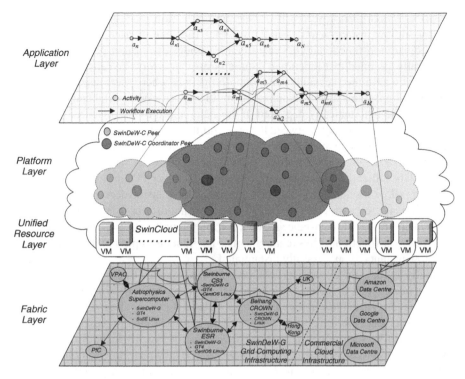

Fig. 13.4 Architecture of SwinDeW-C

cloud workflow specifications (consist of such as task definitions, process structures and QoS constraints). After workflow specifications are created (static verification tools for such as structure errors and QoS constraints may also be provided), they will be submitted to any one of the coordinator peers on the platform layer. Here, an ordinary SwinDeW-C peer is a cloud service node which has been equipped with specific software services similar to a SwinDeW-G peer. However, while a SwinDeW-G peer is deployed on a standalone physical machine with fixed computing units and memory space, a SwinDeW-C peer is deployed on a virtual machine of which the computing power can scale dynamically according to task request. As for the SwinDeW-C coordinator peers, they are super nodes equipped with additional workflow management services compared with ordinary SwinDeW-C peers. Details about SwinDeW-C peers will be introduced in the next section.

At the run-time instantiation stage, the cloud workflow specification can be submitted to any of the SwinDeW-C coordinator peers. Afterwards, the workflow tasks will be assigned to suitable peers through peer to peer based communication between SwinDeW-C peers. Since the peer management such as peer join, peer leave and peer search, as well as the p2p based workflow execution mechanism, is the same as in SwinDeW-G system environment. Therefore, the detailed introduction is omitted here but can be found in (Yang et al., 2007). Before workflow execution, a coordinator peer will conduct an evaluation process on the submitted cloud workflow instances to determine whether they can be accepted or not given the specified non-functional QoS requirements under the current pricing model. It is generally assumed that functional requirements can always be satisfied given the theoretically unlimited scalability of cloud. In the case where users need to run their own special programs, they can upload them through the Web portal and these programs will be automatically deployed in the data centre by the resource manager. Here, a negotiation process between the user and the cloud workflow system may be conducted if the user submitted workflow instance is not acceptable to the workflow system due to the unacceptable offer on budgets or deadlines. The final negotiation result will be either the compromised QoS requirements or a failed submission of the cloud workflow instance. If all the task instances have been successfully allocated (i.e. acceptance messages are sent back to the coordinator peer from all the allocated peers), a cloud workflow instance may be completed with satisfaction of both functional and non-functional QoS requirements (if without exceptions). Hence, a cloud workflow instance is successfully instantiated.

Finally, at run-time execution stage, each task is executed by a SwinDeW-C peer. In cloud computing, the underlying heterogeneous resources are virtualised as unified resources (virtual machines). Each peer utilises the computing power provided by its virtual machine which can easily scale according to the request of workflow tasks. As can be seen in the unified resource layer of Fig. 13.4, the SwinCloud is built on the previous SwinGrid infrastructure at the fabric layer. Meanwhile, some of the virtual machines can be created with external commercial IaaS (infrastructure as service) cloud service providers such as Amazon, Google and Microsoft. During cloud workflow execution, workflow management tasks such as QoS management, data management and security management are executed by the coordinator peers in order to achieve satisfactory system performance. Users can get access to the final results as well as the running information of their submitted workflow instances at any time through the SwinDeW-C Web portal.

13.4.3 Architecture of SwinDeW-C Peers

In this section we will introduce the architecture of a SwinDeW-C peer. As we described above, SwinDeW-C is developed based on SwinDeW-G, where a SwinDeW-C peer has inherited most of the SwinDeW-G peer's components, including the components of task management, flow management, repositories, and the group management. Hence the SwinDeW-G peer plays as the core of a SwinDeW-C peer, which provides the basic workflow management components and communication components between peers. However, some improvements are also made on SwinDeW-C peers to accommodate the cloud computing environment. The architecture of the SwinDeW-C peers is depicted in Fig. 13.5.

Firstly, different from a SwinDeW-G peer, a SwinDeW-C peer runs on the cloud platform. The cloud platform is composed of unified resources, which means the

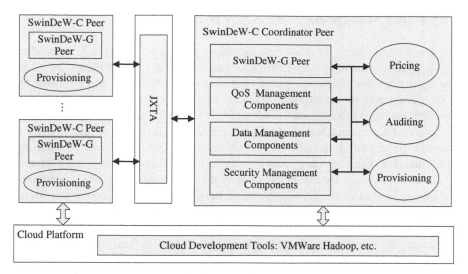

Fig. 13.5   Architecture of SwinDeW-C peers

computation and storage capabilities a SwinDeW-C peer can dynamically scale up or down based on the applications' requirements. Unified resources are offered by cloud service providers and managed in resource pools, hence every SwinDeW-C peer has a provisioning component to dynamically apply and release the cloud resources.

Secondly, in cloud computing environment, different cloud service providers may have different cost model, hence we have to set up a coordinator peer within every cloud service provider. The coordinator peer has the pricing and auditing components, which can coordinate the resource provisioning of all the peers that reside in this service provider.

Last but not least, the coordinator peer of SwinDeW-C also has new functional components related to cloud workflow management. As introduced in Section 13.2.2, the system has new requirements for handling the large scale workflow applications. To meet these new requirements, components of QoS management, data management and security management are added to the SwinDeW-C coordinator peer. More detailed descriptions of these components will be given in the following section.

13.5 New Components in SwinDeW-C

In this section, we introduce the new components in SwinDeW-C. As the three system requirements presented in Section 13.2.2, the three new functional components including QoS Management, Data Management and Security Management are introduced.

13.5.1 QoS Management in SwinDeW-C

The basic requirement for delivering satisfactory temporal QoS (as well as other QoS constraints) includes three basic tasks: the setting of QoS constraints, the monitoring of workflow execution against QoS constraint violations, and the handling of QoS constraint violations. Here, take temporal QoS constraints for example, the new QoS management component in SwinDeW-C is introduced.

Temporal Constraint Setting: In SwinDeW-C QoS management component, a probabilistic strategy is designed for setting temporal QoS constraints at workflow build time (Liu, Chen, & Yang, 2008). Specifically, with a probability based temporal consistency model, the one global or several coarse-grained temporal constraints are assigned based on the negotiation result between clients and service providers. Afterwards, fine-grained temporal constraints for individual workflow activities can be derived automatically based on these coarse-grained ones.

Checkpoint Selection and Temporal Verification: At workflow run time, a checkpoint selection strategy and a temporal verification strategy are provided to monitor the workflow execution against the violation of temporal constraints. Temporal verification is to check the temporal correctness of workflow execution states (detecting temporal violations) given a temporal consistency model. Meanwhile, in order to save the overall QoS management cost, temporal verification should be conducted only on selected activity points. In SwinDeW-C, a minimum time redundancy based checkpoint selection strategy (Chen & Yang, 2010, 2007b) is employed which can select only necessary and sufficient checkpoints (those where temporal violations take place).

Exception Handling: After a temporal violation is detected, exception handling strategies are required to recover the error states. Unlike functional errors which are normally prevented by duplicated instances or handled by roll back and re-execution, non-functional QoS errors such as temporal violations can only be recovered by compensation, i.e. to reduce or ideally remove the current time delays by decreasing the durations of the subsequent workflow activities. Since the previous activities have already been finished, there is no way in the real world that any action can reduce their running time. In SwinDeW-C, for minor temporal violations, the TDA (time deficit allocation) strategy (Chen & Yang, 2007a) is employed which can remove the current time deficits by borrowing the time redundancy of the subsequent activities. As for major temporal violations, the ACOWR (ant colony optimisation based two stage workflow local rescheduling) strategy (Liu et al., 2010) is employed which can decrease the duration of the subsequent workflow segments through ant colony optimisation based workflow rescheduling.

In SwinDeW-C, by constant monitoring of the workflow instance and effective handling of temporal violations along workflow execution, satisfactory temporal QoS can be delivered with low violation rates of both global and local temporal constraints. Similar to temporal QoS management, the management tasks for other QoS constraints are being investigated. Meanwhile, since some of them such as cost and security are partially addressed in the data management and security management components, some functions will be shared among these components.

13.5.2 Data Management in SwinDeW-C

Data management component in SwinDeW-C consists of three basic tasks: data storage, data placement and data replication.

Data Storage: In this component, a dependency based cost-effective data storage strategy is facilitated to store the application data (Yuan et al., 2010). The strategy utilises the data provenance information of the workflow instances. Data provenance in workflows is a kind of important metadata, in which the dependencies between datasets are recorded (Simmhan, Plale, & Gannon, 2005). The dependency depicts the derivation relationship between the application datasets. In cloud workflow systems, after the execution of tasks, some intermediate datasets may be deleted to save the storage cost, but sometimes they have to be regenerated for either reuse or reanalysis (Bose & Frew, 2005). Data provenance records the information of how the datasets have been generated. Furthermore, regeneration of the intermediate datasets from the input data may be very time consuming, and therefore carry a high computation cost. With data provenance information, the regeneration of the demanding dataset may start from some stored intermediated datasets instead. In a cloud workflow system, data provenance is recorded during workflow execution. Taking advantage of data provenance, we can build an Intermediate data Dependency Graph (IDG) based on data provenance (Yuan et al., 2010). All the intermediate datasets once generated in the system, whether stored or deleted, their references are recorded in the IDG. Based on the IDG, we can calculate the generation cost of every dataset in the cloud workflows. By comparing the generation cost and storage cost, the storage strategy can automatically decide whether a dataset should be stored or deleted in the cloud system to reduce the system cost, no matter this dataset is a new dataset, regenerated dataset or stored dataset in the system.

Data Placement: In this component, a data placement strategy is facilitated to place the application data that can reduce the data movement during the workflows' execution. In cloud computing systems, the infrastructure is hidden from users (Weiss, 2007). Hence, for application data, the system will decide where to store them. In the strategy, we initially adapt the k-means clustering algorithm for data placement in cloud workflow systems based on data dependency (Yuan et al., in press). Cloud workflows can be complex, one task might require many datasets for execution; furthermore, one dataset might also be required by many tasks. If some datasets are always used together by many tasks, we say that these datasets are dependant on each other. In our strategy, we try to keep these datasets in one data centre, so that when tasks were scheduled to this data centre, most, if not all, of the data needed are stored locally. Our data placement strategy has two algorithms, one for the build-time stage and one for the run time stage of scientific workflows. In the build-time stage algorithm, we construct a dependency matrix for all the application data, which represents the dependencies between all the datasets. Then we use the BEA algorithm (McCormick, Sehweitzer, & White, 1972) to cluster the matrix and partition it that datasets in every partition are highly dependent upon each other. We distribute the partitions into k data centres, which are initially as the partitions of the k-means algorithm at run time stage. At run time, our clustering algorithm deals with the newly generated data that will be needed by other tasks. For every newly generated dataset, we calculate its dependencies with all k data centres, and move the data to the data centre that has the highest dependency with it.

Data Replication: In this component, a dynamic data replication strategy is facilitated to guarantee data security and the fast data access of the cloud workflow systems. Keeping some replicas of the application data is essential for data security in cloud storage. Static replication can guarantee the data reliability by keeping a fixed number of replicas of the application data, but in a cloud environment, different application data have different usage rates, where the strategy has to be dynamic to replicate the application data based on their usage rates. In large scale workflow applications, many parallel tasks will simultaneously access the same dataset on one data centre. The limitation of computing capacity and bandwidth in that data centre would be a bottleneck for the whole cloud workflow system. If we have several replicas in different data centres, this bottleneck will be eliminated. Hence the data replication will always keep a fix number of copies of all the datasets in different data centres to guarantee reliability and dynamically add new replicas for each dataset to guarantee data availability. Furthermore, the placement of the replicas is based on data dependency, which is the same as the data placement component, and how many replicas a dataset should have is based on the usage rate of this dataset.

13.5.3 Security Management in SwinDeW-C

To address the security issues for the safe running of SwinDeW-C, the security management component is designed. As a type of typical distributed computing system, trust management for SwinDeW-C peers is very important and plays the most important role in security management. Besides, there are some other security issues that we should consider from such as user and data perspective. Specifically, there are three modules in the security management component: trust management, user management and encryption management system.

Trust management: The goal of the trust management module is to manage the relations between one SwinDeW-C peer and its neighbouring peers. For example, to process a workflow instance, a SwinDeW-C peer must cooperate with its neighbouring peers to run this instance. Due to the high level QoS requirements of large scale workflow applications, peer management in SwinDeW-C should be supported by the trust management during workflow run time. The trust management module acts like a consultant. This module can evaluate some tasks and give some advices about the cooperated relation between one peer and other peers for each instance of a specific task. Firstly, peer evaluation makes trust assessment of other neighbouring peers. Secondly, task evaluation makes assessment of re-assignment of the task to other peers. Then the two evaluation scores will be combined by the trust evaluation to reach the conclusion whether this neighbouring peer has adequate trust to take this task. Besides, we design a rule base. For instance, a specific task must not be assigned to one specific neighbouring peer, and this is a simple rule. The rule base is a complement to the previous value-based trust evaluation to fit the real situation (Hess, Holt, Jacobson, & Seamons, 2004).

User management: the user management module is an essential piece in every system. In SwinDeW-C, a user base is a database which stores all user identity and log information that submit service requests. In addition, an authority manager controls the permissions for users to submit some special service requests.

Encryption management System: Given SwinDeW-C peers are located within different geographical local networks, it is important to ensure the data security in the process of data transfer by encryption. In SwinDeW-C, we choose the PGP tool GnuPG (http://www.gnupg.org) to ensure secure commutation.

To conclude, besides the above three new functional components, SwinDeW-C also includes the common cloud functional components such as brokering, pricing, auditing and virtual machine management. Detailed description can be found in (Calheiros, Ranjan, De Rose, & Buyya, 2009) and hence omitted in this paper.

13.6 SwinDeW-C System Prototype

Based on the design discussed above, we have built a primitive prototype of SwinDeW-C. The prototype is developed in Java and currently running on the SwinCloud simulation environment. In SwinDeW-C prototype, we have inherited most of SwinDeW-G functions, and further implemented the new components of SwinDeW-C, so that it can adapt to the cloud computing environment. Furthermore, we have built a Web portal for SwinDeW-C, by which users and system administer can access the cloud resources and manage the applications of SwinDeW-C. The Web portal provides many interfaces to support both system users and administrators with the following tasks, specifically for the system user:

(a) browse the existing datasets that reside in different cloud service providers' data centres;
(b) upload their application data to and download the result data from the cloud storage;
(c) create and deploy workflows to SwinDeW-C system using the modelling tools;
(d) monitor the workflows' execution.

For system administers:

(a) coordinate the workflows' execution by triggering the scheduling strategies;
(b) manage the application datasets by triggering the data placement strategies;
(c) handle the execution exceptions by triggering the workflow adjustment strategies.

Some interfaces of the Web portal are shown in Fig. 13.6.

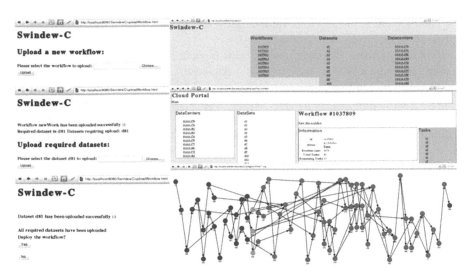

Fig. 13.6 SwinDeW-C web portal

13.7 Related Work

Since the research on cloud workflow management systems is at its initial stage, it is difficult to conduct direct comparison between SwinDeW-C with others at present. Most of the current projects are either on the general implementation of cloud computing or focus on some specific aspects such as data management in the cloud. There exists some research into data-intensive applications on the cloud (Moretti, Bulosan, Thain, & Flynn, 2008), such as early experiences like Nimbus (Keahey, Figueiredo, Fortes, Freeman, & Tsugawa, 2008) and Cumulus (Wang, Kunze, & Tao, 2008) projects. Comparing to the distributed computing systems like cluster and grid, a cloud computing system has a cost benefit (Armbrust et al., 2009). Assunção et al. (2009) demonstrate that cloud computing can extend the capacity of clusters with a cost benefit. Using Amazon clouds' cost model and BOINC volunteer computing middleware, the work in (Kondo, Javadi, Malecot, Cappello, & Anderson, 2009) analyses the cost benefit of cloud computing versus grid computing. In terms of the cost benefit, the work by Deelman, Singh, Livny, Berriman, & Good (2008) shows that cloud computing offers a cost-effective solution for data-intensive applications, such as scientific workflows (Hoffa et al., 2008). The work in (Hoffa et al., 2008) explores the use of cloud computing for scientific workflows, focusing on a widely used astronomy application-Montage. The Cloudbus project (http://www.gridbus.org/cloudbus/) being conducted in the CLOUDS Laboratory at the University of Melbourne are working on a new generalised and extensible cloud simulation framework named CloudSim (Calheiros et al., 2009) which can enable seamless modelling, simulation, and experimentation of cloud computing infrastructures and management services.

With the existing projects for many grid workflow systems developed in recent years, it is agreed by many researchers and practitioners that cloud workflow systems might be built on grid computing environments rather than from scratch. For example, the CloudSim toolkit used in the Cloudbus project is implemented by programmatically extending the core functionalities exposed by the GridSim used in the Gridbus project (2010). Therefore, in this chapter, we review some representative grid workflow system and focus on the related features discussed in this paper such as workflow scheduling architecture, QoS, data and security management. Specifically, we investigate Gridbus (2010), Pegasus (2010), Taverna (2010), GrADS (2010), ASKALON (2010), GridAnt (2010), Triana (2010), GridFlow (2010) and Kepler (2010). For the architecture of the workflow scheduling, Pegasus, Taverna, GrADS, and Kepler use a centralised architecture; Gridbus and GridFlow use a hierarchical architecture; ASKALON and Triana use a decentralised architecture. It is believed that centralised schemes produce more efficient schedules and decentralised schemes have better scalabilities, while hierarchical schemes are their compromises. Similar to SwinDeW-G, SwinDeW-C uses a structured decentralised scheme for workflow scheduling. SwinDeW-G aims at using a performance-driven strategy to achieve an overall load balance of the whole system via distributing tasks to least loaded neighbours.

As far as QoS (quality of service) constraints are concerned, most grid workflow systems mentioned above do not support this feature. Gridbus supports QoS constraints including task deadline and cost minimisation, GrADS and GridFlow mainly use estimated application execution time, and ASKALON supports constrains and properties specified by users or predefined. Right now, SwinDeW-C supports QoS constraints based on task deadlines. When it comes to fault tolerance, at the task level, Gridbus, Taverna, ASKALON, Karajan, GridFlow and Kepler use alternate resource; Taverna, ASKALON and Karajan use retry; GrADS uses rescheduling. At the workflow level, rescue workflow is used by ASKALON and Kepler; user-defined exception handling is used by Karajan and Kepler. Pegasus, GridAnt and Triana use their particular strategies respectively. As a comparison, SwinDeW-C uses effective temporal constraint verification for detecting and handling temporal violations.

As for data management, Kepler has its own actor-oriented data modelling method that for large data in the grid environment. It has two Grid actors, called FileFetcher and FileStager, respectively. These actors make use of GridFTP to retrieve files from, or move files to, remote locations on the Grid. Pegasus has developed some data placement algorithms in the grid environment and uses the RLS (Replica Location Service) system as data management at runtime. In Pegasus, data are asynchronously moved to the tasks on demand to reduce the waiting time of the execution and dynamically delete the data that the task no longer needs to reduce the use of storage. In Gridbus, the workflow system has several scheduling algorithms for the data-intensive applications in the grid environment based on a Grid Resource Broker. The algorithms are designed based on different theories (GA, MDP, SCP, Heuristic), to adapt to different use cases. Taverna proposed a new process definition language, Sculf, which could model application data in a dataflow. It considers workflow as a graph of processors, each of which transfers a set of data inputs into a set of data outputs. ASKALON is a workflow system designed for scheduling. It puts the computing overhead and data transfer overhead together to get a value "weight". It dose not discriminate the computing resource and data host. ASKALON also has its own process definition language called AGWL. Triana is a workflow system which is based on a problem-solving environment that enables the data-intensive scientific application to execute. For the grid, it has an independent abstraction middleware layer, called the Grid Application Prototype (GAP), enables users to advertise, discover and communicate with Web and peer-to-peer (p2p) services. Triana also uses the RLS to manage data at runtime. GridFlow is a workflow system which uses an agent-based system for grid resource management. It considers data transfer to computing resources and archive to storage resources as kinds of workflow tasks.

As for security management, Globus uses public key cryptography (also known as asymmetric cryptography) as the basis for its security management, which represents the main stream in the grid security area. Globus uses the certificates encoded in the X.509 certificate format, an established standard data format. These certificates can be shared among public key based software, including commercial Web browsers from Microsoft and Netscape. The International Grid Trust Federation (IGTF) (http://www.igtf.net/) is a third-party grid trust service provider which aims to establish common policies and guidelines between its Policy Management Authorities (PMAs) members. The IGTF does not provide identity assertions but ensures that within the scope of the IGTF charter, the assertions issued by accredited authorities of any of its PMAs member can meet or exceed an authentication profile relevant to the accredited authority. The European GridTrust project (http://www.gridtrust.eu/gridtrust/) is a novel and ambitious project, which provides new security services at the GRID middleware layer. GridTrust is developing a Usage Control Service to monitor resource usage in dynamic Virtual Organisations (VO), enforce usage policies at run-time, and report usage control policy violations. This service brings dynamic usage control to Grid security in traditional, rigid authorisation models. Other services of the security framework include a Grid Security Requirements editor to allow VO owners and users to define security policies; a Secure-Aware Resource Broker Service to help create VOs based on services with compatible security policies; and a sophisticated Reputation Manager Service, to record past behaviour of VO owners and users as reputation credentials.

13.8 Conclusions and Feature Work

Large scale sophisticated workflow applications are commonly seen in both e-Business and e-Science areas. Workflow systems built on high performance computing infrastructures such as cluster, p2p and grid computing are often applied to support the process automation of large scale workflow applications. However, two fundamental requirements including scalable resources and decentralised management have not been well addressed so far. Recently, with the emergence of cloud computing which is a novel computing paradigm that can provide virtually unlimited, easy-scale computing resources, cloud workflow system is a promising new solution and thus deserves systematic investigation. In this chapter, SwinDeW-C, a novel peer-to-peer based cloud workflow system has been presented. SwinDeW-C is not built from scratch but on its predecessor SwinDeW-G (a p2p based grid workflow system). In order to accommodate the cloud computing paradigm and facilitate the management of large scale workflow applications, significant modifications have been made to the previous SwinDeW-G system. Specifically, the original fabric layer of SwinDeW-G is inherited with the extension of external commercial cloud service providers. Meanwhile, the underlying resources are virtualised at the unified resource layer; functional components including QoS management, data management and security management are added or enhanced at the platform layer to support the management of large scale workflow applications; the user interface is modified to support Internet (Web browser) based access.

This chapter has described the system architecture of SwinDeW-C and its new features for managing instance and data/computation intensive workflow applications. The SwinDeW-C prototype system has been demonstrated but still under further development. In the future, more functional components will be designed and deployed to enhance the capability of SwinDeW-C. Meanwhile, comparison will also be conducted between SwinDeW-C and other workflow systems based on the statistics of performance measurements such as success rate, temporal violation rate, system throughput and others.

Acknowledgment This work is partially supported by Australian Research Council under Linkage Project LP0990393.

References

Ardagna, D., & Pernici, B. (2007). Adaptive service composition in flexible processes. *IEEE Transactions on Software Engineering, 33*(6), 369–384.

Armbrust, M., Fox, A., Griffith, R., Joseph, A. D., Katz, R. H., Konwinski, A., et al. (2009). *Above the clouds: A Berkeley view of cloud computing* (Tech. Rep., University of California, Berkeley).

de Assuncao, M. D., di Costanzo, A., & Buyya, R. (2009). Evaluating the cost-benefit of using cloud computing to extend the capacity of clusters. *Proceedings of the 18th ACM International Symposium on High Performance Distributed Computing, Garching, Germany*, 1–10.

Bhargav-spantzel, A., Squicciarini, A. C., & Bertino, E. (2007). Trust negotiation in identity management. *IEEE Security & Privacy, 5*(2), 55–63.

Bose, R., & Frew, J. (2005). Lineage retrieval for scientific data processing: A survey. *ACM Computing Surveys, 37*(1), 1–28.

Buyya, R., Yeo, C. S., Venugopal, S., Broberg, J., & Brandic, I. (2009). Cloud computing and emerging IT platforms: Vision, hype, and reality for delivering computing as the 5th utility. *Future Generation Computer Systems, 25*(6), 599–616.

Calheiros, R. N., Ranjan, R., De Rose, C. A. F., & Buyya, R. (2009). *CloudSim: A novel framework for modeling and simulation of cloud computing infrastructures and services* (Tech. Rep., Grid

Computing and Distributed Systems (GRIDS) Laboratory, Department of Computer Science and Software Engineering,The University of Melbourne).

Chen, J., & Yang, Y. (2007). Multiple states based temporal consistency for dynamic verification of fixed-time constraints in grid workflow systems. *Concurrency and Computation: Practice and Experience (Wiley), 19*(7), 965–982.

Chen, J., & Yang, Y. (2008). A taxonomy of grid workflow verification and validation. *Concurrency and Computation: Practice and Experience, 20*(4), 347–360.

Chen, J., & Yang, Y. (2010). Temporal dependency based checkpoint selection for dynamic verification of temporal constraints in scientific workflow systems. *ACM Transactions on Software Engineering and Methodology*, to appear. Retrieved 1st February 2010, from http://www.swinflow.org/papers/TOSEM.pdf.

Chen, J., & Yang, Y. (2007). Adaptive selection of necessary and sufficient checkpoints for dynamic verification of temporal constraints in grid workflow systems. *ACM Transactions on Autonomous and Adaptive Systems, 2*(2), Article 6.

Chen, J., & Yang, Y. (2008). Temporal dependency based checkpoint selection for dynamic verification of fixed-time constraints in grid workflow systems. *Proceedings of the 30th International Conference on Software Engineering (ICSE 2008), Leipzig, Germany*, 141–150.

Chervenak, A., Deelman, E., Livny, M., Su, M. H., Schuler, R., Bharathi, S., et al. (2007). Data placement for scientific applications in distributed environments. *Proceedings of the 8th Grid Computing Conference*, 267–274.

Deelman, E., Gannon, D., Shields, M., & Taylor, I. (2008). Workflows and e-Science: An overview of workflow system features and capabilities. *Future Generation Computer Systems, 25*(6), 528–540.

Deelman, E., & Chervenak, A. (2008). Data management challenges of data-intensive scientific workflows. *Proceedings of the IEEE International Symposium on Cluster Computing and the Grid*, 687–692.

Deelman, E., Singh, G., Livny, M., Berriman, B., & Good, J. (2008). The cost of doing science on the cloud: The montage example. *Proceedings of the ACM/IEEE Conference on Supercomputing, Austin, TX*, 1–12.

Erl, T. (2008). *SOA: Principles of service design*. Upper Saddle River, NJ: Prentice Hall.

Foster, I., Zhao, Y., Raicu, I., & Lu, S. (2008). Cloud computing and grid computing 360-degree compared. *Proceedings of the Grid Computing Environments Workshop, 2008, GCE '08*, 1–10.

Hadoop (2009). Retrieved 1st September 2009 from http://hadoop.apache.org/.

Hess, A., Holt, J., Jacobson, J., & Seamons, K. E. (2004). Content-triggered trust negotiation. *ACM Transactions on Information and System Security, 7*(3), 428–456.

Hoffa, C., Mehta, G., Freeman, T., Deelman, E., Keahey, K., Berriman, B., et al. (2008). On the use of cloud computing for scientific workflows. *Proceedings of the 4th IEEE International Conference on e-Science*, 640–645.

Keahey, K., Figueiredo, R., Fortes, J., Freeman, T., & Tsugawa, M. (2008). Science clouds: Early experiences in cloud computing for scientific applications. *Proceedings of the First Workshop on Cloud Computing and its Applications (CCA'08)*, 1–6.

Kondo, D., Javadi, B., Malecot, P., Cappello, F., & Anderson, D. P. (2009). Cost-benefit analysis of cloud computing versus desktop grids. *Proceedings of the IEEE International Symposium on Parallel & Distributed Processing, IPDPS'09*, 1–12.

Lin, C., Varadharajan, V., Wang, Y., & Pruthi, V.t. (2004). Enhancing grid security with trust management. *Proceedings of the 2004 IEEE International Conference on Services Computing (SCC04)*, 303–310.

Liu, K., Chen, J. J., Yang, Y., & Jin, H. (2008). A throughput maximization strategy for scheduling transaction-intensive workflows on SwinDeW-G. *Concurrency and Computation: Practice and Experience, 20*(15), 1807–1820.

Liu, K., Jin, H., Chen, J., Liu, X., Yuan, D., & Yang, Y. (2010). A compromised-time-cost scheduling algorithm in SwinDeW-C for instance-intensive cost-constrained workflows on cloud computing platform. *International Journal of High Performance Computing Applications*.

Liu, X., Chen, J., Liu, K., & Yang, Y. (2008). Forecasting duration intervals of scientific workflow activities based on time-series patterns. *Proceedings of the 4th IEEE International Conference on e-Science (e-Science08), Indianapolis, IN, USA,* 23–30.

Liu, X., Chen, J., Wu, Z., Ni, Z., Yuan, D., & Yang, Y. (2010). Handling recoverable temporal violations in scientific workflow systems: A workflow rescheduling based strategy. *Proceedings of the 10th IEEE/ACM International Symposium on Cluster, Cloud and Grid Computing (CCGrid10), Melbourne, Australia.*

Liu, X., Chen, J., & Yang, Y. (September 2008). A probabilistic strategy for setting temporal constraints in scientific workflows. *Proceedings of the 6th International Conference on Business Process Management (BPM08), Lecture Notes in Computer Science, Vol. 5240, Milan, Italy,* 180–195.

McCormick, W. T., Sehweitzer, P. J., & White, T. W. (1972). Problem decomposition and data reorganization by a clustering technique. *Operations Research, 20,* 993–1009.

Moore, M. (2004). An accurate parallel genetic algorithm to schedule tasks on a cluster. *Parallel Computing, 30,* 567–583.

Moretti, C., Bulosan, J., Thain, D., & Flynn, P. J. (2008). All-Pairs: An abstraction for data-intensive cloud computing. *Proceedings of the IEEE International Parallel and Distributed Processing Symposium, IPDPS'08,* 1–11.

Askalon Project (2010). Retrieved 1st February 2010, from http://www.dps.uibk.ac.at/projects/askalon.

GrADS Project (2010). Retrieved 1st February 2010, from http://www.iges.org/grads/.

GridBus Project (2010). Retrieved 1st February 2010, from http://www.gridbus.org.

Kepler Project (2010). Retrieved 1st February 2010, from http://kepler-project.org/.

Pegasus Project (2010). Retrieved 1st February 2010, from http://pegasus.isi.edu/.

Taverna Project (2010). Retrieved 1st February 2010, from http://www.mygrid.org.uk/tools/taverna/.

Triana Project (2010). Retrieved 1st February 2010, from http://www.trianacode.org/.

Raghavan, B., Ramabhadran, S., Yocum, K., & Snoeren, A. C. (2007). Cloud control with distributed rate limiting. *Proceedings of the 2007 ACM SIGCOMM, Kyoto, Japan,* 337–348.

SECES (May 2008). *Proceedings of the 1st International Workshop on Software Engineering for Computational Science and Engineering, in conjuction with the 30th International Conference on Software Engineering (ICSE2008), Leipzig, Germany.*

Simmhan, Y. L., Plale, B., & Gannon, D. (2005). A survey of data provenance in e-Science. *SIGMOD Rec. 34*(3), 31–36.

VMware (2009). Retrieved 1st September 2009, from http://www.vmware.com/.

Wang, L. Z., Kunze, M., & Tao, J. (2008). Performance evaluation of virtual machine-based grid workflow system. http://doi.wiley.com/10.1002/cpe.1328, 1759–1771.

Wang, L. Z., Jie, W., & Chen, J. (2009). *Grid computing: Infrastructure, service, and applications.* Boca Raton, FL: CRC Press, Talyor & Francis Group.

Weiss, A. (2007). Computing in the cloud. *ACM Networker, 11*(4), 18–25.

Winsborough, W. H., & Li, N. H. (2006). Safety in automated trust negotiation. *ACM Transactions on Information and System Security, 9*(3), 352–390.

Yang, Y., Liu, K., Chen, J., Lignier, J., & Jin, H. (December 2007). Peer-to-peer based grid workflow runtime environment of swinDeW-G. *Proceedings of the 3rd International Conference on e-Science and Grid Computing (e-Science07), Bangalore, India,* 51–58.

Yang, Y., Liu, K., Chen, J., Liu, X., Yuan, D., & Jin, H. (December 2008). An algorithm in swinDeW-C for scheduling transaction-intensive cost-constrained cloud workflows. *Proceedings of the 4th IEEE International Conference on e-Science (e-Science08), Indianapolis, IN, USA,* 374–375.

Yu, J., & Buyya, R. (2005). A taxonomy of workflow management systems for grid computing. *Journal of Grid Computing,* (3), 171–200.

Yuan, D., Yang, Y., Liu, X., & Chen, J. (2010). A cost-effective strategy for intermediate data storage in scientific cloud workflow systems. *Proceedings of the 24th IEEE International*

Parallel & Distributed Processing Symposium, Atlanta, GA, USA, to appear. Retrieved 1st February 2010, from http://www.ict.swin.edu.au/personal/yyang/papers/IPDPS10-IntermediateData.pdf.

Yuan, D., Yang, Y., Liu, X., & Chen, J. A data placement strategy in cloud scientific workflows. *Future Generation Computer Systems,* in press. http://dx.doi.org/10.1016/j.future.2010.02.004.

Part III
Services

Chapter 14
Cloud Types and Services

Hai Jin, Shadi Ibrahim, Tim Bell, Wei Gao, Dachuan Huang, and Song Wu

14.1 Introduction

The increasing popularity of Internet services such as the Amazon Web Services, Google App Engine and Microsoft Azure have drawn a lot of attention to the Cloud Computing paradigm. Although the term "Cloud Computing" is new, the technology is an extension of the remarkable achievements of grid, virtualization, Web 2.0 and Service Oriented Architecture (SOA) technologies, and the convergence of these technologies. Moreover, interest in Cloud Computing has been motivated by many factors such as the prevalence of multi-core processors and the low cost of system hardware, as well as the increasing cost of the energy needed to operate them. As a result, Cloud Computing, in just three years, has risen to the top of the IT revolutionary technologies as shown in Fig. 14.1, and has been announced as the top technology to watch in the year 2010 (Gartner – Gartner Newsroom, 2010).

The name "Cloud Computing" is a metaphor for the Internet. A Cloud shape is used to represent the Internet in network diagrams to hide the flexible topology and to abstract the underlying infrastructure. Cloud Computing uses the internet to deliver different computing services including hardware, programming environments and software while keeping users unaware of the underlying infrastructure and location.

Despite the popularity and interest in cloud computing, a lot of confusion remains about what it is, and there is no formal definition of Cloud Computing. Two of the main definitions that are being used by the Cloud community have been provided by Ian Foster and Jeff Kapalan. Ian Foster gives a detailed definition of the term

H. Jin (✉), S. Ibrahim, W. Gao, D. Huang, and S. Wu
Services Computing Technology and System Lab; Cluster and Grid Computing Lab, Huazhong University of Science and Technology, Wuhan, China
e-mails: {hjin@hust.edu.cn; shadi@hust.edu.cn; gaowei715@gmail.com; hdc1112@gmail.com; wusong@mail.hust.edu.cn}

T. Bell
Department of Computer Science and Software Engineering, University of Canterbury, Christchurch, New Zealand
e-mail: tim.bell@canterbury.ac.nz

B. Furht, A. Escalante (eds.), *Handbook of Cloud Computing*,
DOI 10.1007/978-1-4419-6524-0_14, © Springer Science+Business Media, LLC 2010

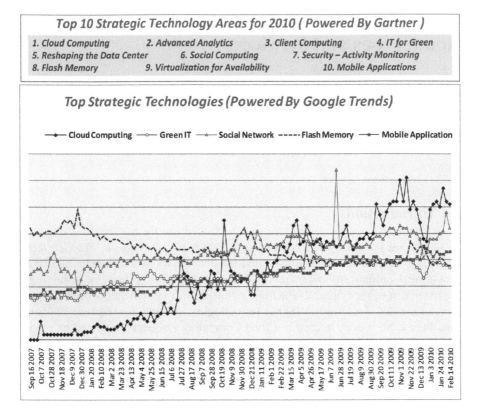

Fig. 14.1 Top 10 strategic technology areas for 2010 and their evolution for the last three Years (based on Gartner (Gartner – Gartner Newsroom, 2010) and Google Trends (2010))

Cloud Computing: "A large-scale distributed computing paradigm that is driven by economies of scale, in which a pool of abstracted, virtualized, dynamically-scalable, managed computing power, storage, platforms, and services are delivered on demand to external customers over the Internet" (Foster, Zhao, Raicu, & Lu, 2008).

Jeff Kaplan views cloud computing as "a broad array of web-based services aimed at allowing users to obtain a wide range of functional capabilities on a "pay-as-you-go" basis that previously required tremendous hardware/software investments and professional skills to acquire. Cloud computing is the realization of the earlier ideals of utility computing without the technical complexities or complicated deployment worries" (Twenty Experts Define Cloud Computing, 2008).

Cloud Computing enables users to access various computing resources simply, including computing cycles, storage space, programming environments and software applications (all you need is a web browser). Moreover, Cloud computing promises to provide other benefits:

- *Less investment.* Clouds provide affordable solutions that handle peaks, or scale easily at a fraction of the traditional costs of space, time and financial investment.
- *Scale.* Cloud vendors have vast data centers full of tens of thousands of server computers, offering computing power and storage of a magnitude never before available – cloud computing promises virtually unlimited resources.
- *Manageability.* The user experience is simplified as no configuration or backup is needed

However, Cloud Computing also raises many concerns, mainly about security, privacy, compliance and reliability. When users move their data to the service provider data center, there is no guarantee that nobody else has access to this data. If the data is being stored in a different country, there can also be issues about jurisdictions for legal rights, and control of the data. Moreover, to date, there are no clearly defined Service Level Agreements (SLA) offered by the cloud providers.

There has been relatively little unification of the Cloud Computing concept. Consequently, it is useful to take a step back, consider the variety of Clouds offered by leading vendors, and describe them in a unified way, putting the different use and types of clouds in perspective; this is the main purpose of this book chapter.

The rest of this chapter is organized as follows: In Section 14.2 we present the different types of Clouds. Then we briefly introduce the three main Cloud service categories. We then describe the variety of Clouds in leading projects in each category, IaaS, PaaS and SaaS, in Sections 14.4, 14.5, and 14.6 respectively. To complete our survey, we present the Amazon cloud family and the different enterprises that are using the Amazon infrastructure, in Section 14.7. A conclusion is provided in Section 14.8.

14.2 Cloud Types

Clouds can be classified in terms of who owns and manages the cloud; a common distinction is Public Clouds, Private Clouds, Hybrid Clouds and Community Clouds (see Fig. 14.2).

14.2.1 Public Cloud

A public cloud, or external cloud, is the most common form of cloud computing, in which services are made available to the general public in a pay-as-you-go manner. Customers – individual users or enterprises – access these services over the internet from a third-party provider who may share computing resources with many customers. The public cloud model is widely accepted and adopted by many enterprises because the leading public cloud vendors as Amazon, Microsoft and Google, have equipped their infrastructure with a vast amount of data centers, enabling users

Fig. 14.2 Cloud types: public, private and hybrid clouds

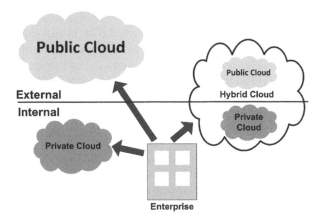

to freely scale and shrink their rented resources with low cost and little management burden. Security and data governance are the main concern with this approach.

14.2.2 Private Cloud

A Private Cloud, or internal cloud, is used when the cloud infrastructure, proprietary network or data center, is operated solely for a business or organization, and serves customers within the business fire-wall. Most of the private clouds are large company or government departments who prefer to keep their data in a more controlled and secure environment. Table 14.1 presents a comparison between public and private clouds.

Table 14.1 Public *vs.* private cloud

	Public cloud	Private cloud
Infrastructure Owner	Third party (Cloud provider)	Enterprise
Scalability	Unlimited and On-Demand	Limited to the installed Infrastructure
Control and Management	Only manipulate the virtual machines, resulting in less management burden	High level of control over the resources, and need more expertise to mange them.
Cost	Lower cost	High cost including: space, cooling, energy consumption and hardware cost
Performance	Unpredictable multi-tenant environment makes it hard to achieve guaranteed performance	Guaranteed performance
Security	Concerns regarding data privacy	Highly secure

14.2.3 Hybrid Cloud

A composition of the two types (private and public) is called a Hybrid Cloud, where a private cloud is able to maintain high services availability by scaling up their system with externally provisioned resources from a public cloud when there are rapid workload fluctuations or hardware failures. In the Hybrid cloud, an enterprise can keep their critical data and applications within their firewall, while hosting the less critical ones on a public cloud.

14.2.4 Community Cloud

The idea of a Community Cloud is derived from the Grid Computing and Volunteer Computing paradigms. In a community cloud, several enterprises with similar requirement can share their infrastructures, thus increasing their scale while sharing the cost (Wikipedia – Cloud Computing, 2010). Another form of community cloud may be established by creating a virtual data center from virtual machines instances deployed on underutilized users machines (Briscoe & Marinos, 2009).

14.3 Cloud Services and Cloud Roles

A Cloud is essentially a class of systems that deliver IT resources to remote users as a service. The resources encompass hardware, programming environments and applications. The services provided through cloud systems can be classified into Infrastructure as a service (IaaS), Platform as a Service (PaaS) and Software as a service (SaaS).

Different enterprises play different roles in building and using cloud systems (Fig. 14.3). These roles range from cloud technology enablers (enabling the

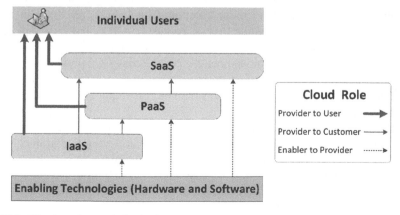

Fig. 14.3 Cloud services and cloud roles

underlying technologies used to build the cloud, such as hardware technologies, Virtualization technology, web services and so on), to cloud providers (delivering their infrastructure and platform to customers), to cloud customers (using the providers' services to improve their web applications), and users (who use the web applications, possibly unaware that it is being delivered using cloud technologies).

14.4 Infrastructure as a Service

Infrastructure as a Service (IaaS) is one of the "Everything as a Service" trends. IaaS is easier to understand if we refer it as Hardware as a Service (i.e. instead of constructing our own server farms, a small firm could consider paying to use infrastructure provided by professional enterprises). Companies such as Google, Microsoft and IBM are involved in offering such services. Large-scale computer hardware and high computer network connectivity are essential components of an effective IaaS.

The IaaS is categorized into: (1) Computation as a Service (CaaS), in which virtual machine based servers are rented and charged per hour based on the virtual machine capacity – mainly CPU and RAM size, features of the virtual machine, OS and deployed software; and (2) Data as a Service (DaaS), in which unlimited storage space is used to store the user's data regardless of its type, charged per GByte for data size and data transfer.

In this section we will describe some popular IaaS systems such as Amazon EC2 (2010), GoGrid (2010), Amazon S3 (2010) and Rackspace (2010). We then compare three widely used CaaS systems (Table 14.2).

14.4.1 Amazon Elastic Compute Cloud (EC2)

Amazon has provided a popular universal and comprehensive solution to Cloud Computing, called the Amazon Elastic Compute Cloud (EC2) (2010). This solution was released as a limited public beta on August 25, 2006, but grew rapidly in the following years. After Amazon added many important and powerful features to EC2, it dropped the beta label on October 23, 2008. Today EC2 provides complete control over a customer's computing resources, so new server instances can be set up and booted in minutes, and their capacity can be scaled quickly through a simple web service interface.

EC2 provides many useful features for customers, including a mature and inexpensive billing system able to charge for computing at a very fine-grained level (memory usage, CPU usage, data transfer, etc.), deployment between multiple locations, elastic IP addresses, connection to a customer's existing infrastructure through a Virtual Private Network, monitoring services by Amazon CloudWatch, and elastic load balancing. EC2 has deployed such fine granularity and precision that it has become a benchmark and model in cloud computing.

Table 14.2 A comparison of three widely used CaaS (prices from Feb 2010)

CaaS	Amazon EC2	GoGrid	Rackspace (cloud server)
Virtualization	Xen	Xen	VMware
OS support	Linux, Windows	Linux, Windows	Linux, Windows
Server RAM	1.7 GB and going up to 68.4 GB	0.5 GB and going up to 8 GB	256 MB and going up to 16 GB
Load Balancer	Amazon Elastic Load Balancer	Free F5 Load Balancer	No
Persistent Block Storage	Yes	Yes	No
Hybrid Hosting	No	Yes	Yes
24/7 Support	No	Yes	Yes
Pricing	Billed $0.085 – $3.18 per hour (vary for different Instance and Regions). The Data Transfer rates vary based on where the data goes out to and comes in from with pricing between $0.00 to $0.15 per GB transferred.	Billed $0.19 per GB of deployed RAM per hour and 60 GB of disk, $0.50 per GB of outbound data transferred, and all inbound data transfer is free.	Billed $0.06 per GB of deployed RAM per Hour and 40 GB of disk, $0.05 per GB of inbound data transfer and $0.22 per GB of outbound data transfer.

Amazon's EC2 provides virtual machine based computation environments. It uses the Xen hypervisor (2010) to manage their Amazon Machine Image (AMI) instance. AMI (Amazon EC2, 2010) is "an encrypted machine image that contains all information necessary to boot instances of your software". Using simple web service interfaces, users can launch, run, monitor and terminate their instances as shown in Fig. 14.4. Moreover they can, on the fly, add any of the abovementioned features to their configuration as they desire.

14.4.2 GoGrid

GoGrid (2010) shares many common characteristics with Amazon in the classic cloud computing areas: supporting multiple operating systems through its own image management, and supporting load balancing, cloud storage, and so on. In addition, GoGrid provides customers with a user-friendly web service interface, easy-to-understand video demonstrations, and a strict but inexpensive billing system. Thus both EC2 and GoGrid provide basic and common features of cloud computing. The difference between the services they provide mainly derives from their respective business models. For example, GoGrid provides free cloud and persistent storage, slightly different from Amazon.

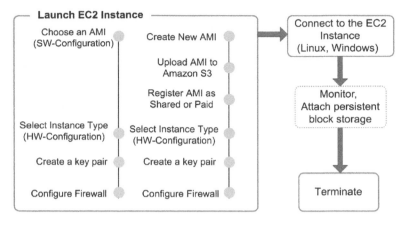

Fig. 14.4 Lifecycle of amazon machine image

GoGrid also provides Hybrid Hosting, which is a distinguishing feature. Many applications simply don't run well in a pure multi-tenant server environment. Databases perform better on a dedicated server where they don't have to compete for input/output resources, and the situation is similar with web server applications. GoGrid provides these special applications with dedicated servers that also have high security assurance.

14.4.3 Amazon Simple Storage Service (S3)

The Amazon Simple Storage Service (2010) (S3) is an online storage web service offered by Amazon Web Services. S3 is accessible to users through web services, REST-style HTTP interfaces,[1] or by involving a SOAP interface. Like other cloud computing services, users can request small or large amounts of storage on the fly, providing a highly scalable storage system.

Amazon S3 organizes the storage space into many "buckets", with each bucket being given a global unique namespace to help locate data addresses, identify the user account for payments, and gathering usage information (ASSSDG, 2010). S3 deals with all type of data as objects and stores them with their metadata into the bucket chosen by the data owner. An object can be accessed through a URL composed of its key and version ID with its bucket namespace as the prefix.

Amazon S3's users are spread across countless fields, for example, SmugMug (2010), Slideshare (2010) and Twitter (2010) use Amazon S3 to host images,

---

[1]REST stands for "Representational state transfer", a software architecture for distributed hypermedia.

Apache Hadoop (2010) uses S3 to store computation data, and online synchronization utilities such as Dropbox (2010) and Ubuntu One (2010) use S3 as their storage and transfer facility.

14.4.4 Rackspace Cloud

Rackspace (2010) Cloud was originally launched on March 4, 2006 under the name "Mosso". In the following three years, it has changed his name from "Mosso LLC" to "Mosso: The Hosting Cloud", and then finally "Rackspace Cloud" on June 17, 2009. This company provides services including a cloud server, cloud files, and cloud site.

The cloud files service is a cloud storage service providing unlimited online storage and a Content Delivery Network (CDN) for media on a utility computing basis. In addition to the online control panel, this company provides an API service that can be accessed over a RESTful API with open source client code. Rackspace solves the security problem by replicating three full copies of data across multiple computers in multiple zones, with every action protected by SSL.

14.5 Platform as a Service

Platform as a Service (PaaS) cloud systems provide a software execution environment that application services can run on. The environment is not just a pre-installed operating system but is also integrated with a programming-language-level platform, which users can be used to develop and build applications for the platform. From the point of view of PaaS clouds' users, computing resources are encapsulated into independent containers, they can develop their own applications with certain program languages, and APIs are supported by the container without having to take care of the resource management or allocation problems such as automatic scaling and load balancing. In this section we introduce three typical PaaS : Google App Engine (2010), Microsoft Azure (2010), and Force.com (2010), and then we compare them in Table 14.3.

14.5.1 Google App Engine

Google App Engine (GAE)'s main goal is to efficiently run users' web applications. As shown in Fig. 14.5, it maintains Python and Java runtime environments on application servers, along with some simple APIs to access Google services.

The front ends spread HTTP requests with load balancing and routing strategies based on the contents. Runtime systems running on application servers deal with the logic processing of applications and provide dynamic web content, while static pages are served by the shared Google infrastructure (Alon, 2009). To decouple the

Table 14.3 Survey of four PaaS provides

PaaS provider	Programming environment	Infrastructure	Hosted application
Google	Python and Java	Google Data Center	Socialwok (2010), Gigapan (2010), LingoSpot (2010)
Azure	.Net (Microsoft Visual Studio)	(Virtual Machine Based) Microsoft Data Centers	Microsoft Pinpoint (2010)
Force.com	Apex Programming and Java1	Saleforce Data Center	EA (2010), Author Solutions (2010), The Wall Street Journal (2010)
Heroku (2010)	Ruby	Amazon EC2 and S3	Übermind (2010), Kukori.ca (2010), act.ly (2010), Cardinal Blue (2010)

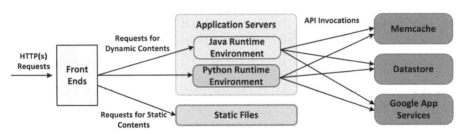

Fig. 14.5 The Architecture of Google App Engine

persistent data from application servers, GAE puts them into the Datastore instead of a local file system. Applications can integrate data services and other Google App Services, such as email, image storage and so on through APIs provided by the GAE.

In addition to the services, Google also provides some tools for developers to help them build web applications easily on GAE. However, since they are tightly connected to the Google infrastructure, there are some restrictions that limit the functionality and portability of the applications.

14.5.2 Microsoft Azure

Microsoft's cloud strategy is to construct a cloud platform that users can move their applications to in a seamless way, and ensure its managed resources are accessible to both cloud services and on-premises applications. To achieve this, Microsoft introduced the Windows Azure Platform (WAP), which is composed of a cloud operating system named Windows Azure, and a set of supporting services, as shown in Fig. 14.6.

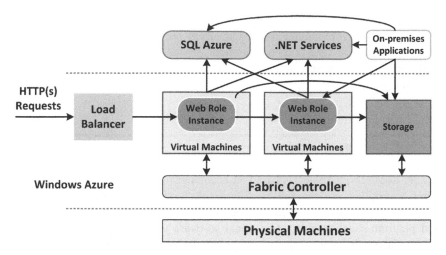

Fig. 14.6 The architecture of the windows azure platform

Windows Azure is the main part of the WAP. It employs virtual machines as its runtime environments. The applications in Microsoft's cloud offerings are divided into two types: Web role instances, which can serve web requests via the internet information services (IIS); and Worker role instances, which can only receive messages from other Web role instances or on-premises applications (David, 2009a). Windows Azure employs a "fabric controller" to manage all virtual machines and storage servers on the physical machines in a Microsoft data center (David, 2009b). Similar to the GAE's datastore, WAP also provides a database service called SQL Azure, to store data in the cloud. One feature of SQL Azure is that it provides a tool for data synchronization across on-premises and/or off-premises databases. Infrastructure services supported by WAP through .NET services currently include access control and exposing services. Both are available for cloud and on-premises applications.

14.5.3 Force.com

Force.com is an enterprise cloud computing platform offered by Salesforce. It helps service venders develop and deliver stable, secure and scalable applications. Two key enabling technologies of Force.com are multi-tenancy and metadata (Force.com, 2009a). The multi-tenancy approach allows different users to share application templates on a public physical computing resource pool, while the application instances are independent from each other. For customized applications, a metadata driven architecture that generates application components according to its own description has been proposed (Force.com, 2009b). Other technologies and services of the Force.com platform include service delivery infrastructure, a database, logic services, user interfaces, and developer tools (Force.com, 2009a).

The idea of the Force.com cloud solution is that it should take care of all common underlying requirements so that users need only focus on the design of their applications. One potential problem of this is that the applications rely heavily on the infrastructure and services of Force.com, which compromises their portability.

14.6 Software as a Service

Software-as-a-Service (SaaS) is based on licensing software use on demand, which is already installed and running on a cloud platform. These on-demand applications may have been developed and deployed on the PaaS or IaaS layer of a cloud platform. SaaS replaces traditional software usage with a Subscribe/Rent model, reducing the user's physical equipment deployment and management costs. The SaaS clouds may also allow users to compose existing services to meet their requirements. This section presents some SaaS clouds and applications.

14.6.1 Desktop as a Service

Desktop as a Service is a special variant of Software as a Service that provides a virtualized desktop-like personal workspace, and sends its image to the user's real desktop. Instead of a local desktop, the user can access their own desktop-on-the-cloud from different places for convenience, and receive the benefit of SaaS at same time.

The "Global Hosted Operating SysTem" (G.ho.st) (2010) is a free and complete Internet-based Virtual Computer (VC) service suite including a personal desktop, files and applications (G.ho.st, 2009). It offers users an operating system image simulated in a web browser using Flash and Javascript, which can be accessed through any browser. The G.ho.st application services are hosted by the Amazon Web Services (AWS) platform, so users can utilize EC2 and S3 resources through their G.ho.st desktops. One limitation of G.ho.st is, as a lightweight desktop service, it only supports on-line applications, and users cannot run legacy programs on it.

Unlike G.ho.st's browser-based desktop image, the Desktone Virtual-D Platform (2010) implements a desktop as a service by encapsulating a virtual machine based desktop, called Virtual Desktop Infrastructure (VDI), into a service. The advantage of VDI is that it can offer the same environment as a native operating system, and allows users to install their own software. The Desktone Virtual-D Platform integrates all desktop virtualization layers and simplifies desktop management, improving security and compliance (The Desktone Virtual-D Platform – Fact Sheet, 2009). This solution delivers desktops as a cost-effective subscription service deployed on cloud.

14.6.2 Google Apps

Google Apps (2010) is a typical SaaS implementation. It provides several Web applications with similar functionality to traditional office software (word processing, spreadsheets etc.), but also enables users to communicate, create and collaborate easily and efficiently. Since all the applications are kept online and are accessed through a web browser, users can access their accounts from any internet-connected computer, and there is no need to install anything extra locally.

Google Apps has several components. The communication components consist of Google mail and Google Talk, which allow for communication through email, instant messaging and voice calls. The office components include docs and spreadsheets, through which users can create online documents that also facilitate searching and collaboration. Google Calendar is a flexible calendar application for organizing meetings and events. With Google's "Web Pages", administrators can easily publish web pages, while "Start Pages" provide users with a rich array of content and applications that can be personalized.

Google Apps has several significant features. First, it provides an easy-to-use control panel which facilitates the most common administration tasks such as enabling/disabling applications, managing accounts, and customizing interfaces. Second, although hosted on Google, the user can control the branding on all interfaces – email addresses will have only the user's domain name with no mention of Google in the message body (Wikipedia, Google App, 2010), and users can customize their web interfaces, layouts and colors on web and start pages. Third, administrators can integrate with existing platforms as well as extend the functionality of the core Google Apps applications with the Application Programming Interfaces (APIs) that are offered. There are APIs available for provisioning, reporting, and migration, as well as manipulating data in Calendar and Spreadsheets, and integrating with Single Sign On (SSO) systems (Google Apps Products Overview, 2010).

14.6.3 Salesforce

Salesforce (2010) is a business SaaS cloud platform that provides customizable applications, mostly Customer Relationship Management (CRM) services, to consumers. There are two major products presented by Salesforce. Sales Cloud is a group of comprehensive applications to improve the convenience and efficiency of business activities; and Service Cloud is provided to integrate social network applications like Facebook and Twitter, to construct a users' customer service community.

Saleforce CRM services are deployed on the Force.com cloud platform, which operates a multi-tenancy oriented metadata-driven architecture (Force.com, 2009b). Multi-tenancy enables sharing the same version of an application among many users, but each user can only access their own private data, which keeps their activities isolated. All applications' functionalities and configurations are described with metadata, so users can customize applications as they want.

Although the shared application model could cause interference between users, the Salesforce SaaS cloud has these advantages: (1) service providers can develop only one version of application, and don't need to worry about heterogeneous execution environments; (2) the sharing of the physical computing resource, operating system and runtime environment lowers the cost of the application service; and (3) service consumers are free to choose their preferred version of the application and customize it to fit their business.

14.6.4 Other Software as Service Examples

As cloud computing technology spreads, more and more Software as a Service implementations have been released. Table 14.4 gives some other SaaS examples. The services cover many fields in addition to personal file processing and business administration.

Table 14.4 Some SaaS examples

SaaS provider	Important services
A2Zapps.com (2010)	Marketing Automation, School Automation (ERP)
Envysion.com (2010)	Video Management
Learn.com (2010)	Training, HR, Online Courses
Microsoft (2010)	Office Live Meeting, Dynamics CRM, SharePoint
OpenID (2010)	Log in Identification
Zoho (2010)	Mail, Docs, Wiki, CRM, Meeting, Business

14.7 The Amazon Family

Since the early stage of Cloud Computing, Amazon has dominated the Cloud market by providing scalable on-demand infrastructure, in particular EC2 and S3, making it easy for enterprises to obtain computing power and storage as a service ("The network world – 10 cloud computing companies to watch", http://www.networkworld.com/supp/2009/ndc3/051809-cloud-companies-to-watch.html).

One of the first success stories about the effectiveness of cloud computing for providing low-cost and fast solutions on-demand for an enterprise was the case of the New York Times. In order to make their articles from 1851 to 1922 available to the public, they were able to create PDF versions of their archives using 100 EC2 instances. This was a large job – in some cases they needed to take several TIFF images and scale and glue them to create one PDF file in less than 24 hours. Once the archive was created, they stored it in the S3, using 4 TB of storage (The New York Times Blog, 2010).

Recently Amazon has equiped their IT infrastructure services with new services as shown in Table 14.5, motivating many businesses, enterprises and academia to

Table 14.5 The Amazon Web services (prices from Feb 2010)

Amazon web service	Brief description	Geographical regions[1]	Pricing range
Amazon Elastic Compute Cloud (Amazon EC2)	"Amazon Elastic Compute Cloud (Amazon EC2) is a web service that provides resizable compute capacity in the cloud" (Amazon EC2, 2010).	US – N. Virginia US – N. California EU – Ireland	$0.085 – $3.18 per hour (vary for different Instance and regions)
Amazon Simple Storage Service (Amazon S3)	"Amazon S3 provides a simple web services interface that can be used to store and retrieve any amount of data, at any time, from anywhere on the web" (Amazon S3, 2010).	US – N. Virginia US – N. California EU – Ireland	0.055 – 0.150 Per GB (vary for different regions)
Amazon SimpleDB (2010)	"Amazon SimpleDB is automatically indexing your data and providing a simple API for storage and access and requiring no schema" (Amazon SimpleDB, 2010).	US – N. Virginia US – N. California EU – Ireland	$0.140 – $0.154 per Hour (vary to different regions)
Amazon CloudFront (2010)	"Amazon CloudFront is a web service for content delivery. It delivers the static and streaming content using a global network of edge locations. By routing the requests for any objects to the nearest edge location, so content is delivered with the best possible performance. Amazon CloudFront works seamlessly with (Amazon S3)" (Amazon CloudFront, 2010).	United States Europe Hong Kong Japan	$0.050 – $0.221 per GB (vary for different Data transfer per month and regions)
Amazon Simple Queue Service (Amazon SQS) (2010)	"Amazon Simple Queue Service (Amazon SQS) offers a reliable, highly scalable, hosted queue for storing messages as they travel between computers. Amazon SQS makes it easy to build an automated workflow, working in close conjunction with the Amazon EC2" (Amazon SQS, 2010).	US – N. Virginia US – N. California EU – Ireland	$0.01 per 10,000 Amazon SQS Requests ($0.000001 per Request)

Table 14.5 (continued)

Amazon web service	Brief description	Geographical regions[1]	Pricing range
Amazon Elastic MapReduce (2010)	"Amazon Elastic MapReduce is a web service that enables businesses, researchers, data analysts, and developers to easily and cost-effectively process vast amounts of data. It utilizes a hosted Hadoop framework running on the web-scale infrastructure of Amazon EC2 and Amazon S3" (Amazon Elastic MapReduce, 2010).	US – N. Virginia US – N. California EU – Ireland	$0.015 – $0.42 per hour (vary for different instance and regions)
Amazon Relational Database Service (Amazon RDS) (2010)	"Amazon Relational Database Service (Amazon RDS) gives you access to the full capabilities of a familiar MySQL database. This means the code, applications, and tools you already use today with your existing MySQL databases work seamlessly with Amazon RDS" ((Amazon RDS, 2010).	US Region	$0.11 – $3.10* (vary for different instances)

[1] Amazon web services provide multiple regions and "availability zones" so customers can connect to the most convenient service, and also choose services in multiple zones to maximize failure-independence.

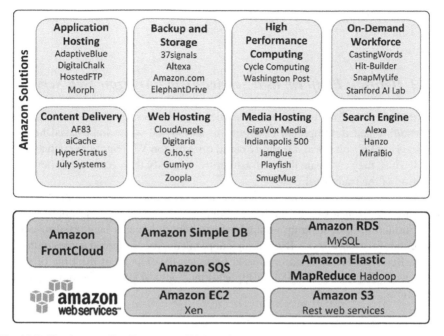

Fig. 14.7 The Amazon Family: Amazon Web services and their Different Solutions and customers (Amazon AWS, 2010)

join the Amazon Web Services using the always-improving IT infrastructure services suite to build their business and applications (Fig. 14.7). Furthermore, some enterprises use the Amazon cloud services to provide new cloud services, including RightScale (Cloud Computing Management Platform by RightScale, 2010) providing IaaS, Heroku (2010) providing PaaS, Animoto (Animoto, 2010) and G.ho.st (2010) providing SaaS.

14.7.1 RightScale: IaaS Based on AWS

Rightscale (2010) is a web based solution for deploying and managing services on top of IaaS cloud providers such as Amazon and GoGrid. RightScale enables users to simply build, monitor and auto-scale their virtual infrastructure. It has been listed among the top 10 companies to watch in cloud computing (Appistry/CloudCamp Survey, 2009) especially after its huge success in managing the Amazon EC2 services for companies such as Animoto (Rightscale, 2010) and G.ho.st (2010). RightScale also provides management tools for deployments across multiple clouds, providing flexibility in the choice of all kinds of services.

Recently, many enterprises have joined Rightscale, running different applications such as scalable websites (ShareThis (2010)), grid applications (Animoto

(RightScale, 2010)), Test & Development, and Social Gaming Applications (PlayFish (2010)).

14.7.2 HeroKu: Platform as a Service Using Amazon Web Service

Heroku (2010) is a Ruby platform-as-a-service, which offers an in-browser Ruby on Rails multi-tenant development environment associated with cloud-based hosting services. Heroku's platform is entirely based on Amazon Web Services such as EC2 and S3. Thus, they can scale their infrastructure to satisfy their customers' demands at fraction of the traditional cost.

In Heroku, the users' code is compiled into self-contained, read only "slugs", which are then run inside a number of "dynos", depending on the application's need. Furthermore, to scale up an application, new dynos can be started in under two seconds for most apps. A Dyno is an independent process spread across multiple servers. Recently, Heroku has been used by many developers; at the time of writing it was hosting more than 45,000 applications, including websites and facebook applications.

14.7.3 Animoto Software as Service Using AWS

Animoto (RightScale, 2010) is a web application that automatically generates fast, free (for video up to 30 s), and unique video pieces from users' photos, video clips and music. It is based around their own patent-pending technology and high-end motion design. Their system is built on top of Amazon Web Services, namely EC2, S3 and SQS. EC2 is used for web servers, application servers, upload servers, "director" servers, and database servers. All the music and photos are stored and served by Amazon S3. Amazon's SQS is used to connect all the operations during the video creation process.

Earlier, Animoto were regularly using a 50 virtual machine instance of EC2, but after the huge success of their Facebook application, they scaled up to 3,500 instances using RightScale, within just three days (at its peak RightScale was launching and configuring 40 new instances per minute (Rightscale Blog, 2010)).

14.7.4 SmugMug Software as Service Using AWS

SmugMug (2010) is a photo sharing company that offers unlimited storage using the Amazon S3. In early 2006, SmugMug (with 15 employees and 1 programmer), moved its storage to S3, and became fully operational on Amazon S3 in one week, with around 100 Terabytes of customer photos (70,000,000 original images and six display copies of each). This saved them roughly $500,000 compared with increasing the space in their data center.

14.8 Conclusion

Cloud computing is a very flexible paradigm for delivering computational power. It will mean many things to many people. For some it means being able to set up a new start-up company knowing that initial resources will be inexpensive but a sudden increase in demand from users won't make the company a victim of its own success, as has happened in some cases in the past where servers have been unable to cope with demand, and the company loses clients as they become unhappy with poor response times. For other people, cloud computing means easier administration, with issues such as licensing, backup and security being taken care of elsewhere. In other cases, cloud computing means having a powerful computational environment available anywhere that the user can access a web browser.

With this flexibility, scalability and ease of maintenance, it is little wonder that cloud computing is being touted as a technology to watch. Of course, there are issues: privacy of data can be a concern, good internet connectivity is required, and some organizations may wish to maintain control over their own resources. However, these problems can usually be addressed, and using a cloud remains a very attractive way to set up a powerful system very quickly.

The various forms of service – infrastructure, platform, and software as a service – provide exciting ways to deliver new products that innovators might come up with. Already there are examples of widely used products and web sites that have sustained remarkable growth because creative ideas could be implemented quickly, and because the subsequent demand could be met easily through the flexibility of cloud computing.

The future seems to be limited only by the imaginations of innovators who can think of applications that will help people communicate, store and process vast quantities of information, whether it is millions of individuals with small collections of personal information, or a single large organization with large collections of data to be processed.

Acknowledgment This work is supported by National 973 Key Basic Research Program under Grant No. 2007CB310900, NSFC under Grant No.60673174, Program for New Century Excellent Talents in University under Grant NCET-07-0334, Information Technology Foundation of MOE and Intel under Grant MOE-INTEL-09-03NSFC, and National High-Tech R&D Plan of China under Grant 2006AA01A115.

References

act.ly Homepage (2010). http://act.ly/. Accessed on July 13, 2010.
A2Zapps.com Homepage (2010). http://www.a2zapps.com. Accessed on July 13, 2010.
Alon, L. (2009). From spark plug to drive train: Life of an app engine request. *Google I/O Developer Conference*, http://code.google.com/events/io/2009/.
Amazon AWS – AWS Solutions (2010). http://aws.amazon.com/solutions/aws-solutions/. Accessed on July 13, 2010.
Amazon Elastic Compute Cloud (Amazon EC2) (2010). http://aws.amazon.com/ec2. Accessed on July 13, 2010.

Amazon Elastic MapReduce (2010). http://aws.amazon.com/elasticmapreduce/. Accessed on July 13, 2010.

Animoto Homepage (2010). http://animoto.com/. Accessed on July 13, 2010.

Amazon Relational Database Service (RDS) (2010). http://aws.amazon.com/rds/. Accessed on July 13, 2010.

Amazon SimpleDB (2010). http://aws.amazon.com/simpledb/. Accessed on July 13, 2010.

Amazon CloudFront (2010). http://aws.amazon.com/cloudfront/. Accessed on July 13, 2010.

Amazon Simple Queue Service (SQS) (2010). http://aws.amazon.com/sqs/. Accessed on July 13, 2010.

Amazon Simple Storage Service (Amazon S3) (2010). http://aws.amazon.com/s3. Accessed on July 13, 2010.

Amazon Simple Storage Service Developer Guide (2010). http://docs.amazonwebservices.com/AmazonS3/latest/index.html?Introduction.html. Accessed on July 13, 2010.

Apache Hadoop (2010). http://hadoop.apache.org/. Accessed on July 13, 2010.

Appistry/CloudCamp Survey (2009). *Inside the cloud.* www.appistry.com/go/inside-the-cloud. Accessed on July 13, 2010.

Author Solutions Homepage (2010). http://www.authorsolutions.com/. Accessed on July 13, 2010.

Briscoe, G., & Marinos, A. (2009). Towards community cloud computting. *Proceedings of the IEEE Digital EcosystemS and Technologies (DEST 2009), Istanbul, Turkey.*

Cardinal Blue Homepage (2010). http://cardinalblue.com/. Accessed on July 13, 2010.

Cloud Computing Management Platform by RightScale (2010). http://www.rightscale.com/. Accessed on July 13, 2010.

David, C. (2009a). *Introducing the windows azure platform.* http://view.atdmt.com/action/mrtyou_FY10AzurewhitepapterIntroWindowsAzurePl_1. Accessed on July 13, 2010.

David, C. (2009b). *Introducing windows azure.* http://view.atdmt.com/action/mrtyou_FY10Azure WhitepaperIntroWindowsAzureSec_1. Accessed on July 13, 2010.

Dropbox Homepage (2010). http://www.dropbox.com/. Accessed on July 13, 2010.

EA Video Games – Electronic Arts (EA) Homepage (2010). http://www.ea.com/. Accessed on July 13, 2010.

Envysion.com (2010). http://www.envysion.com. Accessed on July 13, 2010.

Force.com homepage (2010). www.salesforce.com/platform/. Accessed on July 13, 2010.

Force.com Whitepaper (2009a). *A comprehensive look at the force.com cloud platform.* http://wiki.developerforce.com/index.php/A_Comprehensive_Look_at_the_Force.com_Cloud_Platform. Accessed on July 13, 2010.

Force.com Whitepaper (2009b). *The Force.com multitenant architecture.* http://www.apexdevnet.com/media/ForcedotcomBookLibrary/Force.com_Multitenancy_WP_101508.pdf. Accessed on July 13, 2010.

Foster, I., Zhao, Y., Raicu, I., & Lu, S. Y. (2008). Cloud computing and grid computing 360-degree compared. *Proceedings of the Grid Computing Environments Workshop (GCE'08), Austin, TX.*

Gartner – Gartner Newsroom (2010). http://www.gartner.com/it/page.jsp?id=1210613. Accessed on July 13, 2010.

G.ho.st Homepage (2010). http://g.ho.st/. Accessed on July 13, 2010.

G.ho.st Whitepaper (2009). The G.ho.st virtual computer. http://g.ho.st/HomePageResources/PDF/GhostWhitePaper.pdf. Accessed on July 13, 2010.

Gigapan Homepage (2010). http://www.gigapan.org/. Accessed on July 13, 2010.

GoGrid homepage (2010). http://www.gogrid.com/index.v2.php. Accessed on July 13, 2010.

Google App (2010). http://www.google.com/apps/intl/en/business/index.html.

Google App Engine (2010). http://code.google.com/appengine/.

Google Apps Products Overview (2010). http://services.google.com/apps/resources/overviews_breeze/Apps/index.html.

Google Trends (2010). http://www.google.com/trends, Accessed on February 23rd 2010.

Heroku (2010). *Ruby cloud platform as a service.* http://heroku.com/. Accessed on July 13, 2010.

Kukori.ca Homepage (2010). http://kukori.ca/. Accessed on July 13, 2010.

Learn.com (2010). http://learn.com. Accessed on July 13, 2010.

LingoSpot Homepage (2010). http://www.lingospot.com/. Accessed on July 13, 2010.

Microsoft Online Services (2010). http://www.microsoft.com/online/products.mspx. Accessed on July 13, 2010.

Microsoft Pinpoint Homepage (2010). http://pinpoint.microsoft.com/en-US/. Accessed on July 13, 2010.

OpenID Homepage (2010). http://openid.net/. Accessed on July 13, 2010.

PlayFish Homepage (2010). http://www.playfish.com/. Accessed on July 13, 2010.

Rightscale Blog – Animoto's Facebook Scale-up (2010). http://blog.rightscale.com/2008/04/23/animoto-facebook-scale-up/. Accessed on July 13, 2010.

Salesforce Homepage (2010). http://www.salesforce.com/crm/. Accessed on July 13, 2010.

ShareThis Homepage (2010). http://sharethis.com/. Accessed on July 13, 2010.

Slideshare homepage (2010). http://www.slideshare.net/. Accessed on July 13, 2010.

Smugmug homepage (2010). http://www.smugmug.com/. Accessed on July 13, 2010.

Socialwok Homepage (2010). http://www.socialwok.com/. Accessed on July 13, 2010.

The Desktone Virtual-D Platform – Fact Sheet (2009). http://www.desktone.com/downloads/public/collateral/Virtual-D-Fact-Sheet.pdf. Accessed on July 13, 2010.

The Desktone Virtual-D Platform Homepage (2010). http://www.desktone.com/platform/index.php. Accessed on July 13, 2010.

The New York Times Blog, – Service (2010). *Prorated super computing fun!* http://open.blogs.nytimes.com/2007/11/01/self-service-prorated-super-computing-fun/. Accessed on July 13, 2010.

The Rackspace Cloud (2010). http://www.rackspacecloud.com/. Accessed on July 13, 2010.

The Wall Street Journal Homepage (2010). http://online.wsj.com. Accessed on July 13, 2010.

Twenty Experts Define Cloud Computing (2008). *SYS-CON Media Inc.*, http://cloudcomputing.sys-con.com/read/612375_p.htm. Accessed on July 13, 2010.

Twitter homepage (2010). http://twitter.com/. Accessed on July 13, 2010.

Übermind Homepage (2010). http://www.ubermind.com/. Accessed on July 13, 2010.

Ubunto One (2010). https://one.ubuntu.com/. Accessed on July 13, 2010.

Wikipedia – Cloud Computing (2010). http://en.wikipedia.org/wiki/Cloud_computing. Accessed on July 13, 2010.

Wikipedia, Google App (2010). http://en.wikipedia.org/wiki/Google_Apps. Accessed on July 13, 2010.

Windows Azure platform (2010). http://www.microsoft.com/windowsazure/. Accessed on July 13, 2010.

Xen Hyper visor (2010). http://xen.org/. Accessed on July 13, 2010.

Zoho Homepage (2010). http://www.zoho.com/. Accessed on July 13, 2010.

Chapter 15
Service Scalability Over the Cloud

**Juan Cáceres, Luis M. Vaquero, Luis Rodero-Merino, Álvaro Polo,
and Juan J. Hierro**

15.1 Introduction

Cloud Computing promises an easy way to use and access to a large pool of virtualized resources (such as hardware, development platforms and/or services) that can be dynamically provisioned to adjust to a variable workload, allowing also for optimum resource utilization. This pool of resources is typically exploited by a pay-per-use model in which guarantees are offered by means of customized SLAs (Vaquero, Rodero-Merino, Caceres, & Lindner, 2009). Therefore, Cloud Computing automated provisioning mechanisms can help applications to scale up and down systems in the way that performance and economical sustainability are balanced. So, what does it mean to scale?

Basically, scalability[1] can be defined "as the ability of a particular system to fit a problem as the scope of that problem increases (number of elements or objects, growing volumes of work and/or being susceptible to enlargement)". For example, increasing system's throughput by adding more software or hardware resources to cope with an increased workload (Schlossnagle, 2007; Bondi, 2000). The ability to scale up a system may depend on its design, the types of data structures, algorithms or communication mechanisms used to implement the system components. A characterization of different types of scalability is reported by Bondi (2000), here we summarize some relevant examples:

- *Load Scalability*: when a system has the ability to make good use of available resources at different workload levels (i.e. avoiding excessive delay, unproductive consumption or contention). Factors that affect load scalability may be a bad

J. Cáceres (✉), L.M. Vaquero, Á. Polo, and J.J. Hierro
Telefónica Investigación y Desarrollo, Madrid, Spain
e-mails: {caceres; lmvg; apv; jhierro}@tid.es

L. Rodero-Merino
INRIA-ENS, INRIA, Lyon, France
e-mail: luis.rodero-merino@ens-lyon.fr

[1] http://en.wikipedia.org/wiki/Scalability

B. Furht, A. Escalante (eds.), *Handbook of Cloud Computing*,
DOI 10.1007/978-1-4419-6524-0_15, © Springer Science+Business Media, LLC 2010

use of parallelism, inappropriate shared resources scheduling or excessive over-heads. For example, a web server maintains a good level of load scalability if the performance of the system is maintained in an acceptable level when the number of threads that executes HTTP requests is increased in a workload peak.

- *Space Scalability*: The system has the ability to keep the consumption of sys-tem's resources (i.e. memory or bandwidth) between acceptable levels when the workload increases. For example, an operating system scales gracefully using a virtual memory mechanism that swaps unused virtual memory pages from phys-ical memory to disk, avoiding physical memory exhaustion. Another example would be when the number of user accounts of a web 2.0 service like a social network grows from thousands to millions.
- *Structural scalability*: The implementation of standards at the system allows for the increase of the number of managed objects or at least it will do so within a given time frame. For example, the size of a data type could affect the number of elements that can be represented (using a 16 bit integer as an entity identifier only allows representing 65,536 entities).

It would be desirable that the scaling capabilities of a system remained both, short and long term; having short term reactivity to respond to high and low rate of incoming works. As important as scaling up is scaling down, this impacts directly in the sustainability of business, reducing exploitation cost of unused resources when the workload decreases avoiding over-provisioning.[2]

The factors that could improve or diminish scalability could be hard to identify and even specific for target system. Sometimes the actions taken to improve one of these capabilities could spoil others. For example, the introduction of compres-sion algorithms to improve Space scalability (i.e. bandwidth reduction compressing messages) impacts on load scalability (i.e, increase of use of processor when compressing messages). The actions to scale may be classified in:

- *Vertical scaling*: by adding more horsepower (more processors, memory, band-width, etc.) to equipments used by the systems. This is the way applications are deployed on large shared-memory servers.
- *Horizontal scaling*: by adding more of the same software or hardware resources. For example, in a typical two-layer service, more front-end nodes are added (or released) when the number of users and workload increases (decreases). This is the way applications are deployed on distributed servers.

Scalability has to be kept in mind from the very beginning when designing the architecture of a system. Although appropriate time-to-market, fast prototyping or targeting small number of users could require quick developments, the architec-ture of the solution should have scalability into account. This implies the system

---

[2]http://rationalsecurity.typepad.com/blog/2009/01/a-couple-of-followups-on-my-edos-economic-denial-of-sustainability-concept.html

could increase the number of users from hundreds to thousands or even to millions or it could increase in complexity. By this doing, the risk of failure and system reimplementation would be minimized.

The Cloud is a computational paradigm which aims, among other major targets, to ease the way a service is provisioned helping service providers by delivering the illusion of infinite underlying resources and automatic scalability. This article describes how Cloud Computing may help to build scalable applications by automating the service provisioning process with IaaS (Infrastructure as a Service) Clouds (reducing the management costs and optimizing the use of resources) and providing PaaS (Platform as a Service) frameworks (with scalar execution environments, service building blocks and APIs) to build Cloud-aware applications in a Software as a Service (SaaS) model.

15.2 Foundations

Having made clearer what is currently understood by scalability and briefly outlined, the Cloud scalability is based in three basic pillars:

- *Virtualization*: it reduces systems complexity, standardizing the hardware platform and then reducing resource management costs.
- *Resource sharing*: sharing computing resources among different applications and/or organizations will allow optimizing their use avoiding sparse or idle occupation times. In this sense, virtualization helps to consolidate server in the same physical machine.
- *Dynamic provisioning*: resources should be provided in an on-demand way, they should be also automatically refigured on the fly. Dynamic provisioning implies the need of monitoring the service performance and automating the decisions and actions to respond to an in/decreasing workload.

Through this section, we will analyze the evolution of Information Technology (IT) services from mainframes to Clouds, which explains how scalability has been tackled, and some dynamic resource allocation techniques that could help to implement automatic scalability.

15.2.1 History on Enterprise IT Services

Since the 60's decade, mainframes (Ebbers et al., 2009) have been present at businesses back-office in finance, health care, insurance, government and other public and private enterprises. Starting processing batch tasks, introduced with punch cards, paper tape or magnetic tapes, during last decades it has evolved adding interactive terminals, and supporting multiple Operating System (OS) instances supported by virtual machines.

Mainframes (Ebbers et al., 2009) are centralized computers designed for handling and processing very large amounts of data quickly with high reliability, availability and serviceability. Mainframes scale vertically by adding more computing, storage or connectivity resources. Thanks to continuing compatibility of hardware and Operating Systems, upgrades in the OS, architecture family or model allow to scale systems without having to change the applications that run on top of it.

Decentralized applications, and more particularly client/server architectures (Orfali, Harkey, & Edwards, 1996), were implemented in the late 80's as an alternative to mainframes. An application is composed by autonomous computational entities with their own memory and communicated by a network using message-passing protocols (such as the Remote Procedure Call Protocol, RPC, which is "itself is a message-passing protocol").[3] A server provides a set of functions to one of many clients which invoke requests for such functions. Distributed systems present some advantages over mainframes:

- *Cost reduction*: A mainframe price was by \$2–3 M (now it cost around \$1 M). Mid-size applications could be deployed with only a few thousands of dollars.
- *Flexibility*: Server nodes, typically hosted in dedicated computers (mostly UNIX) can be developed, configured and tested separately from the rest of the system and plugged when ready.
- *Latency reduction*: Server nodes can be distributed among different datacenters to be as close of end-users as possible.
- *Interactive services*: Initially mainframes were batch processing-oriented but client/server is mainly interactive.
- *Unlimited resource addition*: Mainframes present platform-dependent limits when adding more resources (CPU, disk, memory...). Distributed systems allow the addition of more servers to increase the whole system capacity.

Client/server architectures are the base model for the network computing such Internet Services (web, mail, ftp, streaming, etc), telecommunication systems (IMS, VoIP, IPtv, etc.) and enterprise applications (information services, databases, etc.). Figure 15.1 shows how client/server architectures (Orfali et al., 1996) have evolved from two-tier (the client and the server hosted in different machines) to three-tier (client, application logic and data layer) and multi-tier (client, presentation logic layer, business logic layer, and data layer). As the applications grow in complexity and most of the business processes have been automated, datacenters have taken up more and more physical space. Distributed Systems can scale vertically (adding more resources to a host node), horizontally (adding new nodes of the same type) or both (from hosting all the service nodes in the same server to distribute them in dedicated servers taking advantage of transparency of location that communication protocols typically grant). Horizontal scale could require redesigning the application

---

[3]http://www.ietf.org/rfc/rfc1831.txt

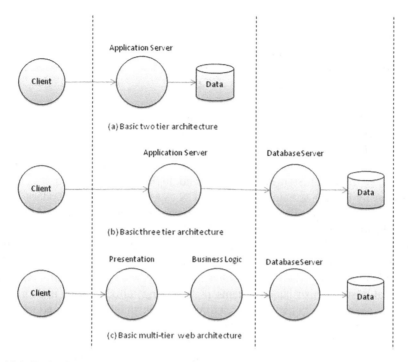

Application Server

Client → Data

(a) Basic two tier architecture

Application Server   Database Server

Client → → Data

(b) Basic three tier architecture

Presentation   Business Logic   Database Server

Client → → → Data

(c) Basic multi-tier web architecture

Fig. 15.1   Basic client/server architectures (Orfali et al., 1996)

to introduce parallelization, load-balancing, etc. But, during the 90's, distributed systems presented some disadvantages as compared to mainframes:

- *Low use of server nodes*: Typically server nodes are dedicated to a single application node, as it requires specific OS, libraries, isolation from other software components, etc.
- *More operation costs*: Distributed systems are more complex, they require more management task performed by human operators.
- *Less energy efficiency*: Wasted power on dedicated servers may increase the energy consumed by the hosts and the cooling systems.
- *More physical space required*: Even grouped in racks or blades servers, they need more space that integrated mainframes.
- *Potentially less performance with I/O*: Data storage is centralized in mainframes, but distributed systems that access through the network to a centralized storage present latency on accessing them.
- *Potentially more difficult to be fault tolerant*: As network and the number of service nodes introduce more failure points. Unless hardware components in mainframes are vulnerable too, they implement redundancy mechanisms to minimize the impact of failures.

- *Inability to share resources in distributed nodes*: spare CPU capacity in one node may not be used by others nodes. Here, the ability to consolidate the number of CPUs in a single mainframe also had a relevant economic impact in terms of middleware software licenses costs savings (since most middleware license prices scale up based on the number of CPUs.)

At the beginning of the 2000's these disadvantages made some IT managers reconsider coming back to renewed mainframes architectures which enabled to consolidate hardware resources while being able to run multiple OS instances. These architectures introduced virtualization technologies (such as IBM's z/Virtual Machine and z/Virtual Storage Extended) and logical partitions which can host UNIX (mostly Linux) or native mainframe OS (VMS, z/OS, etc.). Modern mainframes can also dynamically reconfigure the resources assigned to a virtual machine or logical partition (processors, memory, device connections, etc.).

In the world of distributed systems, clustering technologies, virtualization (VMware[4] or Xen[5]) and datacenter automation tools (IBM's Tivoli[6] or HP's OPSware[7]) make datacenters much easier to manage, minimizing complexity and operational costs of distributed systems provisioning (allocation of network and computing resources, OS and software components installation, etc.) and management (OS patches, software updates, monitoring, power, etc.). Modern datacenters are organized in clusters. A cluster is a group of commodity computers, working together closely so that in many respects they form a single computer. The components of a cluster are commonly, but not always, connected through fast local area networks. Clusters are usually deployed to improve performance (by scaling the available computing power) and/or availability over that of a single computer, while typically being much more cost-effective than single computers of comparable speed or availability.[8] However, putting commodity (often computational) resources together in a cluster often resulted not to be enough for the high computational demand required by some applications and experiments (Bote Lorenzo, Dimitriadis, & Gómez Sánchez, 2004). Further scaling was needed to absorb increasing demand. The need to aggregate sparse and separately managed computational resources gave rise to the concept of Virtual Organizations. These are separate administrative domains that set the means for using unused resourced belonging to other collaborating organizations as if they were locally located. Indeed, virtualization allows consolidating different Virtual Machines in the same physical host reducing waste of computing power and energy.

Indeed, from the computer architecture point of view, the evolution of computers makes not very clear the difference between dedicated servers and mainframes, as

---

[4]http://www.vmware.com

[5]http://www.xen-source.com

[6]http://www.ibm.com/software/tivoli

[7]http://www.hp.com

[8]http://en.wikipedia.org/wiki/Cluster_%28computing%29

mainframes could be consider the topmost model of a computer family. Then, the key issue in an IT Service strategy can be either to scale horizontally with low-cost servers (in the way of clusters or small servers) or vertically with "big" shared-memory servers. As usual, there is no silver bullet and the decision may depend on the application. Anyway, by introducing virtualization the underlying physical hardware layer is transparent for service nodes that potentially may scale vertically to the maximum capacity of the physical machines and horizontally to the maximum capacity of the datacenter. Some studies, see (Michael, Moreira, Shiloach, & Wisniewski, 2007; Barroso & Hölzle, 2009) for example, conclude that horizontal scaling offers better price/performance, although at an increase in management complexity, for Web-centric applications. In addition, horizontal scaling is deemed by some authors as the only solution for supercomputers.[9]

15.2.2 Warehouse-Scale Computers

Internet companies, such as Google, Amazon, Yahoo and Microsoft's online services division, have transformed their datacenters using Warehouse-scale Computers (WSCs) (Barroso & Hölzle, 2009) which differ from traditional datacenters on:

- Datacenters belong to a single company.
- Use of relatively homogeneous hardware and software platforms.
- Large cluster considered as a single computing unit, not only a set of wired individual servers.
- Often much of the software platform (applications, middleware and system software) is build in-house and adapted to the services they provide (search engines, media warehouses, e-commerce, etc.) instead of using third-party software (standard application servers, middleware, OS, etc.).
- They run a smaller number of very large applications.
- Use of a common management layer that flexibly controls the deployment of applications among shared resources.
- High availability achieved assuming large number of component faults with little or no impact on service level performance.

Figure 15.2 shows the typical warehouse-scale systems architecture (Barroso & Hölzle, 2009) which basic elements are low-cost 1U or blade enclosure servers mounted within a rack. Servers are interconnected by a rack-level 1–10 Gbps Ethernet switches with uplink connections to one or more cluster or datacenter level Ethernet switches. Disk drives can be managed by a Network Attached Storage (NAS) connected directly to the cluster switches, or be connected to each individual

---

[9]http://www.top500.org

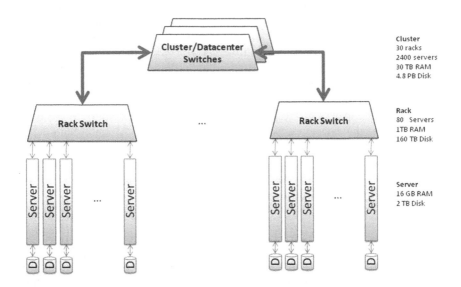

Fig. 15.2 Warehouse-scale system architecture (Barroso & Hölzle, 2009)

Table 15.1 Mainframe vs. datacenters scaling points

Scale type	Mainframes	Clusters
Vertical	• Add more CPUs, Memory or Disk • Upgrade to a bigger model • Add/improve software concurrency, compression, etc.	• Add more CPUs, Memory or Disk to servers • Add more bandwidth to network • Upgrade to a bigger server model • Redistribute Application components • Add/improve software concurrency, compression, etc.
Horizontal	None	• Add more server nodes • Add/improve parallelization algorithms

server and managed by a global distributed file system such as Google's File System (GFS).[10]

As shown above, resource management tools are key to control the dynamic resource provisioning and application deployment, and hence, to scale in a graceful manner. Again, specific distributed application architectures and technologies are used as well to scale the systems: distributed file systems, parallelization algorithms, message passing, etc. Datacenters can be replicated among different geographies to reduce user latency and improve the service performance.

Table 15.1 summarizes scaling solutions used in modern datacenters.

---

[10]http://labs.google.com/papers/gfs.html

15.2.3 Grids and Clouds

In contrast to warehouse-scale systems, Grid and Cloud technologies have emerged to allow resource sharing between organizations. Also called Service-Oriented Infrastructures, their aim is more "general-purpose" as they have to host a number of different applications from different domains and types of organizations (research, governance or enterprise). Anyway, most of functions they should provide are in common, and "public" functions could be adopted by "private" infrastructures and viceversa.

The Grid is one of such technologies extending the scale of computing systems by aggregating commodity resources belonging to different administrative domains into one or more virtual organizations. More formally, the Grid is defined as a "system that coordinates resources which are not subject to centralized control, using standard, open, general-purpose protocols and interfaces to deliver nontrivial qualities of service" (Foster, 2002). More recent definitions emphasize the ability to combine resources from different organizations for a common goal (Bote Lorenzo et al., 2004). In Kurdi, Li, and Al-Raweshidy (2008) and Stockinger (2007) the concern is about coordination of resources from different domains and how those resources must be managed.

Apart from aggregating more organizations into a single virtual organization, a given service was hard to scale and the Grid, as it is traditionally conceived, offered no mechanisms for helping Grid Service developers to scale their systems in accordance with changes in demand. In this regard, the Grid was no different to previous information technology (IT) systems: an administrator should detect service overloads and manually scale the system relying on performance metrics relevant for the service in question. Also, the number of nodes a Virtual Organization comprises may be well below the number needed to accomplish the intended task.

Recently, another paradigm, the Cloud, came into the scene to help increase the scalability provided to the end user. However, the distinctions are not clear maybe because Clouds and Grids share similar visions: reducing computing costs and increase flexibility and reliability by using third-party operated hardware (Vaquero et al., 2009). For example, Grids enhance fair sharing of resources across organizations, whereas Clouds provide the required resources on demand, making the impression of a single dedicated resource. Hence, there is no actual sharing of resources due to the isolation provided through virtualization. Nevertheless, virtualization technologies are also being used to help Grids scale at the vertical level (e.g. adding more re-sources to a virtual machine). Other important difference between traditional Grids and the Cloud has to do with the employed programming model. For example, a Cloud user can deploy Enterprise Java Beans-based applications just as he can deploy a set of Grid services instead. The Cloud will treat them both equally. However, by definition, Grids accept only "gridified" applications (Vaquero et al., 2009; Stockinger, 2007), thus imposing hard requirements to developers. Although virtual organizations share the hardware (and its management) costs, they are still higher than "renting" the required capacity right when it is needed. This is

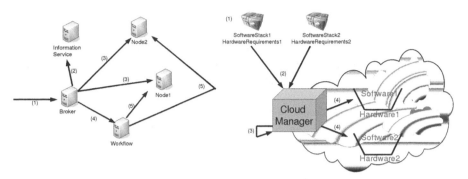

Fig. 15.3 Grid vs cloud provisioning models

indicating that the provisioning model is a very important element when it comes to determine the potential scalability of our IT system.

In the Grid, left hand side of the Fig. 15.3 below, a job request (1) was match-maker by a broker to some available resources found in an information service (2), resources are reserved to guarantee appropriate execution (3) and arranged in a workflow system (4) controlling the coordinated action of the allocated resources (5). On an infrastructure as a service Cloud, the user is in charge of providing the packaged software stack (1), and the Cloud manager finds available resources (2,3) to host the virtual machines containing the software stack (4). These represent two completely different provision and management philosophies.

Cloud computing paradigm shifts the location of computing resources to the network to reduce the costs associated with the management of hardware and software resources. On-demand provision of resources and scalability are some essential features of the Cloud. The Cloud offers many of the typical scaling points that an application may need including servers, storage and networking service underlying resource resizing (in an Infrastructure as a Service, IaaS, Cloud), or advanced development and maintenance platforms for reduced service time to market (in Platform and Software as a Service, PaaS/SaaS, Clouds). Thus, the Cloud could arguably be defined not as a technological advance, but as a model for dynamic service provision, where a service is anything that can be offered as a networked service (XaaS) (Vaquero et al., 2009).

The on-demand nature of Cloud computing combined with the aforementioned pay-per-actual-use model means that as application demand grows, so can the resources you use to service that demand. In this situation, the system eventually reaches a balance and allocated capacity equals demand as long as your application was designed properly and its architecture is amenable to appropriate scaling. Ideally applications in IaaS Cloud deployments should operate in terms of high-level goals and not present specific implementation details to administrators (Eyers, Routray, Zhang, Willcocks, & Pietzuch, 2009). Existing strategies require developers to rewrite their applications to leverage the on-demand resource utilization, thus locking applications to a specific Cloud infrastructure. Some approaches structure servers into a hierarchical tree to achieve scalability without significantly restructuring the code base (Song, Ryu, & Da Silva, 2009). Also, profiles are used to capture experts' knowledge of scaling different types of applications. The profile-based approach automates the deployment and scaling of applications in Cloud without binding to specific Cloud infrastructure (Yang, Qiu, & Li, 2009; Wang, 2009). Similar methods can be employed that analyze communication patterns among service operations and the assignment of the involved services to the available servers, thus optimizing the allocation strategy to improve the scalability of composite services (Wu, Liang, & Bertino, 2009).

The need for the above strategies and code rewriting clearly indicates the difficulties in scaling an application in the Cloud. Although some remarkable attempts have been made that try to add automatic scaling capabilities to service-based systems (see (Poggi, Moreno, Berral, Gavaldà, & Torres, 2009) for example), these are often hard to develop, too dependent on the specific application, and hardly generalizable to be offered as a general-purposed service. Thus, current commercial Grid and Cloud systems rely also on user know-how on the 'maximum' capacity and build out to that capacity. This implies that the system is usually in one of two modes: either under-buying or overbuying (see Fig. 15.4). Unlike the Grid and previous provisioning methods for IT systems, the Cloud allows service providers to simply add capacity as needed, typically with lead times of minutes. The pay-per-actual-use model lets one pay only for what is actually provisioned.

In spite of the flexible provision and billing models, the degree of automation and integration with underlying monitoring systems offered by most existent

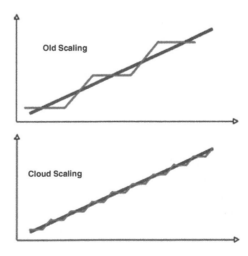

Fig. 15.4 Cloud computing techniques to reduce over- and under-provisioning

Cloud systems are to be further developed. The key to effective on-demand scaling is accurate utilization metrics. At the forefront, Amazon's Cloud Watch and Auto-scale[11] features allow some integration of the underlying monitoring system (providing infrastructure metrics, such as CPU usage) and a service for reacting on user-defined conditions. Also, RightScale[12] allows automating some trigger actions based on infrastructure metrics or on user-defined scripts put in the deployed servers. However, the degree of customization of the scaling rules by users is based on and application–level metrics are hard to be included, resulting in low automation for high level rules or metrics (e.g. service requests over a time period), which are closer to the Cloud user mindset.

Besides, adding more machines on demand, the underlying virtualization technologies inherent to the Cloud, allow one to include vertical scaling (more resources can be added on the fly to a virtual machine). Unfortunately, this very desirable feature is no yet supported by most operating systems (they need a fixed CPU and memory size at boot time).

Summing up, although IaaS Clouds have taken system scalability a step further (see Table 15.2), defining automatic scalability actions based on custom service metrics is not supported as of today. In fact, no Cloud platform supports the configuration of certain business rules, as for example limits on the maximum expenses the SP is willing to pay so she does not go bankrupt due for example to Economic Denial of Sustainability (EDoS) attacks.

Table 15.2 Scaling potential of current grid and IaaS clouds

Grid	Cloud
• Adding new organizations (and their shared resources) to the virtual organization. • Manual service scaling.	• "Renting" resources on-demand. • Preliminary automation means to enforce scaling rules. • Horizontal scaling thanks to hypervisor technology.

While IaaS Clouds and Grids can be different with regard to their scalability potential (see Table 15.2), PaaS/SaaS[13] Clouds still need to make huge progress towards helping to increase the required levels of automation and abstraction.

Just-in-time scalability is not achieved by simply deploying applications in the Cloud. SaaS scaling is not just about having a scalable underlying (virtual) hardware, but also about writing scalable applications. Valuable rules of thumb have been provided by PaaS platforms such as Google's App Engine.[14] These include minimizing work, paging through large datasets, avoiding datastore contention,

---

[11] http://aws.amazon.com/

[12] http://www.rightscale.com/

[13] http://cloudtechnologies.morfeo-project.org/archives/a-cloud-architecture-how-do-we-fill-the-gap-of-real-scalability/lng/es

[14] http://code.google.com/appengine/articles/scaling/overview.html

sharding counters, and effective memory cache. These techniques indicate that, as of today, traditional techniques for helping applications scale need to arise from programmers crafting the scaling points.

More sophisticated means are being called upon for avoiding customers to have to become experts in thread management or garbage-collection schemes. The Cloud, itself, should provide such facilities for programmers as a service (PaaS). One of the big advantages of the Cloud is that it could be used to split up a program so that different instructions can be processed at the same time. But it is hard to write the code needed to do that with most programming languages. For instance, C and other programming languages offered MPI APIs for performing the parallelization of instructions within a computational cluster. However, traditional programming environments provide inadequate tools, because they place the burden to developers who should operate at a too low level of abstraction.

One such effort is the BOOM (Berkeley Orders Of Magnitude) project, which aims to build easily scalable distributed with less code including GFS, Map/Reduce and Chubby, and then allow these components to be reconfigured and recomposed to enable new distributed systems to be easily constructed at low cost (Boom09, 2009).

Lacking such mechanisms results in particular solutions that can hardly be generalized and offered as a service. More general approaches tackle application domain knowledge to increase application scalability by minimizing the changes need to be done to the application's code. This has been applied, for example, to online social networks by getting advantage of the graph structure. Groups are separated in different servers and the nodes belonging to several groups are replicated in all the group servers (Pujol, Siganos, Erramilli, & Rodríguez, 2009).

15.2.4 Application Scalability

The example above (Pujol et al., 2009) clearly indicates that not all the applications are well suited to scale in the Cloud. While this is an ideal environment for Web applications, transactional applications, cannot be "cloudified" in such an easy manner.

Web applications can be scaled horizontally and vertically and spread over several datacenters without firewalls (or any other network related) problems. Web applications are usually stateless, which implies that migrating services from a location to another does not imply any shortcoming for application performance. Also, as new replicas are added (horizontal scaling), load balancers can evenly reroute requests to any available replica.

Databases cannot be ported to the Cloud so easily, though. They do not rely on Internet-ready protocols (such as HTTP), so problems are to be found in a highly distributed and multi-tenant environment such as the Cloud. Also, they are inherently stateful and rollbacks and commits are needed features for an appropriate service behavior. This latter premise implies that services cannot be migrated or located anywhere. Legal restrictions also restrict the migration and actual location of some very sensitive data (although these not technical implications fall beyond this Chapter's scope). Database replication is usually done by expert administrators and is very often dependent on the specific data model. The replication strategy affects load balancing, which further complicates providing transactional applications with automated scalability.

New transactional-like SaaS applications should rely on some basic programming concepts that web applications have been using to achieve high performance or high availability in large-scale deployments (Barroso & Hölzle, 2009), not trying to emulate traditional transactional architectures:

- *Data Replication* (this makes updates more complex)
- *Sharding* (partitioning) data set into smaller fragments and distributing them among a large number of servers.
- *Dynamic Load Balancing* by biasing the sharding policy to equalize the workload per node.
- *Health checking and watchdog timers*: it is critical to know if a server is too slow or unreachable to take actions as soon as possible. Timeouts to remote requests and heartbeat techniques should be used.
- *Integrity checks* to avoid data corruption
- *Application-specific compression*
- *Eventual Consistency*: large-scale systems relax data consistency during limited periods that will return eventually to a stable consistent state.

15.2.5 Automating Scalability

Automatic scalability at the application level can be implemented in several manners. The two most significant ones are:

1. Users provide a set of rules in a well-defined language (e.g. see Galán et al. (2009)) that feeds an "application controller" acting on behalf of the user and in charge of enforcing the specified scalability rules.
2. Design and implement application-specific algorithms or statistical methods for the controller to know when the application should scale without having to resort to users direct actions in an expert system.

As for the first approach above, a very relevant example will be given below (see Section 15.3.2). Suffice to say here that rule-based systems provide meaningful descriptions on the appropriate conditions to scale and datamining techniques can be employed to extract relevant rules and help users (service providers) to gain runtime knowledge on their application performance and optimization techniques. On the other hand, algorithmic techniques (by this, we also mean statistical tools, neural

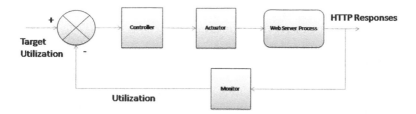

Fig. 15.5   Control loop for a web server process (Abdelzaher et al., 2002)

networks, or traditional control theory techniques) provide a degree of control for real-time response that cannot be attained by rule-based systems yet. This section is focused on some of these systems' features.

Figure 15.5 describes a QoS (Quality-of-Service) management solution for a Web application based in Control-Theory (Abdelzaher, Shin, & Bhatti, 2002). Extending these concepts to an application in an IaaS Cloud, the control loop that tries to avoid system overloads and meet the individual response time and throughput guarantees would be:

- *Service Application*: the service application to be executed in one or more virtual machines in the Cloud.
- *Monitor*: it provides feedback about resources utilization based on available measures such as CPU, disk memory and network bandwidth.
- *Controller*: given the difference between the desirable QoS and resources utilization (as measured by the monitor component), the controller has to decide on the corrective actions to meet the QoS target. Thus, control theory offers analytic techniques for closing the loop such us modelling the system as a gain function that maximizes the resource utilization between acceptable margins.
- *Actuator*: it translates abstract controller outputs into concrete actions taken by the Cloud middleware to scale up/down the service components to change its load.

Here, we highlight an open challenge for dynamic scalability in the cloud: finding a controller that could model a service deployed in a Cloud and make the system to scale as expected.

In Roşu, Schwan, Yalamanchili, and Jha (1997) the authors proposed the use of Adaptative Resource Allocation (ARA) in real-time for embedded system platforms. ARA mechanisms promptly adjust resource allocation (vertical scalability) to changes in runtime variations of the application needs whenever there is a risk of not meeting its timing constrains avoiding "over-sizing" real-time systems to meet the worst-case scenario. ARA models an application as a set of interconnected software components which execution is driven by event streams:

- *Resource Usage Model*: describes an application's expected computational and communication needs and the runtime variations of them.

- *Adaption Model*: acceptable configurations in terms of expected resource needs and application-specific configuration overheads.

This model could be generated by static and dynamic profiling tools that analyze the source code and the application runtime under different workloads, respectively. The ARA controller, then, detects a risk of not meeting performance targets, so that it calculates an acceptable configuration and a more appropriate resource allocation. Resource needs estimations are done based on node characteristics (processor speed factor, communication links speed, communications overhead) and the application static (parallelism level, execution time, number of interchanged messages and processor speed factor) and dynamic (execution factor, intra-component message exchange factors) resource usage models.

Similar real-time techniques based on mathematical application profiling and negative feedback loops, such us ARA, could be adapted to the Cloud scalability automation problem and represent future research challenges. Indeed, time series prediction is a very complex process and very specific for certain application domains (or even very specific applications).

15.3 Scalable Architectures

15.3.1 General Cloud Architectures for Scaling

Several architectures have been proposed to structure the Cloud and its applications. Buyya et al. proposed a QoS-enabling architecture for the Cloud, which presented accounting, pricing, admission control and monitoring as key enablers for its realization (Buyya, Yeo, & Venugopa, 2008). Based on this one, a more complete architectural model has recently been proposed by Brandic, Music, Leitner, and Dustdar (2009). Although these architectures were too QoS oriented (a crucial element for Clouds to become a real option for big enterprises), some essential shared elements have been identified (Lenk, Klems, Nimis, Tai, & Sandholm, 2009). Based on these common elements, an integrated Cloud computing stack architecture to serve as a reference point has been proposed. Based on this one, we further generalize it (see Fig. 15.6) to give a clear overview.

In spite of its widespread acceptance, this architecture is too general for illustrating the major points raised so far. Throughout this section we have been dealing with the problems of scalability at the different layers of the Cloud. Here we summarize the major mechanisms for Cloud-enabled scalability as of today (see Table 15.3 below).

While scalability and basic automation low level mechanisms have already been implemented for IaaS clouds (and many more are to be developed towards full automation and appropriate abstraction level). At the PaaS level, a "Cambrian" period is still to happen in order to increase the number of techniques to help scale

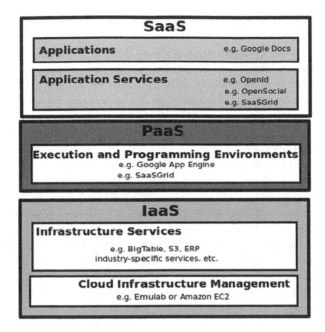

Fig. 15.6   General architecture of the cloud

Table 15.3   Architectural comparison of the cloud scaling status

	Scalability today	Scalability evolution
IaaS	Infrastructure level metrics	Service level metrics
	Limited automation	Full automation
	Virtual machine rules	Whole service rules
PaaS	Few available techniques	Great expansion potential by providing advanced development environments helping to develop Cloud applications.
		Help to exploit inherent parallelism in the developed code that can benefit from distributed architectures.
SaaS	Traditional development techniques. Huge burden for programmers	Services built on top of the PaaS-offered scaling techniques

the applications in the Cloud in a seamless backwards-compatible manner. These techniques are expected to be implemented by the developer herself, rather than helping to build really advanced scaling services by relying on platform-offered services. This change will shift current SaaS scaling efforts from programmers to the underlying platform.

15.3.2 A Paradigmatic Example: Reservoir Scalability

Resources and Services Virtualization without Barriers is a European Union FP7 co-funded project enabling massive scale deployment and management of complex services across administrative domains (Rochwerger et al., 2009). RESERVOIR[15] is structured in a layered architecture that can essentially be comprised in the IaaS layer above (managing virtualized and physical resources and cluster of these resources). One of the essential innovations of this effort with regard to scalability is the definition of techniques and technology for federation several independently-operated RESERVOIR sites. Federation allows a given Cloud provider to outsource some services and "rent" external resources in a seamless manner.

However, this scalability strategy is somewhat similar to that employed by Clouds as compared to Grids. Additional means are required to enable an appropriate architecture for networked services scalability. In order to overcome current IaaS scalability limitations (see Table 15.2 above), RESERVOIR proposes a new abstraction layer closer to the lifecycle of services that allows for their automatic deployment and escalation depending on the service status (not only on the infrastructure status). These features exceed pure IaaS capabilities, but remain essential for actual scalability across layers.

RESERVOIR Service Management (SM) layer automatically handles the service lifecycle and automatically scales services depending on service-level scaling "rules" and metrics. This is done with an abstraction closer to that managed by the service programmer. In the example below, a rule consists of a condition (number of tasks per executor node greater than 50, and there are less than 3 executors running) and an associated action if the rule is meet (create a new replica). This is specified in a to-be standard manner (Galán et al., 2009):

```
<rsrvr:Rule>
   <rsrvr:RuleName>rule2</rsrvr:RuleName>
   <rsrvr:RuleType>AgentMeasureEvent</rsrvr:RuleType>
   <rsrvr:Trigger checkingPeriod="5000ms"
condition="(@{kpis.QueueLength}/(@{components.VEEBigExecutor.
replicas.amount}*3 + @{components.VEEExecutor.replicas.
amount} +1)> 50) && (@{components.VEEExecutor.
replicas.amount} < 3)"/>
   <rsrvr:Action run="createReplica(components.VEEExecutor)"/>
</rsrvr:Rule>
```

In spite of these significant advances for IaaS Clouds, RESERVOIR does not deal with PaaS nor SaaS scalability, letting an important part of the problem to be solved by future research efforts.

---

[15] http://www.reservoir-fp7.eu

15.4 Conclusions and Future Directions

Scalability has followed a twisted line being applied to mainframes, distributed systems, back to mainframes and back to a "centralized" Cloud which is distributed and heterogeneous, but seen as a single entity by edge devices accessing the Cloud through standardized interfaces in order to execute services. Scaling capabilities have, thus, meandered between horizontal and vertical scalability, following the preponderant trend to systems design and implementation. This Chapter has very briefly reviewed the major features with regard to scalability as offered by some of the most salient centralized and distributed systems appeared as of today.

Here we have shown some of the most prominent examples of Cloud-enabled scalability at the different Cloud layers. It is important to note that this scalability is offered in a transparent manner for the end user (either a service provider or a service consumer).

Virtual Machines in IaaS Clouds can scale horizontally (by adding more service replicas to a given service) or vertically (by assigning more resources to a single Virtual Machine). IaaS Clouds themselves can be scaled vertically (by adding more clusters or network resources) or by attaining federation features with other externally managed datacenters (Cloud federation). Users are kept totally unaware of these scaling features and delivered with the illusion of infinite resources.

However, IaaS scalability is still too service-level oriented, meaning that scaling decisions are made on the basis of pure infrastructural metrics. Thus, user's involvement in service management is still required (not quite so with regard to Virtual Machine management in which a relevant degree of automation has been reached). Full automation and application of scalability rules (or load profile–based models) to control services in a holistic manner are granted for future developments on top of IaaS Clouds. These advanced high-level management automation capabilities lay close to the aforementioned PaaS features. However, they deal with deployment and runtime service lifecycle stages only. More elements helping to reduce services' time to market and provide further support for application design, development, debugging, versioning, updating, etc. are still very much needed. For example, some benchmarks specific for Cloud environments are already under way (Wang, 2009). Specific benchmarks for the scaling potential of applications on the Cloud are also to be developed. Generally speaking, the scaling applications in the Cloud will face some "old-fashioned challenges", detecting code parallelism (which could be offered as a PaaS Cloud service), distributing application components in clusters and service operations in cores (for multi-core architectures) will remain subject to massive research in future Clouds supporting actually scalable applications on the Cloud.

These whole service lifecycle facilities, including service scalability, will help to deliver advanced services in a shorter time and minimizing management burdens.

Some other issues remain to be solved, though. In order to scale Cloud services reliably to millions of service developers and billions of end users the next generation cloud computing and datacenter infrastructure will have to follow an evolution similar to the one that led to the creation of scalable telecommunication networks (Mikkilineni & Sarathy, 2009).

References

Abdelzaher, T. F., Shin, K. G., & Bhatti, N. (January 2002). Performance guarantees for web server end-systems: A control-theoretical approach. *IEEE Transactions on Parallel and Distributed Systems, 13*(1), 80–96.

Barroso, L. A., & Hölzle, U. (2009). The datacenter as a computer: And introduction to the design of warehouse-scale machines. *Synthesis Lectures on Computer Architecture, Morgan & Claypool Publishers*. ISBN: 9781598295573, URL http://www.morganclaypool.com/doi/abs/ 10.2200/S00193ED1V01Y200905CAC006.

Boom09 (2009). *BOOM: Data-centric programming in the datacenter* (UC Berkeley EECS Tech. Rep. No. UCB/EECS-2009-113 August 11, 2009).

Bondi, A. B. (September 2000). Characteristics of scalability and their impact on performance. *ACM WOSP'00 Proceedings of the 2nd International Workshop on Software and Performance, New York, NY.*

Bote Lorenzo, M. L., Dimitriadis, Y., & Gómez Sánchez, E. (February 2004). Grid characteristics and uses: A grid definition. (Postproceedings extended and revised version) *First European Across Grids Conference, ACG'03, Springer, LNCS 2970, Santiago de Compostela, Spain,* 291–298.

Buyya, R., Yeo, C. S., & Venugopa, S. (2008). Market-oriented cloud computing: Vision, hype, and reality for delivering it services as computing utilities. *Proceedings of the 10th IEEE International Conference on High Performance Computing and Communications, IEEE CS Press, Oslo, Norway.*

Brandic, I., Music, D., Leitner, P., & Dustdar, S. (2009). Vieslaf framework: Enabling adaptive and versatile SLA-management. *The 6th International Workshop on Grid Economics and Business Models 2009 (Gecon09). In Conjunction with Euro-Par 2009, Delft, The Netherlands,* 25–28.

Ebbers, M., Ketner, J., O'Brien, W., Ogden, B., Ayyar, R., Duhamel, M., et al. (August 2009). *Introduction to the new mainframes: z/OS Basics.* IBM Redbooks. http://www.redbooks. ibm.com/redbooks.nsf/RedbookAbstracts/sg246366.html.

Eyers, D. M., Routray, R., Zhang, R., Willcocks, D., & Pietzuch, P. (2009). Towards a middleware for configuring large-scale storage infrastructures. *Proceedings of the 7th International Workshop on Middleware for Grids, Clouds and E-Science, Urbana Champaign, IL, MGC '09,* November 30–December 01.

Foster, I. (2002). What is the grid? – A three point checklist, *Grid Today.*

Galán, F., Sampaio, A., Rodero-Merino, L., Loy, I., Gil, V., Vaquero, L. M., et al. (June 2009). Service specification in cloud environments based on extensions to open standards. *Fourth International Conference on COMmunication System softWAre and middlewaRE (COMSWARE 2009), Dublin, Ireland.*

Kurdi, H., Li, M., & Al-Raweshidy, H. (2008). A classification of emerging and traditional grid systems. *IEEE DS Onlinearch.*

Lenk, A., Klems, M., Nimis, J., Tai, S., & Sandholm, T. (2009). What's inside the cloud? An architectural map of the cloud landscape. *Proceedings of the 2009 ICSE Workshop on Software Engineering Challenges of Cloud Computing, IEEE Computer Society.* 23–31.

Michael, M., Moreira, J. E., Shiloach, D., & Wisniewski, R. W. (2007). Scale-up x scale-out: A case study using Nutch/Lucene. *IEEE International Parallel and Dis-tributed Processing Symposium, IPDPS, Rome, Italy.*

Mikkilineni, R., & Sarathy, V. (2009). Cloud computing and the lessons from the past. *Wetice, 18th IEEE International Workshops on Ena-bling Technologies: Infrastructures for Collaborative Enterprises, Paris, France,* 57–62.

Orfali, R., Harkey, D., & Edwards, J. (1996). *The essential client/server survival guide* (2nd ed.). New York, NY: Wiley.

Poggi, N., Moreno, T., Berral, J. L., Gavaldà, R., & Torres, J. (July 2009). Self-adaptive utility-based web session management. *Computer Networks, 53*(10), 1712–1721.

Pujol, J. M., Siganos, G., Erramilli, V., & Rodríguez, P. (2009). Scaling online social net-works without pains, NetDB 2009. *5th International Workshop on Networking Meets Databases, Co-located with SOSP 2009, Big Sky, MT.*

Rochwerger, B., Galis, A., Levy, E., Cáceres, J. A., Breitgand, D., Wolfsthal, Y., et al. (June 2009). RESERVOIR: Man-agement technologies and requirements for next generation service oriented infrastructures. *11th IFIP/IEEE International Symposium on Inte-Grated Management, New York, NY, USA*, 1–5.

Roşu, D., Schwan, K., Yalamanchili, S., & Jha, R. (1997). On adaptive resource allocation for complex real-time applications. *Proceedings of the 18th IEEE Real-Time Systems Symposium (RTSS '97), San Francisco, CA.*

Schlossnagle, T. T. (2007). Scalable Internet Architectures. Sams Publishing. ISBN: 0-672-32699-X.

Song, S., Ryu, K. D., & Da Silva, D. (2009). Blue eyes: Scalable and reliable system management for cloud computing. *Ipdps, IEEE International Symposium on Parallel&Distributed Processing, Rome, Italy*, 1–8.

Stockinger, H. (October 2007). Defining the grid: A snapshot on the current view. *Journal of Super-Computing, 42*(1), 3–17.

Vaquero, L., Rodero-Merino, L., Caceres, J., & Lindner, M. (2009). A break in the clouds: Towards a cloud definition. *ACM SIGCOMM Computer Communications Review, 39*(1), 50–55.

Wang, C. (2009). EbAT: Online methods for detecting utility cloud anomalies. *Proceedings of the 6th Middleware Doctoral Symposium, Urbana Champaign, IL.*

Wu, J., Liang, Q., & Bertino, E. (2009). Improving scalability of software cloud for composite web services. *Cloud, IEEE International Conference on Cloud Computing, Honolulu, Hawaii*, 143–146.

Yang, J., Qiu, J., & Li, Y. (2009). A profile-based approach to just-in-time scalability for cloud applications. *Cloud, IEEE International Conference on Cloud Computing, Bucharest, Romania*, 9–16.

Chapter 16
Scientific Services on the Cloud

David Chapman, Karuna P. Joshi, Yelena Yesha, Milt Halem, Yaacov Yesha, and Phuong Nguyen

16.1 Introduction

Scientific Computing was one of the first every applications for parallel and distributed computation. To this date, scientific applications remain some of the most compute intensive, and have inspired creation of petaflop compute infrastructure such as the Oak Ridge Jaguar and Los Alamos RoadRunner. Large dedicated hardware infrastructure has become both a blessing and a curse to the scientific community. Scientists are interested in cloud computing for much the same reason as businesses and other professionals. The hardware is provided, maintained, and administrated by a third party. Software abstraction and virtualization provide reliability, and fault tolerance. Graduated fees allow for multi-scale prototyping and execution. Cloud computing resources are only a few clicks away, and by far the easiest high performance distributed platform to gain access to. There may still be dedicated infrastructure for ultra-scale science, but the cloud can easily play a major part of the scientific computing initiative.

Scientific cloud computing is an intricate waltz of compute abstract programming models, scientific algorithms, and virtualized services. On one end, highly compute intensive scientific data algorithms are implemented upon cloud programming platforms such as MapReduce and Dryad, while on the other, service discovery and execution implement the bigger picture with data product dependencies, service chaining, and virtualization.

The cloud of science services is very tightly knit. It is difficult to make meaningful scientific discoveries from only a single data product. Yet even individual data products are produced from other products, which in turn require even more different products for calibration. Service chaining is essential to the scientific cloud, just as much as with the business cloud. However, the distinctive feature of scientific cloud computing is data processing and experimentation; a compute elephant hiding

D. Chapman (✉), K.P. Joshi, Y. Yesha, M. Halem, Y. Yesha, and P. Nguyen
Computer Science and Electrical Engineering Department, University of Maryland, Baltimore County, MD, USA
e-mails: {dchapm2@umbc.edu; kjoshi1@umbc.edu; yeyesha@csee.umbc.edu; halem@umbc.edu; yayesha@umbc.edu; phuong3@umbc.edu}

B. Furht, A. Escalante (eds.), *Handbook of Cloud Computing*,   379
DOI 10.1007/978-1-4419-6524-0_16, © Springer Science+Business Media, LLC 2010

underneath the service oriented architecture. Both the service lifecycle, and the processing platforms are key ingredients to a successful scientific cloud computation. We discuss both fronts from the perspective of an atmospheric cloud computing system Service Oriented Atmospheric Radiances (SOAR). We also make sure to touch on many related cloud technologies, even if they were not necessarily the best fit for our SOAR system.

16.1.1 Outline

There are two ends to a scientific cloud, the back end and the front end. We briefly describe the *Service Oriented Atmospheric Radiances (SOAR)* system in the next section. The third section on *Scientific Programming Paradigms (back end)* describes how programming platforms affect the scientific algorithms. The fourth section discusses the *Scientific Computing Services* that form the front end, describing in detail how service virtualization affects scientific repositories.

We have found MapReduce and Dryad to be highly effective platforms for more than our own algorithms. We summarize our own and others' work to apply these paradigms to science related problems.

We also describe a five phase service lifecycle, but in the perspective of scientific applications, and address some of the unique challenges that set science apart from other service domains.

16.2 Service Oriented Atmospheric Radiances (SOAR)

SOAR, a joint project between NASA, NOAA, and UMBC, is a scalable web service set of tools that provides complex gridding services on-demand for atmospheric radiance data sets from multiple temperature and moisture sounding sensors. SOAR accepts input through an online Graphical User Interface (GUI), or directly from other programs. The server queues these requests for a variety of complex science data services in a database tracking the various requested workflows. It uses large data sets collected by NASA, NOAA and DOD. These datasets contain satellite readings for temperature and moisture from the last three decades. SOAR uses the cloud Bluegrit at University of Maryland Baltimore County (UMBC) to apply data transformations such as gridding, sampling, subsetting and convolving in order to generate derived data sets from diverse atmospheric radiances (Halem et al., 2009).

Satellite remote sensing instruments orbit the Earth sun-synchronously to observe temperature, moisture, and other atmospheric structure and properties. SOAR facilitates climate oriented experiments by providing geospatial computations and transformations. This puts SOAR in a unique position, as it must chain with remote servers to acquire data, but as a facilitator, would be well placed within an even deeper chain for complicated scientific experiments.

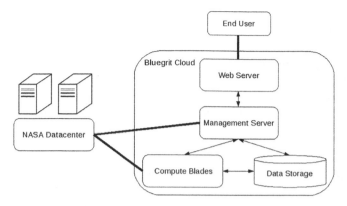

Fig. 16.1   SOAR deployment system

Figure 16.1 is a diagram of the SOAR deployment system. It makes use of the compute and data resources of the Bluegrit cloud. End users could be individual scientists, or other data service centers, and could make use of our graphical or SOAP interface provided by Bluegrit's Web server. Service requests encapsulate various climate related experiments such as tracking easterly cloud motion, or generating high resolution planetary images.

Management Server is a task driver for the various compute subsystems. It is responsible for scheduling tasks on the various compute blades. These tasks include precisely geo-located gridding, and singular value decomposition.

The input data for various service computations may not be locally available at the time of request. Management Server and Compute Blades must interact with various NASA data centers to acquire various data products for the requested scientific computations. Additionally Management Server routinely schedules jobs to compute and cache generic intermediate results, such as daily gridded average radiances.

16.3 Scientific Programming Paradigms

One of the biggest hurdles to unleashing the cloud onto science, is understanding its compute paradigms. The cloud provides a layer abstraction above and beyond the bare system configuration. Cloud abstraction typically arises from distributed middleware and centralized task scheduling. Programming paradigms empower the middleware, and change the way that we program; they force us to think in parallel. The remainder of this chapter is designed to bend our minds into understanding how to program the cloud for scientific applications. We discuss two programming strategies, MapReduce, and Dryad, and various scientific related problems, and how they could be implemented in a cloud environment.

MapReduce is a simple programming paradigm for distributed cloud computing. Google started MapReduce as a parallel processing solution for its indexing pipeline, and quickly realized that MapReduce was useful for many more parallel processing chores within the scope of Internet data retrieval. Google now hosts thousands of MapReduce applications. Although the intended purpose of MapReduce was text analysis and machine learning, it is also useful for many scientific computations, provided however, that they follow certain conditions (Ekanayake, Pallickara, & Fox, 2008). This makes MapReduce a very relevant topic for scientific computing, because it can make problems easier to program, yet it can potentially make hard problems even harder, if the problem does not fit the paradigm.

Dryad is a flexible programming model based on Directed Acyclic Graphs (DAGs). The nodes represent computation and the edges represent data flow direction. Dryad was developed by Microsoft as a generic paradigm for cloud computing problems, and as an alternative to Google's MapReduce. The Microsoft developers quickly found that Dryad was much more flexible than MapReduce, and as evidence were able to implement relational database, Map and Reduce, and many other software paradigms all completely encapsulated within Dryad's framework. Dryad is well rounded, and perfectly suitable for compute intensive, data intensive, dense, sparse, coupled, and uncoupled tasks. Dryad provides a very flexible solution, and is a good alternative for problems that might not fit well in simpler paradigms such as MapReduce. On the flip side, Dryad is relatively complicated, and may not be necessary when easier solutions are possible.

16.3.1 MapReduce

The original MapReduce paper, by Dean and Ghemawat (2004) describes the programming paradigm of MapReduce very concisely as follows.

> The computation takes a set of input key/value pairs, and produces a set of output key/value pairs. The user of the MapReduce library expresses the computation as two functions: Map and Reduce. Map, written by the user, takes an input pair and produces a set of intermediate key/value pairs. The MapReduce library groups together all intermediate values associated with the same intermediate key I and passes them to the Reduce function.
>
> The Reduce function, also written by the user, accepts an intermediate key I and a set of values for that key. It merges together these values to form a possibly smaller set of values. Typically just zero or one output value is produced per Reduce invocation. The intermediate values are supplied to the user's reduce function via an iterator. This allows us to handle lists of values that are too large to fit in memory.

The run-time system that implements the MapReduce handles the detail of scheduling, load balancing and error recovery. MapReduce implementation by Google uses Google Distributed File System where data storages and data placement in this file system is built based on the assumption that terabyte datasets will be distributed over thousands of disks inserted to computer nodes (Ghemawat, Gobioff, & Leung, 2003). In this kind of system, hardware failures occur often and replication strategies become important. In case of hardware failures, the system will redo jobs based on failed hardware only so that they do not have to redo the whole process. The main idea of the MapReduce implementation is that it performs computation where data are located so that the network latency of moving data around is optimized. Google's researchers have recently devised programming support for distributed data structure, called BigTable which provides capabilities similar to those in database systems whereas MapReduce is purely a functional notation for generating new files from old ones. Data recorded in the table by users are then stored and managed by the systems. Although BigTable does not provide a complete set of operations supported by relational databases, it strikes a balance between expressive and the ability to scale for very large databases in a distributed environment (Chang et al., 2006).

An open source framework Hadoop is an implementation of MapReduce by Apache providing abstract to programmers to let them write applications which can access and process massive data on the distributed file system. Essentially, the framework distributes data and processing map/sort/shuffle/reduce functions parallel locally at each node. In this way, massively parallel processing can be achieved using clusters that comprise thousands of nodes. In addition, the supporting runtime environment provides transparent fault tolerance by automatically duplicating data across nodes and detecting and restarting computations that fail on a particular node. Hadoop becomes bigger and better to provide open-source software for reliable, scalable, distributed computing. It includes subprojects such as HDFS a distributed file system that provides high throughput access to application data, HBASE is a scalable, distributed database that supports structured data storage for large tables, and it is an implementation of Google BigTable. Unfortunately, the current version 0.20.0 only allows for text files and does not support binary input files, a common format for science data processing. Apache does promise an updated version in a future release.

Other stages can be added to extend MapReduce paradigm such as sort and shuffle in Hadoop implementation. As one can see, the paradigm has only two user specified functions: Map and Reduce. A great way to become more familiar with MapReduce is by example. Word counting is the canonical example, with pseudocode given from Dean and Ghemawat (2004).

```
map(String key, String value):
  // key: document name
  // value: document contents
  for each word w in value:
    EmitIntermediate(w, "1");
reduce(String key, Iterator values):
  // key: a word
  // values: a list of counts
  int result = 0;
  for each v in values:
    result += ParseInt(v);
  Emit(AsString(result));
```

The map function emits each word plus an associated count of occurrences (just '1' in this simple example). The reduce function sums together all counts emitted for a particular word.

Thus, by performing Map, to report the occurrence of each word, followed by reduce to sum the number of occurrences for each word, the result of this example is the word count for each distinct word in the document.

By constructing new and different Map and Reduce functions, MapReduce can be used to solve many problems in addition to word counting. The processing can be performed in parallel, because both the Map and Reduce functions can be performed in parallel. Map acts in parallel on each input element. Reduce acts in parallel on separate KV groups for each distinct key.

16.3.1.1 MapReduce Merge

Yang, Dasdan, Lung-Hsiao, and Parker (2007) have introduced an improvement to Map-Reduce called Map-Reduce-Merge. This improvement enables better handling of joins on multiple heterogeneous databases, compared with using Map-Reduce. The authors point out that Map-Reduce is good for homogeneous databases. They discuss the problem in performing joins on multiple heterogeneous databases efficiently and mention that Pike, Dorward, Griesemer, and Quinlan (2005) point out that there is quite a lack of fit between Map-Reduce and such joins. In Yang et al. (2007), the authors also mention the importance of database operations in search engines. They also described how Map-Reduce-Merge can be applied to relational data processing

Map-Reduce (Dean & Ghemawat, 2004) is described in Yang et al. (2007) as follows:

For each (key,value) pair, Map produces a list of pairs of the form (key',value').Then Reduce is applied to the pairs created by Map as follows:For every key key'' that appears in the output of Map as a key, Reduce applies user defined logic to all the values $value''$ such that (key'',$value''$) is one of those pairs, and creates a list of values $value'''$

Map-Reduce-Merge is described in Yang et al. (2007) in the context of lineages. In Map-Reduce-Merge, Map is modified to operate on each lineage separately. Reduce is modified to operate on each lineage separately, and further modified to create a list of pairs (key'', $value'''$) rather than a list of values $value'''$.

Also, Merge is added as a third step. Merge is applied to the output of Reduce in two lineages. From the list of values associated with a key key'' in one lineage, and the list of values associated with a key key''' in another lineage, Reduce creates a list of pairs of the form (key'''',$value''''$). All the pairs created by Reduce form a new lineage.

16.3.2 Dryad

Dryad is a programming paradigm and software framework designed around the ideas of task scheduling and data flow. The programmer must create a Directed

Acyclic Graph (DAG) that represents the processing task. The graph nodes are compute kernels that run on various processors, and the graph edges represent the data flow dependence. Each graph node becomes available for computation as soon as all input data is available. A centralized job manager schedules available graph nodes onto idle machines. The machine executes the kernel computation, and upon completion, the node passes its output down to its children, and the machine becomes idle once again. The child graph nodes become available for computation as soon as all input data is available from its deceased parents. The computation continues in this manner until the entire DAG is executed and the program terminates.

Isard et al. describe their Job scheduling system with a concise diagram (Fig. 16.2) (Isard, Budiu, Yu, Birrell, & Fetterly, 2007).

> The job manager (JM) consults the name server (NS) to discover the list of available computers. It maintains the job graph and schedules running vertices (V) as computers become available using the daemon (D) as a proxy. Vertices exchange data through files, TCP pipes, or shared-memory channels. The shaded bar indicates the vertices in the job that are currently running

A common Dryad dilemma is that there is often more than one DAG that will satisfy computation of a particular problem. Which DAG is the fastest? Should one implement a very fine DAG with a high degree of parallelism, or a coarse DAG with low scheduling overhead? Sometimes the choice is clear, such as scheduling one node per machine, but often the choice is much more difficult to understand at first glance. Sometimes the best DAG depends on the design of your compute cluster; network and IO hardware design may play a critical role in determining potential bottlenecks in your data flow DAG. These low level issues may begin to contrast the philosophy that the cloud should be completely abstracted from its underlying hardware. Dryad makes it relatively easy for programmers to play around with the structure of the DAG, until they design one that runs efficiently on their target machine.

Additionally, Microsoft has attacked the graph tweaking dilemma head on with a number of automatic graph pruning and optimization algorithms. These techniques execute at runtime on the job scheduler, and thus can make decisions based on up to date profiles and resource availability. One such algorithm can make the DAG coarser by encapsulating a smaller subgraph to within a single node with serial

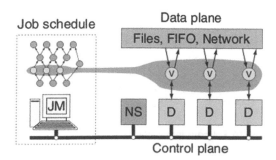

Fig. 16.2 Dryad

execution. Although encapsulation makes the system less parallel, it can greatly improve performance in situations where the graph was designed too finely. Another technique is to automatically make data reductions hierarchical. This can greatly improve performance, by reducing the data volume before sending packets to other machines and across racks.

16.3.3 Remote Sensing Geo-Reprojection

Atmospheric gridding and geo-reprojection is a great example of a single pass scientific problem, which is well suited for the MapReduce, and Dryad programming paradigms. Satellite remote sensing instruments measure blackbody radiation from various regions on earth, to determine weather and climate related forecasting, as well as supply atmospheric models with raw data for assimilation. The geo-reprojection algorithm is one of the first major compute steps along the chain of any satellite atmospheric prediction. The satellite observes a surface temperature, 316 K (43°C ~109°F) that's hot! But where is it? New Mexico? Libya? Geo-reprojection solves this task by producing a gridded map of Earth with average observed temperatures or radiances.

The measured region on Earth is a function of the instrument's position, and the direction that it is observing. Figure 16.3 is a diagram of the NASA Atmospheric Infrared Sounder (AIRS) satellite instrument. The satellite has a sun synchronous

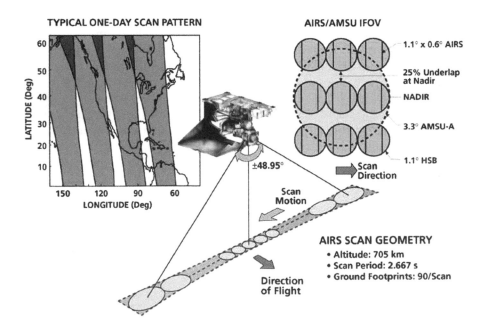

Fig. 16.3 Geo-Reprojection

orbit while the Earth is rotating, so on a Lat Lon, projection, the joint scan pattern is illustrated by the blue striping on the left. The instrument measures many observations during flight, as the sensor quickly oscillates between −48.95 and 48.95 degrees taking 90 observations every 2.7 s.

The map of Earth is uniformly divided into a number of regions, or grid cells. The goal is to have a measured temperature or radiance for each cell on the grid. When the satellite measures the same grid cell more than once, the resulting temperatures are averaged together.

16.3.3.1 Remote Sensing Geo-Reprojection with MapReduce

The MapReduce program for Geo-reprojection is similar in structure to the canonical word counting problem, provided we ignore details about computational geometry and sensor optics. Rather than counting words, we are averaging grid cells. Averaging is only slightly harder than summing, which in turn is only slightly harder than counting. The program has unchanged structure but novel details.

```
map(int timestamp, Measurement measurements):
  // key: timestamp
  // value: a set of instrument observations
  for each measurement m in measurements
    Determine region r containing measurement m
    EmitIntermediate(r, m.value);
reduce(int region, Iterator measurements):
  // key: a region ID
  // value: a set of measurement
  //      values contained in that region
  double result = 0.0;
  for each m in measurements:
    result += m;
  //divide by the total number of measurements
  result = result / measurements.size;
  Emit(result);
```

All of the sensor optics and geometry to determining the appropriate region are glossed over in the above pseudocode, with the line "Determine region r containing measurement m". For more detail see Wolf, Roy, and Vermote (1998). In this simple example, we assume there is only one region per measurement. However, in more realistic re projections, the observation may overlap multiple regions. In such an event, the Map would need to emit partial measurements for each region, and reduce would remain unchanged. Notice, that the final major step of reduce is to divide by the number of measurements (measurements.size). This division transforms the distributed summation to a distributed average, to derive the average measurement of the region.

16.3.3.2 Remote Sensing Geo-reprojection with Dryad

With Dryad, the remote geo-reprojection task may be computed somewhat differently than with MapReduce. We will assume that the output grid is comparably of lower resolution than the input data set. This assumption is usually valid for the problem, because the sounding instrument typically observes overlapping regions multiple times within hours to days of observation. Also, for climate related applications, fine grain grids are often not required, allowing for even further data reduction.

Problems that greatly reduce the volume of data are typically well described by a reduction type of graph (Isard et al., 2007). The basic generic reduction graph is shown in Fig. 16.4.

Notice, that there are many reduction graphs that all produce the same result as the one listed above. An example is the one listed in Fig. 16.5, which features a two level hierarchy. Partial reduction nodes r enumerate partial centroids an then pass the result onto the final reduction node R. This approach is more parallel, because there are more independent nodes working to do part of the reduction. Unfortunately, there is also more overhead in this hierarchical approach, because there are more nodes that need to be scheduled.

Fig. 16.4 Generic reduction graph

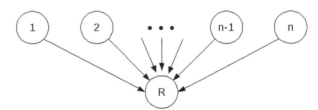

Fig. 16.5 Reduction graph with two level hierarchy

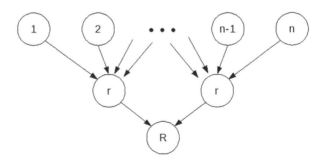

The following pseudocode could be used for the reduction DAGs described in this section. Start() represents nodes *1-n*, and reduce() represents nodes *r* and *R*.

```
start():
  // input0: instrument measurements
  Measurement []measurements = ReadChannel(0);
  // make an empty array of gridcell regions
```

```
Region []regions = new EmptyRegions();
//put each measurement in the region
for each measurement m in measurements
   Determine region r containing measurement m
   r.result += m;
   r.count   += 1;
//write the region array out to the channel zero
WriteChannel(newCentroids, 0);

reduce():
  // input0-n: region arrays
  Region [][]regions;
  for every input channel i
     regions[i] = ReadChannel(i);
  //a single region array for the results
  Region []results = new EmptyRegions();
  //accumulate all of the regions together
  for every input channel i
     for every region j in regions[i]
        results[j].result += regions[i][j].result
        results[j].count  += regions[i][j].count
  //divide, to perform the averaging
  for every region j in results
     results[j].result /= results[j].count
     //don't double divide if we reduce multiple times
     results[j].count = 1
  //We're done, write results to output channel 0
  WriteChannel(results, 0);
```

16.3.4 K-Means Clustering

Clustering is an essential component of many scientific computations, including gene classification (Alon et al., 1999), and N body physics simulations (Marzouk & Ghoniem, 2005). The goal of clustering is to separate a number of multidimensional data points into N groups, based on their positions relative to the other points in the dataset. K-Means clustering uses the concept of the centroid, or average position of all of the points in this group, to define the cluster. Initially, the points are grouped randomly into clusters. K-Means iteratively refines these clusters until it converges to a stable clustering.

K-Means uses the cluster centroid (average position), to determine the cluster grouping. A single iteration of K-Means is as follows:

1. Compute cluster centroid (average all points) for each cluster
2. Reassign all points to the cluster with the closest centroid
3. Test for convergence

16.3.4.1 K-Means Clustering with MapReduce

K-Means clustering is an iterative process that is a good candidate for MapReduce. MapReduce would be used for the centroid and clustering computation performed within each iteration. One would call MapReduce inside "if a" loop (until convergence), in order to compute iterative methods such as K-Means clustering.

The primary reason why K-Means is a reasonable MapReduce candidate, is because it displays a vast amount of data independence. The centroid computation is essentially distributed average, which is a small variation on the distributed summation, as exemplified by canonical word counting. Distributed summations require straightforward list reduction operations. The reassignment of points to clusters, requires only the current point and all cluster centroids; the points can be assigned independently of one another. Below is pseudo-code for K-Means clustering using MapReduce.

The Map function takes as input a number of points, and the list of centroids, and from this it produces a list of partial centroids. These partial centroids are aggregated in the reduce function, and used in the next iteration of the MapReduce

```
map(void, {Centroid []centroids, Point []datapoints}):
    // key: not important
    // value: list of centroids and datapoints
    Centroid []newCentroids;
    Initialize newCentroids to zero
    for each point p in datapoints
        Determine centroid centroids[idx] closest to p
        //accumulate the point to the new centroid
        newCentroids[idx].position += p;
        //we added a point, so remember the
        //total for averaging
        newCentroids[idx].total += 1;
    for each centroid newCentroids[idx] in newCentroids
        //send the intermediate centroids for accumulation
        EmitIntermediate(idx, newCentroids[idx]);

reduce(int index, Centroid []newCentroids):
    // key: centroid index
    // value: set of partial centroids
    Centroid result.position = 0;
    //accumulate the position and total for a grand total
    for each centroid c in newCentroids
        result.position += c.position;
        result.total += c.total;
    //Divide position by total to compute
    // the average centroid
    result.position = result.position / result.total
    Emit(result);
```

16.3.4.2 K-Means Clustering with Dryad

K-Means can equally well be implemented in the Dryad paradigm and mindset. The main task of programming with the Dryad paradigm is to understand the data flow of the system. A keen observations about K-Means, is that typically, each cluster has many points. In other words, there are far fewer clusters than there are points. Thus on each iteration the total amount of information is greatly reduced, when determining centroids from the set of points.

A reduction DAG is a good choice for K-Means, because the data volume is reduced. The following pseudocode would be used to perform the graph reduction operations for the K-Means algorithm using Dryad. In the graph reduction diagrams given the section "Remote Sensing Geo-reprojection with Dryad", start() is the function for nodes *1-n*, and reduce() is the function for nodes *r* and *R*.

```
start():
    // input0: Complete list of centroids from the prior run
    Centroid []centroids = ReadChannel(0);
    // input1: Partial list of datapoints
    Point []datapoints   = ReadChannel(1);
    // make a new list of centroids
    Centroid []newCentroids;
    Initialize newCentroids to zero
    for each point p in datapoints
        Determine centroid centroids[idx] closest to p
        //accumulate the point to the new centroid
        newCentroids[idx].position += p;
        //we added a point, so remember the
        //total for averaging
        newCentroids[idx].total += 1;
    //send the intermediate centroids for accumulation
    //channel zero is the only output channel we have
    WriteChannel(newCentroids, 0);
reduce():
    // input0-n: list of intermediate centroids
    Centroid [][]newCentroids;
    for every input channel i
        newCentroids[i] = ReadChannel(i);
    // make a list of result centroids
    Centroid []results;
    for every input channel i
        results[i].position = 0;
        //accumulate the position and total for a grand total
        for each centroid c in newCentroids[i]
            result[i].position += c.position;
            result[i].total += c.total;
```

```
//Divide position by total to compute
// the average centroid
results.position = result.position / result.total
//don't double divide if we reduce multiple times
   results.total = 1
//write our the results to channel 0
WriteChannel(results, 0);
```

16.3.5 Singular Value Decomposition

Singular value decomposition (SVD) can also be parallelized with cloud computing paradigms. The goal of SVD is very similar to matrix diagonalization. One must describe how a matrix, M, can be represented as the product of three matrices under the conditions as described below:

$$M = U\Sigma V^T$$

Where M is the original m-by-n matrix, U is an m-by-m orthogonal matrix, V^T is a n-by-n orthogonal matrix, and Σ is a m-by-n diagonal matrix.

The idea behind the Jacobi method is to start with the identity $M = IMI$ and attempt to slowly transform this formula into $M = U\Sigma V^T$ by a series of rotations designed to zero out the off-diagonal elements one at a time. Unfortunately, zeroing out one element may un-zero another. However, if this approach is repeated sufficiently, matrix M will converge to the diagonal matrix Σ.

Since the focus of this book is on cloud computing and not matrix algebra, we will not go into detail about the formulas required by the Jacobi and related methods. For further reading, refer to (Klema & Laub, 1980; Soliman, Rajasekaran, & Ammar, 2007; Rajasekaran & Song, 2006).

The One Sided JRS and Jacobi algorithms, to zero out a single element require modification of two rows within the matrix. A single sweep, for each pair of rows in the matrix, one must compute the dot product with the other rows, and use this value to modify both of the rows. For an n-by-n matrix, there are n(n-1)/2 such pairs of rows (Soliman et al., 2007).

It is thus natural to partition the matrix into rows for a parallel implementation. Rajasekaran and Song (2006) propose a round robin approach where each machine stores two blocks, computes all pairs of rows within each block, and then computes all pairs of rows between the two blocks. Then, the blocks are shuffled to other machines as illustrated in Fig. 16.6.

Although the aforementioned data access pattern can be implemented straightforwardly on grid systems with message passing, it is equally straightforward to implement with cloud computing paradigms MapReduce and Dryad. The cloud paradigms still provide additional benefits such as fault tolerance and data abstraction.

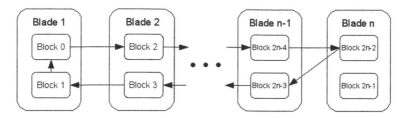

Fig. 16.6 Round robin SVD block communication pattern

16.3.5.1 Singular Value Decomposition with MapReduce

The round robin block data pattern described in Fig. 16.5 can be implemented with MapReduce, but with a single caveat. The difference is that MapReduce prefers to control how data is distributed based on the key/value pair of the block. Thus the key can be used as a *virtual* machine ID, rather than a *physical* ID. Each block is a key value pair. The reduce operation accepts two key value pairs (blocks) modifies them, and emits both of them back as results. This map reduce procedure must be performed iteratively until convergence.

```
map(int blockPos, Block block):
  // key: the current position of the block
  // value: 2D block
  int newBlockPos;
  if (blockPos == 0)
     newBlockPos = 1;
  else if (blockPos = 2n-1)
     newBlockPos = blockPos;
  else if (blockPos == 2n-2)
     newBlockPos = 2n-3;
  else if (blockPos % 2 == 1)
     newBlockPos += 2;
  else // (blockPos % 2 == 0)
     newBlockPos -= 2;
  //key must be equal to the virtual machine ID.
  //however, the slot (top or bottom) is also necessary
  //to disambiguate the block slots
  int machineID = floor(newBlockPos/2);
  BlockValue blockVal;
  blockVal.block = block;
  blockVal.slot  = newBlockPos % 2;
  EmitIntermediate(machineID, blockVal);

reduce(int machineID, BlockValue (Marzouk & Ghoniem, 2005)
  blockVals):
```

```
// key: virtual machine ID
// value: structure with the slot (top or bottom)
//          and the block data
//use a convention arrang the blockvals by slot
if (blockVals[0].slot == 1)
    swap (blockVals[0], blockVals (Alon et al., 1999);
//perform rotations in slot 0
Block block0 = blockVals[0];
for i=0 to block0.numRows-1
    for j=i to block0.numRows-1
        rotate(block0.row[i], block0.row[j]);
//perform rotations in slot 1
Block block1 = blockVals (Alon et al., 1999);
for i=0 to block1.numRows-1
    for j=i to block1.numRows-1
        rotate(block1.row[i], block1.row[j]);
//perform rotations across both slots
for i=0 to block0.numRows-1
    for j=0 to block1.numRows-1
        rotate(block0.row[i], block1.row[j]);
//remember the block position for the next iteration
int blockPos0 = 2*machineID;
int blockPos1 = 2*machineID + 1;
Emit(blockPos0, block0);
Emit(blockPos1, block1);
```

16.3.5.2 Singular Value Decomposition with Dryad

Much like MapReduce, Dryad likes to control how data is distributed via task scheduling. For this reason, it is equally important to use *virtual* machine IDs for the round robin SVD data access pattern. For Dryad, the nodes in the DAG represent virtual machines, and the edges represent the data distribution.

The graph in Fig. 16.6 is *cyclic*. It must be made *acyclic* for use with Dryad. To do so, we must unroll the graph. Unfortunately, the resulting graph would be of infinite length, as one does not know how much iteration must be performed before convergence falls below some error threshold. Fortunately, it is sufficient to make a large finite graph, and simply terminate early upon convergence.

The connectivity of the graph is somewhat intricate in order to achieve the pattern shown in Fig. 16.7. Notice how in Fig. 16.7, every node has two blocks: one on top, and one on bottom. We define that channel 0 reads input to the top block, and input channel 1 reads input from the bottom block. We also define that output channel 0 writes output from the top block, and output channel 1 writes output from the bottom block. At every timestep, both blocks are read, modified, and written to their appropriate channels.

Fig. 16.7 Acyclic SVD
round robin block
communication pattern

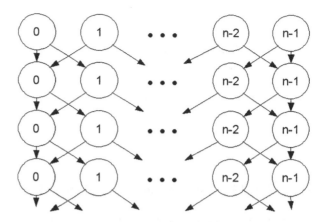

There are several corner cases for the connectivity of the graph. These cases
would be handled in graph construction, and are thus not listed in the pseudocode
of this section. Tables 16.1 and 16.2 show the specific rules of connectivity.

Table 16.1 Standard cases
for connectivity

From		To	
Node	Chan	Node	Chan
x	0	x+1	0
x	1	x−1	1

Table 16.2 Corner cases for
connectivity

From		To	
Node	Chan	Node	Chan
0	1	0	0
n−1	0	n−2	1
n−1	1	n−1	1

The pseudocode within a node to perform the block rotations is listed below.

```
node():
  // input0: Block for top slot
  Block block0 = ReadChannel(0);
  // input1: Block for bottom slot
  Block block1 = ReadChannel(1);
  //perform rotations in slot 0
  for i=0 to block0.numRows-1
      for j=i to block0.numRows-1
          rotate(block0.row[i], block0.row[j]);
  //perform rotations in slot 1
```

```
for i=0 to block1.numRows-1
   for j=i to block1.numRows-1
      rotate(block1.row[i], block1.row[j]);
//perform rotations across both slots
for i=0 to block0.numRows-1
   for j=0 to block1.numRows-1
      rotate(block0.row[i], block1.row[j]);
WriteChannel(block0, 0);
WriteChannel(block1, 1);
```

16.4 Delivering Scientific Computing services on the Cloud

Extant methodologies for service development do not account for a cloud environment, which includes services composed on demand at short notice. Currently, the service providers decide how the services are bundled together and delivered to service consumers. This is typically done statically by a manual process. There is a need to develop reusable, user-centric mechanisms that will allow the service consumer to specify their desired security or quality related constraints, and have automatic systems at the providers end control the selection, configuration and composition of services. This should be without requiring the consumer to understand the technical aspects of services and service composition.

Service Oriented Atmospheric Radiances (SOAR), demonstrates many examples of the forewarned paradox. Climate scientists want to study the earth's atmospheric profile, and they need satellite observations of sufficient quality for the experiments. It would be futile for them to learn every data processing step required, down to the algorithm version numbers, and the compute architectures used to produce every datum they require. Yet, they care that such a tool chain is well documented somewhere. If a colleague were to disagree with his findings years later, when all of the old data, algorithms and hardware have been upgraded, does the scientist even know the tool chain that produced his old experiments? It has been said that science without reproducibility is not science, yet in a world where data and computations are passed around the intricate cloud, provenance is all to easy to lose track of.

We have proposed a methodology for delivering virtualized services via the cloud (Joshi, Finin, & Yesha, 2009). We divide the IT service lifecycle on the cloud into five phases. In sequential order of execution they are requirements, discovery, negotiation, composition, and consumption. Figure 16.8 illustrates our proposed service lifecycle. Detailed lifecycle illustrating the sub-phases and is available at ((Joshi et al., 2009). We have also developed ontology in OWL for the service lifecycle which can be accessed at (Joshi).

16.4.1 Service Requirements

In the service requirements phase the consumer details the technical and functional specifications that a scientific service needs to fulfill. While defining the

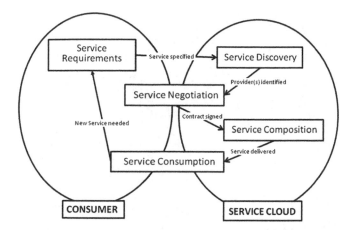

Fig. 16.8 The Service lifecycle on a scientific cloud comprises five phases: requirements, discovery, negotiation, composition and consumption

service requirements, the consumer specifies not just the functionality, but also non-functional attributes such as constraints and preferences on data quality, service compliance and required security policies for the service. Depending on the service cost and availability, a consumer may be amenable to compromise on the service data quality. For example, a simple service providing images of the Earth might deliver data as images of varying resolution quality. Depending on their requirements, service consumers may be interested in the high resolution images (higher quality) or might be fine with lower image resolution if it results in lower service cost.

Such explicit descriptions are of use not just for the consumer of the service, but also the provider. For instance, the cost of maintaining the service data quality can be optimized depending on the type of data quality requested in the service. The advantage for the service provider is that they will not need to maintain the lower quality data with the same efficiency as the higher quality data; but they would still be able to find consumers for the data. They can separate the data into various databases and make those databases available on demand. As maintainability is a key measure of quality, low maintenance need of the service data will result in improved quality of the service.

The service requirement phase consists of two main sub-phases: service specification and request for service.

Service Specification: In this sub-phase, the consumer identifies the detailed functional and technical specifications of the service needed. Functional specification describe in detail what functions/tasks should a service help automate and the acceptable performance levels of the service software and the service agent (i.e. the human providing the service). The technical specifications lay down the hardware, operating system, application standards and language support policies that a service should adhere to. Specifications also list acceptable security levels, data quality and performance levels of the service agent and the service software. Service compliance

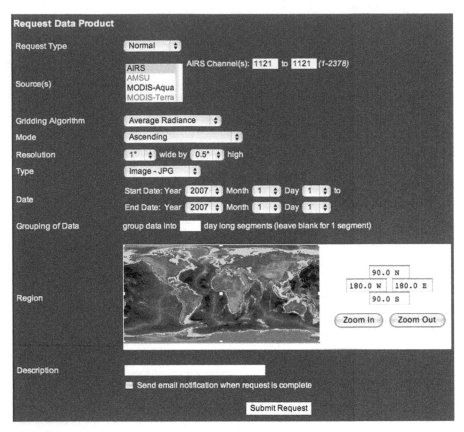

Fig. 16.9 Screenshot of the SOAR system GUI for the gridded average brightness temperature web service.

details like required certifications, standards to be adhered to etc. are also identified. Depending on the requirements, specifications can be as short or as detailed as needed.

Figure 16.9 shows all of the Quality of Services (QoS) parameters available to the user for the SOAR Gridded Average Brightness Temperature web service. GUIs make a good human readable compliment to the machine readable description languages used by SOAP and REST style web services. User-centric QoS parameters include variable resolution, and result type (jpg image, or hdf and binary data files). The service providers make their own decisions about the compute resources and related parameters. For example, a single service job may be distributed across a number of nodes, but this type of distributed computing is not necessary for small services that complete very quickly. It is up to the provider to decide how the task is scheduled, and what resources to designate.

Request for Service: Once the consumers have identified and classified their service needs, they will issue a "Request for Service" (RFS). This request could be made by directly contacting the service providers, such as the "Submit Request" button on Fig. 16.8. This direct approach bypasses any sort of discovery mechanism, but fails when the consumer is unaware of the service provider. Alternatively, consumers can utilize a service search engine on the cloud to procure the desired service, as long as the service is registered with some discovery engine.

Service requirement is a critical phase in service lifecycle as it defines the "what" of the service. It is a combination of the "planning" and "requirements gathering" phases in a traditional software development lifecycle. The consumers will spend the maximum effort in this phase and so it has been depicted entirely in the consumer's area in the lifecycle diagram. The consumer could outsource compilation of technical and functional specifications to another vendor, but the responsibility of successful completion of this phase resides with the consumer and not the service cloud. Once the RFS has been issued, we enter the discovery phase of the service lifecycle.

16.4.2 Service Discovery

In this phase, service providers that offer the services matching the specifications detailed in the RFS are searched (or discovered) in the cloud. The discovery is constrained by functional and technical attributes defined, and also by the budgetary, security, data quality and agent policies of the consumer.

If the consumer elects the option to search the cloud instead of sending the RFS to a limited set of providers, then the discovery of services is done by using a services search/discovery engine. This engine runs a query against the services registered with a central registry or governing body and matches the domain, data type, functional and technical specifications and returns the result with the service providers matching the maximum number of requirements listed at the top.

The discovery phase may not provide successful results to the consumers and so they will need to either change the specifications or alter their in-house processes to be able to consume a service that meets their needs the most.

If the consumers find the exact scientific service within the budget that they are looking for, they can begin consuming the service. However, often the consumers will get a list of providers who will need to compose a service to meet the consumer's specifications. The consumer will then have to begun negotiations with the service providers which is the next phase of the lifecycle. Each search result will also return the primary provider who will be negotiating with the consumer. It will usually be the provider whose service meets most of the requirement specifications.

16.4.3 Service Negotiation

Service negotiation phase covers the discussion and agreement that the service provider and consumer have regarding the service delivered and its acceptance criteria. The service delivered is determined by the specifications laid down in the

RFS. Service acceptance is usually guided by the Service Level Agreements (SLA) that the service provider and consumer agree upon. SLAs define the service data, delivery mode, agent details, compliance policy, quality and cost of the service. While negotiating the service level with potential service providers, consumers can explicitly specify service quality constraints (data quality, cost, security, response time, etc.) that they require.

Of course, scientists love to work with the most accurate and precise data available, until they run in to practical problems with it. The fine dataset might take too much storage, computation may take too much time, or perhaps out of place algorithms such as principal component analysis may require impractical amounts of RAM. These sorts of problems force scientists to reconsider the level of quality that they actually need for their desired experiment. SLAs will help in determining all such constraints and preferences and will be part of the service contract between the service provider and consumer.

Negotiation requires feedback, and the server must be able to estimate the cost involved with the service and deliver these statistics to the consumer before the service is executed. Although it is possible to negotiate automatically using hill-climbing and related optimization algorithms, SOAR was a service system designed primarily for human end users. Depending on the service and the QoS, the desired SOAR transaction could take seconds, minutes, or hours to compute. It is not necessary to give a precise time estimate, but at least a rough guess must be presented. For example, a warning "This transaction could take hours to compute" would be very helpful.

In addition to warning users about high cost services, the provider must also explain some strategy for reducing the cost. Of course, reducing QoS will reduce cost, but the degree may vary. Some service parameters may have a dramatic effect on the service cost, whereas others have little effect at all. For example, in the SOAR Gridded Average Brightness Temperature service, "resolution" has a tremendous effect on performance, whereas "type" (format) does not affect performance very much. Unfortunately, if the consumer doesn't know which parameters to vary, he may find himself toggling options randomly, or give up on the service entirely out of frustration. Although algorithms can happily toggle options systematically, humans would prefer to know what they are doing. SOAR documents the best known ways to cut cost. Preemptive GUIs may even be possible, that flag a user as soon as he selects a high cost option.

At times, the service provider will need to combine a set of services or compose a service from various components delivered by distinct service providers in order to meet the consumer's requirements. For example, SOAR services often need to interact with NASA datacenters to obtain higher resolution data for processing. The negotiation phase also includes the discussions that the main service provider has with the other component providers. When the services are provided by multiple providers (composite service), the primary provider interfacing with the consumer is responsible for composition of the service. The primary provider will also have to negotiate the Quality of Service (QoS) with the secondary providers to ensure that SLA metrics are met.

Thus the negotiation phase comprises of two critical sub phases, the negotiation of SLAs and negotiation of QoS. If there is a need for composite service, iterative discussions takes place between the consumer and primary provider and the primary provider and component providers. The final product of the negotiation phase is the service contract between the consumer and primary provider and between the primary provider and the component (or secondary) providers. Once the service contract is approved, the lifecycle goes to the composition phase where the service software is compiled and assembled.

16.4.4 Service Composition

In this phase one or more services provided by one or more providers are combined and delivered as a single service. Service choreography determines the sequence in the execution of the service components. Often the composition may not be a static endeavor, and may even depend on the QoS parameters used in the RFS.

The SOAR gridded average brightness temperature service is a great example of a dynamic composition workflow. Raw daily observations are cached at coarse resolutions, so only local parallel processing is necessary to service such requests. However, if the data specified is not available in cache, then it must be obtained from NASA's Goddard and MODAPS datacenters. Such is an example of service chaining with foreign NASA entities. Depending on the request, this remote execution could take hours to compute, and thus necessitates asynchronous execution. However, by comparison, the coarse local services could be completed in seconds. This is a great example of how the QoS and RFS specification could have such a huge impact on the resulting workflow, performance, and resources.

Many times what is advertised as a single service by a provider could in turn be a virtualized composed service consisting of various components delivered by different providers. The consumer needs to know that the service is composite for accounting purposes only. The provider will have to monitor all the other services that it is dependent on (like database services, network services etc.) to ensure that the SLAs defined in the previous phase are adhered to, and recorded for scientific reproducibility.

Once the service is composed, it is ready to be delivered to the consumer. The lifecycle then enters the final phase of service consumption.

16.4.5 Service Consumption and Monitoring

The service is delivered to the consumer based on the delivery mode (synchronous/asynchronous, real-time, batch mode etc.) agreed upon in the negotiation phase. After the service is delivered to the consumer, payment is made for the same. The consumer then begins consuming the service. An important part of the consumption phase includes performance monitoring using automated tools.

In this phase, consumer will require tools that enable quality monitoring and service termination if needed. This will involve alerts to humans or automatic termination based on policies defined using the quality related ontologies that need to be developed. The service monitor sub-phase measures the service quality and compares it with the quality levels defined in the SLA. This phase spans both the consumer and cloud areas as performance monitoring is a joint responsibility. If the consumer is not satisfied with the service quality, s/he should have the option to terminate the service and stop service payment. If the service is terminated, the consumer will have to restart the service lifecycle by again defining the requirements and issuing a RFS.

The performance monitoring tool used in SOAR is shown in Fig. 16.10. The tool not terribly complicated, but is very effective. It shows not only completed tasks, but also those currently in progress. The tool provides a timestamp for submission, so it is easy to monitor how long the request is taking. The user can select a task and remove it, even if it is currently in progress. This allows the user to cancel tasks that are taking too much time. Task cancellation is also accessible via SOAP, and can be used by a machine interface as well as a human interface.

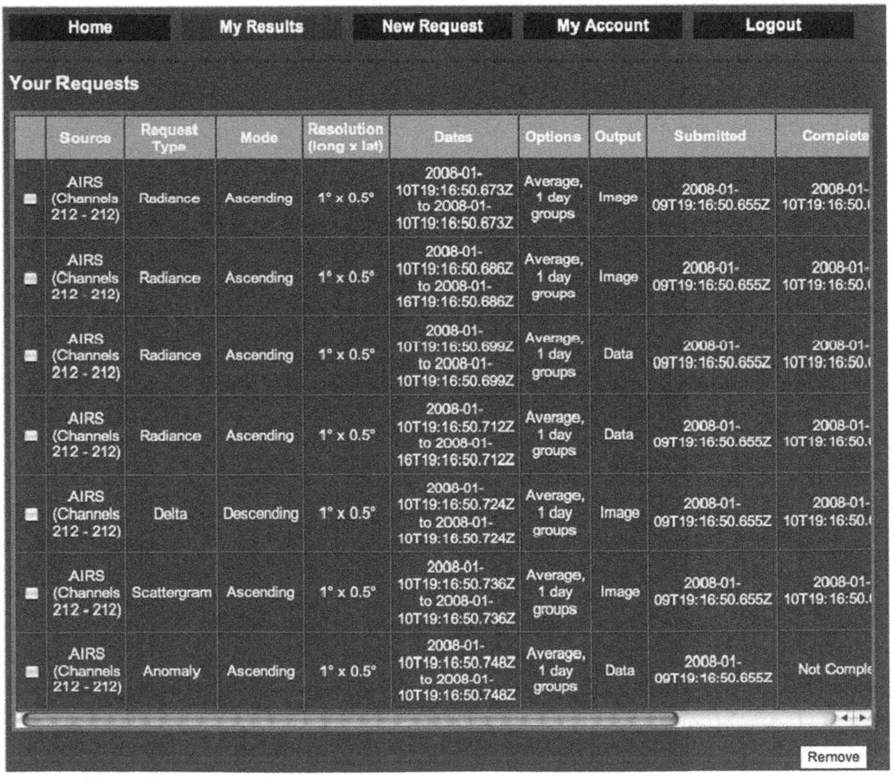

Fig. 16.10 Your Requests tool provided by SOAR

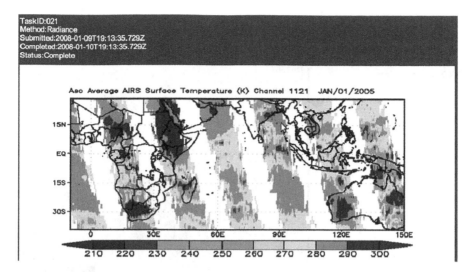

Fig. 16.11 SOAR generated image

An example of a gridded brightness temperature image is shown in Fig. 16.11. This is an example of a simple result generated from the SOAR system. The brightness temperature image is derived from the Atmospheric Infrared Sounder (AIRS) over the Indian Ocean on Jan 1st 2005. The striping on display is due to the sun synchronous polar orbit, and would be resolved if the client data request were for a longer time period. The warmest regions are in South Africa and Ethiopia. The blue dots around Madagascar, New Guinea and other locations are due to clouds. These types of plots, which show clouds very clearly, can be used in Hovmoller (running mean) diagrams to track cloud movement over time and monitor processes such as the Madden Julian Oscillation, as shown in Fig. 16.12.

16.5 Summary/Conclusions

In summary, we believe that cloud computing and in particular the MapReduce parallel programming paradigm and its generalizations offer a new and exciting approach for scientific computing. The compute and services paradigms at first glance may appear counterintuitive. However, once mastered they unleash the benefits cloud provides in terms of flexible compute resources, virtualized web services, software abstraction, and fault tolerance. MapReduce and Dryad are paradigms that can aid with scientific data processing on the back end, while service discovery and delivery provide not only a front end, but an entire compute work flow including data selection and quality of service. SOAR is an example of a complete scientific cloud application, and we hope that the lessons we have learned from this and other systems can help others in their scientific ventures.

Fig. 16.12 Hovmoller showing easterly cloud movement over the Pacific from Dec 15 2006 to Jan 17 2007

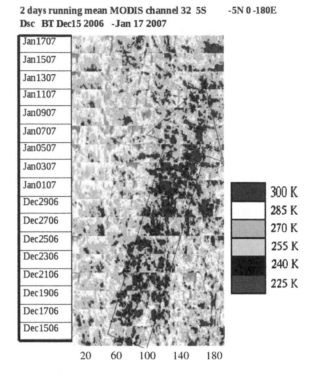

References

Alon, U., Barkai, N., Notterman, D. A., Gish, K., Ybarra, S., Mack, D., et al. (June 1999). Broad patterns of gene expression revealed by clustering analysis of tumor and normal colon tissues probed by oligonucleotide array. *Proceedings of the National Academy of Sciences of the United States of America, 96*(12), 6745–6750.

Chang, F., Dean, J., Ghemawat, S., Hsieh, W. C., Wallach, D. A., Burrows, M., et al. (November, 2006). Bigtable: A distributed storage system for structured data. *OSDI'06, Seattle, WA.*

Dean, J., & Ghemawat, S. (2004). MapReduce: Simplified data processing on large clusters. *Proceedings of OSDI,* 137–150.

Ekanayake, J., Pallickara, S., & Fox, G. (December 2008). Map-reduce for data intensive scientific analysis. *Proceedings of the IEEE International Conference on e-Science.*

Ghemawat, S., Gobioff, H., & Leung, S.-T. (2003). The Google file system. 29–43.

Halem, M., Most, N., Tilmes, C., Stewart, K., Yesha, Y., Chapmam, D., et al. (2009). Service oriented atmospheric radiances (SOAR): Gridding and analysis services for multi-sensor aqua IR radiance data for climate studies. *IEEE Transactions on Geoscience and Remote Sensing, 47*(1), 114–122.

Isard, M., Budiu, M., Yu, Y., Birrell, A., & Fetterly, D. (2007). Dryad: Distributed data-parallel programs from sequential building blocks. *Proceedings of the 2nd ACM SIGOPS/EuroSys European Conference on Computer Systems, Pacific Groove, CA.*

Joshi, K. P., Finin, T., & Yesha, Y. (October 2009). Integrated lifecycle of IT services in a cloud environment. *Proceedings of the 3rd International Conference on the Virtual Computing Initiative (ICVCI 2009), Research Triangle Park, NC.*

Joshi, K. *OWL ontology for lifecycle of IT services on the cloud.* http://www.cs.umbc.edu/
~kjoshi1/IT_Service_Ontology.owl.

Klema, V., & Laub, A. (April 1980). The singular value decomposition: its computation and some
applications. *IEEE Transactions on Automatic Control, 25*(2).

Marzouk, Y., & Ghoniem, A. (August 2005). K-means clustering for optimal partitioning and
dynamic load balancing of parallel hierarchical N-body simulations. *Journal of Computational
Physics, 207*(2), 493–528.

Pike, R., Dorward, S., Griesemer, R., & Quinlan, S. (2005). Interpreting the data: Parallel analysis
with Sawzall. *Scientific Programming Journal,* 227–298.

Rajasekaran, S., & Song, M. (2006). A novel scheme for the parallel computation of SVDs.
Proceedings of High Performance Computing and Communications, 4208(1), 129–137.

Ran, S. (2003). A model for web services discovery with QoS. *ACM SIGecom Exchanges, 4*(1),
1–10.

Soliman, M., Rajasekaran, S., & Ammar, R. (September 2007). A block JRS algorithm for highly
parallel computation of SVDs. *High Performance Computing and Communications, 4782*(1),
346–357.

Wolfe, R., Roy, D., & Vermote, E. (1998). MODIS land data storage, gridding, and compositing
methodology: Level 2 grid. *IEEE Transactions on Geoscience and Remote Sensing, 36*(4),
1324–1338.

Yang, H.-C., Dasdan, A., Hsiao, R.-L., & Scott Parker, D. (June 2007). Map-reduce-merge:
Simplified relational data processing on large clusters. *Proceedings of the 2007 ACM SIGMOD
International Conference on Management of Data, Beijing, China,* 1029–1040.

Chapter 17
A Novel Market-Oriented Dynamic Collaborative Cloud Service Platform

Mohammad Mehedi Hassan and Eui-Nam Huh

17.1 Introduction

In today's world the emerging Cloud computing (Weiss, 2007) offer a new computing model where resources such as computing power, storage, online applications
and networking infrastructures can be shared as "services" over the internet. Cloud
providers (CPs) are incentivized by the profits to be made by charging consumers
for accessing these services. Consumers, such as enterprises, are attracted by the
opportunity for reducing or eliminating costs associated with "in-house" provision
of these services.

However, existing commercial Cloud services are proprietary in nature. They are
owned and operated by individual companies (public or private). Each of them has
created its own closed network, which is expensive to setup and maintain. In addition, consumers are restricted to offerings from a single provider at a time and hence
cannot use multiple or collaborative Cloud services at the same time (Coombe,
2009). Besides, commercial CPs make specific commitments to their customers
by signing Service Level Agreements (SLAs) (Buyya et al., 2009). An SLA is a
contract between the service provider and the customer to describe provider's com-
mitment and to specify penalties if those commitments are not met. For example,
the Cloud "bursting" (using remote resources to handle peaks in demand for an
application) may result in SLA violation and end up costing the provider.

One approach for reducing expenses, avoid adverse business impact and to support collaborative or composite Cloud services, is to form a dynamic collaboration
(DC) (Yamazaki, 2004) platform among CPs. In a DC platform: (i) each CP can
share its own local resources/services with other partner CPs and hence can get
access to much larger pools of resources/services, (ii) each CP can maximize their
profit by offering existing service capabilities to collaborative partners, so they may
create a new value added collaborative service by mashing up existing services.

M.M. Hassan (✉) and E.-N. Huh
Department of Computer Engineering, Kyung Hee University, Global Campus, South Korea
e-mails: {hassan; johnhuh}@khu.ac.kr

B. Furht, A. Escalante (eds.), *Handbook of Cloud Computing*,
DOI 10.1007/978-1-4419-6524-0_17, © Springer Science+Business Media, LLC 2010

These capabilities can be made available and tradable through a service catalog for easy mash-up, to support new innovations and applications; and (iii) the reliability of a CP is enhanced as a result of multiple redundant clouds that can efficiently tackle disaster condition and ensure business continuity. However, the major challenges in adopting such an arrangement include the followings:

When to collaborate? The circumstances under which a DC arrangement should be performed. A suitable auction-based market model is required that can enable dynamic collaboration of Cloud capabilities, hiring resources and assembling new services and commercialized it.

How to collaborate? The architecture that virtualizes multiple CPs. Such architecture must specify the interactions among entities and allow for divergent policies among participating CPs.

How to minimize conflicts when negotiating among CPs? A large number of conflicts may occur in a market-oriented DC platform when negotiating among winning CPs. One reason is that each CP must agree with the resources/services contributed by other CPs against a set of its own policies in DC (Nepal & Zic, 2008; Nepal, Zic, & Chan, 2007). Another reason is due to the inclusion of high collaboration costs (e.g. network establishment, information transmission, capital flow etc.) by the CPs with their bidding prices in the market as they do not know to whom they need to collaborate with after winning an auction.

This chapter discusses the aforementioned challenges to form a DC platform among CPs and present candidate solutions to them. The main contributions of this chapter are as follows:

- We present a novel combinatorial auction (CA) based Cloud market model called CACM with a new auction policy that enables a virtual organization (VO) based DC platform among CPs. The CACM model can help consumers to get collaborative services from different CPs that suits their requirements as well as provide incentives for CPs to share their services.
- To address the issue of conflicts minimization among CPs, the existing auction policy of the CA model (Bubendorfer, 2006; Das & Grosu, 2005) is modified. The existing policy allows each bidder (CP) to publish separate bids to partially fulfill the composite service requirements as it cannot provide all the services. After the bidding, the winning bidders need to negotiate with each other to provide the composite service that results in a large number of conflicts mentioned earlier. The new auction policy in the CACM model allows a CP to dynamically collaborate with suitable partner CPs to form a group before joining the auction and to publish their group bids as a single bid to completely fulfill the service requirements, along with other CPs, who publishes separate bids to partially fulfill the service requirements. This new approach can create more opportunities to win auctions for the group since collaboration cost, negotiation time and conflicts among CPs can be minimized.
- To find a good combination of CP partners required for making groups and reducing conflicts, a multi-objective (MO) optimization model of quantitatively

evaluating the partners using their individual information (INI) and past collaborative relationship information (PRI) (Cowan et al., 2007) is proposed. In the existing approaches (Ko, Kim, & Hwang, 2001; Kaya, 2009) on partner selection, the INI is mostly used, PRI of partners is overlooked.

- Also to solve the model, a multi-objective genetic algorithm (MOGA) that uses INI and PRI called MOGA-IC is developed. A numerical example is presented to illustrate the proposed MOGA-IC in the CACM model. In addition, we developed MOGA-I (Multi-Objective Genetic Algorithm using Individual information), an existing partner selection algorithm, to validate the proposed MOGA-IC performance in the CACM model in terms of satisfactory partner selection and conflicts minimization.

The rest of this chapter is organized as follows: Section 17.2 describes the related works. Section 17.3 outlines the necessity of designing a DC platform among CPs. Section 17.4 introduces the proposed CACM model including the basic market model and key components of a CP to form a DC platform. In Section 17.5, we present the model of CP partner selection and the proposed MOGA-IC. Section 17.6 presents the experimental case study and analysis to show the effectiveness of the CACM model and the MOGA-IC. And finally Section 17.7 presents a conclusion.

17.2 Related Works

Cloud computing evolved rapidly during 2008 and now it is one of the hot topics for research. But no work has been found in the literature regarding the establishment of dynamic collaboration platform among CPs. There are a few approaches proposed in the literature regarding the Cloud market model. In Buyya et al. (2009), authors present a vision of 21st century computing, describe some representative platforms for Cloud computing covering the state-of-the-art and provide the architecture for creating a general auction-based Cloud market for trading Cloud services and resource management. But this market model cannot be directly applicable in creating a DC platform among CPs since the DC platform deals with a combinatorial allocation problem.

There are three types of auctions- one-sided auction (e.g. First Price and Vickrey auctions), double-sided auction (e.g. Double auction) and combinatorial auction (CA) (Bubendorfer, 2006; Bubendorfer & Thomson, 2006; Das & Grosu, 2005; Grosu & Das, 2004; Wolski et al., 2001). To enable the DC platform among CPs, CA is the appropriate market mechanism. In CA-based market model, the user/consumer can bid a price value for a combination of services, instead of bidding for each task or service separately and each bidder or service provider is allowed to wisely compete for a set of services.

But existing CA based market model is not fully suitable to meet the requirements for the CACM model since it cannot address the issue of conflicts minimization among CPs that usually happen when negotiating among providers in the

DC platform (Nepal & Zic, 2008; Nepal et al., 2007). The current approaches to handle conflicts are to design eContract delivery sequences (Nepal & Zic, 2008; Nepal et al., 2007). An eContract (Chen et al., 2008) is used to capture the contributions as well as agreements among all participants. But the main problem of these approaches is that auctioneer may choose an improper set of service providers (competing or rival companies). Then no matter how the eContract delivery sequence is arranged, a large number of conflicts cannot be prevented from happening. So we propose to modify the existing auction policy of CA that allows the CPs to publish their bids collaboratively as a single bid in the auction by dynamically collaborate with suitable partners. This approach can help to minimize the conflicts and collaboration cost among CPs as they know each other very well in the group and also creates more chances to win the auction.

However, collaborator or partner selection problem (PSP) is a complex problem, which usually needs a large quantity of factors (quantitative or qualitative ones) simultaneously, and has been proved to be NP-hard (Ko et al., 2001) or NP-complete (Wu, & Su, 2005). For CP partner selection, for instance, cost and quality of service are the most important factors. Also PSP for CPs in the CACM model is different from other PSP problems in areas like manufacturing, supply chain or virtual enterprise (Ko et al., 2001; Wu & Su, 2005; Wang et al., 2009; Buyukozkan, Feyzioglu, & Nebol, 2008; Ip et al., 2003; Fuqing, Yi, & Dongmei, 2008; Cheng, Ye, & Yang, 2009; Chen, Cheng, & Chuang, 2008; Sha & Che, 2005; Amid, Ghodsypour, & Brien, 2006; Chang et al., 2006; Chen, Lee, & Tong, 2007; Saen, 2007; Huang et al., 2004; Gupta & Nagi, 1995; Fischer, Jahn, & Teich, 2004; Kaya, 2009) since a large number of conflicts may occur among CPs due to dynamic collaboration. In the existing studies on partner selection, the individual information (INI) is mostly used, but the past collaborative relationship information (PRI) (Cowan, 2007) between partners, is overlooked. In fact, the success of past relations between participating CPs may reduce uncertainty and conflicts, shorten the adaptation duration, and help with performance promotion. Thus the existing methods cannot be applied directly to solve the PSP problem of CPs. Therefore, an appropriate MO optimization model using INI and PRI and effective MOGA called MOGA-IC to solve the MO optimization problem are proposed. Although, many MOGAs are available in the literature (Wang et al., 2009; Ip et al., 2003; Chen et al., 2008; Kaya, 2009; Zitzler et al., 2001), but all of these also do not consider PRI for partner selection.

17.3 A Dynamic Collaborative Cloud Services Platform

Dynamic collaboration is a viable business model where each participant within a DC shares their own local resources (services) with other participants by contributing (in a controlled policy driven manner) them to the collaboration. To make Cloud computing truly scalable and to support interoperability issues, a DC platform among CPs is very important.

A dynamic collaborative Cloud service platform can help CPs to maximize their profits by offering existing services capabilities to collaborative business partners. These capabilities can be available and tradable through a service catalog for easy mash-up to provide new value-add collaborative Cloud services to consumers. Also the DC platform can enable a CP to handle Cloud bursting by redirecting some load to collaborators. The Fig. 17.1 shows a formed dynamic collaborative Cloud service platform.

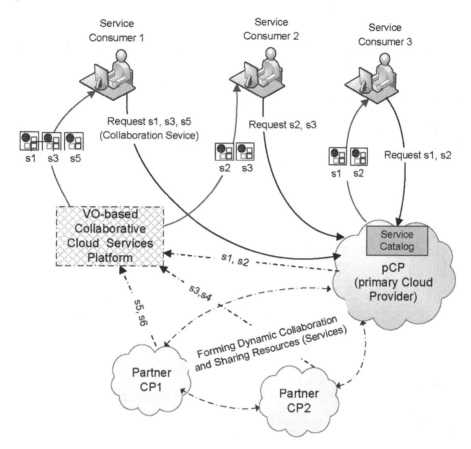

Fig. 17.1 A formed VO based cloud services collaboration platform

Formation of a DC is initiated by a CP, which realizes a good business oppor- tunity which is to be addressed by forming DC with other CPs for providing a set of services to various consumers. The initiator is called a primary CP (pCP), while other CPs who share their resources/services in DC are called collaborating or part- ner CPs. Users interact transparently with the VO-based DC platform by requesting services through a service catalog of the pCP. The CPs offer capabilities/services to consumers with a full consumption specification formalized as a standard SLA. The requested service requirements (single, multiple or collaborative Cloud services) are served either directly by the pCP or by any collaborating CPs within a DC. Let us consider, pCP can provide two services s1 and s2 and CP1 and CP2 can provide services s3, s4 and s5, s6 respectively as shown in Fig. 17.1. The request for collaborative services like s1, s3, s5 or s2, s3 can be served by VO-based DC platform. In case of services s1 and s2, the pCP can directly delivers the services. To enable this DC platform and make it commercialized, a CA-based Cloud market (CACM) model is described in the next section.

17.4 Proposed Combinatorial Auction Based Cloud Market (CACM) Model to Facilitate a DC Platform

17.4.1 Market Architecture

The proposed CACM model to enable a DC platform among CPs is shown in Fig. 17.2. The existing auction policy of the CA is modified in the CACM model to address the issue of conflicts minimization among providers in a DC platform.

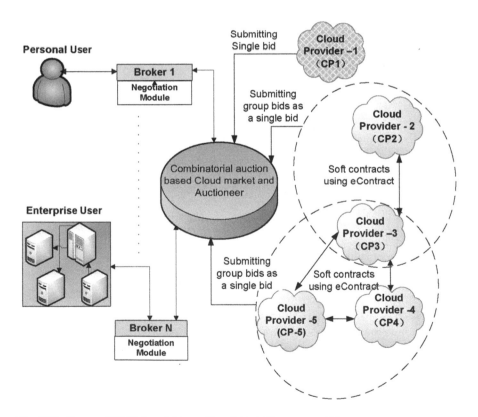

Fig. 17.2 Proposed CACM model to enable a DC platform among CPs

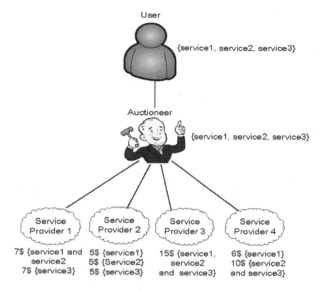

Fig. 17.3 Existing auction policy of CA

The existing and new auction policy for the CA model is shown in Figs. 17.3 and 17.4 respectively. The CACM model allows any CP to dynamically collaborate with appropriate partner CPs to form groups and to publish their group bids as a single bid to completely fulfill the consumer service requirements while also supporting the other CPs to submit bids separately for a partial set of services. The main

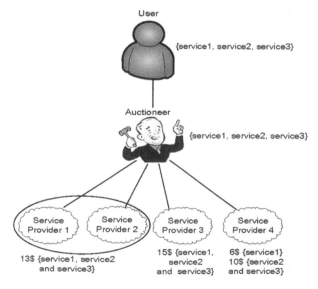

Fig. 17.4 New auction policy in CACM

participants in the CACM model are brokers, users/consumers, CPs and auctioneers as shown in Fig. 17.2.

Brokers in the CACM model mediate between consumers and CPs. A broker can accept requests for a set of services or composite services requirements from different users. A broker is equipped with a negotiation module that is informed by the current conditions of the resources/services and the current demand to make its decisions. Consumers, brokers and CPs are bound to their requirements and related compensations through SLAs. Brokers gain their utility through the difference between the price paid by the consumers for gaining resource shares and that paid to the CPs for leasing their resources.

The users/consumers can be enterprise user or personal user. Consumers have their own utility functions that cover factors such as deadlines, fidelity of results, and turnaround time of applications. They are also constrained by the amount of resources that they can request at any time, usually by a limited budget. The users can bid a single price value for different composite/collaborative Cloud services provided by CPs.

The CPs provide Cloud services/resources like computational power, data storage, Software-as-Service (SaaS), computer networks or infrastructure-as-a Service (IaaS). A CP participates in an auction based on its interest and profit. It can publish bid separately or collaboratively with other partner CPs by forming groups to fulfill the consumers' service requirements.

The responsibility of an auctioneer includes setting the rules of the auction and conducting the combinatorial auction. The auctioneer first collects bids (single or group bids) from different CPs participating in the auction and then decides the best combination of CPs who can meet user requirements for a set of services using a winner determination algorithm. We utilize secured generalized Vickrey auction (SGVA) (Suzuki & Yokoo, 2003) to address the CACM model problem and use dynamic graph programming (Yokoo & Suzuki, 2002) for winner determination algorithm.

17.4.2 Additional Components of a CP to Form a DC Platform in CACM

To achieve DC using the CACM model, in our architecture, a CP should possess the additional components described as follows:

Price Setting Controller (PSC) – A CP is equipped with a PSC which sets the current price for the resource/service based on market conditions, user demand, and current level of utilization of the resource. Pricing can be either fixed or variable depending on the market conditions.

Admission and Bidding Controller (ABC) – It selects the auctions to participate in and submits single or group bid based on an initial estimate of the utility. It needs market information from the information repository (IR) to make decisions which auction to join.

Information Repository (IR) – The IR stores the information about the current market condition, different auction results and consumer demand. It also stores INI (price, quality of service, reliability etc.) and PRI (past collaboration experiences) of other CPs collected from each CPs website, market and consumers feedback about their services.

Collaborator Selection Controller (CSC) – It helps a CP to find a good combination of collaborators to fulfill the consumer requirements completely by running a MOGA called MOGA-IC (described later in Section 17.5.3) utilizing the INI and PRI of other CPs.

Mediator (MR) – The MR controls which resources/services to be used for collaborative Cloud services of the collaborating CPs, how this decision is taken, and which policies are being used. When performing DC, the MR will also direct any decision making during negotiations, policy management, and scheduling. A MR holds the initial policies for DC formation and creates an eContract and negotiates with other CPs through its local Collaborating Agent (CA).

Service Registry (SR) – The SR encapsulates the resource and service information for each CP. In the case of DC, the service registry is accessed by the MR to get necessary local resource/service information. When a DC is created, an instance of the service registry is created that encapsulates all local and delegated external CP partners' resources/services.

Policy Repository (PR) – The PR virtualizes all of the policies within the DC. It includes the MR policies and DC creation policies along with any policies for resources/services delegated to the DC as a result of a collaborating arrangement. These policies form a set of rules to administer, manage, and control access to DC resources and also helps to mash-up Cloud services. They provide a way to manage the components in the face of complex technologies.

Collaborating Agent (CA) – The CA is a policy-driven resource discovery module for DC creation and is used as a conduit by the MR to exchange eContract with other CPs. It is used by a primary CP to discover the collaborating CPs (external) resources/services, as well as to let them know about the local policies and service requirements prior to commencement of the actual negotiation by the MR.

17.4.3 Formation of a DC Platform in CACM Model

The DC creation steps are shown in Fig. 17.5 and are explained as follows-

Step 1: A pCP finds a business opportunity in the market from IR and wants to submit collaborative bids as a single bid in the auction to address consumer requirements as it cannot provide all the service requirements.

Step 2: The CSC is activated by the pCP to find a set of Pareto-optimal solutions for partner selection and it chooses any combination from the set to form groups and send this information to the MR.

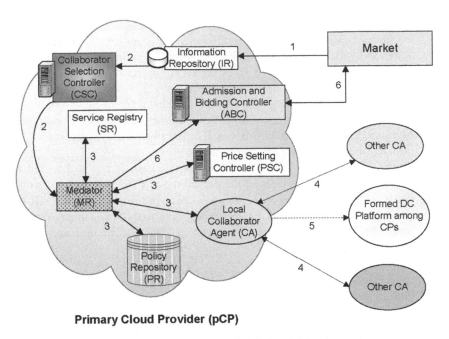

Primary Cloud Provider (pCP)

Fig. 17.5 The formation of a DC platform among CPs before joining the auction

Step 3: The MR obtains the resource/service and access information from the SR, whilst SLAs and other policies from the PR. It generates a eContract that encapsulates its service requirements on the pCP's behalf based on the current circumstance, its own contribution policies, prices of services (generated by PSC) and SLA requirements of its customer(s) and passes this eContract to the local Collaborating Agent (CA).

Step 4: The local CA of pCP carries out negotiations with the CAs of other identified partner CPs using the eContract. As the group members know each other very well, the number of conflicts will be less. So when all CPs (including the pCP) agree with each other, they make a soft contract among them. A soft contract guarantees that resources/services will be available if the group wins the auction.

Step 5: When pCP acquires all services/resources from its collaborator to meet SLA with the consumer, a DC platform is formed. If no CP is interested in such arrangements, DC creation is resumed from Step 2 with another Pareto-optimal solution.

Step 6: After the DC platform creation, the MR of pCP submits collaborative bids as a single bid to the market using the admission and bidding controller (ABC). If this group wins the auction, a hard contract is performed among each group members to firm up the agreement in DC. A hard contract ensures that the collaborating CPs must provide the resources/services according to the SLAs with consumers.

If some CPs win the auction separately for each service (few chances are available), the steps 3–5 are follows to form a DC platform among providers. But they make the hard contract in step 4 and in this case, a large number of conflicts may happen to form the DC platform.

An existing DC may need to either disband or re-arrange itself if any of the following conditions hold: (a) the circumstances under which the DC was formed no longer hold; (b) collaborating is no longer beneficial for the participating CPs; (c) an existing DC needs to be expanded further in order to deal with additional load; or (d) participating CPs are not meeting their agreed upon contributions.

17.4.4 System Model for Auction in CACM

For the convenience of analysis, the parameters and variables for the auction models are defined as follows:

$R = \{R_j \,|j = 1...n\}$: a set of n service requirements of consumer where $S \subseteq R$

$P = \{P_r \,|r = 1...m\}$: a set of m Cloud providers who participate in the auction as bidders where $G \subseteq P$

$P_{rj} = $ a Cloud provider r who can provide service j

$S(P_r) = $ a set of services ($S_{j=1...n}$) provided by any CP r where $S(P_r) \subseteq R$

$\Omega_{\max}(R, Q) = $ payoff function of the user where R is the service requirements and Q defines SLAs of each service.

17.4.4.1 Single and Group Bidding Functions of CPs

Let M be the service cost matrix of any CP P_r. We assume that each CP can provide at most two services. The matrix M includes costs of P_r provider's own services as well as the collaboration costs (CC) between services of its own and other providers. Figure 17.6 illustrates the matrix M. We assume that P_r provides two services – CPU and Memory. Let $a_{ii}(i = 1...n)$ be the cost of providing any service in M independently, $a_{ij}(i, j = 1...n, i \neq j)$ be the CC between S_i and S_j services ($S_i, S_j \in S(P_r)$) and $a_{ik}(i, k = 1...n, i \neq k)$ be the CC between S_i and S_k services ($S_i \in S(P_r)$ and $S_k \notin S(P_r)$). We set nonreciprocal CC between $S(P_r)$ services in M which is practically reasonable.

If CP P_r knows other providers or have some past collaboration experience with others, it can store true CC of services with other providers. Otherwise it can set a high CC for other providers. The CC of services with other providers in matrix M is updated when the providers finish negotiation and collaboratively provide the services of consumers in the DC platform.

Now the *Bidding Function* of any CP say P_r who submits bid separately to partially fulfill the customer service requirements without collaborating with other CPs can be determined as follows: $\phi_{S(P_r)} = C_{S(P_r)} + \gamma(P_r)$ where $C_{S(P_r)}$ is the total cost

Fig. 17.6 Cost matrix M

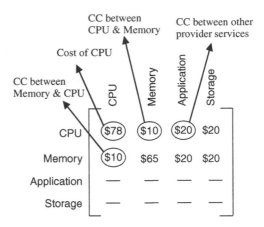

incurred by CP P_r to provide $S(P_r)$ services ($S(P_r) \subseteq R$) and $\gamma(P_r)$ is the expected profit of provider P_r. The total cost $C_{S(P_r)}$ is calculated as follows by using the matrix M:

$$C_{S(P_r)} = \sum_{S_i \in S(P_r)} a_{ii} + \sum_{S_i \in S(P_r)} \sum_{S_j \in S(P_r)} a_{ij} + \sum_{S_i \in S(P_r)} \sum_{S_k \notin S(P_r)} a_{ik} \qquad (17.1)$$

where, $i, j, k = 1...n$ and $i \neq j \neq k$

The first term in the Eq. (17.1) is the cost of providing services $S(P_r)$. The second term is the total collaboration cost between $S(P_r)$ services and third term refers to the total collaboration cost between services of different CPs with whom provider P_r needs to collaborate. As provider P_r does not know to whom it will collaborate after winning the auction, the true cost of a_{ik} cannot be determined. Therefore, P_r may set a high collaboration cost in a_{ik} in order to avoid potential risk in collaboration phase.

Now the *Bidding Function* of a group of CPs, who submit their bids collaboratively as a single bid to fulfill the service requirements completely, can be determined as follows: Let P_r forms a group G by selecting appropriate partners where $S(P_G)$ be the set of services provided by G and $S(P_G) \subseteq R, G \subseteq P$. For any provider like $P_r \in G$, the total cost of providing $S(P_r)$ services is:

$$C_{S(P_r)}^G = \sum_{S_i \in S(P_r)} a_{ii} + \sum_{S_i \in S(P_r)} \sum_{S_j \in S(P_r)} a_{ij} + \sum_{S_i \in S(P_r)} \sum_{S_g \in S(P_G)\backslash S(P_r)} a_{ig} + \sum_{S_i \in S(P_r)} \sum_{S_k \notin S(P_G)} a_{ik}$$
$$(17.2)$$

where, $i, j, g, k = 1...n$ and $i \neq j \neq g \neq k$

We can see from equation (2) that the term $\displaystyle\sum_{S_i \in S(P_r)} \sum_{S_k \notin S(P_r)} a_{ik}$ of Eq. (17.1) is now divided into two terms in $C_{S(P_r)}^G$: $\displaystyle\sum_{S_i \in S(P_r)} \sum_{S_g \in S(P_G)\backslash S(P_r)} a_{ig}$ and $\displaystyle\sum_{S_i \in S(P_r)} \sum_{S_k \notin S(P_G)} a_{ik}$.

The term $\sum\limits_{S_i \in S(P_r)} \sum\limits_{S_g \in S(P_G)\setminus S(P_r)} a_{ig}$ denotes the total collaboration cost of services of provider P_r with other providers in the group. The term $\sum\limits_{S_i \in S(P_r)} \sum\limits_{S_k \notin S(P_G)} a_{ik}$ refers to the total collaboration cost between services of other CPs outside of the group with whom provider P_r needs to collaborate. This term can be zero if the group can satisfy all the service requirements of consumer. Since P_r knows other group members, it can find the true value of the term $\sum\limits_{S_i \in S(P_r)} \sum\limits_{S_g \in S(P_G)\setminus S(P_r)} a_{ig}$. Moreover, if P_r applies any good strategy to form the group G, it is possible for P_r to minimize $\sum\limits_{S_i \in S(P_r)} \sum\limits_{S_g \in S(P_G)\setminus S(P_r)} a_{ig}$. Hence, this group G has more chances to win the auction as compare to other providers who submit separate bids to partially fulfill the service requirements. So the *Bidding Function* for the group G can be calculated as follows:

$$\phi^G_{S(P_G)} = \sum (C^G_{S(P_r)} + \gamma^G(P_r)), \forall P_r \in G, r = 1...l \qquad (17.3)$$

where l is the no. of providers in G and $\gamma^G(P_r)$ is the expected profit of any provider r in the group.

17.4.4.2 Payoff Function of the User/Consumer

With the help of broker user generates the payoff function. During auction, user uses the payoff function $\Omega_{max}(R, Q)$ to internally determine the maximum payable amount that it can spend for a set of services. If the bid price of any CP is greater than the maximum payable amount Ω_{max}, it will not be accepted. In the worst case, auction terminates when the bids of all Cloud service provider is greater than Ω_{max}. In such case, user modifies its payoff function and the auctioneer reinitiates auction with changed payoff function.

17.4.4.3 Profit of the CPs to form a Group

Let $\phi^G_{S(P_r)}$ be the price of the provider r when it forms a group G where $C^G_{S(P_r)}$ is the cost of its services in the group. So the expected profit for the P_r in the group is $\gamma^G(P_r) = \phi^G_{S(P_r)} - C^G_{S(P_r)}$. We know that the expected profit for the provider r, who submits bid separately, is $\gamma(P_r) = \phi_{S(P_r)} - C_{S(P_r)}$. We argue that if any CP forms a group using a good partner selection strategy; it can increase its profit rather than separately publishing the bid. To calculate the increased profit, we consider the following assumptions:

$$C^G_{S(P_r)} \le C_{S(P_r)} \text{ and } \gamma^G(P_r) = \gamma(P_r)$$

Since CP r can collaboratively publish the bid, it may minimize its collaboration cost by selecting good partners, that is, $C^G_{S(P_r)}$ should be less than or equal to $C_{S(P_r)}$. However, $\gamma^G(P_r) = \gamma(P_r)$ means the expectation of profit does not change.

Consequently, we can also deduce the following:

$$\phi^G_{S(P_r)} \leq \phi_{S(P_r)} \tag{17.4}$$

That is the provider who collaboratively publishes bid can provide lower price for its services while maintaining the same expected profit. Thus it has more chances to win the auction. To determine the increased profit for P_r, let $\phi^{2LP}_{S(P_r)}$ be the second lowest price that will be paid to P_r for $S(P_r)$ services if it wins the auction. Now if P_r attends any auction and apply separate and collaborative bidding strategy alternatively, the increased profit $\gamma^I(P_r)$ for P_r can be calculated as follows:

$$\gamma^I(P_r) = \alpha \left(\phi^{2LP}_{S(P_r)} - C^G_{S(P_r)} \right) - \beta \left(\phi^{2LP}_{S(P_r)} - C_{S(P_r)} \right) \tag{17.5}$$

where
$$\alpha = \begin{cases} 1 \text{ if provider } r \text{ collaboratively wins the auction} \\ 0 \text{ otherwise} \end{cases}$$
$$\beta = \begin{cases} 1 \text{ if provider } r \text{ separately wins the auction} \\ 0 \text{ otherwise} \end{cases}$$

From Eq. (17.5), we can figure out that if provider r collaboratively wins the auction, it can always get the increased profit. Otherwise, no increased profit will be achieved. So a good partner selection strategy is required for a CP to make groups. In the next section, we will describe an effective MO optimization model for a good combination of partner selection.

17.5 Model for Partner Selection

17.5.1 Partner Selection Problem

A primary/initiator CP (pCP) identifies a business opportunity which is to be addressed by submitting a bid for a set of services for the consumer. It needs to dynamically collaborate with one or more CP partners to form groups to satisfy the consumer service requirements completely as it cannot provide all the services. We assume that each CP can provide one or at most two services and each service has one or more providers. Also each CP can organize other groups simultaneously. This process of CP partner selection can be presented in Fig. 17.7.

Figure 17.7 shows that the pCP ($P_{1,1}$) can provide s1 service and needs other 4 CP partners among 12 candidate CP partners to provide total 5 kinds of consumer service requirements (s1, s2, s3, s4 and s5). We also assume that the pCP has the INI and PRI of all the other providers for each service. The INI includes price and quality information of services of other providers which are the most important factors. The PRI includes number of projects/auctions accomplished/won by other providers among themselves and also with pCP. The pCP can get all of these information from each CPs website, market and also from consumers feedback about their services.

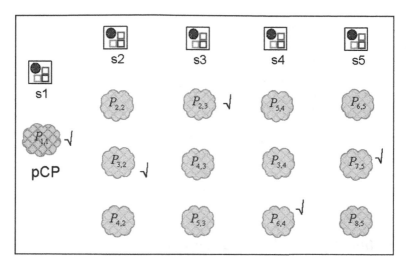

Fig. 17.7 Partner selection process for the pCP

17.5.2 MO Optimization Problem for Partner Selection

The parameters for MO partner selection are defined as follows:

ϕ_{rj} = the price of CP r for providing service j independently

Q_{rj} = the quality value for service j of CP r (qualitative information can be expressed by the assessment values from 1 to 10 (1: very bad, 10: very good))

$W_{rj,xi}$ = the value of past collaboration experience (i.e. the number of times collaboratively wining an auction) between provider r for service j and provider x for service i where $(r, x = 1....m; i, j = 1....n; i \neq j)$

$U = \{U_{rj} \,|\, r = 1.....m, j = 1....n\}$: a decision vector of partner selection

where $U_{rj} = \begin{cases} 1 \text{ if choose } P_{rj} \\ 0 \text{ otherwise} \end{cases}$ and $U_{rj}U_{xi} = \begin{cases} 1 \text{ if choose } P_{rj} \text{ and } P_{xi} \\ 0 \text{ otherwise} \end{cases}$

The optimized goal is selecting a group of CP partners who collaboratively win auctions many times (maximizing past relationship performance values) and making the individual price the lowest and quality value of service the highest. In the most situations, it is impossible that there is a candidate provider group that can make all the goals optimized. So to solve the partner selection problem of a pCP using the INI and PRI, a multi-objective (MO) optimization model to minimize total price and maximize total collaborative past relationship (PR) performance and service quality values can be expressed mathematically as follows:

$$\text{Minimize Obj\_1} = \sum_{j=1}^{n} \sum_{r=1}^{m} \phi_{rj} U_{rj} \qquad (17.6)$$

$$\text{Maximuze Obj\_2} = \sum_{j=1}^{n} \sum_{r=1}^{m} Q_{rj} U_{rj} \tag{17.7}$$

$$\text{Maximize Obj\_3} = \sum_{\substack{i,j=1 \\ i \neq j}}^{n} \sum_{r,x=1}^{m} W_{rj,xi} U_{rj} U_{xi} \tag{17.8}$$

subjec to

$$U_{rj} = \begin{cases} 1 \text{ if choose } P_{rj} \\ 0 \text{ otherwise} \end{cases}$$

$$U_{rj} U_{xi} = \begin{cases} 1 \text{ if choose } P_{rj} \text{ and } P_{xi} \\ 0 \text{ otherwise} \end{cases}$$

17.5.3 Multi-objective Genetic Algorithm

In the CP partner selection problem, MO optimization is preferable because it provides a decision-maker (pCP) with several trade-off solutions to choose from. Actually the CP partner selection problem has multiple conflicting objectives- minimization of the price of service while maximization of past relationship performance and service quality values. Multiple objective formulations are practically required for concurrent optimization that yields optimal solutions that balance the conflicting relationships among the objectives. MO optimization yields a set of Pareto optimal solutions, which is a set of solutions that are mutually non-dominated (Deb et al., 2002). The concept of non-dominated solutions is required when comparing solutions in a multi-dimensional feasible design space formed by multiple objectives.

A solution is said to be Pareto-optimal if it is not dominated by any other solution in the solution space. The set of all such feasible non-dominated solutions in a solution space is termed the Pareto optimal solution set. For a given Pareto-optimal solution set, the curve made in the objective space is called the Pareto front. When two conflicting objectives are present there will always be a certain amount of sacrifice in one objective to achieve a certain amount of gain in the other when moving from one Pareto solution to another. So often it is preferred to use a Pareto optimal solution set rather than being provided with a single solution, because the set helps effectively understand the trade-off relationships among conflicting objectives and make informed selections of the optimal solutions.

MO optimization difficulties can be alleviated by avoiding multiple simulation runs, doing without artificial aids such as weighted sum approaches, using efficient population-based evolutionary algorithms, and the concept of dominance. The use of multi-objective GAs (MOGAs) provides a decision-maker with the practical means to handle MO optimization problems. When solving PSP for CPs using MOGA techniques, one important issue need to be addressed: how to find an appropriate diversity preservation mechanism in selection operators to enhance the yield of Pareto optimal solutions during optimization, particularly for the CP partner selection problems having multiple conflicting objectives. So we develop the MOGA-IC using the non dominated sorting genetic algorithm (NSGA-II) (Deb et al., 2002) which includes an excellent mechanism for preserving population diversity in the selection operators. In this section, the MOGA-IC is designed for the proposed model of CP partner selection as follows:

Natural number encoding is adopted to represent the chromosome of individual. A chromosome of an individual is an ordered list of CPs. Let $y =$ $[y_1, y_2, ..., y_j......y_n]$ $(j = 1, 2, ... n), y_j$ be a gene of the chromosome, with its value between 1 and m (for service j, there are m CPs for a response). If $m = 50$ and $n = 5$, there may be 10 CPs that can provide each service j. Thus a total of 10^5 possible solutions are available. In this way the initial populations are generated. For the selection of individual, the binary tournament selection strategy is used. We employ a two-point crossover. In the case of mutation, one provider is randomly changed for any service.

The multi-objective functions (Obj_1, Obj_2 and Obj_3) are considered as fitness functions when calculating the fitness values. NSGA-II is employed to calculate the fitness values of individual. Any two individuals are selected and their corresponding fitness values are compared according to the dominating-relationships and crowding-distances in the objective space. Then all the individuals are separated into the non-dominated fronts. The individuals in the same fronts do not dominate each other and we call this non-dominated sorting. Now the MOGA-IC is presented step by step as follows:

Step 1: Initialize the input parameters which contain the number of requirements (R), providers (m) and maximum genetic generations (G), population size (N), crossover probability (p_c) and mutation probability (p_m).

Step 2: Generate the initial parent population $P_t, (t = 0)$ of size N_P.

Step 3: Apply binary tournament selection strategy to the current population, and generate the offspring population O_t of size $N_O = N_P$ with the predetermined p_c and p_m.

Step 4: Set $S_t = P_t \cup O_t$, apply a non-dominated sorting algorithm and identify different fronts $F_1, F_2...F_a$.

Step 5: If the stop criterion ($t > G$) is satisfied, stop and return the individuals (solutions) in population P_t and their corresponding objective values as the Pareto-(approximate) optimal solutions and Pareto-optimal fronts.

Step 6: Set new population $P_{t+1} = 0$. Set counter $i = 1$. Until $|P_{t+1}| + |F_i| \leq N$ set $P_{t+1} = P_{t+1} \cup F_i$ and $i = i + 1$.

Step 7: Perform the crowding-sort procedure and include the most widely spread ($N - |P_{t+1}|$) solutions found using the crowding distance values in sorted F in P_{t+1}.

Step 8: Apply binary tournament selection, crossover and mutation operators to P_{t+1} to create offspring population O_{t+1}.

Step 9: Set $t = t + 1$, then return to Step 4.

17.6 Evaluation

In this section, we present our evaluation methodology and simulation results for the proposed CACM model with new auction policy and the MOGA-IC for CP partner selection. First we present a simulation example of PSP for a pCP in the CACM model. It is used to illustrate the proposed MOGA-IC method. Then NSGA-II is utilized to develop the MOGA-IC. Also we implement the existing MOGA that uses only INI called MOGA-I for CP partner selection and analyze its performance with MOGA-IC in the proposed CACM model. We implement the CACM model (winner determination algorithm) with new auction policy as well as the MOGA-IC in Visual C++.

17.6.1 Evaluation Methodology

One of the main challenges in the CACM model and the PSP of CP is the lack of real-world input data. So we conduct the experiments using synthetic data. We generate the input data as follows:

Many CPs ($m = 100$) with different services and also some consumer requirements ($R = 3$–10) are generated randomly. We assume that each CP can provide at most 2 services so that they have to collaborate with others to fulfill the service requirements R. Each service may have one or more CPs. Based on R, CPs are selected. So it is possible that every CP may not provide the required R. Also the cost of providing any independent service is randomly generated from \$80 to \$100. The ranges of collaboration cost (CC) of services as well as the profit are set within \$10 – \$30 and \$10 – \$20 respectively. Quality and collaborative performance values of providers are randomly selected from 1-10 and 0-10 respectively. If any provider has more collaboration experience with other providers, the CC can be minimized. We use the following formula to calculate the CC between any provider P_{rj} and P_{xi}:

$$CC_{rj,xi} = CC_{min} + (CC_{max} - CC_{min}) \times \frac{1}{e^{W_{rj,xi}}} \qquad (17.9)$$

where

CC_{min} = the minimum CC between services (here \$10)
CC_{max} = the maximum CC between services (here \$30)
$W_{rj,xi}$ = the value of number of collaboration experience between P_{rj} and P_{xi}. If it is zero, the highest CC is set between providers. Thus the final price of services is generated for each provider and it is varied based on CC in different auctions.

17.6.1.1 Simulation Examples

Table 17.1 shows the three simulation examples with MOGA-IC parameters of PSP in the CACM model. For each simulation example, MOGA-IC is developed based on NSGA-II. Also in each simulation example, two INI (price and quality of services) and one PRI (number of auctions collaboratively won by other providers among themselves and also with pCP) of candidate CPs are considered. Both the information is presented in Tables 17.2 and 17.3 in normalized forms for the first simulation example. For normalization, the method proposed by Hwang and Yoon (1981) is utilized. We can see that total 21 CPs are found from 35 candidate CPs who can provide 5 randomly generated consumer service requirements. We assume that provider number 1 is the pCP who can provide service no. 7. The number of generations G in the first simulation example is set to 20 as the example search space is quite small.

Table 17.1 The three simulation examples with MOGA-IC parameters

Simulation examples	m	R	N/E	G	P_c	P_m
1	35	5	50	20	0.9	0.1
2	100	5	100	50	0.9	0.1
3	100	5	100	100	0.9	0.1

Table 17.2 The normalized INI of pCP and other candidate CPs

Service no.	Provider no.	Price of service	Quality value of service
2	28	0.17	0.99
2	10	038	0.88
3	10	0.99	0.88
3	15	0.81	0.3
3	6	0.66	0.01
3	32	0.07	0.65
3	14	0.55	0.23
3	20	0.88	0.72
4	9	0.00	0.54
4	33	0.17	0.4
4	18	0.84	0.62
4	17	0.89	0.02
4	34	0.5	0.66
4	26	0.57	0.00
7	**1**	**0.4**	**1**
8	2	0.83	0.19
8	21	0.73	0.48
8	11	0.63	0.06
8	23	0.94	0.22
8	19	0.81	0.63
8	32	0.88	0.82

M.M. Hassan and E.-N. Huh

Table 17.3 The normalized PRI of pCP and other candidate CPs

	P 28, 2	P 10, 2	P 10, 3	P 15, 3	P 6, 3	P 32, 3	P 14, 3	P 20, 3	P 9, 4	P 33, 4	P 18, 4	P 17, 4	P 34, 4	P 26, 4	P 1, 7	P 2, 8	P 21, 8	P 11, 8	P 23, 8	P 19, 8	P 32, 8
P 28, 2	–	–	0.51	0.5	0.79	0.25	0.79	0.14	0.69	0.66	0.24	0.54	0.18	0.3	0.29	0.28	0.36	0.42	0.96	0.97	0.72
P 10, 2	–	–	0.28	0.62	0.7	0.52	0.51	0.48	0.31	0.0	0.81	0.22	0.74	0.94	0.79	0.17	0.4	0.03	0.4	0.39	0.77
P 10, 3	–	–	–	–	–	–	–	–	0.49	0.67	0.28	0.13	0.41	0.63	0.93	0.66	0.17	0.01	0.70	0.26	0.96
P 15, 3	–	–	–	–	–	–	–	–	0.65	0.57	0.04	0.98	0.18	0.08	0.13	0.83	0.66	0.84	0.63	0.20	0.23
P 6, 3	–	–	–	–	–	–	–	–	0.48	0.38	0.28	0.18	0.38	0.27	0.81	0.11	0.77	0.79	1.0	0.29	0.96
P 32, 3	–	–	–	–	–	–	–	–	0.39	0.04	0.09	0.84	0.87	0.35	0.79	0.16	0.43	0.87	0.11	0.80	0.25
P 14, 3	–	–	–	–	–	–	–	–	0.68	0.54	0.29	0.32	0.21	0.44	0.85	0.09	0.18	0.666	0.19	0.52	0.74
P 20, 3	–	–	–	–	–	–	–	–	0.05	0.41	0.81	0.33	0.04	0.01	0.90	0.28	0.0	0.03	0.67	0.82	0.01
P 9, 4	–	–	–	–	–	–	–	–	–	–	–	–	–	–	0.17	0.51	0.56	0.26	0.07	0.56	0.93
P 33, 4	–	–	–	–	–	–	–	–	–	–	–	–	–	–	0.07	0.22	0.50	0.59	0.87	0.16	0.77
P 18, 4	–	–	–	–	–	–	–	–	–	–	–	–	–	–	0.68	0.57	0.41	0.91	0.88	0.04	0.87
P 17, 4	–	–	–	–	–	–	–	–	–	–	–	–	–	–	0.64	0.21	0.50	0.04	0.73	0.02	0.67
P 34, 4	–	–	–	–	–	–	–	–	–	–	–	–	–	–	0.51	0.09	0.11	0.13	0.75	0.49	0.77
P 26, 4	–	–	–	–	–	–	–	–	–	–	–	–	–	–	0.27	0.32	0.68	0.57	0.31	0.07	0.05
P 1, 7	–	–	–	–	–	–	–	–	–	–	–	–	–	–	–	0.22	1.0	0.8	0.47	0.58	0.96

17.6.2 Simulation Results

17.6.2.1 Appropriate Approach to Develop the MOGA-IC

In solving the first simulation example problem of CP partner selection, the best Pareto front among the 10 trials of 20 generations are selected as the final solution. The 16 Pareto-optimal solutions of first front of MOGA-IC with NSGA-II for simulation example 1 are presented in Table 17.4. Also the graphical representations are shown by Fig. 17.8.

Table 17.4 Pareto-optimal solutions of MOGA-IC with NSGA-II for example 1

Pareto-optimal solutions					Optimal objective function values		
$y = (y_7$	y_4	y_3	y_2	$y_8)$	Obj_1	Obj_2	Obj_3
1	18	6	10	32	3.16	3.32	7.63
1	18	10	10	32	3.49	4.2	7.33
1	34	10	28	32	2.94	4.35	6.24
1	9	32	28	21	1.37	3.66	4.93
1	9	10	28	32	2.44	4.23	6.65
1	9	32	28	19	1.45	3.81	5.49
1	9	14	28	32	2.00	3.58	6.82
1	34	32	10	32	2.23	4.01	6.97
1	18	10	28	32	3.28	4.31	6.44
1	9	32	10	32	1.73	3.89	5.88
1	34	32	28	32	2.02	4.12	5.59
1	34	6	10	32	2.82	3.36	7.39
1	34	32	10	19	2.16	3.82	6.48
1	34	10	10	32	3.15	4.24	7.12
1	9	32	28	32	1.52	4.00	5.44
1	18	14	10	32	3.05	3.55	7.27

Figures 17.9 and 17.10 show plots of Pareto optimal solution sets of the first fronts obtained by MOGA-IC using NSGA-II when solving the simulation examples 2 and 3 respectively. Here, we have just provided the graphical representations of the Pareto-optimal solutions for the algorithm as the input data tables are very large.

Figures 17.11 show the average optimized values of three objective functions in the first fronts during 50 generations using MOGA-IC with NSGA-II for the simulation examples 2.

It is seen from Fig. 17.11 that NSGA-II Pareto front moves towards the low-cost region without preserving each generation's extreme solutions. Instead, the entire Pareto front shifts as new solution sets are obtained. In other words, MOGA-IC with NSGA-II distributes solutions in a more focused manner. In MOGA-IC with NSGA-II, dominance ranking is used when forming the fronts of individuals and these fronts are first used to populate the external set, based on ranking, a strategy that allows a set of close-neighbor individuals in the same front to be included in the next generation.

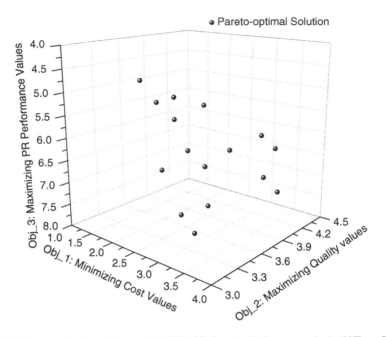

Fig. 17.8 Pareto-optimal solutions of MOGA-IC for simulation example 1 ($N/E = 50$ and $G = 20$) obtained by NSGA-II

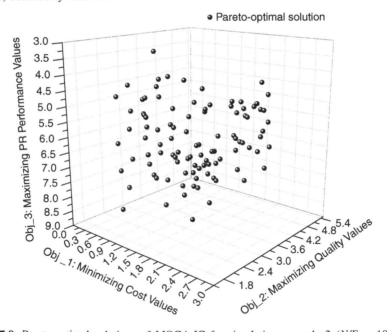

Fig. 17.9 Pareto-optimal solutions of MOGA-IC for simulation example 2 ($N/E = 100$ and $G = 50$) obtained by NSGA-II

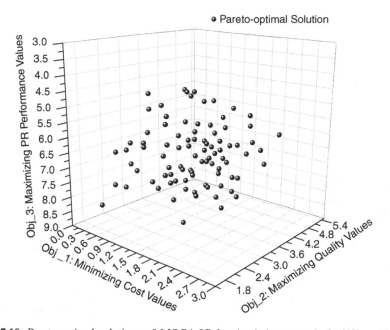

Fig. 17.10 Pareto-optimal solutions of MOGA-IC for simulation example 3 (*N/E* = 100 and *G* = 100) obtained by NSGA-II

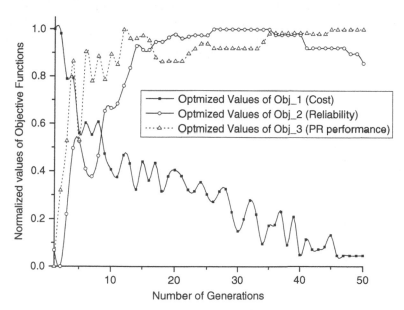

Fig. 17.11 Average optimized values of different objective functions in the first front of MOGA-IC with NSGA-II for 50 generations

The MOGA-IC with NSGA-II is more focused when exploring the search space and generating Pareto solution sets. The MOGA-IC with NSGA-II uses crowding distance when the size of non-dominated solutions exceeds the archive size. So we found that NSGA-II is the appropriate algorithm to develop the MOGA-IC for CP partner selection problem. Thus the pCP can select any combination of CP partners from the Pareto-optimal solution sets obtained from MOGA-IC based on NSGA-II.

17.6.2.2 Performance comparison of MOGA-IC with MOGA-I in CACM Model

In order to validate the proposed MOGA-IC model for CP partner selection in CACM model, we develop another MOGA called MOGA-I based on NSGA-II that uses INI for CP partner selection. We analyze the performance of pCP that uses both MOGA-IC and MOGA-I algorithms to make groups and joins various auctions in CACM model. We assume that initially no collaborative information of other CPs is available to the pCP.

In each auction, for a set of service requirements, first all providers including pCP form several groups using MOGA-I and submit several group bids as single bids. The winner determination algorithm proposed in Yokoo and Suzuki (2002) is used to find the winners. Next in the same auction with the same set of services, the winner determination algorithm is executed again but this time pCP uses the proposed MOGA-IC and others use MOGA-I approach and join the auctions and winners are determined. In our simulation, 1000 auctions are generated for different user requirements. After each 100 auctions, we count the number of auctions pCP wins using both the algorithms. The experimental result is shown in Fig. 17.12.

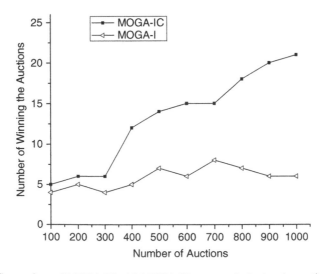

Fig. 17.12 Comparison of MOGA-IC with MOGA-I in terms of winning the auctions

It is seen from Fig. 17.12 that using the MOGA-IC approach, pCP wins more auctions in comparison to MOGA-I approach. The reason is that the past collaborative performance values increase as the number of auctions increases and as a result the MOGA-IC finds good combination of partners for pCP.

We also validate the performance of MOGA-IC as compared to MOGA-I in terms of conflicts minimization among CP providers. We assume that conflicts may happen between providers P_{rj} and P_{xi} with the probability –

$$
p_{\text{conflicts}} = \begin{cases} \frac{1}{\delta \times e^{W_{rj,xi}}}, & \text{if } W_{rj,xi} \neq 0 \\ \frac{1}{\delta} & \text{otherwise} \end{cases} \quad i \neq j, r \neq x, \delta > 1 \qquad (17.10)
$$

where δ is a constant. We set $\delta = 20$ assuming that there is 5% chance of conflicts between any two providers P_{rj} and P_{xi} if they have no past collaborative experience. Like the previous experiment, 1000 auctions are generated. For each auction when pCP uses both the algorithms and forms groups, we count the total number of conflicts that may happen among the group members for various services using the probability $p_{\text{conflicts}}$. The experimental result is shown in Fig. 17.13. We can see from Fig. 17.13 that the MOGA-IC can reduce a sufficient number of conflicts among providers as compared to the MOGA-I algorithm since it can utilize the PRI to choose partners along with INI.

Fig. 17.13 Comparison of MOGA-IC with MOGA-I approach in terms of conflicts minimization

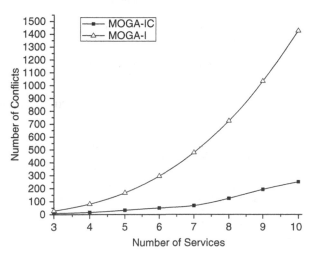

17.7 Conclusion and Future Work

The paper presents a novel combinatorial auction based Cloud market model called CACM that enables a DC platform among CPs. The CACM model uses a new auction policy that allows CPs to dynamically collaborate with other partners and form groups and submit their bids for a set of services as single bids. This policy can help to reduce collaboration costs as well as conflicts and negotiation time among CPs in DC and therefore creates more opportunities to win the auctions for the group. A new multi-objective optimization model of partner selection using the individual and past collaborative information is also proposed. An effective MOGA called MOGA-IC with NSGA-II is then developed to solve the model. In comparison with the existing MOGA-I approach; MOGA-IC with NSGA-II shows better performance results in CP partner selection as well as conflicts minimization among CPs in the CACM model.

There are two important aspects of the CACM model that require further future research efforts. One such aspect is how to measure the level of satisfaction of partners forming the group or allocate optimal resources among partners for the collaborative service- often it is ignored in the existing partner selection problem. Other important aspect is how the aggregate profit/payoff should be divided among partners so that no one leaves the group and thus the group becomes stable. We need to analyze these aspects in future using game theory.

References

Amid, A., Ghodsypour, S. H., & Brien, C. O. (2006). Fuzzy multi-objective linear model for supplier selection in a supply chain. *International Journal of Production Economics, 104,* 394–407.

Bubendorfer, K. (2006). Fine grained resource reservation in open grid economies. *Proceedings of the 2nd IEEE International Conference on e-Science and Grid Computing, Vol. 1, Washington, DC,* 81–81.

Bubendorfer, K., & Thomson, W. (2006). Resource management using untrusted auctioneers in a grid economy. *Proceedings of the 2nd IEEE International Conference on e-Science and Grid Computing, Vol. 1, Prentice Hall, NJ,* 74–74.

Buyukozkan, G., Feyzioglu, O., & Nebol, E. (2008). Selection of the strategic alliance partner in logistics value chain. *International Journal of Production Economics, 113,* 148–158.

Buyya, R., Yeo, C. S., Venugopal, S., Broberg, J., & Brandic, I. (2009). Cloud computing and emerging IT platforms: Vision, hype, and reality for delivering computing as the 5th utility. *Future Generation Computer Systems, 25,* 599–616.

Chang, S. L., Wang, R. C., & Wang, S. Y. (2006). Applying fuzzy linguistic quantifier to select supply chain partners at different phases of product life cycle. *International Journal of Production Economics, 100,* 348–359.

Chen, Y. L., Cheng, L. C., & Chuang, C. N. (2008). A group recommendation system with consideration of interactions among group members. *Expert Systems with Applications, 34,* 2082–2090.

Chen, H. H., Lee, A. H. I., & Tong, Y. (2007). Prioritization and operations NPD mix in a network with strategic partners under uncertainty. *Expert Systems with Applications, 33,* 337–346.

Chen, S., Nepal, S., Wang, C. C., & Zic, J. (2008). Facilitating dynamic collaborations with eContract services. *Proceeding of 2008 IEEE International Conference on Web Services, Vol. 1, Miami, FL,* 521–528.

Cheng, F., Ye, F., & Yang, J. (2009). Multi-objective optimization of collaborative manufacturing chain with time-sequence constraints. *International Journal of Advanced Manufacturing Technology, 40,* 1024–1032.

Cowan, R., Jonard, N., & Zimmermann, J. B. (2007). Bilateral collaboration and the emergence of innovation networks. *Management Science, 53,* 1051–1067.

Das, A., & Grosu, D. (2005). Combinatorial auction-based protocol for resource allocation in grid. *Proceedings of the 19th IEEE International Parallel and Distributed Processing Symposium, Denver, CA.*

Deb, K., Pratap, A., Agarwal, S., & Meyarivan, T. (2002). A fast and elitist multi-objective genetic algorithm: NSGA-II. *IEEE Transactions on Evolutionary Computation, 6,* 182–197.

Fischer, M., Jahn, H., & Teich, T. (2004). Optimizing the selection of partners in production networks. *Robotics and Comput-Integrated Manufacturing, 20*(5), 593–601.

Fuqing, Z., Yi, H., & Dongmei, Y. (2008). A multi-objective optimization model of the partner selection problem in a virtual enterprise and its solution with genetic algorithms. *International Journal of Advanced Manufacturing Technology, 37,* 1220.

Grosu, D., & Das, A. (2004). Auction-based resource allocation protocols in grids. *Proceeding of the 16th IASTED International Conference on Parallel and Distributed Computing and Systems, Vol. 1, Los Angeles, CA,* 20–27.

Gupta, P., & Nagi, R. (1995). *Optimal partner selection for virtual enterprises in agile manufacturing.* Submitted to IIE Transactions on Design and Manufacturing, Special Issue on Agile Manufacturing.

Huang, X. G., Wong, Y. S., & Wang, J. G. (2004). A two-stage manufacturing partner selection framework for virtual enterprises. *International Journal of Computer Integrated Manufacturing, 17*(4), 294–304.

Hwang, C. L., & Yoon, K. (1981). *Multiple attribute decision making: Methods and applications.* Berlin: Springer.

Coombe, B. (2009). Cloud Computing- Overview, Advantages and Challenges for Enterprise Deployment. *Bechtel Technology Journal, 2*(1).

Ip, W. H., Huang, M., Yung, K. L., & Wang, D. (2003). Genetic algorithm solution for a risk-based partner selection problem in a virtual enterprise. *Computers and Operations Research, 30,* 213–231.

Kaya, M. (2009). MOGAMOD: Multi-objective genetic algorithm for motif discovery. *Expert Systems with Applications, 36,* 2.

Ko, C. S., Kim, T., & Hwang, H. (2001). External partner selection using Tabu search heuristics in distributed manufacturing. *International Journal of Production Research, 39*(17), 3959–3974.

Nepal, S., & Zic, J. (2008). A conflict neighboring negotiation algorithm for resource services in dynamic collaboration. *Proceedings of the IEEE International Conference on Services Computing, 2,* 7–11.

Nepal, S., Zic, J., & Chan, J. (2007). A distributed approach for negotiating resource contributions in dynamic collaboration. *Proceedings of the 8th IEEE International Conference on Parallel and Distributed Computing Applications and Technologies, Vol. 1, Genoa, Italy,* 82–86.

Saen, R. F. (2007). Supplier selection in the presence of both cardinal and ordinal data. *European Journal of Operational Research, 183,* 741–747.

Sha, D. Y., & Che, Z. H. (2005). Virtual integration with a multi-criteria partner selection model for the multi-echelon manufacturing system. *International Journal of Advanced Manufacturing Technology, 25,* 793–802.

Suzuki, K., & Yokoo, M. (2003). Secure generalized vickery auction using homomorphic encryption. *Proceedings of 7th International Conference on Financial Cryptography, LNCS, Springer, Vol. 2742, Hong-Kong, China,* 239–249.

Wang, Z.-J., Xu, X.-F., & Zhan, D. C. (2009). Genetic algorithms for collaboration cost optimization-oriented partner selection in virtual enterprises. *International Journal of Production Research, 47*(4), 859–881.

Weiss, A. (December 2007). Computing in the clouds. *netWorker, 11*(4), 16–25.

Wolski, R., Plank, J. S., Brevik, J., & Bryan, T. (2001). Analyzing market-based resource allocation strategies for the computational grid. *The International Journal of High Performance Computing Applications, 15*(3), 258–281.

Wu, N. Q., & Su, P. (2005). Selection of partners in virtual enterprise paradigm. *Robotics Computer-Integrated Manufacturing, 21*(5), 119–131.

Yamazaki, Y. (2004). Dynamic collaboration: The model of new business that quickly responds to changes in the market through 'The Integrated IT/Network Solutions' provided. *NEC Journal of Advanced Technology, 1*(1), 9–16.

Yokoo, M., & Suzuki, K. (2002). Secure multi-agent dynamic programming based on homomorphic encryption and its application to combinatorial auctions. *Proceedings of the First Joint International Conference on Autonomous Agents and Multi-agent Systems, ACM Press, Vol. 1, New York, NY,* 112–119.

Zitzler, E., Laumanns, M., & Thiele, L. (2001). *SPEA2: Improving the strength Pareto evolutionary algorithm* (TIK Rep. No. 103, Swiss Federal Institute of Technology, 2001).

Part IV
Applications

Chapter 18
Enterprise Knowledge Clouds: *Applications and Solutions*

Jeff A. Riley and Kemal A. Delic

18.1 Introduction

With the evolution of cloud computing in recent times we have proposed *Enterprise Knowledge Clouds* (EKC) as the next generation Enterprise Knowledge Management systems (Delic & Riley, 2009).

The Enterprise Knowledge Cloud abstracted architecture is shown in Fig. 18.1. This architecture interconnects business partners and suppliers to company customers and consumers, and uses future cloud technologies to harvest, process and use internal knowledge. Each of the clouds shown in Fig. 18.1 is an autonomous entity, existing for its own purpose and capable of collecting, warehousing, managing and serving knowledge to its own group of users. However, while the clouds are independent, they will be capable of interconnection, overlap and knowledge-sharing so that, for example, customers and consumers might have access to appropriate internal enterprise knowledge, or even partner/supplier knowledge through the Enterprise Knowledge Cloud.

The emergence of these clouds and their coalescence into the Enterprise Knowledge Cloud allows, indeed encourages, the collective intelligences formed within each cloud to emerge and cooperate with each other. As an example, internal IT operations will use private clouds, Sales and Marketing would operate on public clouds, while Outsourcing businesses may reside on the partner clouds – each having different types of users and customers. The interaction and cooperation of the user groups, their knowledge, and the collective intelligences across the three clouds shown in Fig. 18.1 provides both the infrastructure for behavioural, structural and strategic adaptation in response to change as well as an environment for knowledge creation and exchange.

In the following sections we discuss Enterprise Knowledge Management (EKM) applications and solutions in terms of cloud computing and the emergence of the

J.A. Riley (✉) and K.A. Delic
Hewlett-Packard Co., New York, NY, USA
e-mails: {jeff.riley; kemal.delic}@hp.com

B. Furht, A. Escalante (eds.), *Handbook of Cloud Computing*,
DOI 10.1007/978-1-4419-6524-0_18, © Springer Science+Business Media, LLC 2010

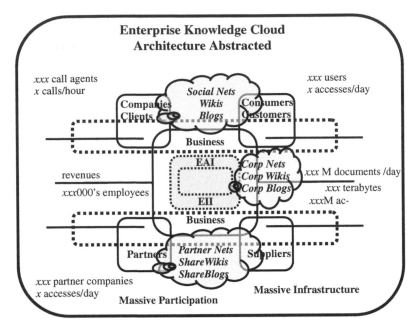

Fig. 18.1 Enterprise knowledge cloud: architectural view

Enterprise Knowledge Cloud. We present what we believe will be user expectations and requirements of cloud-based Knowledge Management Applications. We also discuss issues that will face developers and providers of KM applications as they migrate existing applications to the cloud platform, or build new applications specifically for the cloud. Finally we present our view of the future direction of Knowledge Management in a cloud computing environment.

18.2 Enterprise Knowledge Management

Unfortunately, even today there is no real consensus at to what "knowledge management" really is – ask ten people to define knowledge management and you will get twelve different answers. Worse than that – ask ten people to define "knowledge" and you will probably walk away with twice that number of definitions.

The "DIKW Hierarchy", or "Knowledge Hierarchy", refers to a representation of the relationships between *Data, Information, Knowledge,* and *Wisdom.* An early description of the DIKW Hierarchy was given by Milan Zeleny in 1987 (Zeleny, 1987)[1] in which he describes *Data* as "know-nothing", *Information* as "know-what" and *Knowledge* as "know-how" (we are not concerned with Zeleny's definition of *Wisdom* here and leave it to interested readers to refer to his paper).

---

[1] The origin of the DIKW hierarchy is far from clear – see [12].

Later, Ackoff (1989) gave the following definitions for *Data*, *Information* and *Knowledge*:

- Data – data is raw. It simply exists and has no significance beyond its existence (in and of itself). It can exist in any form, usable or not. It does not have meaning of itself. In computer parlance, a spreadsheet generally starts out by holding data.
- Information – information is data that has been given meaning by way of relational connection. This "meaning" can be useful, but does not have to be. In computer parlance, a relational database makes information from the data stored within it.
- Knowledge – knowledge is the appropriate collection of information, such that its intent is to be useful. Knowledge is a deterministic process. When someone "memorizes" information (as less-aspiring test-bound students often do), then they have amassed knowledge. This knowledge has useful meaning to them, but it does not provide for, in and of itself, integration such as would infer further knowledge

We can see from these definitions that not all data is useful information, and not all information is useful knowledge. It follows then that Knowledge Management is not simply the management of data or information – just providing keyword searching capabilities over one or more information repositories is not Knowledge Management.

Referring to the definitions of knowledge proposed by Zeleny and Ackoff, we define Knowledge Management as the process by which organisations create, identify and extract, represent and store, and facilitate the use of knowledge and knowledge-based assets to which they have access.

18.2.1 EKM Applications

Knowledge can be categorized as either explicit or tacit. Explicit knowledge is knowledge that can be codified – knowledge that can be put in a form which is easily transferred from one person to another. Examples of explicit knowledge are patents, recipes, process documentation, operational plans, marketing strategies and other such knowledge-based hard assets. Tacit knowledge, on the other hand, is knowledge which hasn't been codified and is in fact difficult to codify and transfer – knowledge that exists as "know-how" in people's heads, often unknown or unrecognised. Examples of tacit knowledge are things such as habits, sometimes cultural and often learned by observation and imitation, and the notion of language.

Knowledge Management applications can be similarly categorised, and while traditional KM applications focused on managing explicit knowledge (knowledge base management systems, workflow management systems etc.), with the advent of *Web2.0*, *social networking* and *groupware*, *mashups*, *wikis*, *blogs* and *folksonomies* etc., there is more opportunity for KM applications focused on managing tacit knowledge to evolve.

Typical Knowledge Management applications can be layered into three essential subsystems:

- Front-end portals that manage interactions with internal users, partners' agents and external users, while rendering various *Knowledge Services*. Different classes of users (e.g. internal vs external) are often presented with slightly different portals allowing access to different knowledge and services.
- A core layer that provides the knowledge base and access, navigation, guidance and management services to knowledge portals and other enterprise applications. The core layer provides the *Knowledge Base Management System* (KBMS), the *Knowledge Feeds* – the means by which knowledge is added to the knowledge base or exchanged with other knowledge management systems or users – as well as the mechanism to distribute and inject appropriate knowledge into business processes throughout the enterprise.

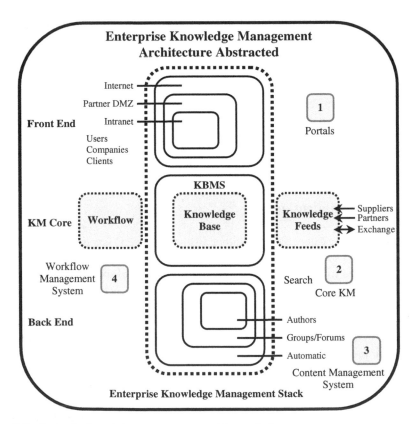

Fig. 18.2 Enterprise knowledge management: architectural view

- The back-end that supplies *Knowledge Content*, and the content management system, from various sources, authors and communities, enabling a refresh of the knowledge base.

Thus, Enterprise Knowledge Management is typically a three-tier enterprise application probably spread over several geographically dispersed data centres, and typically interconnected or integrated with enterprise portals, content and workflow management systems (Fig. 18.2).

The Enterprise Workflow System captures interactions with users and provides necessary context for the Enterprise Knowledge Management system. Various feeds enable flow and exchange of knowledge with partners and suppliers. Today these feeds are mainly proprietary, while we expect that they will evolve into standards-based solutions for large-scale content flows (RESTful services, RSS, ATOM, SFTP, JSON, etc.). To indicate the scale and size of the typical corporate knowledge management system, we presume that the knowledge base contains several million knowledge items, and users number in the hundreds of thousands. Enterprise knowledge management is considered a high-end, mission-critical corporate application which resides in the corporate data centre. High availability and dependability are necessary engineering features for such global, always-on, always-available systems. The transformation of EKM into the enterprise cloud will increase the scale and importance by orders of magnitude.

18.3 Knowledge Management in the Cloud

Knowledge Management in a cloud computing environment requires a paradigm shift, not just in technology and operational procedures and processes, but also in how providers and consumers of knowledge within the enterprise think about knowledge. The *knowledge-as-a-service*, "on-demand knowledge management" model provided by the cloud computing environment can enable several important shifts in the way knowledge is created, harvested, represented and consumed.

18.3.1 Knowledge Content

Collective intelligence is a phenomenon that emerges from the interaction – social, collaborative, competitive – of many individuals. By some estimates there are more than eighty million people worldwide writing web logs (*"blogs"*). The blogs are typically topic-oriented, and some attract important readership. Authors range from large company CEOs to administrative assistants and young children. When taken together, the cloud computing infrastructure which hosts *"blogospheres"* is a big social agglomeration providing a kind of collective intelligence. But it is not just blogs that form the collective intelligence – the phenomenon of collective intelligence is nurtured and enhanced by the social and participatory culture of the internet, so all content developed and shared on the internet becomes part of the collective intelligence. The internet then, and the content available there, appears as an omnipresent, omniscient, giant infrastructure – as a new form of knowledge management. This same paradigm applies, albeit on a smaller scale, to the enterprise cloud – the socialisation and participatory culture of today's internet is mirrored in the microcosm of the enterprise.

Today this represents collaboration of mostly people only, but very soon in the future we may envisage intelligent virtual objects and devices collaborating with people. Indeed, this is already beginning to happen to some extent with internet-attached devices starting to proliferate. Thus, rescaling from the actual ∼1.2 billion users to tens or even hundreds of billions of real-world objects having a data representation in the virtual world is probably realistic.

It is important to note here that content will no longer be located almost solely in a central knowledge repository on a server in the enterprise data centre. Knowledge in the cloud is very much distributed throughout the cloud and not always residing in structured repositories with well-known query mechanisms. Knowledge management applications offered in the cloud need to be capable of crawling through the various structured and ad hoc repositories – some perhaps even transient – to find and extract or index knowledge, and that requires those applications be capable of recognising knowledge that might be useful to the enterprise knowledge consumers. Furthermore, we believe that over time multimedia content will become dominant over ordinary text, and that new methods for media-rich knowledge management will need to be devised.

Even in the smaller world of the enterprise, a real danger, and a real problem to be solved by knowledge management practitioners, is how to sort the wheat from the chaff – or the knowledge from the data and information – in an environment where the sheer amount of data and information could be overwhelming.

18.3.2 Knowledge Users

Leaving aside administrative tasks, there will be two categories of users within the Enterprise Knowledge Cloud: *Knowledge Providers* and *Knowledge Consumers*.

While sketching the architecture of future enterprise knowledge management applications, serious consideration needs to be given to aspects and dimensions of future users – the evolution of technology in consumer and corporate domains has created a new type of user that will be very different from contemporary knowledge consumers.

The younger generation – the so-called "*Generation Y*" or "*Millennial Generation*" – seems to have developed a way to quickly exchange information snippets, being either very short text messages or particular multimedia content. Members of this generation also typically have a much better ability to multitask naturally while not losing or intermixing communication threads – probably a natural consequence of their exposure to electronic gaming and new work and living styles. This new generation of knowledge consumer will drive the on-demand nature of knowledge management in the cloud. Moreover, they will require their knowledge served up to them in *"BLATT"* format – with the "bottom line at the top" – and with media-rich content.

Knowledge providers will need to develop applications that recognise the salient knowledge in response to queries, and deliver a synopsis of that salient material ahead of the detail. Delivering a list of documents for the user to read through and identify and extract the knowledge themselves will no longer be acceptable. Furthermore, knowledge management applications in the cloud will need to be capable of presenting media-rich results to the consumer concurrently with more traditional text-based results. Knowledge consumers that have grown up with the internet know that knowledge is more than just text in a data base, and when they seek knowledge they will want to view the video as well as read the text. The new, internet-savvy knowledge consumers will want short, sharp, to-the-point responses to their queries – but responses which are complete and include audio and visual content where appropriate.

18.3.3 Enterprise IT

From our experience, the best domain for Enterprise Knowledge Management is in the Enterprise IT domain, as it is a domain under huge cost pressure but one which is essential for strategic development.

From a highly abstracted view, the Enterprise Knowledge Management IT domain consists of problem solving, monitoring, tuning and automation, business intelligence & reporting, and decision making tasks (Fig. 18.3).

The tasks of problem solving, monitoring, tuning and automation, business intelligence and reporting, and decision making are the most promising areas for the future deployment of Enterprise Knowledge Clouds. The knowledge available to both IT administrators and automated management agents via the Enterprise

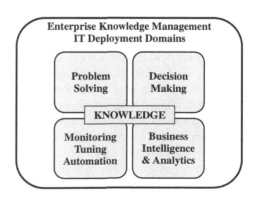

Fig. 18.3 Enterprise knowledge management: IT deployment domains

Knowledge Cloud will help drive the development of a slew of new technologies addressing the problems which previous computing facilities couldn't resolve.

Currently, the majority of the indicated IT tasks include people, while we suggest that this balance will be changed in the future through automation, ultimately leading to self-managing enterprise IT systems (Delic & Faihe, 2007). When mapped into more precise form, this conceptual drawing (Fig. 18.3) will evolve into the enterprise-scale knowledge management application stack discussed earlier (Fig. 18.2).

18.3.3.1 Problem Solving

Problem solving, especially in the Enterprise Knowledge Management IT domain, is the task for which knowledge management techniques and systems are most commonly deployed. The proliferation of knowledge management systems for problem analysis and solving is many and varied, spanning the gamut from knowledge capture, representation and transformation, through to recognition, extraction and reuse. Knowledge from all sources, including human expertise, in the form of plain text, models, visual artefacts, executable modules, etc. is used by intelligent knowledge management systems to enable users to solve problems without reference to scarce, and often expensive, human experts.

With the advent of the Enterprise Knowledge Cloud and associated interfaces to knowledge bases and knowledge feeds, software agent-based problem solving and guidance becomes much more of a realistic proposition. Intelligent software agents, able to be trained by "watching" and learning from experienced and expert human support engineers, can be deployed across the cloud to assist less experienced, sometimes novice, engineers and end users. These intelligent software agents will have knowledge of and access to all knowledge repositories and knowledge feeds across the Enterprise Knowledge Cloud.

The potential for collaborative problem solving is expanded in the Enterprise Knowledge Cloud, with the social networking aspect of the cloud environment facilitating both greater interaction between the end user and support engineer and amongst support engineers and others with relevant knowledge – including software agents, and eventually intelligent software agents collaborating with other software agents.

The Enterprise Knowledge Cloud provides a platform not only for collaborative problem solving, but also for distributed problem solving. Distributed problem solving is not new, and there are established grid middleware and resource brokering technologies that facilitate the development of distributed applications (e.g. Foster & Kesselman, 1999; Frey, Tannenbaum, Livny, Foster, & Tuecke, 2004; Venugopal, Buyya, & Winton, 2006), but more recently distributed problem solving applications developed specifically for the Enterprise Cloud have been described. In Vecchiola, Kirley, and Buyya (2009), Vecchiola, Kirley and Buyya describe a network-based, multi-objective evolutionary algorithm to be deployed on an Enterprise Cloud for solving large and complex optimisation problems – for example, the tuning and management of computer systems and networks.

18.3.3.2 Monitoring, Tuning and Automation

In recent years a wide variety of Artificial Intelligence (AI) techniques and heuristics have been deployed in knowledge management systems in an effort to make the systems smarter and more responsive. These smarter knowledge management systems are particularly well suited to automation and self-management tasks, where the goal is to provide automated monitoring of system use and predictive tuning of system parameters to achieve automatic system scale out.

Delic et al. describe a system for the self-management of Apache Web Servers using a hierarchical control system well-suited to implementation on an Enterprise Knowledge Cloud platform (Delic, Riley, Bartolini, & Salihbegovic, 2007a). Their hybrid approach harvests and combines knowledge from different sources and utilises machine learning techniques to learn and implement heuristics to manage the web server (Fig. 18.4).

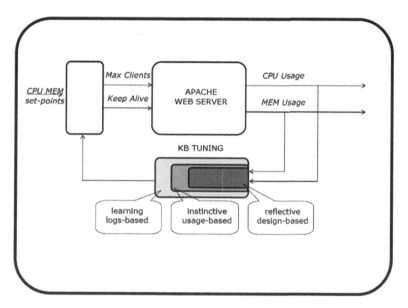

Fig. 18.4 Control system for self-management of apache web servers

The self-managing Enterprise IT system is becoming more achievable as knowledge management systems become smarter and knowledge more accessible. The automation of system monitoring, problem recognition, analysis and, when necessary, routing to a human expert is a critical step in architecting self-managing systems. Figure 18.5 shows an architecture for a state-based, adaptive, self-managing Enterprise IT system (Delic, Riley, & Faihe, 2007b). The system shown in Fig. 18.5 recognises the system state and applies solutions to known-problem where applicable, or routes the problem to an expert to solve as necessary. Problems routed to experts are solved, and knowledge of the solutions is fed back into the system to allow that knowledge to be applied as "known-problem solutions" in the future.

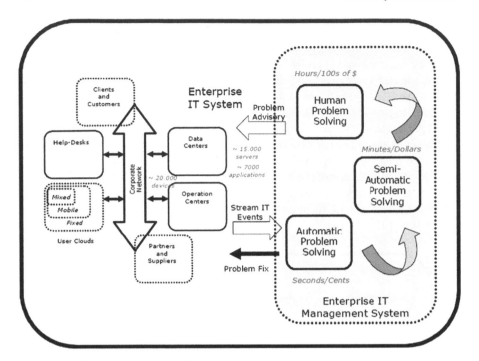

Fig. 18.5 Self-managed enterprise IT system

System state and known-problem recognition are learned in an "offline" mode by the system – facilitated by the integrated knowledge management systems of the Enterprise Knowledge Cloud in a way that has not been possible, or is vastly more difficult, in a non-cloud environment.

18.3.3.3 Business Intelligence and Analytics

Business Intelligence (BI) refers to a range of methodologies, technologies, skills, competencies, and applications businesses implement and utilize in order to better understand their commercial context. Typically business intelligence systems are knowledge management systems that provide current and predictive views of the business based on historical and current data relating to the business itself and the commercial environment in which it exists. Business Intelligence reporting is more than the simple reporting of data gathered – BI reporting uses a wide range of AI techniques to extract relevant knowledge from incoming data streams and repositories and provide observations, hints and suggestions about trends and possible futures.

Enterprise analytic systems provide different classes of users a wide range of analytics from holistic views of the enterprise state and subsystems to the tasks of optimization and forecasting (Delic & Dayal, 2003). These systems cover different domains, address varying value-at-risk entities, and require different analytic arte-facts to be provided to users in order to improve and accelerate decisions. Highly abstracted, the enterprise architecture can be dissected into event, transaction and analytic layers, each having different purposes, objectives and design constraints (Fig. 18.6).

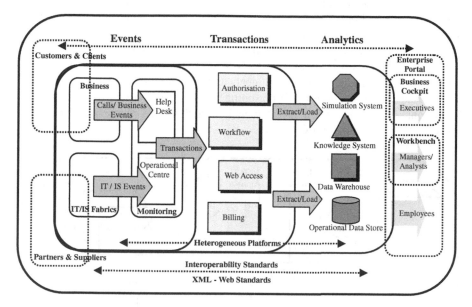

Fig. 18.6 Abstracted enterprise architecture

Different types of users of enterprise analytics use different artefacts: senior executives and analysts typically require interactive/active content, whereas pas-sive/static content is used by larger communities of employees. Each generic type of user also exhibits different usage patterns: analysts typically need a powerful client/server setup; managers need browser access to analytics from a portable system; lower-level employees might need analytics right on their desktops or accessible via temporarily connected PDAs, mobile phones, or other portable devices.

The Enterprise Knowledge Cloud brings together the technologies and knowl-edge repositories underpinning today's business intelligence and analytics systems.

18.3.3.4 Decision Making

Decision making is most often done by humans after consuming the results of the business intelligence and analytics reporting, but with the volume of business intelligence available to analysts increasing almost exponentially it is becoming more and more difficult for humans to make sensible, rational and timely decisions. For this reason more responsibility for the decision making task is being given to knowledge-driven Decision Support Systems (DSS) which typically employ some sort of artificial intelligence tuned to the environment of the DSS deployment.

Delic et al. describe a real-time Decision Support System for a typical IT situation: help-desk support for enterprise customers (Delic, Douillet, & Dayal, 2001). Such help-desk support deals with IT equipment usage problems (usually desktop workstations), provides operations centre support to manage the network, servers and applications. Figure 18.7 depicts a help-desk system which uses a workflow system to capture "*problem cases*" (phone call records) and "*trouble tickets*" (events initiated from the equipment or/and applications). Two kinds of knowledge bases support IT help-desk operations: case knowledge bases containing problem-solving knowledge (often in the form of documents and search tools, diagnostic tools such as Bayesian networks, case-based reasoning tools, etc.); and real-time knowledge bases containing events, event management knowledge (e.g., *event-condition-action* rules) and enterprise topology.

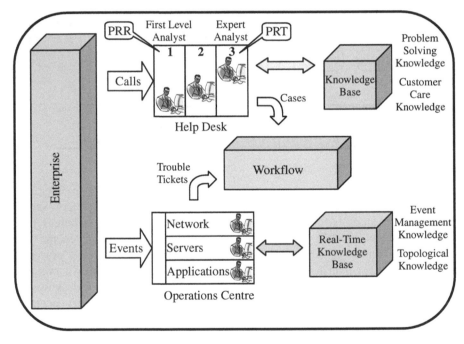

Fig. 18.7 Real-time decision support system for IT help desk

A help-desk is usually staffed by first call contact analysts who use the knowledge bases and tools available to solve less complex problems, and dispatch the more complex problems to second and third line analysts for resolution.

A manager responsible for IT help-desk support operations typically wants to see a variety of performance metrics, including Problem Resolution Rate [PRR] for the front-line analysts, and Problem Resolution Time [PRT] for the second and third line of support. Traditionally the manager would typically get monthly or weekly reports from the workflow system, but with more current information the manager could react more quickly to changes in operating conditions. For example, if the help-desk manager was able to observe *in real time* that PRR is going down and PRT is going up, he or she could decide quickly what corrective action to take. The causes of the problem could be varied and interdependent, masking the prevailing cause or disabling proper diagnosis and decision-making. The manager may need additional information to determine the actual causes, so a real-time Decision Support System is critical in this case.

The Enterprise Knowledge Cloud, with its integrated knowledge repositories and Knowledge Management Systems is an ideal environment for a knowledge-based DSS. A knowledge-driven Decision Support System can be viewed in general terms as the combination of a problem solving system and business intelligence and analytics reporting. The knowledge-driven DSS applies specialist problem solving capabilities to the analytics reported by the enterprise analytics and business intelligence systems and recommends decisions and actions to users of the system. In some cases more advanced and intelligent systems will actually perform recommended actions or implement recommended decisions.

18.3.4 The Intelligent Enterprise

Business enterprises today use the existing internet infrastructure to execute various business operations and provide a wide variety of services. As we see the shift of all non-physical operations versus the internet, we observe a new type of enterprise emerging: the *Intelligent Enterprise* (Delic & Dayal, 2010).

The Intelligent Enterprise is able to interact with its environment, change its behaviour, structure and strategy – behaving actually as an intelligent entity. It is able to adapt to rapid changing market circumstances, gradually change its business model and survive into the next market cycle. The Intelligent Enterprise as we see it is characterized by its ability to learn from and adapt to changes in its environment and reinvent itself, sometimes with surprising results. In order to keep up with the rapidly changing demands of doing business, most enterprises implement increasingly complex IT solutions. Although implemented to make the enterprise more efficient, coupled with the organizational complexity of such large enterprise business, the technical complexity introduced by the many and varied IT solutions helps create pockets of inefficiencies within the organization. We see future Intelligent Enterprises deriving efficiencies through the automation of their core business processes, and the exploitation of knowledge inherent in their organization. Their ability to respond quickly to changes will improve significantly as the knowledge base and "intelligence density" within the enterprise grows and problem-solving capabilities improve dramatically. Intelligent Enterprises will form dynamic partnerships with other enterprises to create dynamic business ecosystems, which will be self-managed, self-configured and self-optimized. In short, future enterprises will become smarter – more intelligent – and by doing so will evolve automatically into organizations more suited to their changing environment.

We postulate that the emergence of collective intelligences in the cloud computing infrastructure will influence markets and established businesses, allowing – even encouraging – Intelligent Enterprises to emerge, and reshape the contemporary approach to Enterprise Knowledge Management.

18.4 Moving KM Applications to the Cloud

Not all applications are suited to cloud computing, so it follows that not all Knowledge Management applications will be good candidates for migration to a cloud platform. A number of issues need to be considered when deciding whether any application should run on a cloud platform, all of which apply to KM applications.

Some considerations when either migrating applications to a cloud environment or designing for the cloud are:

- *Security and Privacy*: while less of a problem in an enterprise cloud than a public cloud or the open internet, security and privacy issues need to be considered when migrating KM applications to the cloud, or when designing new KM applications for the cloud.
- *Latency*: if the speed with which knowledge queries are answered is critical, then careful consideration should be made as to whether the knowledge application serving those queries should be moved to, or developed on, the enterprise cloud. Individual business, even within larger enterprise business, may be able to provide faster access to dedicated knowledge bases than is possible in a cloud environment.
- *Transaction Management*: data integrity, concurrency and latency are all issues for transaction management in the cloud. There are many strategies for coping with issues in this area – the right ones need to be chosen for each application.
- *Criticality and Availability*: the criticality of the underlying cloud infrastructure should match the criticality of the applications to be run on that infrastructure. Mission critical KM applications, for example, should not be deployed in a non-mission critical cloud environment.

18.5 Conclusions and Future Directions

Cloud technology is driving a new paradigm in knowledge management. The participatory and collaborative nature of the internet and cloud computing is both creating more knowledge and providing access to knowledge that was hitherto not generally accessible. There is much more data and information to sort through to find the gems of knowledge – some of the data and knowledge is transient, and some of it not recognisable as knowledge. New technologies will be developed to cope with an almost overwhelming volume of data, information and knowledge.

Knowledge content today is mostly text-based, but for the future we see an evolution towards multimedia and active content. Users today are either fixed or mobile: tomorrow we expect they will be virtual, and later will take personalities of "avatars" to protect privacy and integrity. And while today's enterprise applications are developed by IT departments, with users becoming more tech and internet savvy we predict a shift towards user-developed applications: mash-ups written in high-level mash-up languages.

Current Enterprise Knowledge Management systems are enterprise applications in data centres, while we expect them to evolve into "Enterprise Grids" on which others envisage the development of "Knowledge Management Grids" (Cannataro & Talia, 2003). Once the technology is stable and markets grow, we predict the development of clouds as the super-structure of Enterprise Grids, interconnecting enterprise data centres providing various functionalities.

Thus, while the architecture of today's Enterprise Knowledge Management systems is built around the enterprise stack, tomorrow's Enterprise Knowledge Management architecture will be distributed and loosely-coupled, and later moving to decoupled, completely pluggable, intelligent knowledge management appliances capable of adapting to interface with Enterprise Knowledge Clouds as required (Table 18.1).

Table 18.1 Evolution of EKM systems

EKM systems	Today	Tomorrow	Beyond
Architecture	Enterprise Stack	Distributed	Decoupled/Pluggable
Infrastructure	Datacentre	Grid	Cloud
Application	IT Controlled	User Produced	On Demand
Content	Mainly Text	Multimedia	Active
Users	Fixed/Mobile	Virtual	Avatars
Standards	3 W.org	Web 2.0	Web 3.0

We are in the midst of important social, technological and market changes where we see some major companies announcing their intention to enter, drive and dominate the field of cloud computing (Hayes, 2008; Staten, 2008; Weiss, 1987). We see this as a precondition for the emergence of the intelligent, adaptive enterprise which was announced in the previous century, but can be created only in the right technological circumstances. We believe that enterprise intelligence will draw its capacities from the Enterprise Knowledge Clouds embedded in the global, dependable fabrics consisting of subjects, objects and devices. Cloud computing will enable massive and rapid rescaling of the content production, consumption and participation of the various groups of cloud users at an unprecedented scale.

Massive collaboration (on content tagging, for example) followed by the emergence of ontologies based on the Semantic Web, and adjusted by the *folksonomies* developed as user-oriented Web 2.0 applications, will embody collective intelligence as the new source of knowledge. To see this happen, we postulate the necessity of massive, global, mega-scale infrastructure in the form of cloud computing (interconnected grids and data centres). We are at the very beginning of important new developments where we expect that the field of Enterprise Knowledge Management will be rescaled by an order of magnitude and will spawn the creation of a new kind of EKM system – the *Enterprise Knowledge Exchange*, enabling trade, exchange and monetisation of knowledge assets.

References

Ackoff, R. L. (1989). From data to wisdom. *Journal of Applied Systems Analysis, 16*, 3–9.

Cannataro, M., & Talia, D. (2003). The knowledge grid. *Communications of the ACM, 46*(1), 89–93.

Delic, K. A., & Dayal, U. (2003). A new analytic perspective. *Intelligent Enterprise Magazine, 6*(11). Available from: http://intelligent-enterprise.informationweek.com/030630/611feat2_1.jhtml.

Delic, K. A., & Dayal, U. (2010). The rise of the intelligent enterprise. *ACM Ubiquity*, (45). Retrieved February 1, 2010, from http://www.acm.org/ubiquity.

Delic, K. A., & Riley, J. (2009). Enterprise knowledge clouds: Next generation KM systems? *Proceedings of the 2009 International Conference Information, Process, and Knowledge Management (eKnow 2009), Cancun, Mexico*, 49–53.

Delic, K. A., Douillet, L., & Dayal, U. (2001). Towards an architecture for real-time decision support systems: Challenges and solutions. *Proceedings of the 2001 International Database Engineering & Applications Symposium (IDEAS'01), Grenoble, France*, 303–311.

Delic, K. A., Riley, J., Bartolini, C., & Salihbegovic, A. (October 2007a). Knowledge-based self-management of apache web servers. *Proceedings of the 21st International Symposium on Information, Communication and Automation Technologies (ICAT'07), Sarajevo, Bosnia and Herzegovena.*

Delic, K. A., Riley, J., & Faihe, Y. (June 2007b). Architecting principles for self-managing enterprise IT systems. *Proceedings of the 3rd International Conference on Autonomic and Autonomous Systems, Athens, Greece*, 60–65.

Foster, I., & Kesselman, C. (1999). Globus: A toolkit-based grid architecture. In I. Foster, & C. Kesselman (Eds.), *The grid: Blueprint for a new computing infrastructure* (pp. 259–278). Los Altos, CA: Morgan Kaufmann.

Frey, J., Tannenbaum, T., Livny, M., Foster, I., & Tuecke, S. (2004). Condor-G: A computation management agent for multi-institutional grids. *Cluster Computing, 5*(3), 237–246.

Hayes, B. (2008). Cloud computing. *Communications of the ACM, 51*(7), 9–11.

Staten, J. (March 2008). Is cloud computing ready for the enterprise?, *Forrester Research*, March 7.

Vecchiola, C., Kirley, M., & Buyya, R. (2009). Multi-objective problem solving with offspring on enterprise clouds. *Proceedings of the Tenth International Conference on High-Performance Computing in Asia-Pacific Region (HPC Asia 2009), Hongkong.*

Venugopal, S., Buyya, R., & Winton, L. (2006). A grid service broker for scheduling e-science applications on global data grids. *Concurrency and Computation: Practice and Experience, 18*(6), 685–699.

Weiss, A. (2007). Computing in the clouds. *netWorker, 11*(4), 16–25.

Zeleny, M. (1987). Management support systems: Towards integrated knowledge management. *Human Systems Management, 7*(1), 59–70.

Chapter 19
Open Science in the Cloud: Towards a Universal Platform for Scientific and Statistical Computing

Karim Chine

19.1 Introduction

The UK, through the e-Science program, the US through the NSF-funded cyber infrastructure and the European Union through the ICT Calls aimed to provide "the technological solution to the problem of efficiently connecting data, computers, and people with the goal of enabling derivation of novel scientific theories and knowledge".[1] The Grid (Foster, 2002; Foster, Kesselman, Nick, & Tuecke, 2002), foreseen as a major accelerator of discovery, didn't meet the expectations it had excited at its beginnings and was not adopted by the broad population of research professionals. The Grid is a good tool for particle physicists and it has allowed them to tackle the tremendous computational challenges inherent to their field. However, as a technology and paradigm for delivering computing on demand, it doesn't work and it can't be fixed. On one hand, "the abstractions that Grids expose – to the end-user, to the deployers and to application developers – are inappropriate and they need to be higher level" (Jha, Merzky, & Fox, 2009), and on the other hand, academic Grids are inherently economically unsustainable. They can't compete with a service outsourced to the Industry whose quality and price would be driven by market forces. The virtualization technologies and their corollary, the Infrastructure-as-a-Service (IaaS) style cloud[2], hold the promise to enable what the Grid failed to deliver: a sustainable environment for computational sciences that would lower the barriers for accessing federated computational resources, software tools and data; enable collaboration and resources sharing and provide the building blocks of a ubiquitous platform for traceable and reproducible computational research. Amazon Elastic Compute Cloud (EC2) (Amazon, Inc., 2006) is an

K. Chine (✉)
Cloud Era Ltd, Cambridge, UK
e-mail: karim.chine@polytechnique.org

[1] http://en.wikipedia.org/wiki/Cyberinfrastructure.

[2] Interview with KateKeahey (2009), *An interview with kate keahey of the nimbus project, a cloud computing infrastructure* [Online]. Available: http://www.nsf.gov/news/news_videos.jsp?cntn_id=114788&media_id=65105.

B. Furht, A. Escalante (eds.), *Handbook of Cloud Computing*,
DOI 10.1007/978-1-4419-6524-0_19, © Springer Science+Business Media, LLC 2010

example of Infrastructure-as-a-Service that anyone can use today. Its considerable success announces the emergence of a new era. However, bringing that era for research and education still requires the software that would bridge the gap between the cloud and the scientists' everyday tools (YarKhan, Dongarra, & Seymour, 2006) and would make Infrastructure-as-a-Service a trivial commodity.

This article describes Elastic-R,[3,4] platform that makes working with R on the cloud as simple as working with it locally. More generally, it aims to be the missing link between the cloud and the most widely used data analysis tools and Scientific Computing Environments (SCEs). Elastic-R synergizes the usage scenarios of those environments with the usage scenarios of the cloud and empowers them with what the cloud has best to offer:

User-friendly and flexible access to Infrastructure-as-a-Service: The cloud interfaces are simple and expose the right abstractions for managing and using virtual appliances within a federated computing environment but the cloud consoles remain tools for computer savvies. Elastic-R offers a simplified façade to the cloud that makes any scientist able to choose and run the virtual machine with the specific scientific computing environment (example: R version 2.9, Scilab,[5] Sage,[6] Root,[7] etc.). The scientist can then have access to the full capabilities of the environment using the Elastic-R Java workbench or from within a standard web browser using the Elastic-R Ajax workbench. The scientist can issue commands, install and use new packages, generate and interact with graphics, upload and process files, download results, create and edit R-enabled server-side spreadsheets, etc. The scientist can disconnect from the engine and reconnect again from anywhere, retrieving his full session including workspace, graphics, etc. and can continue working from where he left off. The virtual machine can be simply shut down when not needed anymore. The user is charged only for the usage time. User's data (working directory content) remains on a virtual disk on the cloud that can be attached once again to a new virtual machine instance.

Collaboration: A virtual machine instance on the cloud has a public IP address and can be seen and used by the owner's collaborators who can be located anywhere. Elastic-R allows the scientist to expose his machine and his R sessions (for example) to his collaborators (Fig. 19.1). All of them can connect their Java or Ajax workbenches to the same R engine and can control that engine and update its environment. The actions of each collaborator in the console (commands issued, chat), on the graphics (plotting, annotating, resizing and slides viewing) or on the spreadsheets (cells updating, cells selecting) are broadcasted to the others and their workbenches show the changes in real time.

---

[3] www.elasticr.net/platform

[4] www.elasticr.net/portal

[5] http://www.scilab.org

[6] http://www.sagemath.org/

[7] http://root.cern.ch/

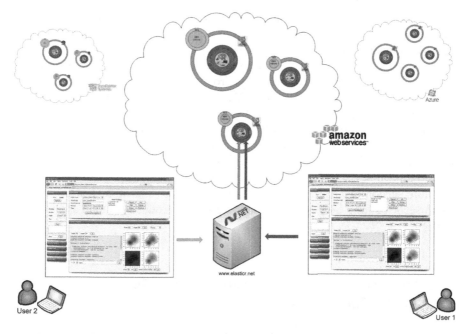

Fig. 19.1 Elastic-R portal: collaborative Virtual Research Environment

On-demand elasticity: Elastic-R exposes to the scientist a major feature of the cloud which is the ability to choose the capacity of the virtual machine instances, such as the number of virtual cores, the memory size and the disk space. He can then run on the cloud analysis or simulations that require more memory than available on his laptop or would take days if run locally. Elastic-R allows the scientist to solve compute-intense problem by starting any number of virtual machines hosting R engines that can process in parallel partial tasks. Those pools of engines can also be used to create Web applications with dynamic analytical content generated by R or any other environment. For those applications, the Elastic-R platform enables cloudbursting: virtual machines can be fired up or shut down (increasing or decreasing the engines pool size) to scale up or scale down according to the load of the application.

Applications deployment flexibility: The cloud can host very easily client-server style applications. Elastic-R is a platform that allows anyone to assemble statistical/numerical methods and data on the server (an Elastic-R virtual machine instance on the cloud) and to visually create and publish, in the form of URLs, interactive user interfaces and dashboards exposing those methods and data. Elastic-R also provides tools that allow anyone to expose those methods (implemented by R functions for example) as SOAP Web services that can be used as computational services on the cloud for data analysis pipelines or as nodes for workflow workbenches.

Recording capabilities: Because the scientific computing environments accessible through Elastic-R are running on virtual machines and because the working directories are hosted by virtual disks, a snapshot of the full updated computational environment can be produced at any point in time. That snapshot can be archived or made available to anyone using the Elastic-R workbenches: an author can share his environment with the reviewers of the journal to which he submitted his paper, a teacher can make the statistical learning environment needed for his course available to his students, a researcher in laboratory A can make his simulation environment accessible to his collaborators in laboratory B, etc.

This chapter is organized as follows: The first section describes the building blocks of the Elastic-R platform and the ecosystem it creates for the interoperability, sharing and reuse of the computing artifacts. The second section describes how the usage scenarios of Elastic-R can be integrated with those of an IaaS. The third section details the major e-Science use cases Elastic-R deals with. The fourth section shows how it can be used as a highly productive cloud applications platform.

19.2 An Open Platform for Scientific Computing, the Building Blocks

R is a language and environment for statistical computing and graphics and it became the *lingua franca* of data analysis (R Development Core Team, 2009).[8] R has a very powerful graphics system as well as cross-platform capabilities for packaging any computational code. Hundreds of available R packages, exponentially growing in number, implement the most up-to-date computational methods and reflect the state-of-the-art of research in various fields. R packages have become a reproducible research enabler because they enable functions and algorithms to be reused and shared. There is no obstacle to a large-scale deployment of R on public clouds and Grids since it is licensed under the GNU GPL (Chambers, 1998). However, R is not multithreaded and does not operate as a server. As a language, it implements the powerful S4 class system but as a library, R has only a low-level non-object oriented Application Programming Interface (API). Graphical User Interfaces (GUIs) development for R remains non standardized. R's potential as a computational back end engine for applications and service-oriented architectures has yet to be fully exploited. While its user base is growing at a high rate, this growth rate would be significantly higher with a user-friendly and rich workbench. Elastic-R brings to the R ecosystem all those missing features which may enable it to be applied in many more situations, in various different ways. Its ambition though goes far beyond the provision of new tools and frameworks. By extending R's logic of openness and extensibility, Elastic-R builds an environment where all the artifacts and resources of computing become "pluggable" and not only the computational component (the R package).

---

[8]http://www.nytimes.com/2009/01/07/technology/business-computing/07program.html

Figure 19.2 shows the key features of Elastic-R. The Java or Ajax Elastic-R workbench allows the scientist, the statistician, the financial analyst, etc. to easily assemble (plug together) synergetic capabilities, described in the following sections:

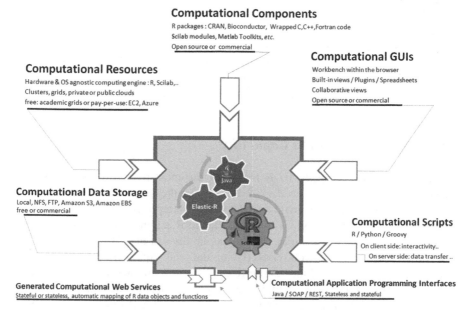

Computational Components
R packages : CRAN, Bioconductor, Wrapped C,C++,Fortran code
Scilab modules, Matlab Toolkits, *etc.*
Open source or commercial

Computational GUIs
Workbench within the browser
Built-in views / Plugins / Spreadsheets
Collaborative views
Open source or commercial

Computational Resources
Hardware & OS agnostic computing engine : R, Scilab,..
Clusters, grids, private or public clouds
free: academic grids or pay-per-use: EC2, Azure

Computational Data Storage
Local, NFS, FTP, Amazon S3, Amazon EBS
free or commercial

Computational Scripts
R / Python / Groovy
On client side: interactivity..
On server side: data transfer ..

Generated Computational Web Services
Stateful or stateless, automatic mapping of R data objects and functions

Computational Application Programming Interfaces
Java / SOAP / REST, Stateless and stateful

Fig. 19.2 The open computational platform's ecosystem

19.2.1 The Processing Capability

By providing a simple URL and credentials, the scientist connects his workbench to an R-based remote computational engine and gain access to the computational resource whether it is a node of Grid, a virtual machine of a cloud, a cluster or his own laptop. The engine is agnostic to its hosting operating system and hardware. IaaS requires such a mechanism for computing to become "like electricity". EC2 (Amazon, Inc., 2006) allows the user to choose the capacity of the virtual machine (number of virtual cores, memory size, etc.) he would like to launch. The Elastic-R workbench exposes that choice to the scientist in its simplified EC2 Console. The state of an Elastic-R engine persists until the computational resource is released (the virtual machine is shut down, the interactive Grid job is killed, the process on the physical server is killed, etc.). The scientist can disconnect form the engine and reconnect again from anywhere: he retrieves his session with all variables, functions, graphics, spreadsheets, etc.

19.2.2 The Mathematical and Numerical Capability

By gaining access to an R session and by importing into his workspace the R packages related to his problem domain, the scientist gathers the functions and mathematical models needed to process data and transform it into knowledge and insight. The R package can be a wrapper of any mathematical library written in C, C++, FORTRAN, etc. R can be considered as a universal framework for computational code and computational toolkits. From within his R session, the scientist can also call Scilab,[5] Sage,[7] Root,[9] etc. and increase the mathematical capability of his environment. An architecture for server-side extensions allows anyone to build java bridges that couple the Elastic-R engine with any software. Such bridges are available for Matlab[10] and OpenOffice.[11]

19.2.3 The Orchestration Capability

The S language implemented by R is one of the most powerful languages ever created for "programming with data".[6] Besides R, the user can orchestrate tasks and control data flow using python and groovy. Interpreters for theses scripting languages are embedded both within the Elastic-R engine (on server side) and within the workbench (client side). The full capabilities of the platform are exposed via SOAP and RESTful front-ends (Computational Application Programming Interfaces in Fig. 19.1) and the Elastic-R engine can be piloted programmatically from Java, Perl, C#, C++, etc. A tool is provided to enable the scientist to generate and deploy SOAP Web services exposing a selection of his R functions (Fig. 19.1: generated computational web services). They can be used as nodes within workflow workbenches. The nodes are dynamically connected to the scientist's R session and the processing of data is done on the cloud if the Elastic-R engine exposing the Web service is on the cloud.

19.2.4 The Interaction Capability

The console views within the Java and Ajax workbenches allow the full control of an R session as well as the use of python, groovy and Linux shells. Besides the consoles, both workbenches have several dockable built-in views including remote directory browsers for the viewing, download and upload of files from and to the remote engine's working directory, syntax-highlighting-enabled code editors, help browsers, viewers for various file formats (PDF, SVG, HTML, etc.), interactive server-side graphic devices with built-in resizing, zooming, scrolling, coordinate

---

[9]http://www.openoffice.org/

[10]http://open.eucalyptus.com/

[11]http://www.mathworks.co.uk/

tracking and annotation capabilities, data inspectors, linked plots, spreadsheets fully integrated with R functions and data.

The workbench's architecture for plugins lets anyone create his own views and dashboards to make workbenches more productive or to expose statistical and numerical models through simple graphical user interfaces. All the views of the workbenches are collaborative: when more than one user is connected to the same Elastic-R engine, the actions of one collaborator are broadcasted to all the others. An example of the Elastic-R Java workbench is shown in Fig. 19.3.

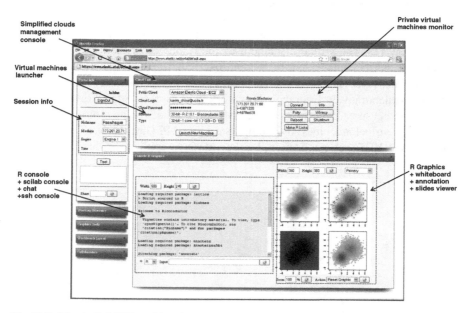

Fig. 19.3   Elastic-R AJAX workbench

19.2.5 The Persistence Capability

The Elastic-R computational engine's working directory can be on a local or a network file system and its content can be easily synchronized with an FTP server or with Amazon S3 (Amazon, Inc., 2006). When the scientist uses the simplified EC2 console to start a Elastic-R-enabled Amazon Machine Image (AMI), an Amazon Elastic Block Store (EBS) is automatically attached to the running AMI. That EBS becomes the working directory of all the Elastic-R engines hosted by the AMI and all the files generated by the scientist including the workspaces serialization, the spreadsheets content, the generated web services, etc. are kept when the AMI is shutdown. The EBS is also a place where the R packages installed by the scientist are stored. Those packages are made available to the Elastic-R engines when a new AMI is started. A snapshot of the EBS can be created by the scientist who can decide to share that snapshot with other EC2 users.

19.3 Elastic-R and Infrastructure-as-a-Service

Elastic-R can be used on any type of infrastructure. However, the platform takes its full dimension only when it is used within an IaaS-style-cloud whether it is Amazon EC2 or a private cloud based on Eucalyptus,[12] OpenNebula[13] or Nimbus (Keahey & Freeman, 2008).

Figure 19.4 shows the role of Elastic-R within an Infrastructure as service environment (the concentric circles). Elastic-R wraps R (in the very centre) with Java Object-oriented layers that can be accessed remotely from anywhere. The R engine created (inside the Java virtual machine circle) is agnostic to the operating system and to the hardware. In this case, it runs within a virtual machine that can be based on any OS. The virtual machine uses the resources of the hardware via the Hypervisor and its management by the end user (start-up, shutdown, etc.) is done using the outer layer (IaaS API). "User1" and "User2" can connect to the R engine and use it collaboratively from their Java workbenches, their Ajax workbenches or from their Excel spreadsheets. The Developer can use the R engine (one or many, on one or many virtual machines) by calling its remote API (Java-RMI, SOAP, REST) from his ASP.NET web application, his Java desktop application, his Perl scripts,

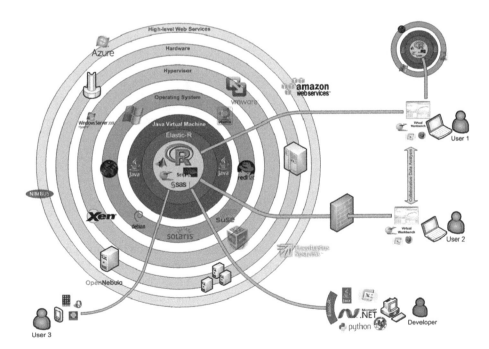

Fig. 19.4 Elastic-R in the IaaS environment

[12] http://www.opennebula.org/

[13] http://www.nimbusproject.org/

his Excel Add-in, etc. "User 3" from his portable device (iPhone, Android-based phone, etc.) can use the Ajax workbench to access the same R engine as "User 1" and "User 2" and show them his data, spreadsheets, slides, etc. interactively: the Ajax workbench gives the same capabilities to "User 1", "User 2" and "User 3". They all can issue commands to R, install and use new packages, generate and interact with graphics, upload and process files, download results, etc.

19.3.1 The Building Blocks of a Traceable and Reproducible Computational Research Platform

Elastic-R on an IaaS-style cloud provides a system so that the computational environment, the data and the manipulations of the data (scripts, applications) can be recorded. These can be used by reviewers, collaborators and anyone wanting to investigate the data. Elastic-R provides an end-to-end solution for traceable and reproducible computational research. Snapshots of computational environments can be created as virtual machine images (AMIs). Snapshots of versioned libraries and working directories can be created as Elastic Block Stores (EBSs). The Elastic-R Java and Ajax workbenches make it possible to all scientists to work with these snapshots (AMIs + EBSs) and produce them easily (Fig. 19.5). By Providing the Elastic-R-enabled AMI identifier, the complementary computational libraries EBS snapshots identifiers and the working directory (data) EBS snapshot identifier that

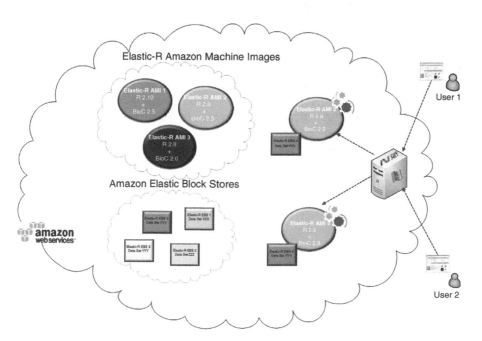

Fig. 19.5 The IaaS-style cloud as a reproducible research platform

have been used for his research, the scientist makes it possible to anyone to rebuild all the data and the computational environment required to process that data.

19.3.2 The Building Blocks of a Platform for Statistics and Applied Mathematics Education

Besides being free and mostly open source and therefore accessible to students and educators, Elastic-R provides education-friendly features that only proprietary software could offer so far (for example the centralized and controlled server-side deployment of the Scientific Computing Environments) and enables new scenarios and practices in the teaching of statistics and applied mathematics. With Elastic-R, it becomes possible for educators to hide the complexity of R, Scilab, Matlab, etc. with User Interfaces such as the Elastic-R plugins and spreadsheets. These are very easy to create and to distribute to students. The User Interfaces reduce the complexity of the learning environment and keep beginning students away from the steep learning curves of R, Scilab or Matlab. Once created by one educator, the User Interfaces can be shared, reused and improved by other educators. Dedicated repositories can be provided to centralize the efforts and contributions of the community of educators and help them sharing the insight gained in using this new environment. One could envisage these methods being used from primary schools to graduate-level studies.

Educators can adapt the Elastic-R virtual machines images to the specific needs of their courses and tutorials. For example, after choosing the most appropriate image, they can add to it the missing R packages, the required data files, install the missing tools, etc. The new image can then be provided to students on USB keys or made accessible on an IaaS-style cloud. In the first case, the students need only to have Java and a virtual machine player (the free VMware player for example) installed on their laptops to run the Elastic-R workbench and to connect to a computational engine on the virtual machine. In the second case, they need only a browser. Once again, a virtual machine prepared by one educator can be shared, reused and improved by other educators.

The virtual machine is fully self-contained: the code needed to run the workbench or the plug-ins prepared by the educator can be delivered by the virtual appliance itself thanks to the Elastic-R code server that runs at startup. The interaction between the student and the SCE as well as the artifacts produced are saved within the Elastic-R-enabled-virtual machine. The educator can retrieve the USB keys used by the students (or connect to the virtual machine instance on the IaaS-style cloud) and checks not only the validity of the different intermediate results they obtained but also the path they followed to get those results.

The collaboration capabilities of the workbench open also new perspectives in distributed learning. The educator can connect anytime to the SCEs of students at any location. He can see and update their environments and guide them remotely. Collaborative problem solving becomes also possible and can be used as a support for learning.

19.4 Elastic-R, an e-Science Enabler

Elastic-R is an e-Science platform that deals with some of the most timely use cases related to the use of Information and Communications Technologies (ICT) in research and education:

19.4.1 Lowering the Barriers for Accessing on-Demand Computing Infrastructures. Local/Remote Transparency

The same application, the Elastic-R workbench (Fig. 19.6), makes it easy to connect to various environments locally or on remote machines whether they are nodes of a Grid or virtual machines of a cloud. Switching from one resource to another (for example from one virtual machine instance on Amazon Elastic Compute Cloud to another or from an interactive Grid job on the European Grid EGI to an interactive job on an intranet cluster) becomes as simple as replacing one URL with another.

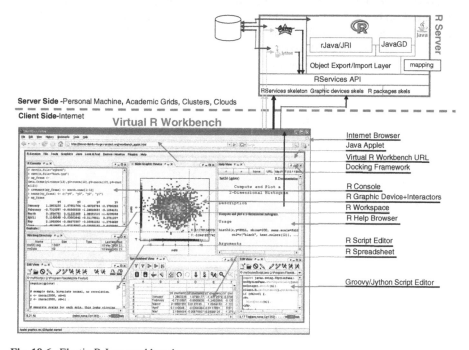

Fig. 19.6 Elastic-R Java workbench

19.4.2 Dealing with the Data Deluge

The data generated by modern science tools can become too large to move easily from one machine to another. This can be an issue for large collaborative projects.

The analysis of such data can't be performed the way it has been so far. The answer to this increasingly acute problem is to take the computation to the data and is what Elastic-R enables users to do: The generic computational engine can run on any machine that has a privileged connectivity with the data storage machine or within the large scale database. This is the case when an Elastic-R EC2 AMI is used to process data that is already on Amazon's Elastic Cloud. The user can connect his virtual workbench (or his scripts using the Elastic-R SOAP clients) to the computational engine, set the working directory to the location of the data (e.g. via NFS) and view or analyze the data using R/Scilab packages.

19.4.3 Enabling Collaboration Within Computing Environments

Users can connect to the same remote engine and work with large scale data collaboratively using broadcasted commands/graphics and collaborative spreadsheets (Fig. 19.7). Every command issued by one of them is seen by all the others. Synchronized R graphics panels allow them to see the same graphics and to annotate them collaboratively. Chatting is enabled. Linked plots views based on a refactored iplots package (Theus & Urbanek, 2008) enable collaborative highlighting and color brushing on a variety of high interaction graphics.

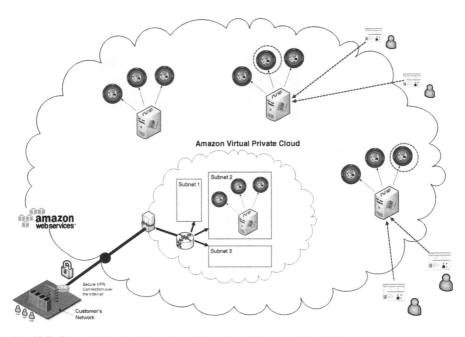

Fig. 19.7 Decentralized collaboration: Elastic-R portal as an EC2 AMI

19.4.4 Science Gateways Made Easy

Web-based interfaces and portals allowing scientists to use federated distributed computing infrastructures to solve their domain specific problems have always been difficult to develop, upgrade and maintain. We should have front ends that are easy to create. Elastic-R proposes a different paradigm for the creation and distribution of such front-ends to HPC/cloud environments with plugins and server-side spreadsheets (see Sections 19.5.1 and 19.5.2)

19.4.5 Bridging the Gap Between Existing Scientific Computing Environments and Grids/Clouds

Once the user's workbench is connected to a remote R/Scilab engine, a RESTful embedded server (local http relay) enables third-party applications such as emacs, OpenOffice Calc or Excel to access and use the Grid/cloud-enabled engine. For example, an Excel add-in enables scientists to use the full capabilities of the Elastic-R platform and reproduce the features of the Elastic-R spreadsheets from within Excel. The bi-directional mirroring of server-side spreadsheets' models into Excel cell ranges is also available. This allows users to overcome some of the Excel flaws (limited capabilities in statistical analysis, inaccurate numerical calculations at the edge of double, inconsistent identification of missing observations...). Excel becomes a front-end of choice to Grid/cloud resources and can then become the universal workbench for different sciences.

19.4.6 Bridging the Gap Between Mainstream Scientific Computing Environments

The platform has a server-side extensions architecture that enables the creation of bridges between the remote computational engine and any third party tool. Besides R and Scilab, several widely used environments can be integrated (Matlab, Root, SAS, etc.). Since R and Scilab are running within the same process (same Java Virtual Machine), it is easy and very fast to exchange data between them. This can be achieved for example by using the Groovy interpreter available as part of the remote engine. The SOAP API can be called from any environment. It enables SciPy users for example to work with Elastic-R engines on the cloud and to call R and Scilab functions.

19.4.7 Bridging the Gap Between Mainstream Scientific Computing Environments and Workflow Workbenches

Elastic-R enables automatic exposure of R functions and packages as Web Services (Fig. 19.8). The generated Web Services are easy to deploy and can use back-end

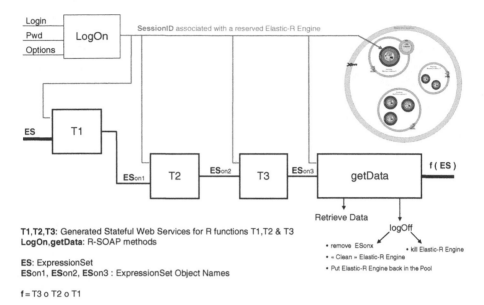

Fig. 19.8 Workflows with generated Stateful SOAP web services

computational engines running at any location. They can be seamlessly integrated as workflows nodes and used within environments such as Knime,[14] Taverna[15] or Pipeline Pilot.[16] They can be stateless (an anonymous R worker performs the computation) or stateful (an R worker reserved and associated with a session ID is used and can be reused until the session is destroyed). The statefulness solves the overhead problem caused by the transfer of intermediate results between workflow nodes.

19.4.8 A Universal Computing Toolkit for Scientific Applications

Elastic-R frameworks and tools make it possible to use R as a Java object-oriented toolkit or as an RMI server. All the standard R objects have been mapped to Java (Fig. 19.9) and user defined R classes can be mapped to Java on demand (Fig. 19.10). R functions can be called from Java as if they were Java functions. The input parameters are provided as Java objects and the result of a function call is retrieved as a Java object. Calls to R functions from Java locally or remotely cope with local and distributed R objects. The full capabilities of the platform are exposed via a SOAP

---

[14]http://www.knime.org/

[15]http://www.taverna.org.uk/

[16]http://accelrys.com/products/scitegic/

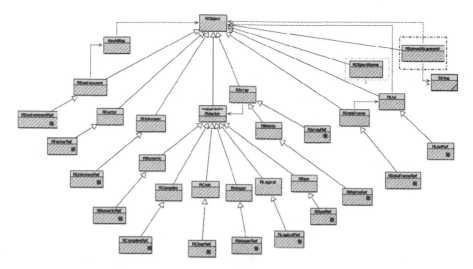

Fig. 19.9 Java classes diagram: mapping of standard R objects

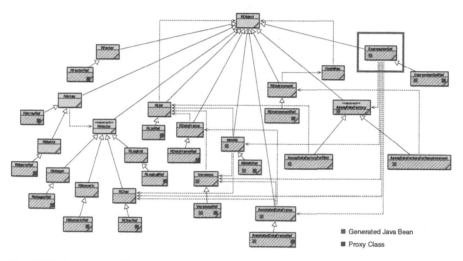

Fig. 19.10 Java classes diagram: generated mapping for ExpressionSet (S4 class)

and RESTful front-ends. Several tools and frameworks are provided to help building analytical desktop/web applications and scalable data analysis pipelines in any programming language (Java, C#, C++, Perl, etc.)

19.4.9 Scalability for Computational Back-Ends

Elastic-R provides a pooling framework for distributed resources (RPF) allowing pools of computational engines to be deployed on heterogeneous nodes/virtual machines instances. These engines are managed and used via a simple borrow/return API for multithreaded web applications and web services, for distributed and parallel computing, for dynamic content on-the-fly generation (analytic results, tables and graphics in various formats for thin web clients) and for computational engines' virtualization in a shared computational resources context (Fig. 19.11). The engines become agnostic to the hosting operating system. Several tools are provided to monitor and manage the pools programmatically or interactively (Supervisor UI). The pooling framework enables transparent cloudbursting: Amazon EC2 virtual machines instances hosting one or many computational engines can be fired up or shut down to scale up or scale down according to the load in a highly scalable web applications deployment for example.

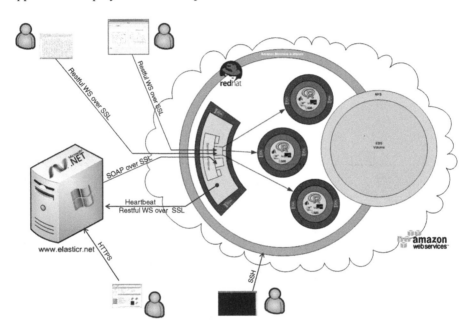

Fig. 19.11 Elastic-R security architecture

19.4.10 Distributed Computing Made Easy

To solve heavily computational problems, there is a need to use many engines in parallel. Several tools are available but they are difficult to install and beyond the technical skills of most scientists. Elastic-R solves this problem. From within a main R session and without installing any extra toolkits/packages, it becomes possible to create logical links to remote R/Scilab engines either by creating new processes or by connecting to existing ones on Grids/clouds (Fig. 19.12). Logical links are variables that allow the R/Scilab user to interact with the remote engines. rlink.console, rlink.get, rlink.put allow the user to respectively submit R commands to the R/Scilab worker referenced by the rlink, retrieve a variable from the

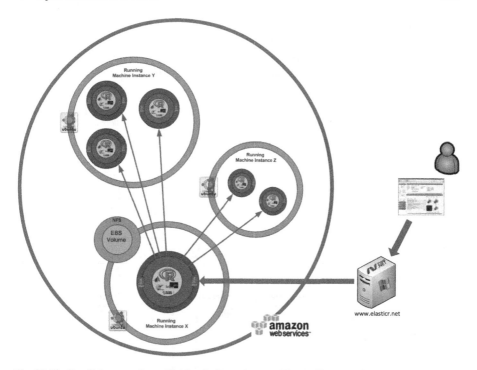

Fig. 19.12 Parallel computing with Elastic-R on Amazon Elastic Compute Cloud

R/Scilab worker's workspace into the main R workspace and push a variable from the main R workspace to the worker's workspace. All the functions can be called in synchronous or asynchronous mode. Several rlinks referencing R/Scilab engines running at any locations can be used to create a logical cluster which enables to use several R/Scilab engines in a coordinated way. For example, a function called cluster.apply uses the workers belonging to a logical cluster in parallel to apply a function to a large scale R data.

19.5 Elastic-R, an Application Platform for the Cloud

Elastic-R is extensible with java components both on client-side (the plugins) and on server-side (the extensions). With those components, anyone can create and deploy his application on the cloud without having any specific knowledge about the infrastructure.

19.5.1 The Elastic-R Plug-ins

The Elastic-R platform defines a contract for creating cross-platform statistical/numerical new interfaces in Swing-Java either programmatically or using visual composition tools like the Netbeans GUI designer. The views can be bundled into zip files and opened by anyone using the Elastic-R workbench (Fig. 19.13). The views receive a Java Interface that allows them to use the R/Scilab engine to which the workbench is connected and that can be running at any location.

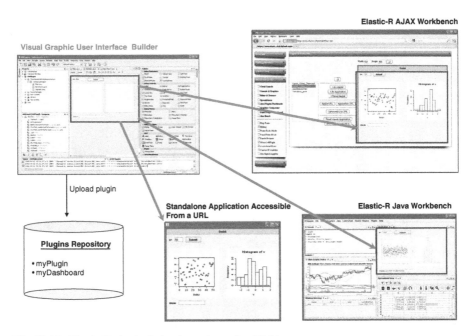

Fig. 19.13 Elastic-R plugins' visual creation and publishing

Three-parts-URLs (Elastic-R's Java Web Start trigger + computational engine's URL parameter + plugin's zip file URL parameter) can be used to deliver those GUIs to the end-user. He retrieves them in one click and the only software required to be preinstalled on his machine is a Java runtime. Instead of requiring a transparent connection to a server-side Grid/cloud-enabled engine, the distribution URLs can be written to trig transparently the creation of a computational engine on the user's machine: a zipped version of R is copied on the user's machine (with or without administrative privileges) and is used transparently by the GUI.

19.5.2 The Elastic-R Spreadsheets

The Elastic-R spreadsheets are Java-based built originally using the OSS jspreadsheet. Unlike jspreadsheet, Calc and Excel's spreadsheets, they have their models on server-side, are HPC and collaboration enabled and are fully connected with the remote statistical/numerical engine's workspace. This enables for example R data import/export from/to cells and R functions use in formula cells. Dedicated R functions (cells.get, cells.put, cells.select, etc.) allow the R user to retrieve the content of cell ranges into the R workspace or to update them programmatically: An R script can reproduce the spreadsheet entirely. A macros system allows the user to define listeners on R variables and on cell ranges and to define corresponding actions as R/Java scripts. Specific macros called datalinks allow the user to bi-directionally mirror R variables with cell ranges. R graphics and User Interface components can be docked onto cell ranges. UI components can be for example: Sliders mirroring R variables; Graphic Panels showing R Graphics (in any format) produced by user defined R scripts and automatically updated in case user-defined R variables have their values change or in case cells within a user-defined cell ranges list are updated; Buttons executing any user-defined R script, etc. (Fig. 19.14). This spreadsheet enables scientists without programming skills to create sophisticated Grid or cloud-based analytical views and dashboards and lowers the barriers for creating science gateways and distributing them.

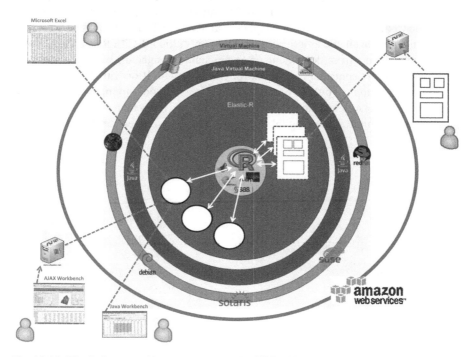

Fig. 19.14 Elastic-R server-side spreadsheet models/GUI widgets

19.5.3 The Elastic-R extensions

Elastic-R extensions are Java components that can be uploaded anytime to an Elastic-R engine's extensions folder. They are dynamically loaded by the engine's Java Virtual Machine and the code they expose can be called from the client.

Extensions allow anyone to build java bridges that couple the Elastic-R engine with any software such as Matlab[10] and OpenOffice.[11]

19.6 Cloud Computing and Digital Solidarity

The FOSS community established Open Source Software as a credible alternative to proprietary software and allowed users from developing countries to have free access to tools of the highest quality. Cloud computing allows anyone today to use over the internet virtual machines of any computing capacity running on federated infrastructures. The combination of sponsored access to public clouds and Software delivered as a service through the internet opens perspectives for reducing the digital divide of an unprecedented scale and enables revolutionary new scenarios for knowledge sharing and digital solidarity.

Elastic-R is the first Software platform to bring to light the tremendous potential of this combination in research and education. It puts the best of the existing Scientific Computing Environments and data analysis tools in the hands of everyone by making them available as a service on Amazon's cloud (EC2). Researchers, educators and students from developing countries can use standard web browsers and low-speed internet connections to work for example with R or with Scilab. African scientists for example would have on demand access not only to machines with any number of processors and any size of memory to perform their research but also to potentially infinite shared digital resources. Those resources can be for example

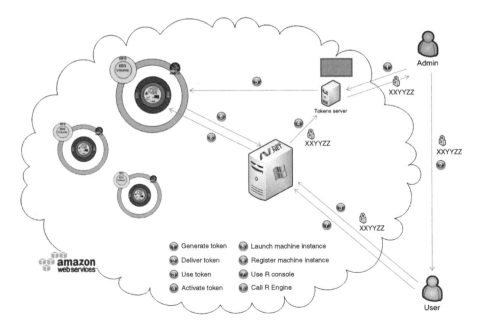

Fig. 19.15 Elastic-R's digital tokens: cloud vouchers for digital solidarity

pre-prepared and ready-to-run virtual machine images with research software tools or public data or analytical applications provided by the world's scientists.

The Elastic-R portal lowers the barriers for anyone to use the cloud. It also provides a mechanism of secure digital tokens (Fig. 19.15) that can be delivered by international organizations and charities to African scientists for example. The tokens allow the scientists to start virtual machines for a specified number of hours and use them for their research. Elastic-R is also a Virtual Research Environment that allows any number of geographically distributed users to work simultaneously and collaboratively with the same virtual machine, the same tool and the same data. It makes it much easier for developing countries' scientists to get more actively involved in large international collaborations, to have real time scientific interaction with their peers in the US and in Europe and to gain access to scientific data and to the computational environments required to process that data. Elastic-R is finally a real-time cloud-based e-Learning system that makes it possible for volunteering educators from developed countries to teach statistics and math interactively to African students without having to be physically in Africa.

19.7 Conclusions and Future Directions

This article described the Elastic-R as a new environment that has the potential to democratize the cloud and to push forward the reproducibility of computational research. Its current availability and easy access on Amazon Elastic Compute Cloud maximizes its chances for uptake and adoption. Academia, Industry and Educational Institutions would benefit from the emergence of a new environment for the interoperability, sharing and reuse of computational artifacts. The creation and sharing of analytical tools and resources can become accessible to anyone (open science). An international portal[4] for on demand computing is being built using the different frameworks provided by Elastic-R and could become a single point of access to Virtualized SCEs on public servers and on virtual appliances that are ready for use on various clouds. There is no question about the need for more usability in the computational landscape. Java, Xen, VMware, EC2, R and Elastic-R prove that the target of a universal computational environment for science and for everyone is definitely within reach.

References

Amazon, Inc. (2006). *Amazon elastic compute cloud* [Online]. Available: aws.amazon.com/ec2.
Amazon, Inc. (2006). *Amazon simple storage service* [Online]. Available: aws.amazon.com/s3.
Chambers, J. M. (1998). *Programming with data: A guide to the S language.* New York, NY: Springer.
Foster, I. (2002). What is the grid? A three point checklist. *Grid Today, 1*(6), 22–25.
Foster, I., Kesselman, C., Nick, J., & Tuecke, S. (2002). *The Physiology of the Grid: An Open Grid Services Architecture for Distributed Systems Integration.* Open Grid Service Infrastructure WG, Global Grid Forum.

Jha, S., Merzky, A., & Fox, G. (2009). Using clouds to provide grids with higher levels of abstraction and explicit support for usage modes. *Concurrency and Computation: Practice and Experience, 21*(8), 1087–1108.

Keahey K., & Freeman, T. (October 2008). *Science clouds: Early experiences in cloud computing for scientific applications*. Chicago, IL: Cloud Computing and Its Applications 2008 (CCA-08).

R Development Core Team (2009). R: A language and environment for statistical computing. *R Foundation for Statistical Computing, Vienna, Austria*. ISBN 3-900051-07-0, URL http://www.R-project.org.

Theus, M., & Urbanek, S. (2008). *Interactive graphics for data analysis: Principles and examples.* CRC Press. ISBN 978-1-5848-8594-8, 2008.

YarKhan, A., Dongarra, J., & Seymour, K. (July 2006). NetSolve to GridSolve: The evolution of a network enabled solver. *IFIP WoCo9 Conference "Grid-Based Problem Solving Environments: Implications for Development and Deployment of Numerical Software," Prescott, AZ.*

Chapter 20
Multidimensional Environmental Data Resource Brokering on Computational Grids and Scientific Clouds

Raffaele Montella, Giulio Giunta, and Giuliano Laccetti

20.1 Introduction

Grid computing has widely evolved over the past years, and its capabilities have found their way even into business products and are no longer relegated to scientific applications. Today, grid computing technology is not restricted to a set of specific grid open source or industrial products, but rather it is comprised of a set of capabilities virtually within any kind of software to create shared and highly collaborative production environments. These environments are focused on computational (work-load) capabilities and the integration of information (data) into those computational capabilities. An active grid computing application field is the fully virtualization of scientific instruments in order to increase their availability and decrease operational and maintaining costs. Computational and information grids allow to manage real-world objects in a service-oriented way using industrial world-spread standards.

Boosted by the rapid advance in the multi-core technology, the approach to grid computing has been changing and has evolved towards the right convergence among stability, effectiveness, easy management, convenient behavior and, above all, a better ratio between costs and benefits. This technology led the distributed computing world to a third generation of grids characterized by the increase of the intra site computing power, due to the multi-core technology, and the terabyte up to petabyte scale storage availability. This approach makes convenient the development of virtualization techniques and allows an effective exploitation of massive server factories. Advanced virtual machine software has a slight impact on performance, e.g. using techniques as paravirtualization (Youseff, Wolski, Gorda, & Krintz, 2006), and it permits to deploy, start, pause, move, stop and un-deploy virtual machines in a high performance, secure and collaborative environment. The software technologies underlying this grid approach, usually defined as cloud computing (Foster

R. Montella (✉) and G. Giunta
Department of Applied Science, University of Napoli Parthenope, Napoli, Italy
e-mails: {raffaele.montella; giulio.giunta}@uniparthenope.it

G. Laccetti
Department of Mathematics and Applications, University of Napoli Federico II, Napoli, Italy
e-mail: giuliano.laccetti@unina.it

B. Furht, A. Escalante (eds.), *Handbook of Cloud Computing*,
DOI 10.1007/978-1-4419-6524-0_20, © Springer Science+Business Media, LLC 2010

et al., 2006), are quite similar to those developed for the second generation of grids offering features in a software as service way. The true novelty is the dynamical deployment of full virtual machines packed around a software running in a local or remote site. The user asks the resource broker for a service. This service could be even not deployed, but available in a catalogue and then packed into the proper virtual machine and eventually made ready to use. Furthermore, the rapid decline of the cost of highly integrated clusters has spurred the emergence of the data center as the underlying platform for a growing class of data-intensive applications (Llorente, 2008a, 2008b).

In this scenario metadata augmented data, both stored or produced by an on line acquisition system, have a key role in overcoming what old fashioned computing grids and data grids were in the past. In particular, in a cloud environment the resource is seen as a high level entity, while brokering on application needs is performed automatically using a self describing service approach.

We developed a grid aware component based on a Resource Broker Service implementing a wrap over the OpenDAP (Open source Project for a Network Data Access Protocol) (Gallagher, Potter, & Sgouros, 2004) scientific data access protocol and ensuring effective and efficient content distribution on the grid. In a previous work (Montella, Giunta, & Riccio, 2007) we described the behavior of a Grads Data Distribution Service (GDDS) that could be considered the ancestor of the Five Dimension Data Distribution Service (FDDDS) presented in this work. This service relies on a different legacy OpenDAP engine and provides wider environmental data format capabilities and better performance.

Most of grid middleware implementations assume instruments as data sources. Instruments can be considered off-line and only processed data are usually published, leveraging on common storage facilities as Replica Location, Reliable File Transfer and GridFTP technologies. With this kind of approach, acquired data are concentrated, rather than distributed, with poor benefits in terms of dynamic load balanced access.

In the field of environmental data acquisition, the latter issue can be crucial when a large number of users require an access to the data during extreme weather events, such as hurricanes, flooding or natural disasters as volcanic eruptions, earthquakes and tsunamis.

The challenge of integrating instruments into a grid environment is strategically relevant and it can lead to a more efficient and effective use of the instruments themselves, a reduction of the general overhead and to an improvement of the throughputs. Usually instruments produce huge amount of data that could be stored elastically using cloudily available storage services, reducing management costs and charging users only for resources that they use. From this point of view, data produced by instruments, and stored on cloud computing services, can be advertised on a grid index serviced and provisioned on-demand using OpenDAP based services.

In the following we discuss the implementation and the integration in the Globus Toolkit 4 of FDDDS and we describe how FDDDS and the Resource Broker (RBS) services work. We also focus on the implementation of the mapping component among the resources exposed by GT4 index service and the Condor

ClassAd representation, and we provide some examples related to weather forecast model evaluation.

Our RB can answer to queries such as the following one (looking for the best 5D environmental dataset fitting specified requirements):

```
[
    Rank=other.ConnectionSpeed;
    Requirements=other.Type="dataset";
    other.Time>='03:07:2008 00:00:00' &&
    other.Time<='09:07:2008 00:00:00' &&
    other.Lat>=36 &&
    other.Lat<=42 &&
    other.Lon>=8 &&
    other.Lon<=20 &&
    other.Variables=u10m,v10m; &&
    other.DataOrigin=="wrf"
]
```

In this ClassAd data oriented query, the user is looking for a dataset containing the east-component and the north-component of the wind vector at 10m from the sea level (u10m and v10m, respectively). The requested data have to be produced by the Weather and Forecast model and related to the Southern Italy domain area, expressed as latitude and longitude ranges. The best dataset provider is selected by ranking the ConnectionSpeed of the hosts producing the web service. Therefore, if two or more data providers are discovered, the best performing one is selected. Finally, the user submits a specific request to FDDDS for sub-setting (i.e. extracting sub-cubes from data cubes) the desired data. Taking into account local data management policies, the sub-set data could be made available as advertised resource, distributed, replicated, cached or eventually even deleted if not used within a time threshold. In a grid based on cloud environment, FDDDS instances could be dynamically deployed in order to fit actual user needs.

If data are directly produced by instruments and stored using a locally provided or cloud available storage, the developed Instrument Service (InS) adverts data sources on the grid index service, so a ClassAd query as the following is possible:

```
[
    Type="DataConsumer";
    Rank=1/gis.getDistance(
          14.22,40.85,other.Longitude,other.Latitude
          );
    Requirements=
          other.Type=="Instrument" &&
          other.Desc=="WeatherStation" &&
          other.Sensor=="windDir,windSpeed"
]
```

In this query the user is looking for a weather station instrument provided by the data channel acquiring wind direction and speed as close as possible to the point located at longitude 14.22° and latitude 40.85°.

In Section 20.2, the design and the behavior of our RBS are briefly described. In Section 20.3 the implementation of GDDS and related issues of performance, flexibility and portability are addressed. In Section 20.4 the FDDDS architecture is introduced as evolution of previously developed systems. Section 20.5 describes the Instrument Service design and implementation. Section 20.6 describes an approach to weather forecast evaluation focusing on a novel algorithm based on time shifted ensembles. Section 20.7 is dedicated to the integration of data provided by FDDDS and data provided by InS. Finally Section 20.8 contains some conclusions and highlights on future work.

20.2 Resource Discovery and Selection Using a Resource Broker Service

The Resource Broker Service (Montella, 2007) is the key component in our grid web service infrastructure. It is based on a Latent Semantic Indices native matchmaking algorithm. We have integrated the resource broker service into the GT4 context, where each resource is published in the Index Service, identified by an EPR (End Point Reference) and selectable using standard GT4 command line tools, such as wsrf-query, and xpath queries.

In the RBS architecture the most important component is the collector process. The generic collector parses index service entry elements and stores them in a host (grid element) oriented format suitable for the native resource brokering algorithm. The collector is charged with managing in the most appropriate way different kinds of resources, such as the Default Index Service, the Managed Job Factory Service and the Reliable File Transfer Factory Service that are parsed and mapped to the Globus.Service.Index, Globus.Service.GRAM and Globus.Service.RTF properties respectively. The collector finds out properties and performs a mapping between one or more properties to new ones and then it stores the results in a local data structure. The collector is fully extensible and customizable using a documented API.

When the Resource Broker service is loaded into the GT4 Web Services container, an instance of the collector component is created and initialized performing queries to the VO main index service.

The RBS accepts queries in native notation; each selection criterion uses expressions like "equal", "different", "greater than", "greater or equal to", "less than", "less or equal to" and "max"/"min" to maximize or minimize a property and "dontcare" to ignore a pre-set condition. The broker returns a match by pointing the consumer directly to the selected resource with an End Point Reference (EPR). The selected resource is tagged as claimed to prevent another resource broker query from selecting the same resource, so that a potential overbooking is avoided. The resource remains claimed until a new update event occurs and the resource status reflects its actual behavior. Notification messages are filtered, to avoid a degradation in performance due to the data renewal event management. In order to automatically control the resource lifetime, the effective update availability is not persistent, but it is renewed whenever a status change event exceeds a given threshold.

20.3 Anagram Based GrADS Data Distribution Service

OpenDAP is a community initiative with the goal to create an open-source solution for serving distributed scientific data and for allowing ocean and atmosphere researchers to access environmental data anywhere on the Internet from a wide variety of new and existing programs. The OpenDAP project can capitalize on years of development of data analysis and display packages that use those APIs, allowing users to continue to use programs which they are already familiar with and to develop network versions of commonly used data access Application Program Interface (API) libraries, such as NetCDF, HDF, JGOFS, and others. The OpenDAP architecture relies on a client/server model, with a client that sends requests for data out onto the network to some server answering with the requested data. This is exactly the model used by the world wide web, where clients submit their requests to web servers for data that make up web pages, even though there is still lack of support to the web service technology and to computational data grids. We previously developed a GT4 based web service in order to provide a web grid service fashioned access to environmental data, leveraging on grid features as the Grid Security Infrastructure and the Reliable File Transfer for secure and high performance data management. Our initial design was focused on serving data provided by the Grid Analysis and Display System (GrADS) (Doty & Kinter, 1995). This software tool is a free and open source interactive desktop tool, developed in ANSI C and ported on different platforms and operating systems, that is widely used for its easy access, manipulation, and visualization of earth science data stored in various formats as binary, GRIB, NetCDF, or HDF-SDS (Scientific Data Sets). GrADS uses a 5-Dimension data environment: longitude, latitude, vertical level, time and parameters (variables). A plain text descriptor file acts as metadata for both station and gridded data distributed over regular, non-linearly spaced, Gaussian, or variable resolution grids. Different datasets may be integrated and graphically overlaid, with their correct spatial and time registration, through a large variety of graphical techniques and, moreover, the output can be exported in either postscript or image formats.

The GrADS-DODS Server (GDS) combines GrADS and OpenDAP to create an open-source solution for serving distributed scientific data (Wielgosz & Doty, 2003). The GDS provides a wide range of clients with remote dataset access via the OpenDAP protocol and with some analysis tool which has been OpenDAP enabled. Metadata and subsets are retrieved transparently from the server as needed. Data extraction and sub-setting are the main features of GDS, but GDS also provides powerful and multi-stage complex analysis tools, using GrADS software as a computing engine component. In order to reuse as much as possible of the standard GDS distribution, we developed an adapting framework over the Anagram engine, a software component which GDS (Wielgosz, 2004) is based on. Our GDDS implements a "same binary" conservative approach along the way we followed in the grid enabling process of a set of environmental models, as for example the Weather Research and Forecast model (WRF) (Giunta, Laccetti, & Montella, 2008). Since the Anagram engine heavily builds on module late binding at runtime and a deep integration of the servlet management into the service implementation, we have developed an adapter framework in order to obtain a better behavior when using it as a component, without modifying the legacy distribution. This solution inherits from GDS the need to invoke an external process, i. e. the GrADS executable, for each data access. Translated in the grid enabled version, this means that at every invocation of the operation provider, the host machine has to configure a scratch directory where to work with a local instance of GrADS. This causes a loss of performance, especially in stressfully working conditions in which many consumers ask for data to be extracted by huge dataset, and a lack of portability because of the need to recompile the GrADS source for specific operating system and architecture. Moreover, the number of data types accessible by GrADS is limited. For instance, only few kinds of NetCDF files (following restricted conventions) can be directly accessed (Fig. 20.1).

20.4 Hyrax Based Five Dimension Distribution Data Service

The Five Dimensional Data Distribution Service relies on the Hyrax OpenDAP server (Gallagher, Potter, West, Garcia, & Fox, 2006). Hyrax is a data server which combines the efforts at UCAR/HAO to build a high performance DAP-compliant data server for the Earth System Grid II project with existing software developed by OpenDAP. This server is based on the Java servlet mechanism to manage web requests from DAP format-specific software. This approach improves the performance for small sized data requests. The servlet front end, instead of launching an external local process as in GDS, looks at each request and formulates a query to a second server, Back End Server (BES), which may be or may be not on the same machine. BES handles the reading of the selected data from the data storage and returns DAP-compliant responses to the front end, which, in turn, could send these responses back to the requestor or might process them to build more complex responses. We notice that this architecture makes possible a better integration in a web grid service ecosystem. In implementing FDDDS we had to extend the client OpenDAP class APIs in order to completely disjoin them from the front end and to provide the needed interfaces to the grid.

We have completely integrated the Back End Service into the web service environment, by reusing any kind of configuration files within the GT4 service architecture. In this way the migration from the classic client server OpenDAP protocol based architecture to the service oriented grid environment application occurs in a really soft fashion.

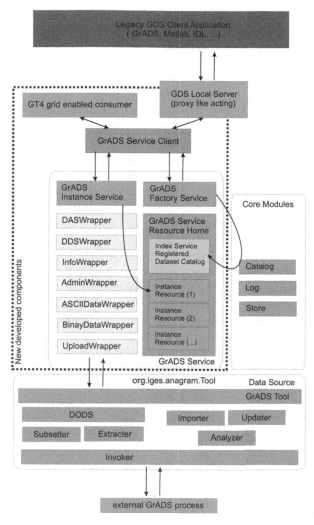

Fig. 20.1 The GrADS Service block diagram

FDDDS has a self decorating dataset metadata which provide a grid oriented data representation and enforce the interaction and the collaboration among different and geographically spread components, such as instruments, and resource brokers. The native BES provides security tools based on X509 certificates. In our implementation we relied on this feature to ensure the access to the BES only from the local host machine and from FDDDS, delegating to this component and to the underlying GSI the task of authentication, authorization and encryption. This allows that only certificates that have been authenticated by grid users can perform OpenDAP operations including new dataset uploading and administration management.

When FDDDS is loaded into the container, it contacts the BES for retrieving the dataset list. For each dataset a web service resource is created, with the properties extracted from the dataset metadata and then published on a specified Index Service. Then the RBS collector component can carry out a ClassAd representation of the resource involved in the matchmaking process. If a suitable dataset resource is found, the resource broker returns the EPR referencing to the best FDDDS service instance, which is selected taking into account the network behavior between the consumer and the producer endpoints. Then the consumer invokes the getData, getDAS or getDDX to retrieve dataset metadata while getData retrieves data in binary format. The getData operation provider accepts data selection and subsetting parameters via a standard OpenDAP query string specifying the variable name, the running dimensions and the step intervals. In the same way, the getData operation provider allows to perform a sort of remote analysis, using the BES engine and specifying datasets and expressions, as well as to choose the (to be returned) data format (DAP or NetCDF).

FDDDS uses SOAP messages only for an operation provider invocation and for returning computed results, while for transferring data a dedicated transfer service is invoked. This prevents a decrease of performance that can be caused by the transport protocol, especially when the requested data size is very large, as in the case of wide space and temporal domains with many variables. The getData operation provider returns an EPR to the resource associated with the result of the sub-setting and allocated in a temporary storage area. It follows that efficient grid oriented file transfer protocols, as GridFTP, can be used to improve the performance of the distribution system. The sub-set data can be made available for request by advertising them on the Index service. Tracking the log for requested data, the RB notifies FDDDS to distribute a popular in queries sub-set dataset over the grid. FDDDS has in charge the management of such a dynamically created dataset, empting caches and destroying it if needed.

The use of the OpenDAP protocol is very common in the environmental scientists community because of a great amount of compliant applications. FDDDS precludes a direct use of distributed data unless upgraded versions of such applications would provide a full WSRF compliant grid web service access. However, this can be achieved through the use of a local server, based on the original Hyrax front end acting as a proxy interface between OpenDAP server compliant applications (such as IDL, Matlab, GrADS, Ferret and more) and the grid world, represented by the resource broker service, the FDDDS service and the Grid Security Infrastructure.

This proxy-like acting local server accepts connections only from the loop-back network interface and translates the standard OpenDAP query string into a direct or resource brokered grid interaction. The local OpenDAP server implements an FDDDS client which handles certificates, EPR and resource broker requests in a fully transparent way (Fig. 20.2).

In a performance analysis, we compared the Grads Data Distribution Service with the Five Dimension Data Distribution Service by setting up a testbed of real data produced by weather forecast operational runs.

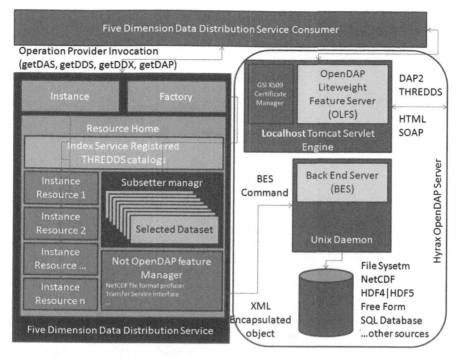

Fig. 20.2 The five dimension data distribution service block diagram

In the current setup, each WRF model run produces a 11 gigabytes data file storing the coarse domain forecast results. The size of the data set on the fine domain amounts to 2 gigabytes for each model's run. The coarse domain output file stores a five-dimension data set, comprised of, 19 vertical levels, 655×594 cells, 27 2D and 3D variables, and 144 time steps.

In order to test the services, a getData operation provider is invoked by requesting the u10m variable for all time steps (144, six days, one time step per hour) and subsetting an area of 2^n square cells with $3 \leq n \leq 9$. The invocation time includes the subsetting and the data transfer via gridFTP. In this conditions, for small sized subsets the performance of both services are comparable, while for more demanding subsetting tasks the FDDDS shows a higher efficiency (Fig. 20.3).

20.5 Design and Implementation of an Instrument Service for NetCDF Data Acquisition

Our ultimate goal is the instruments sharing on computing grids. In order to abstract different kinds of instruments (such as weather stations, surface current radars and wind profilers), with a wide variety of hardware interfaces and acquisition data

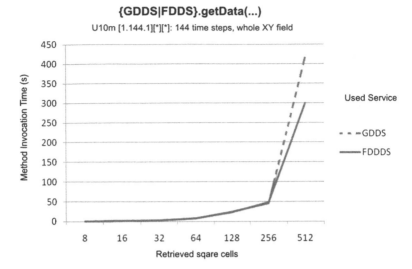

Fig. 20.3 Performance analysis results

rates, a suitable design leveraging on a plug-in based framework is needed. We have developed an Abstract Instrument Framework (AIF) to decouple the acquisition Java interface from the grid middleware technology (Montella, Agrillo, & Di Lauro, 2008) and to introduce a streaming protocol after the SOAP interaction started the communication (Fig. 20.4). Environmental data acquisition instruments

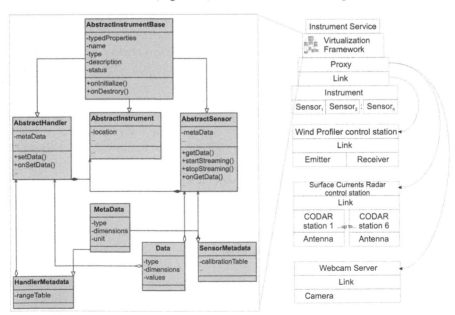

Fig. 20.4 The Abstract Instrument Framework class diagram and the instrument service architecture

interact with their proxy hardware in different ways. Weather station data loggers could work in real-time or in batch mode depending on the type of data link. Wind profilers and surface currents radars operate in a similar way using a power workstation as a proxy machine. Wind profile equipment workstations are a sort of a very powerful data logger in which vertical profile integrated values are stored. The sea surface current instruments use workstations acting as an automatic integration and post processing computing node producing surface time integrated values. We used a layered approach relying on virtualization. The instrument framework defines the behavior of a generic instrument built assembling sensors for data acquisition and actuators for instrument handling leveraging on the component abstraction technique.

Each abstract component leaves unimplemented each low level hardware interaction with the instrument and it provides tools for data delivery as SOAP typed values or streaming protocols.

Weather stations are usually equipped with an autonomous simple data logger with remote interaction capabilities for instrument setup, sensor calibration and data retrieving in both batch and real-time mode. We operate different kind of weather stations built on different sensors and data loggers. Weather stations fit straightforwardly in this framework model, because they are a collection of sensors. Data channels may be operated in two ways, depending on the hardware sensor and the required measurement. The real sensor components implement the real data access either with a real-time request or retrieving them form a data logger which specifies the type and the nature of the acquired data.

Typically sea surface currents radar control stations are powerful workstations where all operations on data and sensors (remote antennas) are managed by a proprietary closed source software. In this case there is no direct access to the sensor, but only to the automatically post-processed data. The sea surface currents radar sensor wraps provide surface data averaged in time e localized on a geo-referenced matrix. The virtualization is obtained through interfacing real sensor components related to sea current speed and direction directly with data produced by the standard equipment. The wind profile instrument works in a very similar manner and the profiled data, averaged in time and distributed along the vertical axis, are stored by the proxy workstation acting as an enhanced data logger which provides data transfer tools based on standard protocols.

The Instrument Service exposes operation providers to start and stop data streaming using a specified protocol (published, for each sensor, on the index service as a metadata field) with a protocol plug in interface designed to easily customize the data distribution. In this case the standard SOAP message exchanging between the web service consumer and the web service producer is performed only to control data transmission, while the actual data transfer is carried out using a more efficient protocol.

The Instrument Service automatically publishes Instruments, Sensors and Handlers metadata – and even data – on the GT4 Index Service, so that a grid service consumer could leverage on the WSRF Notification feature to interact with a data acquisition device. RBS component interacts with the Index Service and allows any instrument shared on the grid to participate in the dynamic resource allocation process. An instrument, as any grid shared data source FDDDS, is discoverable and selectable via ClassAd queries.

A matchmaker algorithm can match ClassAds to different kinds of entities in a specified manner: The expression other.Type=Instrument is evaluated as true if the ClassAd which is matched with contains a property named Type with the string value equal to Instrument. In the matchmaking protocol, two ClassAds match if each one has an attribute requirement that evaluates to true in the context of the other ClassAd; then matched grid elements activate a separate claiming protocol to confirm the match and to establish the allocation of resources.

The representation of an instrument as a grid resource is automatically done by RBS and considered as a sort of dynamically varying stored data. The following example shows the ClassAd representation of a weather station:

```
[
    Type="Instrument";
    Desc="WeatherStation";
    Area="Napoli";
    Longitude="14.32";
    Latitude="40.50";
    TimeStamp="23-12-2007 03:30:00";
    Sensor="windDir,windSpeed,airTemp,airPressure";
    Values="335,3.280,14,1013";
    Units="°N,ms-1,°K,HPa";
    Rank=1;
    Requirements=other.Type=="DataConsumer"
]
```

This instrument is defined as a weather station resource in the Napoli area located at longitude 14.32°, latitude 40.50°. The Instrument Service framework publishes the Sensor metadata, then it collects the instrument features and finally publishes on the grid index service the instrument geographical position, the data currently acquired by each sensor and the actuators feedback parameters.

20.6 A Weather Forecast Quality Evaluation Scenario

The assessment of the quality of an operational weather forecasting model is a crucial process for stating the reliability of the overall operational forecasting system.

A forecast is an estimate of the future state of the atmosphere. It is created by estimating the current state of the atmosphere using observations, and then calculating how this state will evolve in time by means of a numerical weather prediction computer model. As the atmosphere is a chaotic system, very small errors in its initial state can lead to large errors in the forecast. This means that one cannot create a perfect forecast system because it is not possible to observe every detail of the atmosphere's initial state. Tiny errors in the initial state will be amplified, so there is always a limit to how far ahead we can predict any detail.

In a weather forecast operational scenario, the ensemble methodology is a common technique for the assessment of the model behavior by means of a sensitivity analysis on initial and boundary conditions. To test how small discrepancies in the initial conditions may affect the outcome of the forecast, an ensemble system can be used to produce many forecasts. The complete set of forecasts is referred to as the ensemble, and any individual forecast as an ensemble member.

The ensemble forecasts give a much better idea on which weather events may occur at a particular time. By comparing the ensemble members the forecaster can decide how likely a particular weather event may be. Shortly, if the forecasts exhibit a large divergence, then one may conclude that there is a large uncertainty about the future weather, while if most of the forecasts are similar then much confidence can be taken in predicting a particular event.

Ensemble modeling requires high computational power since a large number of simulations must be run concurrently, so that grid computing technology is a key and widely adopted solution (Ramakrishnan et al., 2006).

Technical issues related to our grid application for operational weather forecasts are discussed in (Ascione, Giunta, Montella, Mariani, & Riccio, 2006). Briefly, a Job Flow Scheduler Service (JFSS) orchestrates the grid application behavior; it relies on a Job Flow Description Language which specifies the relations among web service consumers and uses the RBS for dynamic resource allocation, data and instrument discovery and selection. With the Five Dimension Data Distribution Service, we set up a grid application which integrates a grid enabled WRF weather model with a set of grid based validating tools leveraging on FDDDS in a operational scenario.

We have developed a weather forecast quality control and validation algorithm aimed at saving computing time and avoiding multiple model runs, as in a standard ensemble simulation. The algorithm uses an approach we called "Time Shift Ensemble". The WRF model is run each day, with initial and boundary conditions provided by an NCEP global model. The output of a daily simulation is a forecast on the next six simulated days. Thus the first simulated day (hours ranging from 0 to 23) has a set of six forecasted data: the 0 h..23 h dataset generated by the current day run; the 24 h..47 h dataset returned by the previous day run and so on, up to the dataset produced by the run of 6th previous day.

We have considered the Geo-potential height (GpH) computed by the model at 500 HPa level as a good atmosphere status descriptor parameter, as usual in the meteorology community (Warren, 2007). The GpH is a vertical "gravity-adjusted" coordinate that corrects the geometric height (elevation above mean sea level) using the variation of gravity with latitude and elevation. GpH is usually referred to a pressure level, which would correspond to the GpH necessary to reach the given pressure.

At each point of the spatial domain, we compute the difference between the hourly forecasted GpH produced by the current run and the GpH values produced by each of the five simulations that we ran at the previous days. These set of values can be considered as the discrepancies of the results of an ensemble built on 6 model's runs with different initial/boundary conditions. The quality of the forecast can be deduced from the behavior of those time series (Montella, Agrillo, Mastrangelo, & Menna, 2008). To be more precise, let $F_{s,f}$ denote the GpH forecast produced by the run at day s ($s = 0$ for the current day, $s = -1$ the previous one, and so on) for the forecasted day f ($f = 0$ denotes the first day of the simulation and $f = 5$ the last simulated day, with a total simulated time equal to 144 h). The current day is forecasted for the first time as the last day of the run executed at the day -5 ($F_{-5,5}$), the second time as $F_{-4,4}$ up to the simulation $F_{-1,1}$ (Fig. 20.5).

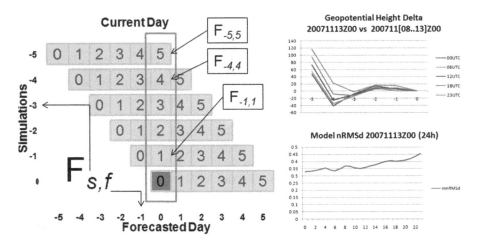

Fig. 20.5 The validation algorithm and some results

Figure 20.5 (right-up) also reports the behavior of the five time series of the computed discrepancies (for a given simulated day and at a specific spatial point) in the case of a reliable simulation, while the right-down side of Fig. 20.5 shows the related time series of normalized Root Mean Square error (nRMSd) between the $F_{-5,.-1}$ and the current day run's values F_0.

20.7 Implementation of the Grid Application

Here we give some insight into relevant implementation details of our grid application. The application workflow starts with an invocation to FDDDS which collects initial and boundary conditions needed for the initialization of WRF (Fig. 20.6).

Data are downloaded from the NCEP ftp server using an automatic process. Since forecasts have to be computed each day on the same domain, another invocation of FDDDS allows to download terrain data previously produced by the GeoGrid

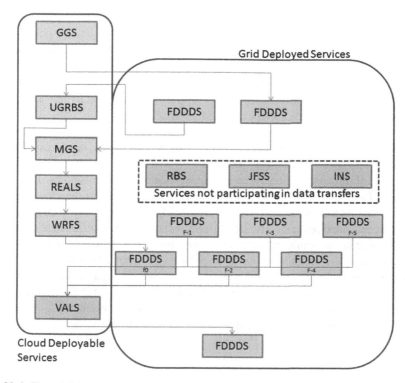

Fig. 20.6 The grid/cloud application schema

Service (GGS), which is a wrap over the WRF geogrid module. Then the Ungrib Service (UGRBS) is invoked to convert initial and boundary conditions in a file format suitable for the metgrid WRF module. The process continues with a call to the MetGrid Service (MGS) which interpolates initial and boundary conditions over the domain and, finally, to the Real Service (RS), that is wrapped on the Real WRF module and that prepares data for the WRF Service, a service shell over the WRF main module. The WRF main model produces results that are stored using FDDDS. This component makes the produced data available on the grid, either using the RBS or directly through the EPR. To perform our validation tests we set up an application with only one domain covering all Europe and part of northern Africa, with 256x120 square cells (30×30 Km^2) centered at longitude 14.22° and latitude 40.85°.

The validation process starts collecting pre-processed data related to the same current simulated day and it invokes RBS and FDDDS to get information on where the data are available. These data are used by the Validation Service (VALS), that wraps around a tool implementing the evaluation algorithm described in the previous section. The VALS component returns a map of the Geopotential Height nRMSd and stores the computed data in a FDDDS service selected via RBS.

Figure 20.6 shows the application architecture from a deployment point of view. The services classified as deployable on a cloud computing infrastructure are the ones characterized by large need of computing power. These services can be conveniently managed by a batch execution using virtual machines (Sotomayor, Keahey, & Foster, 2008), where the computing power, required by the web services underlying the application modules of the grid application (WRF: GGS, UGRBS, REALS, WRFS; Evaluation: VALS, METS), is dynamically allocated on demand and eventually released. We deployed the computing power demanding services on a public infrastructure as a service oriented computing cloud.

The services classified as deployed on the grid are data oriented. In this case the need for the elasticity provided by the cloud infrastructure is addressed to the computing intensive part of the application.

The use of the cloud infrastructure is convenient because resources, i.e. virtual machines, have to be instanced only for the needed time. Using the cloud for data storing is technically possible, but some drawbacks can raise when a large amount of data are allocated permanently in the cloud without the need of storage elasticity; we notice that in our application the needed amount of storage could be estimated in advance. The evaluation algorithm requires 6 files of about 11 gigabyte each. After any run of the model, the oldest dataset is deleted and the last produced dataset is added to the storage. However, the application requirements for the storage result in a total cost of ownership which is less critical than the requirement for computing power, so that a self hosted storage turns out to be more effective than a cloud storage solution.

20.8 Conclusions and Future Work

We have described some recent results in distributing 5D environmental data using GT4 WSRF compliant tools in a high performance grid environment. The application has been deployed in an Infrastructure as a Service public cloud. We have briefly discussed how the computing and storage resources can be made available on a computing grid or dynamically allocated on a computing cloud. In particular, the application we discussed here has a critical issue in computing power needs, so that the WRF related services can be deployed in the cloud. On the other hand, storage does not call for an elastic solution and hosting data on the cloud appeared unsuitable.

We have shown that a suitable integration of legacy software, i.e. the Back End Server of the Hyrax OpenDap server suite and the gridFtp parallel file transfer protocol, can improve the operational throughput and can allow a useful, on demand data sub-setting and data collecting process. The Five Dimension Distribution Service relies on a previously designed resource broker collector and matchmaking algorithm, which use the grid ClassAd notation automatically extracted from metadata. This approach also opens a wide range of applications in the field of the environmental model validation, where issues like component integration and security are crucial since a validation algorithm must involve several community tools in order to achieve scientific relevance.

We are aware that a more extensive performance analysis of the application has to be performed, for instance regarding the cache behavior in a real grid environment and the effect on the overall performance of the FDDDS dynamical deployment on a cloud system. Our next goal is to develop a GT4 WSRF compliant web service, wrapping the MET (Holland, Gotway, Brown, & Bullock) model evaluation tool. The MET offers grid-to-point, grid-to-grid and advanced spatial forecast verification techniques in an unified and modular toolkit, that builds on capabilities available in other verification systems. Tools provided by MET can be grouped by function to describe the overall structure of the MET: data handling, statistical calculations and data analysis. This architecture is suitable for our web service infrastructure in which grid applications are straightforwardly arranged using the JFDL or other similar tools. The idea is to build up a double evaluation system based on VALS and on the service wrapped over MET, which can be able to operate with observed data either downloaded from distribution services or directly sampled via InS, and with data stored using FDDDS.

In an operational scenario, such as that one producing the results shown in Fig. 20.7, the model evaluation has to be performed automatically in order to provide thematic maps about the forecast's confidence. This result could be achieved using machine learning algorithms, which can return decision support rules. The choice of the most suitable classification algorithm and the related issues of its tuning and assessment will be addressed in a future work.

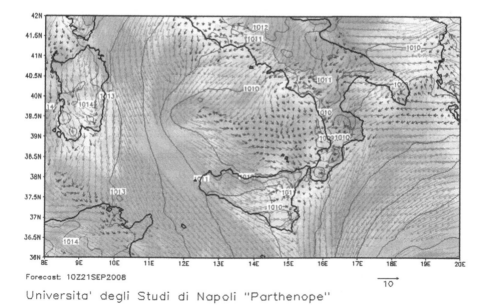

Fig. 20.7 Weather forecast on a Southern Italy domain

References

Ascione, I., Giunta, G., Montella, R., Mariani, P., & Riccio, A. (2006). A grid computing based virtual laboratory for environmental simulations. *Proceedings of 12th International Euro-Par 2006, Dresden, Germany*, August/September 2006. LNCS 4128, Springer 2006.

Doty, B. E., & Kinter, J. L., III (1995). Geophysical data analysis and visualization using GrADS. In E. P. Szuszczewicz & J. H. Bredekamp (Eds.), *Visualization techniques in space and atmospheric sciences* (pp. 209–219). Washington, DC: NASA.

Foster, I., Freeman, T., Keahey, K., Scheftner, D., Sotomayor, B., & Zhang, X. (May 2006). *Virtual clusters for grid communities*. Singapore: CCGRID.

Gallagher, J., Potter, N., & Sgouros, T. (2004). *DAP data model specification DRAFT*. Retrieved November 6, 2004, Rev., 1.68, from www.opendap.org.

Gallagher, J., Potter, N., West, P., Garcia, J., & Fox, P. (2006). OPeNDAP's Server4: Building a high performance data server for the DAP using existing software. *AGU Meeting in San Francisco, San Francisco, CA.*

Giunta, G., Laccetti, G., & Montella, R. (2008). A grid-based service oriented environmental modeling laboratory for research and production applications. In R. Wyrzykowski, J. Dongarra, K. Karczewski, & J. Wasniewski (Eds.), *Chapter of Parallel Processing and Applied Mathematics 2007* (pp. 951–960). Lecture Notes in Computer Science n. 4967. New York, NY: Springer.

Holland, L., Gotway, J. H., Brown, B., & Bullock, R. A. *Toolkit for model evaluation* (National Center for Atmospheric Research Boulder, CO 80307).

Llorente, I. M. (July 2008). Towards a new model for the infrastructure grid. *Panel From Grids to Cloud Services in the International Advanced Research Workshop on High Performance Computing and Grids, Cetraro, Italy.*

Llorente, I. M. (July 2008). Cloud computing for on-demand resource provisioning. *International Advanced Research Workshop on High Performance Computing and Grids, Cetraro, Italy.*

Montella, R. (May 2007). Development of a GT4-based resource broker service: An application to on-demand weather and marine forecasting (Vol. LNCS 4459 of LNCS). New York, NY: Springer.

Montella, R., Giunta, G., & Riccio, A. (June 2007). Using grid computing based components in on demand environmental data delivery. *ACM Proceedings About Upgrade Content Network HPDC2008 Workshop. Monterey Bay, CA, USA.*

Montella, R., Agrillo, G., & Di Lauro, R. (April 2008a). *Abstract instrument framework: Java interface for instrument abstraction* (DSA Tech. Rep.) Napoli.

Montella, R., Agrillo, G., Mastrangelo, D., & Menna, M. (2008b). A globus toolkit 4 based instrument service for environmental data acquisition and distribution. *Proceedings of Upgrade Content Workshop HPDC2008, Boston, MA.*

Ramakrishnan, L., Blanton, B., Lander, H., Luettich, R., Reed, D., & Thorpe, S. (2006). Real-time storm surge ensemble modeling in a grid environment.

Sotomayor, B., Keahey, K., & Foster, I. (June 2008). Combining batch execution and leasing using virtual machines. *ACM/IEEE International Symposium on High Performance Distributed Computing 2008 (HPDC 2008), Boston, MA.*

Warren, R. (2007). Development and illustrative outputs of the community integrated assessment system (cias), a multi-institutional modular integrated assessment approach for modelling climate change. *Environmental Modelling and Software, 20*, 1–19.

Wielgosz, J., & Doty, J. A. B. (2003). The grads-dods server: An open-source tool for distributed data access and analysis.

Wielgosz, J. (2004). Anagram: A modular java framework for high-performance scientific data servers.

Youseff, L., Wolski, R., Gorda, B., & Krintz, C. (December 2006). Paravirtualization for HPC systems XHPC. *Workshop on XEN in High-Performance Cluster and Grid Computing.*

Chapter 21
HPC on Competitive Cloud Resources

Paolo Bientinesi, Roman Iakymchuk, and Jeff Napper

Abstract Computing as a utility has reached the mainstream. Scientists can now easily rent time on large commercial clusters that can be expanded and reduced on-demand in real-time. However, current commercial cloud computing performance falls short of systems specifically designed for scientific applications. Scientific computing needs are quite different from those of the web applications that have been the focus of cloud computing vendors. In this chapter we demonstrate through empirical evaluation the computational efficiency of high-performance numerical applications in a commercial cloud environment when resources are shared under high contention. Using the Linpack benchmark as a case study, we show that cache utilization becomes highly unpredictable and similarly affects computation time. For some problems, not only is it more efficient to underutilize resources, but the solution can be reached sooner in realtime (wall-time). We also show that the smallest, cheapest (64-bit) instance on the studied environment is the best for price to performance ration. In light of the high-contention we witness, we believe that alternative definitions of efficiency for commercial cloud environments should be introduced where strong performance guarantees do not exist. Concepts like *average, expected performance* and execution time, expected cost to completion, and variance measures—traditionally ignored in the high-performance computing context—now should complement or even substitute the standard definitions of efficiency.

21.1 Introduction

The *cloud computing* model emphasizes the ability to scale compute resources on demand. The advantages for users are numerous. Unlike conventional cluster

P. Bientinesi (✉) and R. Iakymchuk
AICES, RWTH, Aachen, Germany
e-mails: {pauldj; iakymchuk}@aices.rwth-aachen.de

J. Napper
Vrije Universiteit, Amsterdam, Netherlands
e-mail: jnapper@cs.vu.nl

B. Furht, A. Escalante (eds.), *Handbook of Cloud Computing*,
DOI 10.1007/978-1-4419-6524-0_21, © Springer Science+Business Media, LLC 2010

systems, there is no significant upfront monetary or time investment in infrastructure or people and ongoing expenses are simplified. When resources are not in use, total cost can be close to zero. Instead of allocating resources according to average or peak load, the cloud user can pay costs directly proportional to current need. Individuals can quickly create and scale-up a custom compute cluster, paying only for sporadic usage. However, there are also some disadvantages. Costs can be divided into different categories that are billed separately: for example, network, storage, and CPU usage. This model can be complex when attempting to minimize costs (Strebel & Stage, 2010). Further, the setup time for computation resources is currently quite long (on the order of minutes for Amazon), and the granularity for billing CPU resources is coarse: by the hour. These two factors imply that resources should be very conservatively scaled in current clouds, reducing some of the benefits of scaling on demand. Finally, in many cloud environments, physical resources are shared among virtual nodes belonging to different users, which can negatively impact performance.

The ability to allocate on demand nodes in current commercial cloud environments implies a new problem in scientific computing: contention. Good high-performance computing clusters balance the CPU, memory, and network usage of applications to maintain efficiency while scaling up resource usage but assume exclusive access to these resources by an application. For example, the RoadRunner supercomputer at Los Alamos National Laboratory can solve a linear system using 100 K cores while maintaining 90% efficiency of CPU usage. This feat requires carefully balancing the application's needs across CPU, memory, and the network. With exclusive access, developers of applications and compute libraries can achieve such balanced computing by reducing the noise in resource usage that prevents efficient synchronization. For example, one node out of a large system might run extra software for monitoring that causes it to reach the end of a computation slower than the other nodes. If all nodes must synchronize at the end of a computation, all other nodes in the system must wait for the slowest node to finish. Small amounts of noise can significantly affect the overall performance of a tightly coupled application (Petrini, Kerbyson, & Pakin, 2003).

Although cloud environments typically provide a small degree of resource availability guarantees, they do not provide a complete facsimile of an unshared virtual node or strong performance guarantees. Contention is visible along almost all resources: memory bandwidth, last-level cache space, network bandwidth and latency (Wang & Ng, 2010), and even CPU time. In addition to introducing the kind of noise discussed above, such contention can severely limit the use of resources shared between virtual environments. For example, memory contention can increase the compute time of a simple matrix multiplication by an order of magnitude.

To run scientific applications efficiently, cloud computing needs to provide resources to scientific applications comparable to current well-balanced, exclusive-access high-performance computing (HPC) systems. Isolation between different virtual machines in the cloud should provide predictable performance for developers of compute-intensive applications and libraries. However, our evaluation of the Amazon EC2 cloud—currently the largest commercial cloud environment—demonstrates that significant work is still needed to isolate virtual machines in the cloud or to adjust HPC libraries to adapt to the dynamic, contentious environment of the cloud. The Amazon EC2 cloud system was built for web service workloads (that is, without extremely fast interconnects) and cannot compare favorably with purpose-built, heavily-engineered HPC supercomputers for many tightly-coupled scientific computations such as dense linear algebra. Although we do not expect comparable performance, the dynamic scalability of the environments holds some promise for smaller workloads.

In order to determine the achievable efficiency of current cloud systems for HPC, we consider the execution of dense linear algebra, which provides a favorable ratio of computation to communication: $O(n^3)$ operations on $O(n^2)$ data. Dense linear algebra algorithms can thus overlap the slow communication of data with quick computations over much more data (Dongarra, Van de Geijn, & Walker, 1994). Cloud systems will generally be limited by slow communication more than specialized HPC systems given their relatively cheap, slow interconnects. We are concerned primarily with the efficiency of shared cloud systems as they scale up and show that the effects of contention can far outweigh (at least in current offerings) the imbalance caused by an underprovisioned interconnect. We thus focus on CPU costs, ignoring possible associated (and significantly less per computation) storage and networking costs.

This chapter empirically explores computational efficiency in a cloud environment when resources are shared under high contention. We consider single node performance of dense linear algebra operations while nodes are shared with unknown other applications on Amazon EC2 (Services, 2010). Further, we analyze how the performance changes on clusters of up to 64 compute cores. Our results show that the performance of single nodes available on EC2 can be as good as nodes found in current HPC systems (TOP500.org, 2010), but the average performance is much worse and shows high variability. In fact, dense linear algebra algorithms do not appear to scale well on cloud systems with high levels of contention. Contention skews the balance of node's CPU and memory resource availability to network capacity to prevent efficient usage of the resources. Cache utilization becomes highly unpredictable and similarly affects computation time. We show that for some problems, not only is it more efficient to underutilize CPU resources, but the solution can even be reached sooner in realtime (wall-time).

In light of the high-contention we witness, we believe that alternative definitions of efficiency for cloud environments should be introduced where strong performance guarantees do not exist. Concepts like *average expected performance* and execution time, expected cost to completion, and variance measures—traditionally ignored in the HPC context—now should complement or even substitute the standard definitions of efficiency. In addition to standard metrics such as GFLOP and efficiency used in HPC, we introduce $/GFLOP(dollars per billions of floating point operations) to analyze in depth the pros and cons of clouds. The $/GFLOP metric allows users to estimate straightforward costs—currently limited to CPU usage—for different applications with respect to computational efficiency.

21.2 Related Work

Cloud computing has been proposed as a service to scale out or share existing scientific clusters. The Eucalyptus project provides an open source cloud management tool (Nurmi et al., 2008). Similarly, the Nimbus project provides cloud management tools for clusters (Keahey, Freeman, Lawret, & Olson, 2007). These tools do not typically provide for multi-tenancy—unlike the commercial cloud environments we study—and thus observe different performance characteristics closer to existing scientific clusters with exclusive access. We are concerned in this chapter with the effects of contention on HPC applications when multiple VMs share resources.

Tikotekar et al. run different HPC benchmarks in virtual machines to determine the effects of virtualization on high-performance computing applications (Tikotekar et al., 2009). The results show isolating the effects of multiple VMs on the same machine is difficult, but in general virtualization has little effect on the performance of compute intensive HPC applications. Similarly, Youseff et al. see little performance impact on the memory hierarchy behavior of linear algebra libraries under virtualization (Youseff, Seymour, You, Dongarra, & Wolski, 2008). However, this study does not consider contention between multi-tenant VMs.

The impact of virtualization on the networking performance of EC2 is analyzed empirically by Wang and Ng (2010). They report heavy network instabilities and delays, especially on small 32-bit nodes. We observe poor performance on the HPL parallel benchmark using multiple instances but are more concerned with the scaling properties. Although the instances do not approach peak performance, the parallel job can scale (for tens of nodes) with poor network performance provided the instances have enough installed RAM due to the high ratio of computation to communication in dense linear algebra. Our experiments (see Fig. 21.8) demonstrate this effect.

The effects of sharing the last-level cache in multi-core processors is discussed by Iyer et al. (2007). The paper discusses different approaches to improving the performance guarantees of each core with respect to cache behavior. Using these techniques could greatly increase the predictability of performance on cloud platforms.

Previously, Edward Walker has compared EC2 nodes to current HPC systems (Walker, 2008). Our results here are similar to his for the small clusters of 4 nodes that he used. We previously reported preliminary results in (Napper & Bientinesi, 2009).

21.3 Background

We perform our experiments on the Amazon Elastic Compute Cloud (EC2) service as a case study for commercial cloud environments (Services, 2010). Although there are competing cloud offerings that were publicly available at the time (Xcalibre, 2010; Hosting, 2010), Amazon's service is the largest that provides highly configurable virtual machines. The nodes allocated by EC2 run a kernel or operating system configured by Amazon, but all software above this level is configured by the user. Many other cloud offerings by other providers limit applications to certain APIs or languages. To use existing highly optimized dense linear algebra libraries, we use Amazon as a case study.

Nodes allocated through EC2 are called *instances*. Instances are allocated from Amazon's data centers according to unpublished scheduling algorithms. Allocations are initially limited to 20 total instances, but this restriction can be lifted upon request. Data centers are combined into entities known as an *availability zone*, Amazon's smallest logical geographic entity for allocation. These zones are further combined into regions, which consist of only the US and Europe at the moment.

After allocation, each instance automatically loads a user-specified *image* containing the proper operating system (in our case Linux) and user software (described below). Images are loaded automatically by Amazon services onto one or more virtualized processors using the Xen virtual machine (VM) (Barham et al., 2003). Each processor is itself multi-core, resulting in a total of 2–8 virtual cores for the instances we reserved. The Terms of Service provided by Amazon do not provide strong performance guarantees. Most importantly for this study, they do not limit Amazon's ability to implement *multi-tenancy;* that is, to co-locate VMs from different customers. We discuss the performance characteristics of different instances below.

Tools written to Amazon's public APIs provide the abilities to allocate extra nodes on demand, release unused nodes, and create and destroy images to be loaded onto allocated instances. Using these tools and developing our own, we built images with the latest compilers provided by the hardware CPU vendors AMD and Intel. We use HPL 2.0 (Petitet, Whaley, Dongarra, & Cleary, 2010) from the University of Tennessee, compiled with GotoBLAS 1.26 (Goto, 2010) from the Texas Advanced Computing Center (TACC), and MPICH2 1.0.8 (Laboratory, 2010) from the Argonne National Laboratory. Using our tools we can allocate and configure variable size clusters in EC2 automatically, including support for MPI applications.

Although we developed tools to automatically manage and configure EC2 nodes for our applications, there are also other publicly available tools for running scientific applications on cloud platforms (including EC2) (Nimbus Science clouds, 2010; Nurmi et al., 2008). Further, as the cloud computing platform matures, we expect much more development for specific applications such as high-performance computing to reduce or eliminate much of the initial learning curve for deploying scientific applications on cloud platforms. Already, for example, public images are available on EC2 supporting MPICH (Gemignani & Skomoroch, 2010).

21.3.1 Overview of Amazon EC2 Setup

Our case study was carried out using various instance types on the Amazon Elastic Compute Cloud (EC2) service from November 2008 through January 2010.

Table 21.1 Information about various instances types: processor type, number of cores per instance, installed RAM (in Gigabytes), and theoretical peak performance (in GFLOP/sec). Prices are on Amazon EC2 as of January, 2010

Instance	Processor	Cores	RAM (GB)	Peak (Gflops)	Price ($/hr)
m1.large	Intel Xeon E5430	2	7.5	21.28	$0.34
m1.large	AMD Opteron 270	2	7.5	8.00	$0.34
m1.xlarge	Intel Xeon E5430	4	15	42.56	$0.68
m1.xlarge	AMD Opteron 270	4	15	16.00	$0.68
c1.xlarge	Intel Xeon E5345	8	7	74.56	$0.68
m2.2xlarge	Intel Xeon X5550	4	34.2	42.72	$1.20
m2.4xlarge	Intel Xeon X5550	8	68.4	85.44	$2.40

Table 21.1 describes the salient differences between the instance types: number of cores per instance, installed memory, theoretical peak performance, and the cost of the instance per hour. We only used instances with 64-bit processors so that we treat the m1.large as the smallest instance although Amazon provides a smaller 32-bit m1.small instance. The costs per node vary by a factor of 7 from $0.34 for the smallest to $2.40 for nodes with significant installed memory. We note that cost scales more closely with installed RAM than with peak CPU performance—the c1.xlarge instance being the outlier. Peak performance is calculated using processor-specific capabilities. For example, the c1.xlarge instance type consists of 2 Intel Xeon quad-core processors operating at a frequency of 2.3 GHz with a total memory of 7 GB. Each core is capable of executing 4 floating-point operations per clock cycle, leading to a theoretical peak performance of 74.56 GFLOP per node. There are additional costs for bandwidth used into and out from Amazon's network and for long-term storage of data, but we ignore these costs in our calculations because they are negligible compared to the costs of the computation itself in our experiments.

In regards to multithreaded parallelism provided by the multi-core processors, extensive testing typically delivered the best performance when we set the Goto BLAS library to use as many threads as available cores per socket—4 and 2, for the Xeon and the Opteron, respectively. We provide the number of threads used to obtain specific results in the following section when presenting peak achieved efficiency. With these settings and using the platform-specific libraries and compilers, we reached 76% and 68% of theoretical peak performance (as measured in GFLOP/sec) for the Xeon E5345 and Opteron 270, respectively, for single node performance on xlarge instances. We thus believe the configuration and execution of LINPACK in HPL on the high-CPU and standard instances is efficient enough to use as an exemplar of compute-intensive applications for the purposes of our evaluation.

All instance types (with Intel or AMD CPUs) execute the RedHat Fedora Core 8 operating system using the 2.6.21 Linux kernel. The 2.6 line of Linux kernels supports autotuning of buffer sizes for high-performance networking, which is enabled by default. The specific interconnect used by Amazon is unspecified (Services, 2010) and multiple instances might even share a single hardware network card

(Wang & Ng, 2010). Therefore, the entire throughput might not be available to any particular instance. In order to reduce the number of hops between nodes to the best of our ability, we run all experiments with cluster nodes allocated in the same availability zone.

21.3.2 Overview of HPL

Our goal is to determine the suitability of commercial cloud environments for certain kinds of scientific applications. We focus on the HPL benchmark (Petitet et al., 2010) as the exemplar of tightly coupled, highly parallel scientific applications. HPL computes the solution of a random dense system of linear equations via LU factorization with partial pivoting and triangular solves. This algorithm requires $O(n^3)$ floating point operations on $O(n^2)$ data; that is, HPL is compute-intensive, and represents a realistic upper bound for the performance of such scientific applications. The actual implementation is driven by more than a dozen parameters, all of which may have a significant impact on the resulting performance and therefore require fine tuning. We describe the HPL parameters that were tuned below:

1. Block size (*NB*). It is determined in relation to the problem size and the performance of the underlying BLAS kernels. We used four different block sizes, namely 192, 256, 512, and 768.
2. Process grid ($p \times q$). This is the number of process rows and columns of the compute grid. As with most clusters, we empirically observed that on EC2 it is better to use process grids where $p \leq q$. This is a product of the data flow of the algorithms used in HPL.
3. Broadcast algorithm (*BFACT*). It depends on the problem size and network performance. Testing suggested that the best broadcast parameters are 3 and 5. For large machines featuring fast nodes compared to the available network bandwidth, algorithm 5 is observed to be best.

With respect to the other HPL settings, we kept them fixed for all the experiments.

21.4 Intranode Scaling

We begin our empirical analysis of EC2 performance for linear algebra by evaluating the consistency of achievable performance on a single node. In order to evaluate the consistency in performance delivered by EC2 nodes, we executed DGEMM—the matrix-matrix multiplication kernel of BLAS—and HPL tests for 24 h, repeating the experiment over different days. We first focus on the DGEMM results, then discuss results from HPL. DGEMM is at the core of the Basic Linear Algebra Subroutines (BLAS) library. Implementing the basic operation of multiplying two matrices, DGEMM is the building block of all the other Level-3 BLAS routines and of virtually every linear algebra algorithm. It is highly optimized for each target architecture and its performance, often in the 90+% range of efficiency, is ordinarily interpreted as the processor peak achievable performance.

21.4.1 DGEMM Single Node Evaluation

In our DGEMM experiment we initialize three square matrices A, B and C of a fixed size and then invoke the GotoBLAS implementation of DGEMM (Goto, 2010). We only time the call to the BLAS library and not the time spent allocating and initializing the matrices. Due to the dense nature of the matrices involved in the experiments—most of the entries are non-zero—we expect little to no fluctuations in the execution time on a single node independent of problem size and the number of cores used.

Figure 21.1 presents the time to complete the same DGEMM computation repeatedly over six hours on EC2. For space reasons we focus on six hours; however, the rest of the time shows similar behavior. In this experiment, only four of the eight cores of a c1.xlarge instance were used. The results show very high variability in execution time with an average of 191.8 seconds and standard deviation of 68.6 seconds (36% of average). There are several possible sources for such variability: (1) the process is not being run for extended, variable periods of time, (2) the threads

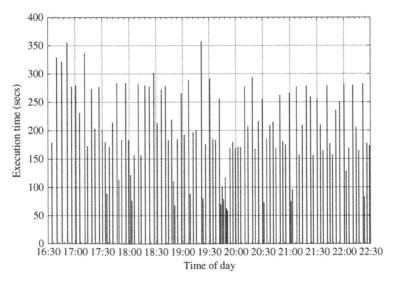

Fig. 21.1 Execution time of repeated DGEMM using 4 (of 8) cores over 6 hours on c1.xlarge instance. Average execution is 191.8 s with a standard deviation of 68.6 s. The input sizes for each DGEMM are identical. On a stand alone node all executions would be identical

are being scheduled on different cores each time (reducing first-level cache performance), and (3) the last-level cache shared by all cores is less available to each thread (because it is being used by another thread on a different core).

Figure 21.2 shows a similar experiment to Fig. 21.1 but using only one of the eight cores. The experiment shows none of the variability when using only one core that is present using more cores. With an average execution time of 227.9 s, the standard deviation is only 0.23 s. The results for using all eight cores shows even higher levels of variability than Fig. 21.1. The reduced variability of a single core demonstrates cause (1) above is unlikely. The process is scheduled similarly, but performance is much more predictable. Cause (2), however, cannot be ruled out because Amazon EC2 provides no mechanism to *pin* threads to particular cores so that the thread always executes on the same physical core of the processor. Finally, we conjecture cause (3) plays at least as significant a role as (2), and we look at different experiments to explore these effects.

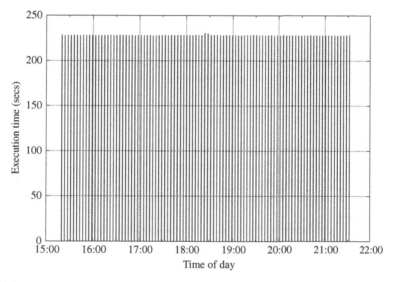

Fig. 21.2 Execution time of repeated DGEMM using 1 (of 8) core over 6 hours on c1.xlarge instance. Average execution is 227.9 s with a standard deviation of 0.23 s. The input sizes to each DGEMM are identical

The effects of using different numbers of cores is easily observed. The average and minimum execution time against the number of cores appear in Fig. 21.3. In this graph, DGEMM was executed over several hours on a matrix of size $n = 10k$ so that the matrices (each 762 MB) cannot fit in the 8 MB last-level cache of the Intel XEON running the c1.xlarge instance. The results presented are average and minimum execution times for the DGEMM. Again, allocation and initialization of memory are not included in the timings. Error-bars show the standard deviation of the average. There are several notable characteristics of the graph:

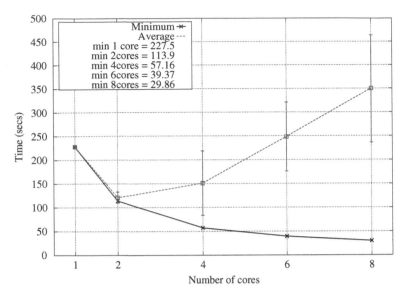

Fig. 21.3 Average execution time of repeated DGEMM with $n = 10k$ on c1.xlarge instance by number of cores used. The matrices cannot fit within the last-level cache. The inputs sizes for each DGEMM are identical. Error bars on average show one standard deviation

1. The best performance in graphs (minimum line) shows that we can reach roughly 90% efficiency. Such performance is close to the optimal achievable. For example, such performance would be expected in a stand alone node. We note that virtualization alone clearly does not have a significant impact on peak performance.
2. The average performance shows that similar computations will not likely ever reach optimal. The average efficiency—as given by the inverse of the spread in graph between the min and average—drops dramatically. The best average performance at two cores is still several times worse than the minimum execution time at eight cores.
3. As the number of cores increases, the average significantly increases along with the standard deviation. The standard deviation increases by four orders of magnitude. Expected performance diverges so significantly from best performance that when using eight cores one would rarely expect to achieve the best performance.
4. Underutilization is a good policy on EC2. Using two cores appears to be the sweet spot in this experiment. Efficiency is good and the average and expected performance are the best and in line with optimal. Using four cores already takes longer on average than using two, although using four is still slightly faster than a single core, and the trend only worsens as the number of cores increases. Note that the fastest expected time is obtained by only using only a quarter of the machine!

While we can attribute the performance degradation when using more than two cores per node in Fig. 21.3 to ever worse cache behavior, we cannot easily distinguish which caches are being missed. In the multicore processors in these instances, cores have individual caches and a large shared last-level cache (LLC). Pinning a thread to a particular core would help the thread maintain cache consistency for the individual caches while reducing memory demand of the threads would reduce contention on the shared LLC. However, the kernels supplied by Amazon EC2 do not provide support to allow us to pin threads. To attempt to distinguish these effects without the ability to fix threads to particular cores, we performed similar DGEMM experiments with smaller matrices that all fit within the LLC of the processor.

Figure 21.4 shows the average and minimum execution times using different numbers of cores for a DGEMM on small matrices $n = 500$ yielding matrices of 1.9 MB that can easily fit within the 8 MB LLC. The error-bars on the average execution times give a standard deviation from the mean. Again, allocation and initialization of memory are not included in the timings. We note several characteristics of the graph:

1. As with the out-of-cache DGEMM, the minimum execution times are consistent with the expected performance of a stand-alone node.
2. Contention is likely for the last-level cache. The two orders of magnitude performance degradation between a single core and eight cores is also the difference in time between accessing the last-level cache and main memory.

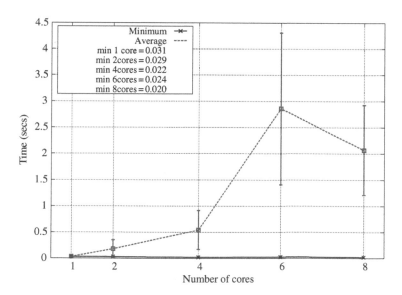

Fig. 21.4 Average execution time of repeated DGEMM with $n = 500$ on c1.xlarge instance by number of cores used. All the matrices fit within the last-level cache at once. The inputs sizes for each DGEMM are identical. Error bars on average show one standard deviation

3. The ability to pin threads to cores would probably help significantly. Using multiple cores always performs worse than with a single core, implying significant overhead from coordination. Since the minimum scales well, we conjecture that the average suffers from poor cache behavior. However, the average time with multiple cores does not become orders of magnitude worse—as would be caused by going to main memory—until all of the cores are used. Lower level cache misses are thus more likely when underutilizing the instance with fewer than eight cores.

21.4.2 HPL Single Node Evaluation

In order to determine whether the effects we have seen using DGEMM scale to multiple nodes, we use the HPL benchmark (Petitet et al., 2010) that solves a system of linear equations via LU factorization. HPL can scale to large compute grids. However, we first consider HPL given similar matrices to the previous DGEMM experiments, namely, solving a system of linear equations on a single node where the data does fit and does not fit into the LLC of the processor. In the following section, we extend HPL to examine internode scaling using parallel algorithms.

In Fig. 21.5 we repeatedly execute HPL with $n = 25$ k on the Amazon EC2 c1.xlarge instance. We plot average and minimum execution times against the number of cores. As with the DGEMM experiments, error bars show one standard

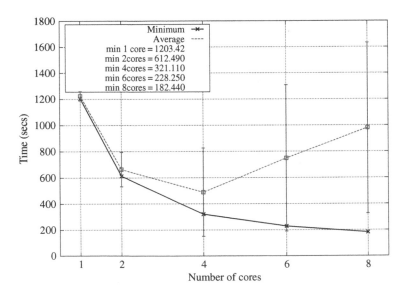

Fig. 21.5 Average execution time of repeated HPL with $n = 25$ k on c1.xlarge instance by number of cores used. The matrices cannot fit within the last-level cache. The input configuration to each execution of HPL is identical. Error bars on average show one standard deviation

deviation. As with the first DGEMM experiment, these matrices do not fit within the LLC of the processor, but do fit within the main memory of a single instance. The results are quite similar to DGEMM, but show slightly different trends. We do not directly compare the DGEMM and HPL experiments because of the use of very different algorithms. We only point out that the trends show similar behavior on a single node.

In the last single node figure, we show HPL with a matrix of size $n = 1$ k so that all data for the computation fits within the LLC. Figure 21.6 plots the average (and one standard deviation) and minimum execution times of repeated runs of the HPL benchmark. The results show similar behavior to the corresponding DGEMM experiment. We could not find a source for the anomalous improved performance at six cores as compared to four or eight. In the following section, we extend HPL to multiple nodes to take a look at the effects of contention on parallel computations in a commercial cloud environment.

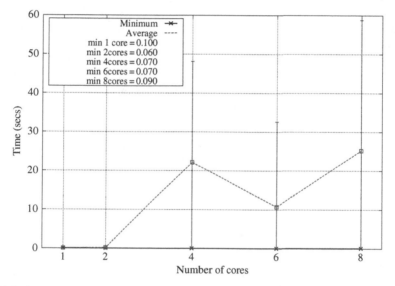

Fig. 21.6 Average execution time of repeated HPL with $n = 1$ k on c1.xlarge instance by number of cores used. The matrices fit within the last-level cache. The input configuration to each execution of HPL is identical. Error bars on average show one standard deviation

From the single node experiments we conclude that the very high variance in performance implies that the best achieved performance is not a good measure of expected performance. The average expected performance can be several orders of magnitude worse depending on the cache behavior of the algorithm in both efficiency and time to solution. In our experiments, the best average expected performance is obtained on Amazon using much less of the machine than is allocated: In an experiment accessing a large amount of main memory, using only a quarter of the machine provided the best expected performance!

21.5 Internode Scaling

The previous section demonstrates the significant effects of contention on single node performance in a commercial cloud environment. In this section, we extend our empirical analysis to parallel multi-node algorithms using the HPL benchmark. The HPL benchmark represents tightly coupled, highly parallel algorithms frequently used in scientific applications. We executed the HPL benchmark on different instance types on Amazon EC2, varying the parameters as described in Section 21.3.2. Using different instance types allowed us to see the effects on performance, efficiency, and cost of varying the amount of RAM available and the problem size.

Tables 21.2, 21.3, 21.4 and 21.5 show the parameters used to obtain the best results over different problem sizes, instance types, and number of cores used per node. To calculate the total number of cores used for a particular execution, use the result of $p \times q \times$ Threads. Due to the numerous parameters of HPL, it was necessary to try different input configurations to maximize performance. To maximize the performance, we minimized reliance on the interconnect by maximizing the memory allocation per node used to solve a particular problem size. Figure 21.7 shows the number of nodes used for different instance types to solve each problem size. Only the m2.4xlarge instance with 68.4 GB of RAM is large enough to solve all problem sizes on a single node. We used this as a reference point in Fig. 21.8 to bound the effects of network usage on the performance.

Table 21.2 Results of HPL benchmarks for matrix size $n = 20$ k. Columns are, in order: Amazon EC2 instance type, CPU type (Xeon or Opteron), block size (NB), process grid ($p \times q$), number of threads used in the BLAS routines, broadcast algorithm (*BFACT*), best elapsed time, and corresponding efficiency using the theoretical peak performance from Table 21.1. A more detailed description of the parameters can be found in Section 21.3.2

Instance	Proc.	Block size	$p \times q$	Threads	Bcast	Time (min:sec)	Efficiency (% peak)
m1.large	Xeon	256	1x1	2	3	06:29.35	64.38
m1.large	Opt.	192	1x2	1	3	12:24.21	89.62
m1.xlarge	Xeon	256	1x1	4	3	05:07.30	40.78
m1.xlarge	Opt.	256	1x2	2	3	06:28.16	85.88
c1.xlarge	Xeon	512	1x1	8	3	01:51.85	63.96
m2.2xlarge	Xeon	256	1x2	2	3	03:19.74	62.50
m2.4xlarge	Xeon	256	1x1	8	3	01:44.55	59.71

In Figs. 21.8, 21.9, 21.10, 21.11, 21.12, and 21.13 we present different aspects of the results of our HPL benchmark experiments given in Tables 21.3, 21.4, and 21.5. We first discuss the best results obtained to give a reasonable lower bound on performance in a contentious commercial cloud environment. In these parallel experiments, the minimum times do not differ as greatly from the average as the single node experiments for several reasons: (1) the elapsed time required (more than an hour for large problem sizes) implies an averaging effect and (2) the overhead for parallel jobs is higher than single node experiments due to network usage. Finally,

Table 21.3 Results of HPL benchmarks for matrix size $n = 30$ k

Instance	Proc.	Block size	p × q	Threads	Bcast	Time (min:sec)	Efficiency (% peak)
m1.large	Xeon	256	1x1	2	3	20:20.82	69.31
m1.large	Opt.	256	1x2	1	3	41:21.01	90.75
m1.xlarge	Xeon	256	1x2	2	3	11:00.73	64.00
m1.xlarge	Opt.	192	1x2	2	3	21:20.46	87.88
c1.xlarge	Xeon	512	1x1	8	3	05:19.10	75.66
m2.2xlarge	Xeon	256	1x2	2	3	11:12.15	62.69
m2.4xlarge	Xeon	512	1x1	8	3	05:46.53	60.80

Table 21.4 Results of HPL benchmarks for matrix size $n = 50$ k

Instance	Proc.	Block size	p × q	Threads	Bcast	Time (hr:min:sec)	Efficiency (% peak)
m1.large	Xeon	512	1x3	2	5	35:55.66	60.56
m1.large	Opt.	256	1x3	2	3	1:09:44.47	83.00
m1.xlarge	Xeon	512	1x2	4	5	28:07.85	58.00
m1.xlarge	Opt.	512	1x4	2	5	1:00:46.59	71.41
c1.xlarge	Xeon	512	3x1	8	5	18:14.36	34.04
m2.2xlarge	Xeon	256	1x2	2	3	51:09.45	63.11
m2.4xlarge	Xeon	256	1x2	8	3	26:16.52	61.87

Table 21.5 Results of HPL benchmarks for matrix size $n = 80$ k[a]

Instance	Proc.	Block size	p × q	Threads	Bcast	Time (hr:min:sec)	Efficiency (% peak)
m1.large	Xeon	768	2x4	2	5	1:01:06.31	54.69
m1.xlarge	Xeon	512	1x4	4	3	1:14:57.67	44.58
m1.xlarge	Opt.	512	2x2	4	5	3:27:46.61	42.78
c1.xlarge	Xeon	768	1x8	8	5	2:17:50.49	0.07
m2.2xlarge	Xeon	512	1x2	4	3	1:50:56.69	60.02
m2.4xlarge	Xeon	512	1x1	8	3	1:46:06.05	62.76

[a]We could not allocate enough m1.large nodes of Opteron type to solve a problem of size $n = 80$ k

we note that in these figures the line for m1.large instances using the Opteron processor does not extend to $n = 80$ k because we could not allocate enough m1.large instances with Opteron CPUs in a single availability zone to solve a problem of size $n = 80$ k.

21.5.1 HPL Minimum Evaluation

Figure 21.8 demonstrates that efficiency is generally better than 60% for problem sizes below 80 k. Here, efficiency is considered to be percentage of theoretical peak

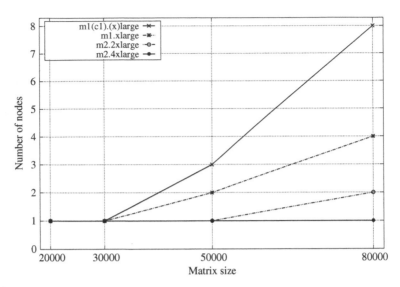

Fig. 21.7 Required number of nodes for different input matrix sizes n. The number of nodes is determined by maximizing usage of RAM on each node. We could not allocate enough m1.large nodes of AMD type to solve a problem of size $n = 80$ k

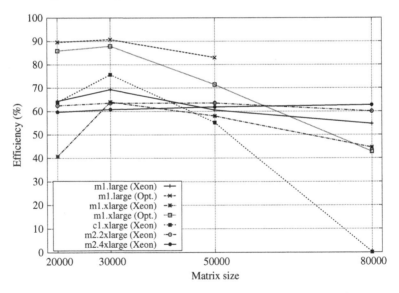

Fig. 21.8 Efficiency scaling by problem size. Efficiency is determined with respect to the theoretical peak given in Table 21.1 multiplied by the number of nodes used

given in Table 21.1 multiplied by the number of nodes used. We consider 60% to be reasonable performance given the relatively slow interconnect provided by Amazon EC2 compared to purpose-built HPC systems. The balance of resources available to the computation is clearly important. For example, the m2.2xlarge and m2.4xlarge

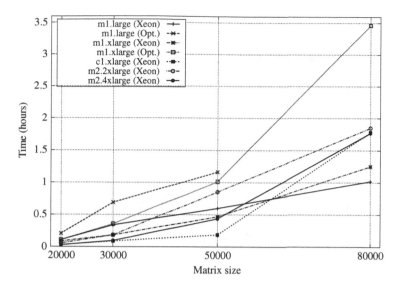

Fig. 21.9 Total time to solution for different instance types by problem size

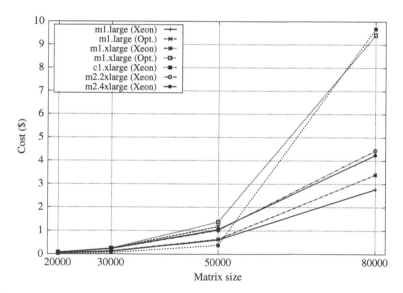

Fig. 21.10 Cost to solution prorated to actual time spent for different instance types by problem size

instances have roughly equal performance at 80 k although HPL needs two of the m2.2xlarge instances and only one m2.4xlarge. In this case the interconnect does not have a significant effect because the nodes are provisioned with enough RAM to keep the CPU busy. The c1.xlarge instances suffer severe performance degradation,

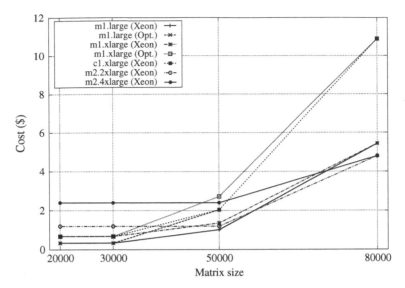

Fig. 21.11 Actual cost to solution for different instance types by problem size. Execution time is rounded to the nearest hour for cost calculation

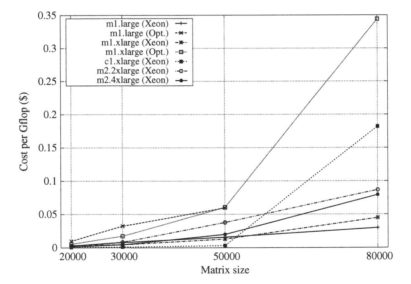

Fig. 21.12 Cost per GFLOP ($/GFLOP) prorated to actual time spent for different instance types by problem size

however. At $n = 80$ k, the c1.xlarge instances perform two orders of magnitude worse than a single node. We conjecture that the 7 GB of RAM of c1.xlarge nodes is insufficient to keep the CPU busy given the same interconnect between instances. In general, nodes with more RAM clearly scale up in problem size much better as

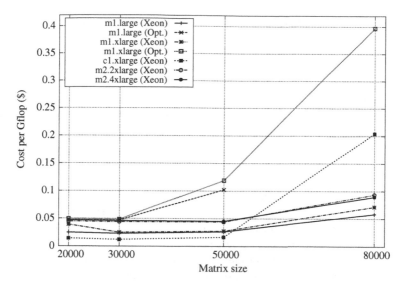

Fig. 21.13   Actual cost to solution per GFLOP ($/GFLOP) for different instance types by problem size. Execution time is rounded to the nearest hour for cost calculation

expected due to the high ratio between the required number of operations and the data size to solve a dense linear system.

While efficiency is important to gauge the scalability of the implementation, for any particular experiment the most important concern is generally time to solution. Fig. 21.9 shows the total time to solution for different instance types by problem size. This graph shows that although the efficiency of larger nodes is somewhat better than that of the smaller ones, in most cases the smaller nodes reach the solution faster. The c1.xlarge instances, recommended for high-compute applications, and the m2.2xlarge and m2.4xlarge instances, recommended for high-memory applications, are generally the slowest.

One of the important concerns when using commercial cloud environments are the differing costs of different instances. Figure 21.10 provides the cost to solution for different instance types by problem size. In this graph, the cost is prorated to the second to illustrate the marginal costs incurred when allocating a cluster to solve several problems. Figure 21.11 provides the comparable actual costs of using the different instance types to solve a single execution each problem size; that is, this graph includes the costs of the remaining hour after the execution completes. The salient difference between the figures is the relation of the largest nodes to the smaller nodes. Using absolute cost, the largest nodes are cheapest for large problems, but using prorated costs they are more expensive. Since the prorated trends are more informative for different problem sizes and for multiple jobs, we conclude the smaller instances m1.large and m1.xlarge are the most cost effective for parallel jobs.

In addition to cost to solution, we also consider a more general cost measure, $/GFLOP, calculated using the ratio between the total GFLOPS returned by the HPL benchmark and the (prorated) cost for the specific computation. This measure allows a rough conversion of expected cost for other problem sizes including other scientific applications characterized by similar computational needs. Figures 21.12 and 21.13 show the results of the HPL benchmarks giving the cost per GFLOP for different instances by problem size. The cost per GFLOP measure magnifies the differences between the instance types, but of course the best instance type in Figs. 21.10 and 21.11—m1.large—is still the best for the price to performance ratio.

The m1.large instance is the fastest by Fig. 21.9 and the best for price to performance by Fig. 21.10, leading us to recommend the smallest (64-bit) instance for most parallel linear algebra compute jobs on the Amazon EC2 cloud environment according to the empirical upper bound on performance. The high-compute and high-memory instances are not worth the extra costs in our experiments. Given the high variability in performance of single nodes, we examine in the following section the expected performance from instances instead of the upper bound to determine if our recommendation holds also for expected average performance.

21.5.2 HPL Average Evaluation

The minimum execution times from the previous section provide a rough upper bound for performance. In this section, we examine the average execution times for HPL to provide a better estimate of the expected performance of applications with dense linear algebra. Figures 21.14, 21.15, 21.16, and 21.17 provide the minimum and average (and standard deviation) execution times for different instances

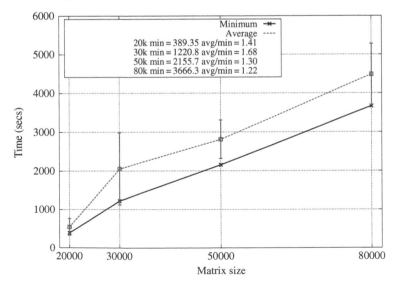

Fig. 21.14 Minimum and average execution time of HPL on m1.large (Xeon) instances by problem size. Error bars on average show one standard deviation

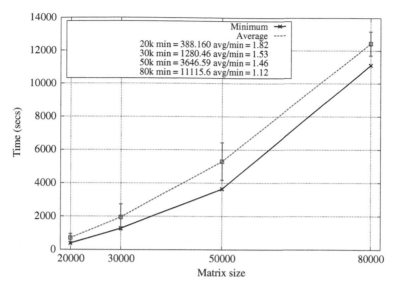

Fig. 21.15 Minimum and average execution time of HPL on m1.xlarge (Opt.) instances by problem size. Error bars on average show one standard deviation

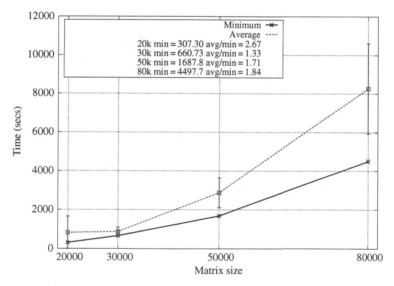

Fig. 21.16 Minimum and average execution time of HPL on m1.xlarge (Xeon) instances by problem size. Error bars on average show one standard deviation

by problem size. We do not show the m2.2xlarge or m2.4xlarge instances due to insufficient data. These large instances are also quite expensive to allocate.

As with the single node experiments, Fig. 21.14, 21.15, 21.16, 21.17 show that the expected performance is worse than the best performance, but by a much smaller

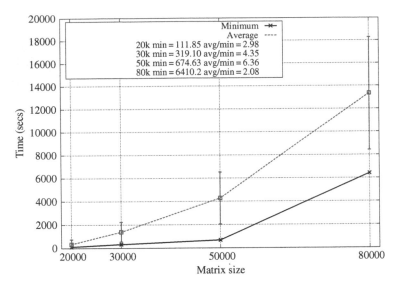

Fig. 21.17 Minimum and average execution time of HPL on c1.xlarge instances by problem size. Error bars on average show one standard deviation

margin. Indeed, for the m1.large instances in Fig 21.14, the expected performance at $n = 80$ k is only 22% worse than the best execution time. Generally, the average for the HPL experiments is between 30% and 2 × worse than the minimum execution times. As we mentioned at the beginning of this section, we believe the smaller differences between the expected execution time and the best times are due to an averaging effect from the length of the experiments and the greater overhead of the HPL benchmark from network traffic (as compared to the DGEMM experiments in Section 21.4.1).

The m1.large instance—the smallest that we tested—remains the best instance on EC2 for tightly coupled, compute intensive, parallel jobs. Although the average expected time is longer than the minimum by 20% for the largest problem size, the expected time is still faster than the next fastest minimum time (m1.xlarge Xeon). Given that the m1.large also costs half as much as the m1.xlarge, the smallest instance is the clear winner for the tightly-coupled, dense linear algebra computation that we evaluated, and the high marginal costs for high-compute or high-memory nodes is not cost effective.

21.6 Conclusions

In this chapter we demonstrated empirically the computational efficiency of high-performance numerical applications in a commercial cloud environment when resources are shared under high contention. Through a case study using the Linpack benchmark, we show that cache utilization becomes highly unpredictable and similarly affects computation time. For some problems, not only is it more efficient to underutilize resources, but the solution can be reached sooner in realtime (walltime). We also show that the smallest, cheapest (64-bit) instance on the Amazon EC2 commercial cloud environment is not only the fastest, but also the best for price to performance ratio.

We presented the *average expected* performance and execution time, expected cost to completion, and variance measures—traditionally ignored in the high performance computing context—to determine the efficiency and performance of the Amazon EC2 commercial cloud environment. Under omnipresent contention for resources, the expected performance in the cloud environment diverges by an order of magnitude from the best achieved performance. We conclude that there is significant space for improvement in providing predictable performance in such environments.

Acknowledgments The authors wish to acknowledge the Aachen Institute for Advanced Study in Computational Engineering Science (AICES) as sponsor of the experimental component of this research. Financial support from the Deutsche Forschungsgemeinschaft (German Research Association) through grant GSC 111 is gratefully acknowledged. Also, support from the XtreemOS project, which is partially funded by the European Commission under contract #FP6-033576 is gratefully acknowledged.

References

Barham, P.T., Dragovic, B., Fraser, K., Hand, S., Harris, T.L., Ho, A., et al. (2003). Xen and the art of virtualization. In *Symposium on operating systems principles,* (pp. 164–177). New York, NY: Bolton Landing, USA.

Dongarra, J., van de Geijn, R., & Walker, D. (1994). Scalability issues affecting the design of a dense linear algebra library. *Journal of Parallel and Distributed Computing, 22*(3), 523–537.

Gemignani, C., & Skomoroch, P. (2010). *Elasticwulf: Beowulf cluster run on Amazon EC2.* Available via the WWW. Retrieved 1 January 2010, from http://code.google.com/p/elasticwulf/.

Goto, K. (2010). *GotoBLAS.* Available from WWW. Retrieved 1 January 2010, from http://www.tacc.utexas.edu/.

Hosting, S. D. (2010). *GoGrid cloud hosting.* Available from WWW. Retrieved 1 January 2010, from http://gogrid.com.

Iyer, R., Zhao, L., Guo, F., Illikkal, R., Makineni, S., Newell, D., et al. (2007). Qos policies and architecture for cache/memory in cmp platforms. *SIGMETRICS Performance Evaluation Review, 35*(1), 25–36. DOI http://doi.acm.org/10.1145/1269899.1254886.

Keahey, K., Freeman, T., Lauret, J., & Olson, D. (2007). Virtual workspaces for scientific applications. *SciDAC 2007 Conference, Boston, MA.*

Laboratory, A. N. (2010). *MPICH2: High-performance and widely portable MPI.* Available via the WWW. Retrieved 1 January 2010, from http://www.mcs.anl.gov/research/projects/mpich2/.

Napper, J., & Bientinesi, P. (2009). Can cloud computing reach the Top500? *Unconventional High-Performance Computing (UCHPC),* Italy.

Nimbus Science Clouds (2010). *Available from WWW.* Retrieved 1 January 2010, from http://www.nimbusproject.org/.

Nurmi, D., Wolski, R., Grzegorczyk, C., Obertelli, G., Soman, S., Youseff, L., et al. (2008). The Eucalyptus open-source cloud-computing system. *Proceedings of Cloud Computing and Its Applications* [online].

Petitet, A., Whaley, R. C., Dongarra, J., & Cleary, A. (2010). *HPL - a portable implementation of the high-performance LINPACK benchmark for distributed-memory computers.* Available from WWW. Retrieved 1 January 2010, from http://www.netlib.org/benchmark/hpl/.

Petrini, F., Kerbyson, D. J., & Pakin, S. (2003). The case of the missing supercomputer performance: Achieving optimal performance on the 8,192 processors of ASCI Q. *SC '03: Proceedings of the 2003 ACM/IEEE conference on Supercomputing*, IEEE Computer Society, Washington, DC, USA, p. 55.

Services, A. W. (2010). *Amazon elastic compute cloud (EC2).* Available from WWW. Retrieved 1 January 2010, from http://aws.amazon.com/ec2.

Strebel, J., Stage, A. "An Economic Decision Model for Business Software Application Deployment on Hybrid Cloud Environments", In: M. Schumann, L.M. Kolbe, M.H. Breitner, A. Frerichs (eds.), Multikonferenz Wirtschaftsinformatik 2010, Universitätsverlag Göttingen, Ottingen, 2010, pp. 195–206.

Tikotekar, A., Vallée, G., Naughton, T., Ong, H., Engelmann, C., & Scott, S. L. (2008). An analysis of HPC benchmarks in virtual machine environments. *Euro-Par 2008 Workshops - Parallel Processing: VHPC 2008, UNICORE 2008, HPPC 2008, SGS 2008, PROPER 2008, ROIA 2008, and DPA 2008, Las Palmas de Gran Canaria, Spain,* August 25–26. Revised Selected Papers, 63–71. Springer, Berlin, Heidelberg (2009). DOI http://dx.doi.org/10.1007/978-3-642-00955-6_8

TOP500.Org (2010). *Top 500 supercomputer sites.* Available from WWW. Retrieved 1 January 2010, from http://www.top500.org/.

Walker, E. (2008). Benchmarking Amazon EC2. LOGIN, 18–23.

Wang, G., & Ng, E. (2010). The impact of virtualization on network performance of Amazon EC2 data center. *INFOCOM '10: Proceedings of the 2010 IEEE Conference on Computer Communications. IEEE Communication Society.* San Diego, CA.

Xcalibre Communications Ltd (2010). *FlexiScale cloud computing.* Available from WWW. Retrieved 1 January 2010, from http://www.flexiscale.com.

Youseff, L., Seymour, K., You, H., Dongarra, J., & Wolski, R. (2008). The impact of paravirtualized memory hierarchy on linear algebra computational kernels and software. *HPDC '08: Proceedings of the 17th International Symposium on High Performance Distributed Computing,* ACM, New York, NY, USA, 141–152. DOI http://doi.acm.org/10.1145/1383422.1383440.

Chapter 22
Scientific Data Management in the Cloud: A Survey of Technologies, Approaches and Challenges

Sangmi Lee Pallickara, Shrideep Pallickara, and Marlon Pierce

22.1 Introduction

Experimental sciences create vast amounts of data. In astronomy, data produced by the Pan-STARRS project (Pan-STARRS project, 2010; Jedicke, Magnier, Kaiser, & Chambers, 2006) is expected to result in more than a petabyte of images every year. In high-energy physics, the Large Hadron Collider will generate 50–100 petabytes of data each year, with about 20 PB of that data being stored and processed on a worldwide federation of national grids linking 100,000 CPUs (Large Hadron Collider project, 2010; Massimo Lammana, 2004).

Cloud computing is immensely appealing to the scientific community, who increasingly see it as being part of the solution to cope with burgeoning data volumes. Cloud computing enables economies-of-scale in facility design and hardware construction. Groups of users are allowed to host, process, and analyze large volumes of data from various sources. There are several vendors that offer cloud computing platforms; these include Amazon Web Services (Amazon Web Services, 2010), Google's App Engine (Google App Engine, 2010), AT&T's Synaptic Hosting (AT&T Synaptic Hosting, 2010), Rackspace (Rackspace, 2010), GoGrid (GoGrid, 2010) and AppNexus (AppNexus, 2010). These vendors promise seemingly infinite amounts of computing power and storage that can be made available on demand, in a pay-only-for-what-you-use pricing model.

The science community has substantial experience in dealing with data management issues in distributed computing environments. Data Grids (Chervenak, Foster, Kesselman, C. Salisbury, & Tuecke, 2001), which is based on the grid computing paradigm, has provided large-scale scientific data storage with support for data discovery and accesses over the grid network (Singh et al., 2003; The Globus toolkit,

S.L. Pallickara (✉) and S. Pallickara
Department of Computer Science, Colorado State University, Fort Collins, CO, USA
e-mails: {sangmi; shrideep}@cs.colostate.edu

M. Pierce
Community Grids Lab, Indiana University, Bloomington, IN, USA
e-mail: mpierce@cs.indiana.edu

B. Furht, A. Escalante (eds.), *Handbook of Cloud Computing*,
DOI 10.1007/978-1-4419-6524-0_22, © Springer Science+Business Media, LLC 2010

2010; Moore, Jagatheesan, Rajasekar, Wan, & Schroeder, 2004; Antonioletti et al., 2006). Similarly, many cyberinfrastructures and gateways provide their own large-scale data management schemes to satisfy domain-specific requirements (Peng & Law, 2004; Plale et al., 2005).

Scientific data management in the cloud differs from prior approaches in several aspects (Gannon & Reed, 2009). First, cloud computing consolidates computing capabilities and data storage in very large datacenters or economics-driven local datacenters. In conventional high performance computing, scientists share massively parallel super computers. Access to these supercomputing resources is managed by batch queue systems. Computational input and output data are staged back and forth from the computing nodes to the data storage, which is located separately. In cloud computing the data storage and the computing capability are in the same place. This leads to leads to a paradigm shift in many of data processing and analysis.

Second, hosting data in a centralized facility can be a catalyst for data sharing across different scientific domains. An example of multidisciplinary data hosting can be found in the SciDB (SciDB, 2010) open source database. SciDB is designed to serve scientists across a variety of disciplines including astronomy, biology, meteorology, oceanography, and physics. This sharing is not only between different fields, but also among scientists from diverse institutions of the same discipline.

Finally, from the perspective of data hosting, cloud computing is inherently more sustainable. In cloud settings data is often replicated to cope with transient and permanent failures in addition to coping with data corruptions caused by the use of commodity hardware. Since data preservation is often a critical aspect in most scientific domains, data management in the cloud is often a far superior alternative to using local storage, which would involve resolving technical problems that include inter alia replication, fault tolerance, and error detection.

This chapter is organized as follows. In Section 22.2 we discuss common characteristics of scientific data processing to help understand the requirements for scientific data management. In Section 22.3 we discuss current data cloud technologies. We describe scientific applications that are adopting data cloud technologies in Section 22.4. We present an analysis of the gaps between current data cloud technologies and scientific data management requirements in Section 22.5. Finally, in Section 22.6 we present our conclusions.

22.2 Data Management Issues Within Scientific Experiments

Data management in scientific computing involves data capture, curation, and analysis of the datasets. A lot of this data is produced by observational or experimental instruments such as survey telescopes, doppler radars, satellites, and particle accelerators such as the Large Hadron Collider. Collecting large amounts of data from these instruments can occasionally cause problems in the data ingest and transfer phases. Integrating data from diverse data sources has also been a challenge because of the differences in data delivery patterns and also the heterogeneity in data formats. Additionally, data is also produced during the computing phase and during simulation runs. Furthermore, in addition to the experimental data (raw, derived and recombined), any results and publications produced because of these experiments are also collected and managed as part of the scientific data (Gray, 2009).

Data analysis and simulation often involve visualization. The collected data is in general stored prior to being accessed by the data analysis or visualization process. The curation process involves efficient information extraction besides data organization including indexing and replication. A related consideration is the long-term preservation of the collected data.

Different stages of scientific data processing execute not only sequentially, but also recursively and interactively with other stages during the scientific experiments. Data processing often involves dynamic sharing between scientists or groups. For example, Unidata publishes atmospheric observational data to the public (Unidata, 2010); and in the biotechnology domain NCBI (NCBI, 2010) publishes information that includes gene sequences and chemical structures besides providing a library for biomedical articles (GenBank, 2010; PubCam, 2010; PubMed, 2010).

22.3 Data Clouds: Emerging Technologies

Peta-scale datasets pose new challenges. Here, the file system has to be able to manage billions of files some of which may themselves be a few terabytes long. To cope with this a synthesis of database systems and file systems has been proposed (Gray et al., 2005); here, the file hierarchy would be replaced with a database cataloging various attributes of each file.

The Map-Reduce (Dean & Ghemawat, 2004) programming model by Google enables concurrent processing of a voluminous dataset on a large number of machines. Computations and the data they operate on are collocated on the same machine. Thus a computation only needs to perform local disk I/O to access its input data. Map-Reduce can also be thought of as an instance of the SPMD model in parallel computing.

The Google File System (GFS) (Ghemawat, Gobioff, & Leung, 2003) fits in rather nicely with the Map-Reduce programming model. GFS disperses a large file onto a set of machines each with its own commodity hard drives: thus, portions of a file reside on multiple machines, which is where Map-Reduce computations would be pushed during subsequent processing. To account for failures that occur fairly often in settings involving commodity components, GFS replicates each file (the default is 3). Failures are assumed to be permanent, and once the system detects a failure it works to ensure that portions of files that were hosted on the failed machine are replicated elsewhere (using copies from other machines) to ensure that the replication levels for the affected files are preserved.

To provide database-like access to the data stored in GFS, Google developed a distributed data store, BigTable (Chang et al., 2006). BigTable provides a multi-dimensional map sorted by {row, column, timestamp}. Files are clustered as tablets that have a size of 100 ~ 200 MB, and BigTable manages more than a few Giga tablets. Google's implementation is available to the public through the Google Application Engine (Google App Engine, 2010). Google also provides a data analysis infrastructure, Sawzall (Pike, Dorward, Griesemer, & Quinlan), which makes effective use of large computing clusters with large datasets. Sawzall allows users to specify execution instructions through statements and expressions that are borrowed from C, Java or Pascal. Sawzall then converts these instructions into a program that can run on Google's highly parallel cluster.

Google's efforts have inspired several open source projects such as Hadoop (Hadoop, 2010) and several derivative projects including HBase (HBase, 2010), Hypertable (Hypertable, 2010) and Hive (Hive, 2010). Hadoop provides Map-Reduce programming environment on top of the Hadoop distributed file system (HDFS). HBase and Hypertable harness Hadoop to implement a distributed store that mirrors the software design of Google's BigTable. Apache's Hive (Hive, 2010) is a data warehouse infrastructure, which allows SQL-like ad hoc querying of data stored in HDFS. Zookeeper (Zoopeeker, 2010) is a high-performance coordination service for distributed applications.

Cassandra (Lakshman, Malik, & Ranganathan, 2008), developed by Facebook and Yahoo, provides a distributed column store for Internet scale data management. Cassandra focuses on fault tolerance to achieve an "always writable" feature, which is critical for several web-based software. Pig (Olston, Reed, Srivastava, Kumar, & Tomkins, 2008) is a platform for analyzing large datasets. Pig is both a high-level data-flow language and execution framework for parallel computation. Pig automatically translates user queries into efficient parallel evaluation plans, and then orchestrates their execution on a Hadoop cluster. Pig utilizes the Pig Latin language, developed by Yahoo, which combines high-level declarative querying in the spirit of SQL, and low-level, procedural programming, MapReduce.

Similarly, Microsoft provides its own software stack and computing facilities to the DataCloud clients. Azure (Microsoft, Windows Azure, 2010) is a large-scale distributed storage system in this software stack. SQL Azure (Microsoft, SQL Azure, 2010) provides a distributed key-value store and its interface extends standard SQL. Microsoft uses the dataflow graph based processing model called Dryad (Isard, Budiu, Yu, Birrel, & Fetterly, 2007) that is designed for Windows based cloud clusters. DryadLINQ (Yu, Gunda, & Isard, 2009; Isard & Yu, 2009) exploits LINQ (Language Integrated Query) to provide a hybrid of declarative and imperative programming. Any LINQ-enabled programming language such as C#, VB, and SQL can be used for distributed computation with DryadLINQ. DryadLINQ built on top on Dryad, provides a sequential data analysis language. The DryadLINQ system transparently translates the data-parallel portions of the program into a distributed execution plan, which is then passed to the Dryad execution platform. Objects in DryadLINQ datasets can be of any .NET type; this makes it easy to work with datasets such as image patches, vectors, and matrices.

In contrast to Google and Microsoft, Amazon web services (Amazon Web Services, 2010) provides a much lower level of computing and data storage infrastructure. Additionally, an Elastic Map-Reduce web service (Amazon Elastic

MapReduce, 2010) based on Hadoop is also provided. Users can run their own virtual cluster with specific batch processing such as Open PBS (2010), Open MPI (2010) or Condor (Thain, Tannenbaum, & Livny, 2005). Similarly, several commercial database management systems are available on the EC2 cluster including, IBM DB2 (Baru et al., 1995), Microsoft SQL Server (Microsoft, SQL Azure, 2010), and Oracle Database 11 g (Oracle Database 11g, 2010). Users can access storage services, such as Simple Storage Service (S3) (Palankar et al., 2008) and Elastic Block Service (EBS) (Amazon EBS, 2010). In S3 users store 1 byte $\sim$ 5 GB objects which are accessed through web service interface. EBS provides a much larger storage of 1 GB $\sim$ 1 TB and this block is mounted on a user's instance. Amazon provides a web service interface to simple database functions such as indexing and querying with SimpleDB (Amazon SimpleDB, 2010). Recently, Amazon introduced a more full-featured SQL like database instances, Relational Database Service (RDS) (Amazon RDS, 2010).

Table 22.1 below summarizes the various data cloud technologies that we have discussed so far.

Many of the technologies enabling the data cloud have been investigated in recent distributed data management projects. The Boxwood project (MacCormick, Murphy, Najork, Thekkath, & Zhou, 2004) from Microsoft provides a distributed data management system which provides distributed locking, clustering, and storage of data based on B-trees. The Boxwood project aims to provide an infrastructure for

Table 22.1 Summarizing the data cloud technologies

Vendors	Execution engine	Distributed data storage (unstructured)	Distributed data storage (Structured)	High-level data analysis
Google	Google Map-reduce	Google File System(GFS)	BigTable	Sawsall
Microsoft	Dryad	Azure, Cosmos	SQL Azure	DryadLINQ
Apache	Hadoop Map-reduce	Hadoop Distributed File System (HDFS)	HBase Hypertable (Zvents)	Hive, Pig Latin, and Pig
Amazon	Elastic Compute Cloud, Elastic MapReduce,	Simple Storage Service (S3), Elastic Block Storage (EBS)	Dynamo, SimpleDB, Relational Database Service (RDS)	
Facebook/ Yahoo			PNUTS, Cassandra	Pig, Hive
Other Efforts	WheelFS, Synaptic Hosting, AppNexus, GoGrid, Rackspace	Synaptic Storage		

building high-level services such as file systems or databases. In the domain of structured peer-to-peer systems, several distributed hash table (DHT) based projects have dealt with similar problems of providing distributed storage or higher-level services over wide area networks. These DHT-based projects include CAN (Ratnasamy, S., Francis, P., Handley, M., Karp, R., and Shenker, 2001), Chord (Stoica, R. Morris, D. Karger, M.F. Kaashoek, and H. Balakrishnan, 2001), Tapestry (Zhao, J. Kubiatowicz, and A. D. Joseph, 2001), and Pastry (Rowstron, and P. Drushel, 2001). In addition, several database vendors have developed parallel databases that can store large volumes of data. Oracle's Real Application Cluster database (Oracle Real Application Cluster, 2010) uses shared disks to store data and a distributed lock manager. IBM's DB2 Parallel Edition (Baru et al., 1995) is based on the shared-nothing architecture (Stonebraker, 1986) similar to BigTable. Each DB2 server is responsible for a subset of the rows in a table that it stores in a local relational database. These approaches provide a complete relational model with transactions.

22.4 Case Studies: Harnessing the Data Cloud for Scientific Data Management

The past decade has seen a new set of challenges emerge in the scientific computing area. Data volumes underpin several of these challenges. It is very common for domain scientists to work with datasets in the order of tends of terabytes. By the same token, it is not uncommon for data volumes to be in the order of petabytes. Problems stem from the fact that the access times and transfer rates for commodity hard drives have not kept pace with improvements in their capacities. Some of this stems from the electro-mechanical nature of these disk drives.

Problems are further exacerbated by the fact that the concomitant processing for these datasets are also becoming computationally intensive. For a set of N data points the processing complexity could be super-linear. The data processing can also entail multiple accesses to the underlying datasets. There are several scientific applications that are adopting data cloud technologies to cope with their data-intensive computing challenges.

22.4.1 Pan-STARRS Data with GrayWulf

The Pan-STARRS project (Pan-STARRS project, 2010; Jedicke et al., 2006) is a large astronomical survey. The project will use a special telescope in Hawaii with a 1.4 gigapixel camera to sample the sky over a period of four years. The large field of view and relatively short exposures will enable the telescope to cover three quarters of the sky 4 times per year in 5 optical colors. This will result in more than a petabyte of images per year. The images will then be processed through an image segmentation pipeline that will identify individual detections, at the rate of 100 million detections per night. These detections will be associated with physical objects in the sky and loaded into the project's database for further analysis and processing. It is expected that the database will contain over 5 billion objects and well over 100 billion detections. The projected size of the database is 30 terabytes by the end of the first year, growing to 80 terabytes by the end of year 4.

As a part of an effort for better analysis of the data produced by Pan-STARRS, astronomers at Johns Hopkins University are partnering with Microsoft External Research to develop a set of software services and design principles known as GrayWulf (Szalay et al., 2009; Simmhan et al., 2009), which is based on the use of commodity hardware, Windows HPC Server 2008, and Microsoft SQL Server 2008. GrayWulf is an extension of the Beowulf cluster and provides access to the cloud from a user's desktop.

GrayWulf provides a shared queryable data store for users who perform analyses on the shared database. For achieving scalability, the shared database is partitioned and layered hierarchically. The lower data layer contains vertical partitions of the tables and includes three types of tables: Detections, Objects, and Metadata. Detection rows correspond to astronomical sources detected in either single or stacked images. Object rows correspond to unique astronomical sources and summarize statistics from both single and stack detections that are stored in different sets of columns respectively. Metadata refers mainly to telescope and image information. In general, this partition of data is associated with a specific portion of the sky. On top of the lower level data store, the loader/merger servers ingest new detections into the database's daily base.

Each user gets their own database (MyDB) on the servers to store intermediate results. Users have full control over their own MyDBs. Data may be uploaded to or downloaded from a MyDB and tables can be shared with other users: this creates a collaborative environment to share results.

GrayWulf uses workflows for several data analysis actions. The data valet workflow infrastructure includes a set of services for workflow visual composition, automatic provenance capture, scheduling, monitoring, and fault handling. Besides the Pan-STARR project, other projects (Li et al., 2008; Budavari, Malik, Szalay, Thakar, & Gray, 2005) have managed their large datasets with GrayWulf.

22.4.2   GEON Workflow with the CluE Cluster

The San Diego Super Conputing Center (SDSC) (San Diego Supercomputing Center, 2010) established its Data Central site, which hosts 27 PB of data and more than 100 specific databases. In 2009, SDSC started investigation on the hosting extremely large data sets on the Academic Cluster Computing Initiative (Zverina, 2010) cluster which is a joint effort by IBM/Google. The research will focus on using the data-parallel GEON LiDAR Workflow application (Jaeger-Frank et al., 2006).

22.4.3 SciDB

SciDB (SciDB, 2010) was initiated in two successive Extremely Large Databases (XLDB) (XLDB, 2007; XLDB, 2008) workshops that were organized in order to address the challenge of designing databases that can support the complexity and scale involved in scientific applications. Data storage and computation scale equally in SciDB. The data storage in SciDB can scale up to a few petabytes. Users can access a 10,000-node cloud using their laptop to process (a subset of) the dataset. Reported use cases of SciDB include domains such as optical astronomy, radio astronomy, earth remote sensing, environmental observation and modeling, and seismology (SciDB, 2010).

SciDB incorporates support for arrays and vectors. These data types come with several built-in optimized operators, which can be categorized as structural or content-dependent. As the name suggests, structural operators operate on the structure of the array independent of the data. Examples of structural operators include Subsample and Reshape. The Subsample operator takes as its input an array A and a predicate specified on the dimensions of A. The operator then generates a new array with the same number of dimensions as A, but where the dimension values satisfy the specified predicate (e.g., every 10th value of the dimension). Reshape is a more advanced structural operator. Reshape can convert an array to a new one with a different number of dimensions possibly with new dimension names, but the same number of cells. Content-dependent operators are those whose result depends on the data that is stored in the input array. Filter is one such operator, which takes as input an array A and a predicate P over the data values that are stored in the cells of A and returns an array with the same dimensions as A.

Additionally, SciDB supports uncertain data; here, "uncertain x" for any data type x uses two values. This feature is particularly useful in computing the result of a location calculation for observed objects commonly used in astronomy and GIS databases, which may contain some approximation errors due to hardware calibration.

SciDB places a special emphasis on the management of provenance for scientific data. Most data is expected to be fed to SciDB from input sources such as scientific measurement devices or sensors. Due to the large volumes of data, the data needs to be written into disk buckets that will contain rectangular chunks of the array. The R-Tree data structure is used to keep track of the location and contents of these buckets.

22.4.4 Astrophysical Data Analysis with Pig/Hadoop

(Loebman et al. 2009) have studied the emerging data management needs of the "N-body Shop" group, which specializes in the development and utilization of large-scale simulations specifically, "N-body tree codes" (Stadel, 2001). The objective of these simulations is to investigate the formation and evolution of large-scale structures in the universe.

The University of Washington's N-body Shop is representative of the current state-of-art in astrophysical cosmological simulation. In 2008, the N-body Shop was the 10th largest consumer of NSF Teragrid time, using 7.5 million CPU hours and generated 50 terabytes of raw data, with an additional 25 terabytes of post-processing information (Teragrid, 2010).

The N-body Shop has tried different approaches to solve issues in the management of voluminous datasets generated during the simulation. The efforts focused on improving the scalability in the RAM and I/O bandwidth. TIPSY (TIPSY, 2010) is one of the popular toolkits in this domain and it includes several scripts written in interpreted languages such as Python, Perl, or Interactive Data Language (IDL) (IDL, 2010). A trait common to these tools is that they operate in main memory. In the astrophysical community, Ntropy (Gardner, A. Connolly, and C. McBride, 2007; Gardner, 2007), a parallel library for the analysis of massive particle datasets has been harnessed to cope with the distributed memory needs. DBMSs and frameworks like Hadoop and Dryad offer similar scalability as application-specific libraries in terms of utilizing distributed memory. However, researchers benefit significantly from declarative languages built on top of these frameworks, such as Pig Latin and DryadLINQ.

Loebman et al's work (Loebman et al., 2009) has evaluated the performance of data analysis queries between DBMSs and Hadoop/Pig environments. For DBMS, this evaluation partitioned the data via an optimization process. For Hadoop/Pig, the data partitioning was done manually. In the experiments, for larger number of nodes, Hadoop/Pig showed shorter response times for the data analysis queries.

Similarly, (Cary et al., 2009) studied the applicability of MapReduce to spatial data processing workloads and validated the excellent scalability of MapReduce in that domain. (Palankar, et al., 2008) recently evaluated Amazon S3 as a feasible and cost effective alternative for hosting scientific dataset, particularly those produced by large community collaborations such as LSST (LSST, 2009).

22.4.5 Public Data Hosting by Amazon Web Services

Amazon Web Services (AWS) hosts Public Data Sets (Amazon Public Datasets, 2010) in their centralized repository of public data sets that can be easily integrated with AWS cloud-based applications. Currently, users can access various scientific datasets such as DNA sequences from GenBank, human genome data, influenza virus including updated Swine Flu sequences published by NCBI, Sloan Digital Sky Survey DR6 subset, and daily global weather measurements (1929–2009). Users can create their own Elastic Block Service (EBS) volume with a snapshot of the selected dataset and access, modify and perform computations on their virtual machine instances.

22.5 A Gap Analysis of Data Cloud Capabilities

In this section we present an analysis of how current capabilities within the data cloud fall short of the current and future needs of scientific data management and processing. In some cases, existing systems have come up with ad hoc solutions to address the shortcomings while in others a lot of research still needs to be done.

22.5.1 The Impedance Mismatch

Database researchers have been dealing with impedance mismatch (Gray et al., 2005), which refers to the mismatch between the programming model and the database capabilities. Data cloud technologies also lag in their inability to match the functionalities required by the scientific application. For example, most of the current data cloud implementations do not support N-dimensional arrays as core data types in their computing model. This issue has been addressed in the SciDB project (SciDB, 2010) by providing a data structure of arrays and vectors along with a set of operators. Indexing schemes based on the appropriate data types for the scientific data objects are another feature that is currently lacking in data cloud technologies. These can cause significant performance issues for scientific data management.

22.5.2 Fault Tolerance

The cloud computing environment is in general built using affordable, commodity hardware; as a result, failures are not uncommon in these settings. The probability of a failure occurring during a long-running data analysis task is thus relatively high. For example, Google reports an average 1.2 failures per analysis job (Deanand & Ghemawat, 2004). Fast failure detections and recovery schemes will provide a more reliable data analysis environment for scientific users.

22.5.3 Scientific Data Format and Analysis Tools

Discovering a sub-array for a computing module is a critical task for scientific applications. Scientific data arrays are often stored as files that are based on scientific file formats such as HDF (HDF, 2010), NetCDF (NetCDF, 2010), and FITS (FITS, 2010). These file formats provide attributes for these files, which in turn give clues to discover the relevant sub-array.

Scientific applications open files to check their attributes. To achieve better performance, scientists often encode key attributes into the file system hierarchy. This set of strings represents not only the path to the file in the file system, but also provides useful filtering functionality to assist the data discovery process. In these settings the users discover data in two steps: (1) the user searches the files relevant to the sub-array by means of the information encoded in the file path. (2) then the application scans the file(s) and searches the right sub-array.

However, as the file system grows to billions of files with petabytes of data volumes, scientific data needs more information to identify and describe itself. Similarly, more database-like features are required to handle files and the information about them. Many scientific data management systems (Peng & Law, 2004; Plale et al., 2005) have addressed this need for metadata – information about the data. As a part of the scientific data management system, the metadata of the data products are stored and managed in a conventional database system. This metadata is generally stored separate from the actual datasets and contains a logical or physical link to the data file.

BigTable like data cloud storage provides characteristics of a distributed file system and databases with parallel data discovery and processing. However, to benefit from this technology, scientific users incur an overhead similar to the one in current scientific data management systems: extracting attributes, storing attributes in the table, and generating queries for them. In general, many of these steps are similar across applications if they use the same data formats. For scientific applications that share the same data format, many of these steps will overlap. Therefore, predefined data types and associated querying mechanism would be a useful technology for scientific users and developers.

22.5.4 Integration with the Object Oriented Programming Model

Integrating database systems with the object oriented programming model can be counted as one of the significant advances in database technology. There are many object-oriented databases that treat any data type as an encapsulated type that can be stored as a value in the field of a record (Ozone, 2010; ZODB, 2010). This approach virtually resolves the problem of the lack of data types in database-style systems. This also provides an easy-to-use interface for the programming environment. Current data cloud systems do not adequately address whether their approaches are a natural fit for the object-oriented programming model.

22.5.5 Working with Legacy Software

A closer look at scientific software reveals that computing or simulation components in many of the applications rely on conventional file systems. Therefore, file path is widely used as an input parameter for executables. Likewise, many of the script files are dependent on the file system. A new method is required which can transform the software interface from one that is file system based to one that is data cloud compatible.

Finally, visualization tools require support for data from the data cloud. Most of the vendors of data clouds develop their applications based on Web browser technology. However, visualization tools are widely used in the scientific community, because general-purpose Web browsers cannot satisfy specific requirements such as 3D graphical rendering, and support for scientific data formats. To provide active access to the data stored in the data cloud, efficient mechanisms to interact with the visualization tools is required.

22.5.6 Real-Time Data

One of the distinctive characteristics in scientific data management is the variety of data sources. Modern sensors and digitized experimental equipments inject the data directly into the data storage. This real-time data often arrives at the storage as streaming data. Stream data processing differs from conventional data processing performed in data cloud implementations, such as crawled web pages, or personal information.

For example, Unidata (Unidata, 2010) provides data collections from observational data sources for meteorology research and weather forecasting. Satellite data is provided every hour with various resolutions. Every day around 140,000 wind and temperature observations are delivered from around 4000 aircrafts. Also, the data from 152 NexRed Doppler radar stations are collected every 5, 6, or 10 minutes. Some of this data must race to be delivered to the simulation component for emergency weather event such as a tornado or hurricane. Efficient data access to real-time data would enable real-time data mining and eventually improve the overall performance of computing significantly.

22.5.7 Programmable Interfaces to Performance Optimizations

Data cloud promises scalability for scientific data management. Scalability is a critical requirement for data management with continually increasing data volumes. However, many scientific research projects also require reasonably low latency to satisfy the on-demand access and computing requirements. Google Earth stores preprocessed imagery (approximately 70 terabytes) on their disk space (Chang et al., 2006). These imageries are indexed into a relatively small ($\sim$500 GB) table. This table must serve tens of thousands of queries per second per data center with low latency. Therefore, this table is hosted across hundreds of tablet servers and contains in-memory column families.

As we see in Google Earth's example, designing well performing data systems is not trivial with data cloud implementations. Unlike conventional database systems that provide a built-in cache, replication scheme, or optimizing schemes, data cloud implementations require programmers and system designers to be involved in the process of performance optimization. Programmable interfaces to performance optimizations is needed.

22.5.8 Distributed Database Issues

Data clouds share many of the issues from distributed database systems. This includes fault tolerance, conflict management, distributed lock, and data integrity. Fault tolerance is critical for high-throughput computing, which can involve processing that can take up to several days to a few weeks to complete a job. Data integrity is essential to ensure the accuracy of the result. Furthermore, data assurance is required for data that is archived for the long term.

22.5.9 Security and Privacy

Scientific users require security and privacy to access their personal data products. Users require secure access to the data for discovery, browsing and computing. Therefore, sensitive data may be encrypted prior to being uploaded to the data cloud storage. To avoid unauthorized access to the sensitive data, any application running in the cloud should not be allowed to directly decrypt the data. However, to decrypt the dataset, moving entire (or large part of) dataset back and forth from the data cloud storage is a very bandwidth/computing intensive task. Thus, (Adabi, 2009) suggests that a data analysis system that can operate directly on encrypted data (Agrawal, Kiernan, Srikant, & Xu, 2004; Hacigumus, Iyer, Li, & Mehrotra, 2002; Ge & Zdonik, 2007; Kantarcoglu & Clifton, 2004; Mykletun & Tsudik, 2006) will improve the performance significantly.

22.6 Conclusions

Cloud computing offers obvious advantages, such as co-locating data with computations and an economy of scale in hosting the services. While these platforms obviously perform very well for their current intended use in search engines or elastic hosting of commercial Web sites, their role in scientific computing is still evolving. In some scientific analysis scenarios, the data needs to be close to the experiment. In other cases, the nodes need to be tightly integrated with a very low latency, while in some cases a high I/O bandwidth is required.

There has been a strong trend to move scientific data to the cloud. We expect this trend to continue and accelerate in the future. As more and more systems start using the data cloud we expect that the issues outlined in the preceding section will become increasingly important, and also be an area where there will be a good deal of research activity.

References

Adabi, D. J. (2009). Data management in the cloud: Limitations and opportunities. *IEEE Data Engineering Bulletin, 32*(1), 4–12.

Agrawal, R., Kiernan, J., Srikant, R., & Xu, Y. (2004). Order preserving encryption for numeric data. *Proceedings of SIGMOD*, 563–574.

Amazon Elastic MapReduce (2010). http://aws.amazon.com/elasticmapreduce/. Accessed on February 20, 2010.

Amazon EBS (2010), http://aws.amazon.com/ebs/. Accessed on February 20, 2010.

Amazon Public Datasets (2010), http://aws.amazon.com/publicdatasets/. Accessed on February 20, 2010.

Amazon RDS (2010), http://aws.amazon.com/rds/. Accessed on February 20, 2010.

Amazon SimpleDB (2010), http://aws.amazon.com/simpledb/. Accessed on February 20, 2010.

Amazon Web Services (2010). http://aws.amazon.com/ Accessed on February 20, 2010.

Antonioletti, M., Krause, A., Paton, N. W., Eisenberg, A., Laws, S., Malaika, S., et al. (2006). The WS-DAI family of specifications for web service data access and integration. *ACM SIGMOD Record, 35*(1), 48–55.

AppNexus (2010). http://www.appnexus.com/ Accessed on February 20, 2010.

AT&T Synaptic Hosting (2010). http://www.business.att.com/enterprise/Family/application-hosting-enterprise/synaptic-hosting-enterprise/ Accessed on February 20, 2010.

Baru, C. K., Fecteau, G., Goyal, A., Hsiao, H., Jhingran, A., Padmanabhan, S., et al. (1995). DB2 parrel edition. *IBM Systems Journal, 34*(2), 292–322.

Budavari, T., Malik, T., Szalay, A. S., Thakar, A., & Gray, J. (2003). SkyQuery – A prototype distributed query web service for the virtual observatory. In H. Payne, R. I. Jedrzejewski, & R. N. Hook (Eds.), *Proceedings of ADASS XII, Astronomical Society of the Pacific, ASP Conference Series* (Vol. 295, p. 31).

Cary, A., Sun, Z., Hristidis, V., & Rishe, N. (2009). Experiences on processing spatial data with mapreduce. *Proceedings of the 21st SSDBM Conference. Lecture notes in computer science, Vol. 5566*, 302–319.

Chang, F., Dean, J., Ghemawat, S., Hsieh, W. C., Wallach, D. A., Burrows, M., et al. (November 2006). Bigtable: A distributed storage system for structured data. *OSDI'06: Seventh Symposium on Operating System Design and Implementation, Seattle, WA*, 205–218.

Chervenak, A., Foster, I., Kesselman, C., Salisbury, C., & Tuecke, S. (2001). The data grid: Towards an architecture for the distributed management and analysis of large scientific datasets. *Journal of Network and Computer Applications, 23*, 187–200.

Dean, J., & Ghemawat, S. (2008). MapReduce: Simplified data processing on large clusters. *Communications of the ACM, Vol. 51(1)*, 107–113.

Deanand, J., & Ghemawat, S. (December 2004). Mapreduce: Simplified data processing on large clusters. *In Proceedings of OSDI, San Francisco, CA*, 137–150.

Dennins D. Gannon, D., & Dan D. Reed, D. (2009). "Parallelism and the cloud." In: T. Hey, S. Hensley, and K. Tolle (Eds.), *The fourth paradigm: Data-intensive scientific discovery, Microsoft research* (pp 131–136), ISBN-10:0982544200.

FITS (2010). http://fits.gsfc.nasa.gov/. Accessed on February 20, 2010.

Gardner, J. (2007). Enabling knowledge discovery in a virtual universe. *Proceedings of TeraGrid '07: Broadening Participation in the TeraGrid*, ACM Press.

Gardner, J. P., Connolly, A., & McBride, C. (2007). Enabling rapid development of parallel tree search applications. *Proceedings of the 2007 Symposium on Challenges of Large Applications in Distributed Environments (CLADE 2007)*, ACM Press, 1–10.

Ge, T., & Zdonik, S. (2007). Answering aggregation queries in a secure system model. *Proceedings of VLDB*, 519–530.

GenBank (2010). http://www.ncbi.nlm.nih.gov/Genbank/ Accessed on February 20.

Ghemawat, S., Gobioff, H., & Leung, S.-T. (October 2003). The google file system. *Appeared in 19th ACM Symposium on Operating Systems Principles, Lake George, NY, 29*–43.

GoGrid (2010). http://www.gogrid.com Accessed on February 20, 2010.

Google App Engine (2010). http://code.google.com/appengine/ Accessed on February 20, 2010.

Gray, J. (2009). Jim gray on eScience: A transformed scientific method. In T. Hey, S. Hensley, & K. Tolle (Eds.), *The fourth paradigm: Data-intensive scientific discovery, Microsoft research* (pp xvii–xxxi), ISBN-10:0982544200.

Gray, J., Liu, D. T., Nieto-Santisteban, M. A., Szalay, A. S., Heber, G., & DeWitt, D. (December 2005). Scientific data management in the coming decade. *SIGMOD Record, 34*(4), 34–41.

Hacigumus, H., Iyer, B., Li, C., & Mehrotra, S. (2002). Executing sql over encrypted data in the database-service-provider model. *Proceedings of SIGMOD*, 216–227.

HDF (2010) http://www.hdfgroup.org/HDF5/. Accessed on February 20, 2010.

Isard, M., & Yu, Y. (July 2009). Distributed data-parallel computing using a high-level programming language. *Proceedings of the International Conference on Management of Data (SIGMOD)*, 987–994.

Isard, M., Budiu, M., Yu, Y., Birrel, A., & Fetterly, D. (March 2007). Dryad: Distributed data-parallel programs from sequential building blocks. *Proceedings of European Conference on Computer Systems (EuroSys), Lisbon, Portugal*, March 21–23, 59–72.

IDL (2010) Interactive Data Language. http://www.ittvis.com/ProductServices/IDL.aspx. Accessed on February 20, 2010.

Jaeger-Frank, E., Crosby, C. J., Memon, A., Nandigam, V., Conner, J., Arrowsmith, J. R., et al. (December 2006). A domain independent three tier architecture applied to Lidar processing and monitoring. *In the Special Issue of the Scientific Programming Journal devoted to WORKS06 and WSES06*, 185–194.

Jedicke, R., Magnier, E. A., Kaiser, N., & Chambers, K. C. (2006). The next decade of solar system discovery with pan-STARRS. *Proceedings of IAU Symposium 236*, 341–352.

Kantarcoglu, M., & Clifton, C. (2004). Security issues in querying encrypted data. *19th Annual IFIP WG 11.3 Working Conference on Data and Applications Security*, 325–337.

Lakshman, A., Malik, P., & Ranganathan, K. (2008). Cassandra, structured storage system over a P2P network. *Keynote Presentation, SIGMOD, Calgary, Canada*, 5–5.

Lammana, M. (November 2004). Nuclear instruments and methods in physics research section A: Accelerators, spectrometers, detectors and associated equipment. *In the Proceedings of the 9th international Workshop on Advanced Computing and Analysis Techniques in Physics Research* (Vol. 534, No. 1–2, pp. 1–6).

Large Hadron Collider project (2010). http://public.web.cern.ch/public/en/LHC/LHC-en.html Accessed on February 20, 2010.

Li, Y., Perlman, E., Wan, M., Yang, Y., Meneveau, C., Burns, R., et al. (2008). A public turbulence database and applications to study lagrangian evolution of velocity increments in turbulence. *Journal of Computational Physics, 9*(31), 1468–5248.

Loebman, S., Nunley, D., Kwon, Y. C., Howe, B., Balazinsk, M., & Gardner, J. P. (2009). Analyzing massive astrophysical datasets: Can pig/hadoop or a relational DBMS help? *Proceedings of the Workshop on Interfaces and Architecture for Scientific Data Storage (IASDS)*, 1–10.

LSST Science Collaborations and LSST Project (2009). *LSST Science Book, Version 2.0, arXiv:0912.0201*, http://www.lsst.org/lsst/scibook.

MacCormick, J., Murphy, N., Najork, M., Thekkath, C. A., & Zhou, L. (December 2004). Boxwood: Abstractions as the foundation for storage infrastructure. *Proceedings of the 6th Symposium on Operating Systems Design and Implementation (OSDI 2004), San Francisco, CA, USA*, 105–120.

Microsoft, SQL Azure (2010). http://www.microsoft.com/windowsazure/sqlazure/ Accessed on February 20, 2010.

Microsoft, Windows Azure (2010). http://www.microsoft.com/windowsazure/ Accessed on February 20, 2010.

Moore, R. W. Moore, R. W., Jagatheesan, A. Jagatheesan, A., Rajasekar, A. Rajasekar, A., et al. (April 2004). "Data grid management systems,". *Proceedings of the 21st IEEE/NASA Conference on Mass Storage Systems and Technologies (MSST), April 13–16, 2004, College Park, Maryland, USA*, April 13–16, 2004.

Mykletun, E., & Tsudik, G. (2006). Aggregation queries in the database-as-a-servicemodel. *IFIP WG 11.3 on Data and Application Security*, 89–103.

NCBI (2010). http://www.ncbi.nlm.nih.gov/guide/ Accessed on February 20, 2010.

NetCDF (2010). http://www.unidata.ucar.edu/software/netcdf/. Accessed on July 16, 2010.

Olston, C., Reed, B., Srivastava, U., Kumar, R., & Tomkins, A. (June 2008). Pig latin: A not-so-foreign language for data processing. *ACM SIGMOD 2008 International Conference on Management of Data, Vancouver, Canada*, 1099–1110.

OpenMPI (2010). http://www.open-mpi.org/. Accessed on February 20, 2010.

OpenPBS (2010). http://www.pbsgridworks.com. Accessed on February 20, 2010.

Oracle Database 11 g (2010), http://www.oracledatabase11g.com/. Accessed on February 20, 2010.

Oracle Real Application Cluster (2010). http://www.oracle.com/technology/products/database/clustering/index.html. Accessed on Febrary 20, 2010

Ozone (2010). http://www.ozone-db.org/frames/home/what.html. Accessed on February 20, 2010.

Palankar, M. R., Iamnitchi, A., Ripeanu, M., & Garfinkel, S. (2008). Amazon S3 for science grids: A viable solution? *DADC '08: Proceedings of the 2008 International Workshop on Data-Aware Distributed Computing*, 55–64.

Pan-STARRS project (2010). http://pan-starrs.ifa.hawaii.edu/public/ Accessed on February 20, 2010.

Peng, J., & Law, K. H. Reference NEESgrid data model (Tech. Rep. NEESgrid-2004-40).

Pike, R., Dorward, S., Griesemer, R., & Quinlan, S. Interpreting the data: Parallel analysis with Sawzall. *Scientific Programming Journal Special Issues on Grids and Worldwide Computing Programming Models and Infrastructure, 13*(4), 227–298.

Plale, B., Gannon, D., Alameda, J., Wilhelmson, B., Hampton, S., Rossi, A., et al. (2005). Active management of scientific data. *IEEE Internet Computing Special Issue on Internet Access to Scientific Data, 9*(1), 27–34.

PubCam (2010). http://pubchem.ncbi.nlm.nih.gov/ Accessed on February 20, 2010.

PubMed (2010). http://www.ncbi.nlm.nih.gov/pubmed/ Accessed on February 20, 2010.

Rackspace (2010). http://www.rackspace.com/index.php Accessed on February 20, 2010.

Ratnasamy, S., Francis, P., Handley, M., Karp, R., & Shenker, S. (August 2001). A scalable content-addressable network. *Proceedings of SIGCOMM*, 161–172.

Rowstron, A., & Drushel, P. (November 2001). Pastry: Scalable, distributed object location and routing for large scale peer-to-peer systems. *Proceedings of Middleware 2001*, 329–350.

San Diego Supercomputing Center (2010), http://www.sdsc.edu/. Accessed on February 20, 2010.

SciDB (2010). http://scidb.org/ Accessed on February 20, 2010.

Simmhan, Y., Barge, R., van Ingen, C., Nieto-Santisteban, M., Dobos, L., Li, N., et al. (2009). GrayWulf: Scalable software architecture for data intensive computing. *Proceedings of the 42nd Hawaii International Conference on System Science*, 1–10.

Singh, G., Bharathi, S., Chervenak, A., Deelman, E., Kesselman, C., Manohar, M., et al. (2003). A metadata catalog service for data intensive applications. *IEEE, ACM, Super Computing the international conference for High Performance Computing, Networking, Storage and Analysis*, 33–50.

Stadel, J. G. (2001). Cosmological N-Body simulations and their analysis. (Doctoral dissertation, University of Washington, 2001).

Stoica, I., Morris, R., Karger, D., Kaashoek, M. F., & Balakrishnan, H. (August 2001). Chord: A scalable peer0to-peer lookup service for internet applications. *Proceedings of SIGCOMM*, 149–160.

Stonebraker, M. (1986). The case for shared nothing architecture. *Database Engineering, 9*(1), 4–9.

Szalay, A., Bell, G., Vandenberg, J., Wonders, A., Burns, R., Fay, D., et al. (2009). GrayWulf: Scalable clustered architecture for data intensive computing. *Proceedings of the 42nd Hawaii International Conference on System Science*, 1–10.

Teragrid (2010), http://www.teragrid.org/. Accessed on February 20, 2010.

Thain, D., Tannenbaum, T., & Livny, M. (February–April 2005). Distributed computing in practice: The condor experience. *Concurrency and Computation: Practice and Experience, 17*(2–4), 323–356.

TIPSY (2010). http://hpcc.astrowaxhington.edu/tools/tipsy/tipsy.html. Accessed on February 20, 2010.

The Academic ClusterComputing Initiative (ACCI 2007). Google and IBM Announce University Initiative to Address Internet-Scale Computing Challegne, Google Official Press Center, http://www.google.com/intl/en/press/pressrel/20071008_ibm_univ.html.

The Globus Toolkit (2010). Data replication service. http://www-unix.globus.org/toolkit/docs/4.0/techpreview/datarep/ Accessed on February 20, 2010.

Unidata (2010). http://www.unidata.ucar.edu/ Accessed on February 20, 2010.

Yu, Y., Gunda, P. K., & Isard, M. (October 2009). Distributed aggregation for data-parallel computing: Interfaces and implementations. *Proceedings of the Symposium on Operating Systems Principles (SOSP)*.

Zhao, B. Y., Kubiatowicz, J., & Joseph, A. D. (April 2001). *Tapestry: An infrastructure for fault-tolerant wide-area location and routing* (Tech. Rep. UCB/CSD-01-1141, CS Division, UC Berkeley).

Zverina, J. (2010) *San Diego supercomputing center begins cloud computing research using the Google IBM clue cluster*. http://www.sdsc.edu/News%20Items/PR021309_clue.html, Accessed on February 20, 2010.

ZODB (2010). http://wiki.zope.org/481zope2/ZODBZopeObjectDatabase. Accessed on February 20, 2010.

Zookeeper (2010). http://wiki.apache.org/hadoop/ZooKeeper. Accessed on February 20, 2010.

Chapter 23
Feasibility Study and Experience on Using Cloud Infrastructure and Platform for Scientific Computing

Mikael Fernandus Simalango and Sangyoon Oh

23.1 Introduction

In the academia, conducting experiments, gathering and processing data are trivial tasks. However, in some circumstances for example when processing large-set of data, method to accomplish the task can be tricky and non-trivial, not to mention problematic. When it comes to numerical float or array, processing a very large set of this kind of data will require superior computing power. Some entities with splendid compute nodes may process the data directly in their own infrastructure. Yet, providing decent infrastructure for processing large data sets or resource-extensive task can be a challenge for other entities. Infrastructure owned by an entity conducting the computation can be insufficient or just enough to execute the tasks above the minimum requirements. A common practice for infrastructure-limited research entity in dealing with such kind of situation is to outsource the compute task to external party with more superior compute power in order to reduce waiting time for producing the result of data processing.

The development of grid computing, however, enables institutions within local vicinity or even remotely separated ones in global scope to collaborate and contribute their compute resources to build a more superior compute facility. This may lessen the gap between less superior entity and more superior entity in processing the scientific data. Nevertheless, the notion that grid is formed by collaboration among several institutes may hinder another body outside the collaboration contract to use the leveraged computing power. Efforts have been made to build grid infrastructure that can be widely used by articulating economic principles (Buyya & Bubendorfer, 2008), but still, they have yet to become mature.

M.F. Simalango (✉)
WISE Research Lab, Ajou University, Suwon, South Korea
e-mail: mikael@ajou.ac.kr

S. Oh
WISE Research Lab, School of Information and Communication Engineering,
Ajou University, Suwon, South Korea
e-mail: syoh@ajou.ac.kr

B. Furht, A. Escalante (eds.), *Handbook of Cloud Computing*,
DOI 10.1007/978-1-4419-6524-0_23, © Springer Science+Business Media, LLC 2010

Cloud computing initially came up to the surface as a buzzword in 2007 (Google Trend for Cloud Computing, http://www.google.com/trends?q=cloud+computing) and has been gaining more popularity these days. Some consider it as marketing hype instead of sound breakthrough in distributed computing (Johnson, 2008; Boulton) Nevertheless, the trend shows that public interests in cloud computing keep growing in the first two years of the cloud computing era. Big enterprises like Amazon through Amazon EC2 (Amazon Elastic Compute Cloud, http://aws.amazon.com/ec2/) and Google with Google App Engine (Google App Engine. http://code.google.com/appengine/) have become the main advocates for introducing and driving the adoption of cloud computing to wider audience.

During the advent of cloud computing, debates were over the definition and scope of cloud computing (Foster, Zhao, Raicu, & Lu, 2008; Mei, Chan, & Tse, 2008; Vouk, 2008; Buyya, Yeo, & Venugopal, 2008). As time progresses, cloud computing starts to find its shape (Rimal, Choi, & Lumb, 2009) and issues have been shifting from debatable opinions over interpretation of corresponding terms in the new paradigm into maturing and nurturing the technology through the execution of prospective researches (Birman, Chockler, & van Renesse, 2009). Through the diversity of the researches, it should also be reasonable to expect the implementation of cloud computing paradigm and techniques in the academia if they are proven to be feasible and can outperform or at least comparable with existing approach.

In a previous article, we raised the issue about the use of cloud computing especially the enterprise cloud for scientific computing (Simalango & Oh, 2009). We revisited characteristics of scientific computing in general and explained common methods used in scientific computing realm. We studied how the enterprise-powered cloud computing could also satisfy the requirements of scientific computing and complement or even replace classical approach. In this article, we will elaborate more about the scientific computing, summarize other researchers' efforts to do computation in the cloud, add our experience in setting up the environment for a tiny internal cloud and scaling up the cloud through the utilization of external enterprise cloud, and also give recommendation for adoption of cloud computing in accomplishing scientific compute tasks.

23.2 Scientific Compute Tasks

Dissecting characteristics of scientific computing is like mapping the genes. At the bigger scope -the double helix structure of DNAs in a chromosome- we notice big affinity of the exterior structure of genes. As we digest more scrupulously to the smaller details, however, it can be noticed that tasks pertaining to and information confined in each gene are different. Similarly, scientific computing tasks may seem indistinguishable for non-practitioners while researchers at the same time perceive the broadness of the tasks. Each field in sciences conducts various compute tasks with different computing strategies. We are then interested more in how a certain strategy is selected for accomplishing a scientific compute task.

Scientific compute tasks are mainly related with two things, processing time and result. A compute task that can be completed in less time is favorable in general sense. With quicker task completion, other jobs in queue can be staged and sequenced to be processed in order to obtain the final result. Implications brought by the final result after series of analyses of the outputted data will determine the state of success of the corresponding compute task. The result, however, depends on the correctness of the underlying logic. Since the logic is formulated by the process designer or researcher, proper and efficient implementation of the logic will yield less processing time and expected result given that the logic is verifiably correct for the task.

Designing implementation of algorithms for a scientific compute task can be trivial. Still, in some cases complexities exist and transform the task to be non-trivial. The complexities are primarily caused by two factors: size of input and algorithms in the processing block. Input to a compute task can be huge hence the time needed to process it completely also becomes lengthier. At the other side, the processing block may also implement complex algorithms which in turn requires hefty amount of resources. This results in the necessity of superior computation capability. Huge inputs and complex algorithms infer intense computing. By basing on compute intensity as criteria, we can categorize scientific compute tasks into two groups, resource-intensive tasks and data-intensive tasks.

A resource-intensive task refers to a task that makes use of plenty of compute resources. Compute resources can be I/O, CPU cycles, memory, temporary storage, and also energy. Examples of resource-intensive task in scientific realm are system modeling, image rendering (especially in 3D), and forecasting. Such kinds of tasks involve chains of complex computations which demand big allocation of resources. Slightly different with the previous one, data-intensive task refers to a task that deals with processing of huge size of input data. An example of task that falls into this category is data mining for huge set of data (iris data in biometrics, protein data in biology, access log data in computer network, etc.). However, for resource-constrained processing node, a data-intensive task can also become a resource-intensive task if proper strategy for processing such data can not be applied.

Enabling parallelism has been a strategy used in the execution and accomplishment of scientific compute tasks especially the complex ones. With parallelism, a complex compute task which requires enormous amount of resources, let say equivalent with the specifications of a minicomputer or mainframe, can be divided into smaller parts, each to be fitted and deployed into less powerful infrastructure, for example PC-class compute nodes. Techniques and paradigm in achieving parallelism have also evolved, from clustering to grid computing and now cloud computing.

In classical clustering, a virtually more superior compute node which is named a compute cluster is formed by several physical compute nodes which share the same software system. The nodes in a compute cluster are also placed in close vicinity and in general case, the hardware specifications are also the same. Hence, a compute cluster is basically built over homogeneous compute nodes. Homogeneity in elements of a compute cluster lessens communication overhead among nodes in cluster. At the same time, however, the homogeneity brings inflexibility in extending the compute capability. Since the requirements are tight, future node provisioning and configuration can be a problem for an institution with limited resources.

Grid computing enables resource pooling from distributed sets of compute nodes. The compute-node set can be a cluster of commodity computers and servers, or simply an individual node. Different with a traditional compute cluster which is built on top of homogeneity, a grid can be formed by diverse systems. A grid federates resources comprising storage, network, and compute which are distributed geographically and generally heterogeneous and dynamic. A grid defines and provides a set of standard protocols, middleware, toolkits, and service to discover and share the componential distributed resources. Upon the creation of a grid, a computation power equivalent to supercomputers and large dedicated clusters can be built and utilized at cheaper price compared to the purchase of a mainframe or supercomputers.

There are two types of grid based on its usage method namely institutional grid and community grid. In institutional grid, utilization of resources in the grid is possible only by the institutions or individuals donating compute resources to the grid. In contrast, community grid model also offers compute resources to public users. Nevertheless, resource utilization is usually tied within a contract. A grid user is allocated certain amount of resources and extra resource allocations will be made possible after approval of proposal of additional resource request (Foster et al., 2008).

A common technique in accomplishing scientific compute task in the grid is through batch processing. This technique is mainly aimed at solving data-intensive tasks. Initially, data are segmented into several sequences and workflows pertinent to the segmentations (Simmhan, Barga, Lazowska, & Szalay, 2008) are then created by a scheduler. A batch process is then executed in order to process all the sequences and output the result. The overall process is depicted in Fig. 23.1. In the picture, we redraw and combine common processes in implementations mentioned in Matsunaga, Tsugawa, and Fortes (2008) and Liu and Orban (2008). In the picture, the job scheduler manages the process of assigning workflows for processing in compute infrastructure, which is the grid. The outputs of the process are chunks of data which have to be merged later to yield the final result.

Based on the scheme in Fig. 23.1, we can see the main idea is implementing parallelization through simultaneous processing of segmented data. Through the creation of multiple workflows and their delegation to worker/compute nodes in the grid, an initially time-consuming process over a large set of data (data-intensive task) can be reduced into sets of smaller processes running in parallel. Consequently, this approach shortens the time to yield the result compared to processing of such data in serial.

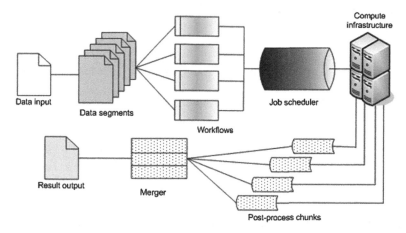

Fig. 23.1 A batch-process scheme for processing large scientific data

23.3 Scientific Computing in the Cloud

If we revisit the question about the feasibility of cloud computing for processing scientific data, the first question can be how scalable the cloud is to compute scientific tasks which can be increasingly complex in their dynamics. Bureaucracy in the grid and the notion that a grid is operated on project basis may have hindered the paradigm to become more popular in use. Conversely, such hurdle may not be found in the cloud computing paradigm which offers the economies of scale and pay-as-you-go principle. Utilizing compute resources offered by cloud service providers by only providing credit card information is tempting and the same time attractive and appealing. Also, extra computational resources can be provisioned immediately to a cloud user (on-demand utilization) thus skipping the hassle that may have been experienced by a grid user. The feature of utility computing in the cloud paradigm, which is paying only the amount of resources used, also brings convenience that makes utilization of the cloud seem to be compelling for the academia.

23.3.1 Cloud Architecture as Foundation of Cloud-Based Scientific Applications

Doing science in the cloud inevitably depends on how cloud architecture can satisfy the requirements of scientific computing. Despite differing in articulation over the definition of cloud computing, there is commonness in cloud architecture adhering to layered architecture as depicted in Fig. 23.2. In the figure, we can see that cloud architecture is constituted of three layers from the bottom up: IaaS (Infrastructure as a Service) layer, PaaS (Platform as a Service) layer, and SaaS (Software as a

Fig. 23.2 Layered cloud architecture

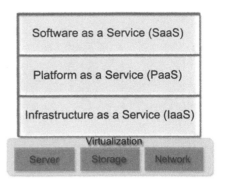

Service) layer respectively. The layers also refer to types of services offered in the cloud. Below the three layers, there exists virtualization technology which handles the logic of transforming physical server, storage, and network into a virtualized environment.

Virtualization technologies architect the creation of virtual machines on physical infrastructure which support the overlaying cloud services. Revisiting its definition, virtualization is defined as the abstraction of the logical resources away from the underlying physical resources thus consequently enabling abstraction of the coupling between hardware and OS. This definition is closed to operating system virtualization if compared with types of virtualizations noted by Sun Microsystems (Cloud Computing Primer. http://www.sun.com/offers/docs/cloud_computing_primer.pdf.). In a virtualized environment, multiple operating systems can run the same physical infrastructure and share the resources. It is made possible by the role of virtualization machine monitor (VMM) or hypervisor placed and running between the hardware and operating systems (also referred to as the guests of VMM). VMM arbitrates accesses to underlying physical host and manages the resource sharing between its guest operating systems.

Capabilities pertinent to virtualization can be categorized into workload isolation, workload consolidation, and workload migration (Uhlig, et al., 2005). Traditionally, an operating system runs various applications under the same system environment. Workload isolation enables isolation of multiple software stacks in their own VMs thus improving security and reliability. Meanwhile, workload consolidation refers to the capability of virtualization to consolidate individual workloads onto single physical platform thus reducing the total cost of ownership. Finally, workload migration makes it possible to decouple the guest from the hardware it is currently running and migrate it to a different platform.

In Fig. 23.1, we reviewed how we can do science, which is processing data-intensive task, in the grid. As we may notice earlier, the basic notion is parallel processing of data chunk in the worker nodes followed with the compilation of the partial result to yield the final output. Referring to the cloud architecture in Fig. 23.2 and the role of virtualization, we consider that science in the cloud implements the same basic principle, parallelism, in different methodology. Through the techniques

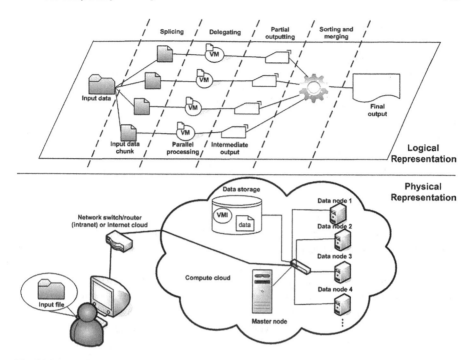

Fig. 23.3 Processing data-intensive scientific task in the cloud

in doing science in the cloud also differ, in general they resemble the process we depict in Fig. 23.3.

Referring to Fig. 23.3, a researcher keeps in hand input data to be processed. The process will be handled by certain package of software applications that will be loaded to the data nodes. Before any data node is deployed, the package is bundled in a virtual machine image (VMI) that is stored in data storage infrastructure in the cloud. The example for the persistent storage service in the enterprise cloud is Amazon S3 (Simple Storage Service) or Amazon EBS (Elastic Block Store). There can be several VMIs stored in the storage infrastructure thus the cloud management application or system should provide the interface to the repository of the images. Cloud management system itself can be located in the master node or hosted in another node outside the cloud. It enables dynamic listing of VMIs and provides other interfaces to cloud management and monitoring. However, in the absence of cloud management system, image deployment to compute nodes and corresponding configurations should be done manually.

Upon getting the list of VMIs from the cloud management interface, the researcher will then choose the suitable VMI and then create a number of instances accordingly. One or some of the instances will act as the master node which will splice the data input into several parts and delegate their processing to the worker nodes. Master node(s) control the executions of jobs, collate the partial output, and notify the researcher when all jobs have been finished and final output is ready to be analyzed. In a bigger picture, this resembles the process in the grid computing paradigm. It is then understandable if some experts also state that the cloud is the evolution of the grid (Foster et al., 2008, Giunta, Lacetti, Montella, 2008).

A recent finding by Dejun, Pierre, and Chi (2009) shows that the performance of EC2 instances is relatively stable. However, one can not predict the performance of a new instance even though statistics of performance of the same instances over some period of time in the past has been collected. The consequence of this finding is two-fold. First, performance stability of a VM instance hints the prospect for conducting lengthy computation since system failure during the compute process can be expected to be minimal. Contrarily, a new resource provisioning mechanism is deemed necessary to ensure that the deployed VM profile matches the requirements of the compute task.

Given that the performance of VM instances in an enterprise cloud service can differ even for those with exactly the same profile, a question about feasibility of entire resource provisioning by a single cloud may be raised. A cloud itself can differ in size thus implying heterogeneity of profiles it may create. A small cloud with less virtualized nodes may support less set of basic profiles. Similarly, a bigger cloud with more resources tends to be more capable of creating various profiles. Meanwhile, another question about the possibility of integrating an internal, private cloud with public, enterprise cloud whenever necessary thus creating a new type of cloud named hybrid cloud or interoperable cloud is also natural to come up to the surface. Adhering to the principle of efficiency and reusability, an entity or organization tends to use its own resource whenever possible instead of paying third parties for renting out and utilizing their resources. These questions end up as a survey of scientific computation cloud infrastructure, platform, and application which is immediately described in the next section.

23.3.2 Emergence of Cloud-Based Scientific Computational Applications

We have surveyed several projects and initiatives resulting as cloud-based applications that are primarily targeted to the scientific domain. The summary is listed as Table 23.1. Our main emphasis is to characterize and categorize the domain of each application, parallelism model, base platform, and cloud scalability in term of support to interconnection and integration with other types of cloud.

There are some other cloud-based scientific applications that are not listed in the table. However, the items put in the table should be representative enough to convey the idea that cloud computing has been touching multi-disciplinary fields. Our table consists of sample applications in bioinformatics, computation biology, climatology and computer science with majority categorized as bioinformatics applications. It was surprising for us to find out the interest of researchers and practitioners outside the computer science field in solving their complex problem with cloud computing paradigm. Success stories from those pioneers may naturally incite curiosity of other

Table 23.1 Some cloud-based projects for scientific purpose

Projects	Domain	Purpose	Parallelism model	Platform	Cloud scalability
CloudBLAST (Matsunaga, Tsugawa, & Fortes, 2008)	Bioinformatics	Protein sequence analysis	MapReduce using BLAST as Mapper	Hadoop	A private cloud in virtual network
CloudBurst (Schatz, 2009)	Bioinformatics	Genomic mapping	MapReduce	Hadoop	A private or public cloud (EC2)
CrossBow (Langmead, Schatz, Lin, Pop, & Salzberg, 2009)	Bioinformatics	DNA sequence alignment and SNP detection	MapReduce using Bowtie and SOAPsnp	Hadoop	A private or public cloud (EC2)
BioVLAB-Microarray (Yang, Choi, Choi, & Pierce, 2008)	Computational biology	Microarray data analysis	Workflow	XBaya	A public cloud (EC2)
Cloud-based Classifiers (Moretti, Steinhaeuser, Thain, & Chawla, 2008)	Computer science	Distributed data mining	Classify	Condor	A private cloud
Satellite data processing (Golpayegani & Halem, 2009)	Geoinformatics	Gridding remote sensing data	MapReduce	Hadoop	A private cloud
MPMD climate application (Evangelinos & Hill, 2008)	Climatology	Simulating atmosphere-ocean models	MPI	LAM, MPICH, and GridMPI	A public cloud (EC2)

parties to implement the paradigm thus bringing cloud computing to broader use especially in the scientific field.

Observing Table 23.1 more scrupulously, we can find out that MapReduce implementation in Hadoop has been the dominant approach in parallelization recently. In our opinion, this is driven by the adequate information, documentation, samples, case studies and also enterprise support for MapReduce. MapReduce model itself was initially proposed and used by Google (Dean & Ghemawat, 2004) and Hadoop provides its open-source Java implementation. This model consists of a map function written by user that takes a set of input key/value pairs in order to generate a set of intermediate key/value pairs, and a reduce function also written by user which merges all intermediate values associated with the same intermediate key. Referring back to Fig. 23.3, the map function is equivalent to delegating, and partial outputting phases whereas the reduce function is to sorting and merging phase.

In Hadoop implementation of MapReduce, map and reduce functions can be replaced by any executable software program. This feature is supported by a utility named Hadoop streaming, which is shipped along with default Hadoop package. For example, CloudBLAST uses Basic Local Alignment Search Tool (BLAST) through NCBI BLAST 2 implementation as the map function. This software program executes the map function in lieu by finding region of local similarity between nucleotide or protein sequences. Another example is CrossBow that reuses Bowtie to enable fast and memory-efficient alignment of short reads to mammalian genomes hence the map function substitute. The reduce function is also replaced by the invocation of SOAPsnp whose task is to provide Single Nucleotide Polymorphism (SNP) calls from short read alignment data. The ability to reuse legacy software may have also contributed to wider adoption of MapReduce model and Hadoop for parallel programming.

We can also notice in Table 23.1 that academia have begun using cloud infrastructure and platform offered by the enterprises. Amazon with its variety of cloud services has been a major testbed cited in corresponding research works. Amazon EC2 and S3 are used for infrastructural cloud services while Amazon Elastic MapReduce is used as a cloud platform for MapReduce-based parallel applications. This finding leads us to a further study of cloud-computing feasibility for scientific purpose through our own experience in building and using a compute cloud.

23.4 Building Cloud Infrastructure for Scientific Computing

In this section, we summarize our early experience in setting up and using cloud infrastructure and platform for scientific computing. We also propose potential future use by transforming our existing systems to the cloud-based infrastructure. Additionally, we share our experience in using the enterprise cloud especially Amazon EC2. We highlight some important remarks which can be useful for other research institutions which are also interested in virtualizing their existing compute environment or simply testing the water by applying cloud computing paradigm.

23.4.1 Setup and Experiment on Tiny Cloud Infrastructure and Platform

Our research lab has diverse types of compute nodes ranging from mobile devices like PDAs, desktops, and, workstations. We run some server programs on the workstations and let the desktop utilized for personal research purpose. We wanted to assess the difficulty level of creating a tiny cloud which comprises small number of compute nodes. Pointing out numbers, the workstations add up to 20 cores in total. We are familiar with the operating systems we use (some distributions of GNU/Linux and Windows) and have been self-managing and self-administering our infrastructure. In our view, if it doesn't take too much time and effort to set up a tiny cloud, other research institutions not associated with computer science/computer engineering field can also transform their legacy system to the cloud by their own.

Since we are interested in MapReduce, we experimented on our workstations by installing and configuring Hadoop in the network and then ran some MapReduce-based applications. We wanted to assess the difficulties of self-installation of Hadoop in a cluster and its performance when running some MapReduce applications. The specifications of our infrastructure can be seen in Table 23.2.

Table 23.2 Infrastructure for running a tiny compute cloud

Part	Item	Description
Hardware	Processors	20 cores total with speed ranging from 1.6 GHz to 3.0 GHz
	RAMs	At least 2 GB on a physical server
Software	OSes	RHEL 4 64-bit, Fedora Linux 12 64-bit (Linux kernel 2.6.31), Windows Server 2003 64-bit
	Other software	Hadoop 0.20, CloudBurst 1.0.1, CrossBow 0.1.3
Network	Ethernet	Gigabit Ethernet
	IP Addresses	Public IPv4 addresses
	Firewall	Two-level firewall, no NAT

In our experience, installing and configuring Hadoop in a cluster may take some time but should not bear a lot of technical difficulties. There exists Hadoop distribution from Cloudera Hadoop Distribution, (http://www.cloudera.com/hadoop) which streamlines the installation of packages and dependencies with the use of yum software program. The challenge can be in configuring the connection between master node and slave node. Since Hadoop requires SSH connection to a slave node with root-level access to start its daemons and other startup scripts, there can be difficulties in configuring the permission and scaling the infrastructure horizontally in a network shielded by NAT. Also, there is some security concern regarding the network configuration of the cluster. Enabling remote root login is potential to bring security breach if variety of software is installed in the machine. Certain software may have security holes or bugs hence once it is exploited, it can affect the whole system.

Virtualization can isolate the workload thus improving the security. It also improves the utilization of idle resources since several VMs are invoked simultaneously. However, our finding indicates that there are still tough difficulties in transforming existing infrastructure into totally virtualized one. We are interested in type 1 hypervisor like Xen Hypervisor. (http://xen.org) that runs directly on computer hardware and manages all its guest operating systems. There are two modes of an operating system running as a Xen guest, dom0 and domU modes. In dom0 mode, an operating system will be assigned as the main guest OS with more privileges. The dom0 guest OS also manages and controls communication between Xen and other guest OSes. In domU mode, the guest OS has less privilege and confined to direct network and I/O access respectively. The major problem was native OS support for running Xen as the virtualization software. We wanted to use the version of Fedora Linux as the dom0 guest but since Xen requires modification to Linux kernel it was not possible to run Xen without manually patching the kernel or downgrading to older version of the OS whose kernel still supports dom0 mode. We consider this as a serious hurdle especially if the kernel patching is to be conducted by somebody with minimum system administration experience. Also, the option to downgrade the OS brings incompatibility problem to other software and software build process.

Beyond the technical difficulties, we noted that Hadoop is potential to be used as a platform for broader scientific applications. Besides the streaming feature, Hadoop also provides built-in web interface for analyzing logs of executed jobs[1] along with its configuration. This can ease the collection of log data for further analysis. For example, in one of our experiments, we ran PatternRecog job which counted the occurrence of certain pattern in several input files. Through the web interface, the total execution time can be seen along with more detailed info like the execution time and details of each phase like Map and Reduce phases. This helped us not only in analyzing and solving some data-intensive problems but also getting to know what happened in each compute node.

23.4.2 On Economical Use of the Enterprise Cloud

If grid computing has been dormant in the advocacy for wider adoption in the academia, it is deemed necessary to question whether cloud computing can take place in the candidacy of scientific computing platform. The first question can be, "How easy and efficient it is to do science in the cloud?" When efficiency and economical use of resources comes in mind, the question can be later followed with "How economical is it to do science in the cloud?" We will provide our perspective based on the experience of using Amazon cloud services.

The term pay-as-you-go can be tempting for an institution with insufficient infrastructural resource but eager to take benefit of enterprise infrastructure offered

---

[1] In Hadoop terminology, a job is divided into several tasks

at low unit price. In Langmead et al. (2009), the cost incurred for analyzing human genome data with 320-CPU cluster in EC2 was $85. According to the authors, the approach is beneficial since it condensed 1000 h of computation into a few hours without requiring the user to own or operate a computer cluster, which cost much more if purchased. We both agree and disagree to this opinion and explain the underlying arguments for the statement.

There are two services offered by Amazon that can be pertinent to scientific compute task, Amazon EC2 and Amazon Elastic MapReduce. Amazon Elastic MapReduce is tailored to applications built on top of MapReduce model. Researchers and developers can conveniently test and deploy their MapReduce-based applications on the platform and scale the amount of resources used only by using a web interface. However, there are also other scientific applications not based on MapReduce model. To deploy such applications in the cloud, bundling and packaging from the ground up is required. An image containing the applications and container OS should be created and referenced every time a new virtual machine instance is launched.

In the absence of suitable images, user should manually configure the basic VMI offered by provider. Upon the creation of an instance, user should manually download the application binaries or build the sources and related dependencies when binaries are not available or compatibility is an issue. As the cloud operates on utility basis, downloading binaries or sources and working remotely to configure a virtual machine is considered as using provider's infrastructure hence user is charged for the data transfer. In Amazon EC2 case for example, by January 2010, a standard Linux large instance in North Virginia (7.5 GB memory, 2 virtual cores, 850 GB local storage, 64-bit architecture) cost $0.34/h with data transfer out starting from $0.15/GB and decreasing progressively, and free to $0.1/GB unit price of data transfer in. A user who is running the instance for one day (24 h) with 1 TB data in and 1 TB data out should pay $158.16 consisting of $8.16 compute instance fee and $150 data transfer fee. If the user activity is configuring the standard virtual machine, he may have not anticipated that he is also being charged for the activity which may potentially contribute significant number to the final amount he has to pay. When he has to configure a decent number of virtual machines, the cost will be multiplied, which is getting worse when failure occurs for several installations, thus configuration has to be done several times. Consequently, this additional cost can reduce the economical benefit of doing computation in the cloud.

To put it into a bigger picture, the current offer by cloud providers is attractive but the hidden cost may hinder the wider use in the academia. Current cloud-based applications either use a private cloud or a public cloud. If scaling can be done more dynamically by integrating virtualized resources in a private cloud with the public cloud by on-demand basis, the cost efficiency can be improved. Also, there should be decent number of VMIs supporting various scientific applications and a mechanism that can self-describe applications bundled within a VMI. We then propose the idea and current effort for the cloud integration and interoperability in the next section.

23.5 Toward Integration Of Private and Public Enterprise Cloud Environment

As an emerging computing paradigm, cloud computing is prospective for various types of research. One of the challenging cloud computing-related researches is dynamic, on-demand scaling of cloud. As can be seen in Fig. 23.4, currently a cloud is a single entity. It is either private or public. Scaling a cloud depends on the whole size of the cloud. A user in a private research institution with a small private cloud can have difficulties in provisioning more resources if a certain computation task grows in complexity and data input size. Current option when such situation occurs is to move the whole computation to a public cloud that supports more resources. This principally lessens the usability and effectiveness of transforming legacy system into a cloud environment.

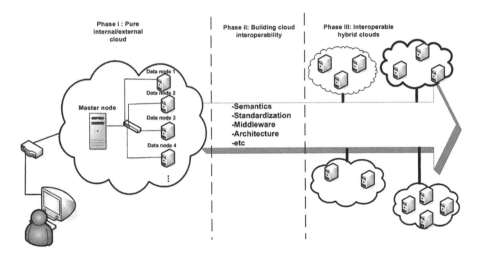

Fig. 23.4 Projected evolution of cloud computing interoperability

Cloud integration and interoperability can be achieved by several methods. We would like to highlight several key methods that can foster the integration and interoperability. Building semantics is one of the methods deemed essential. Currently, each cloud provider goes with its own way in defining the abstraction for the resources. Amazon, for example, uses manifest XML file to describe a virtual machine image which is called Amazon Machine Image (AMI). By complying with certain semantics instead, it is possible to create a directory service that contains list of VMIs from various Independent Software Vendors (ISVs) located across different domains along with description about each image. This will assist in the enablement of auto VMI discovery and deployment across multiple authorities. This idea is similar with WSDL and service registry in web service SOA.

Standardization is also an important aspect to consider. Cloud services in each provider use different interfaces thus without the help of middleware or a cloud service broker, it is difficult to integrate the heterogeneous services. Currently, there has been effort from Open Grid Forum (OGF) to standardize the interfaces for IaaS cloud computing facilities (OGF Open Cloud Computing Interface Working Group, http://www.occi-wg.org/doku.php). The standardization will cover the development of APIs for cloud consumers, integrators, aggregators, providers, and vendors.

Early effort for the integration can be seen in the development of middleware. Eucalyptus (Nurmi et al., 2009) and OpenNebula (Vozmediano, Montero, Llorente, 2009) are samples of middleware for cloud management. Eucalyptus provides the same tools and interfaces with Amazon EC2. It consists of node controller which controls the execution, inspection, and termination of VM instances, cluster controller which gathers information about and schedules VM execution as well as manages virtual instance network, storage controller which provides mechanism for storing and accessing VMI and user data, and cloud controller for high-level scheduling decisions and displaying information about resources. Eucalyptus is basically a middleware for managing a private cloud infrastructure but recent development enables the management of a hybrid cloud infrastructure.

On the other hand, OpenNebula tries to provide a uniform cloud management layer based on service type. Its basic principle is to separate resource provisioning from the service management. By implementing this principle, it enables elastic cluster capacity through on-demand deployment or shutdown of virtual worker nodes which can be located in a private or public cloud, cluster partitioning through segregation and isolation of worker nodes based on service type, and heterogeneous configurations which enables a service to have multiple software configurations. OpenNebula supports various virtualization technologies including Xen, KVM, and VMWare which are usually used in a private cloud and interfaces to public IaaS-cloud like AmazonEC2 and ElasticHosts.

Still, existing efforts can accelerate the progress if there is common consensus about the architecture of a cloud computing system. As cloud computing nowadays is much driven by the enterprises, it is inevitable to see various cloud systems with each implementing different interfaces, abstractions, layering, and management scheme. With common architecture, an ecosystem consisting of private clouds and public clouds interacting dynamically for elastic resource provisioning and service exchange will be closer to come.

23.6 Conclusion

Growing interests in cloud computing raises questions about leveraging this computing paradigm in scientific field. This article addressed the issue by highlighting the features and characteristics of scientific computing and how it can be realized within cloud computing paradigm. Several scientific applications built on top of cloud infrastructure were described along with our experiences in setting up a private cloud and utilizing public cloud infrastructure. Based on our experience, an organization can transform their infrastructure into a cloud environment and take benefit of better resource utilization but should also be aware that some difficulties can be faced. Also, institution with tight budget-spending policy should be cautious when deciding to do massive computation on a public cloud.

Due to its infant state, there are various researches to be done in cloud computing. We consider research in cloud interoperability as one of the important ones. We provide some projection and recommendations based on our finding that can be useful by other researchers having similar interests.

References

Birman, K., Chockler, G., & van Renesse, R. (2009, June). *Toward a Cloud Computing Research Agenda. ACM SIGACT News, 40*(2), 68–80.

Boulton C. Oracle CEO Larry Ellison Spits on Cloud Computing Hype, http://www.eweek.com/c/a/IT-Infrastructure/Oracle-CEO-Larry-Ellison-Spits-on-Cloud-Computing-Hype/.

Buyya, R., & Bubendorfer, K. (2008). Market Oriented Grid and Utility Computing. Wiley, New York.

Buyya, R., Yeo, C. H., &Venugopal, S. *Market Oriented Cloud Computing: Vision, Hype, and Reality for Delivering IT Services as Computing Utilities.* IEEE International Conference on High Performance Computing and Commnunications (HPCC).

Dean, J., & Ghemawat, S. (2004, December). *MapReduce: Simplified Data Processing on Large Clusters.* Proceedings of 6th Symposium on Operating Systems Designs and Implementations.

Dejun, J., Pierre, G., & Chi, C.-H. (2009, November). *EC2 Performance Analysis for Resource Provisioning of Service-Oriented Applications.* Proceedings of 3rd Workshop on Non-Functional Properties and SLA Management in Service-Oriented Computing.

Evangelinos, C., & Hill, C. N. (2008). *Cloud Computing for Parallel Scientific HPC Applications: Feasibility of Running Coupled Atmosphere-Ocean Climate Models on Amazon's EC2.* Cloud Computing and its Application (CCA).

Foster, I., Zhao, Y., Raicu, I., & Lu, S. (2008, December). "Cloud Computing and Grid Computing 360-Degree Compared," Grid Computing Environment Workshop.

Giunta, G., Lacetti, G., Montella, R. (2008). *Five Dimension Environmental Data Resource Brokering on Computational Grids and Scientific Clouds.* IEEE Asia-Pasific Services Computing Conference.

Golpayegani, N., & Halem, M. (2009). *Cloud Computing for Satellite Data Processing on High End Compute Clusters.* IEEE International Conference on Cloud Computing.

Johnson, B. (2008). Cloud Computing is a Trap, Warns GNU Founder Richard Stallman," http://www.guardian.co.uk/technology/2008/sep/29/cloud.computing.richard.stallman.

Langmead, B., Schatz, M. C., Lin, J., Pop, M., & Salzberg, S. L. (2009). Searching for SNPs with cloud computing. *Genome Biology, 10*(11), R134.

Liu, H., & Orban, D. (2008). *GridBatch: Cloud Computing for Large-Scale Data-Intensive Batch Applications.* Eighth IEEE International Symposium on Cluster Computing and the Grid.

Matsunaga, A., Tsugawa, M., & Fortes, J. (2008). *CloudBLAST: Combining MapReduce and Virtualization on Distributed Resources for Bioinformatics Applications.* Fourth IEEE International Conference on eScience.

Mei, L., Chan, W. K., & T. H. Tse, (2008). *A Tale of Clouds: Paradigm Comparison and Some Thoughts on Research Issues.* IEEE Asia-Pasific Services Computing Conference.

Moretti, C., Steinhaeuser, K., Thain, D., & Chawla, N. V. (2008). *Scaling Up Classifiers to Cloud Computers.* Eighth IEEE International Conference on Data Mining.

Nurmi, D. et al. (2009, May). *The Eucalyptus Open-Source Cloud-Computing System.* 9th IEEE/ACM International Symposium on Cluster Computing and the Grid.

Rimal, B. P., Choi, E., & Lumb, I. (2009). *A Taxonomy and Survey of Cloud Computing Systems.* Fifth International Joint Conference on INC, IMS, and IDC.

Schatz, M. C. (2009, June). CloudBurst: Highly sensitive read mapping with mapreduce. *Bioinformatics, 25*(11), 1363–1369.

Simalango, M. F., & Oh, S. (2009, July). *On Feasibility of Enterprise Cloud for Scientific Computing.* Korea Computer Congress.

Simmhan, Y., Barga, R., Lazowska, E., & Szalay, A. (2008). *On Building Scientific Workflow Systems for Data Management in the Cloud.* Fourth IEEE International Conference on eScience.

Uhlig, R. et al. (2005, May). Intel virtualization technologies. *Computer, 38*(5), 48–56.

Vouk, M. A. (2008). *Cloud Computing – Issues, Research, and Implementations.* Proceedings of 30th International Conference on Information Technology Interfaces.

Vozmediano, R. M., Montero, R. S., & Llorente, I. M. (2009). *Elastic Management of Cluster-Based Services in the Cloud.* Proceedings of 1st Workshop on Automated Control for Datacenters and Clouds, ICAC.

Yang, Y., Choi, J. Y., Choi, K., Pierce, M. (2008). *BioVLAB-Microarray: Microarray Data Analysis in Virtual Environment.* Fourth IEEE International Conference on eScience.

Chapter 24
A Cloud Computing Based Patient Centric Medical Information System

Ankur Agarwal, Nathan Henehan, Vivek Somashekarappa, A.S. Pandya, Hari Kalva, and Borko Furht

24.1 Introduction

This chapter discusses an emerging concept of a cloud computing based Patient Centric Medical Information System framework that will allow various authorized users to securely access patient records from various Care Delivery Organizations (CDOs) such as hospitals, urgent care centers, doctors, laboratories, imaging centers among others, from any location. Such a system must seamlessly integrate all patient records including images such as CT-SCANS and MRI'S which can easily be accessed from any location and reviewed by any authorized user. In such a scenario the storage and transmission of medical records will have be conducted in a totally secure and safe environment with a very high standard of data integrity, protecting patient privacy and complying with all Health Insurance Portability and Accountability Act (HIPAA) regulations.

The sharing of medical records, specifically radiology imaging databases with CDOs will have potential to drastically reduce medical redundancies, exposure to radiations, costs to patients. In addition such system can empower the patients with the automated ownership of their secure personal medical information. It is essential to use the cloud computing in this application since it would allow the CDOs to address the challenge of sharing medical data that is overly complex and highly expensive to address with traditional technologies. In addition to providing community of care, proposed system can also serve as a valuable tool in clinical research, medical decision-making, epidemiology, evidence-based medicine, and in formulating public health policy. Figure 24.1 shows a high level simplified overview of

A. Agarwal (✉), A.S. Pandya, H. Kalva, and B. Furht
Department of Computer Science and Engineering, FAU, Boca Raton, FL, USA
e-mails: {ankur@cse.fau.edu; pandya@fau.edu; hari.kalva@fau.edu; bfurht@fau.edu}

N. Henehan
Senior Software Developer, NACS Solutions, Oberlin, OH, USA
e-mail: nhenehan@gmail.com

V. Somashekarappa
Senior Software Developer, Armellini Inc., Palm City, FL, USA
e-mail: vsomashekar@gmail.com

B. Furht, A. Escalante (eds.), *Handbook of Cloud Computing*,
DOI 10.1007/978-1-4419-6524-0_24, © Springer Science+Business Media, LLC 2010

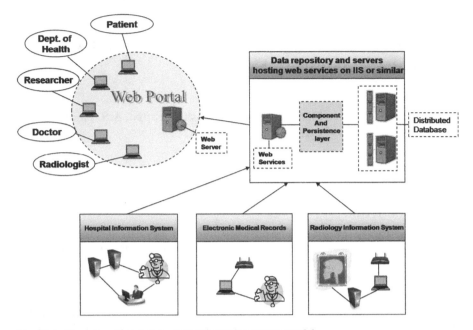

Fig. 24.1 Overview of global medical information system model

the proposed system. This defines the following specific objectives of the chapter below:

1. To discuss an approach that would allow us to shift from institute centered hospital information system towards a regional/global medical information system by developing standards based Service-Oriented-Architecture (SOA) for interfacing heterogeneous medical information systems such that it would allow real-time access to all medical records from one medical information system to another.

2. To discuss a brief architecture and implementation plan to develop a "Lossless Accelerated Presentation Layer" that will allow one to view all radiology images (Digital Imaging and Communication in Medicine (DICOM) objects) that reside in a cloud based distributed database. This component would further allow the radiologist to annotate the image through a web-based viewer and store it back into the distributed cloud based database. Such layer would provide an instantaneous lossless access to all DICOM objects thereby, eliminating the download time. The current solutions do not provide the lossless view of the DICOM objects.

3. To discuss the architecture for web-based interface that will provide a holistic view of all medical records to every patient. The proposed environment will be scalable and reside on the cloud. Such a system would empower all patients with the automated ownership of their secure personal medical records. This interface can be extended to provide an anonymous view for all medical records for research purposes to scientific community and organizations such as department of health.

4. To discuss the strategies and architecture for the design of a distributed architecture that would ensure data consistency/integrity. The proposed architecture would provide autonomous scalability to allow dynamic growth of the cloud based medical record system.

Figure 24.2 shows the layered view of the proposed system architecture and how it relates to stated project objectives. The detailed architecture of each layer is discussed in Section 24.4, "Project Plan and Research Methods".

Fig. 24.2 Layered architecture for the proposed medical information system

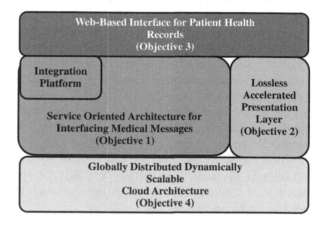

24.2 Potential Impact of Proposed Medical Informatics System

This project focuses on the development of an architecture for integrating heterogeneous medical information systems such as (Hospital Information System, Radiology Information System, and Electronic Medical Records among others). These systems in their current form do not transfer information from one system into another outside a network. The proposed approach for a global medical informatics system would allow all medical records to be completely portable. In the current system a significant amount of delay is involved in transmitting medical records from one CDO to another, leading to repetitive medical testing and increased cost of healthcare to the patient, insurance companies and federal government. The development of a cloud based service oriented architecture that will provide all patients with an interactive view of all their medical records. Such a system would provide all patients with the ownership of their medical records, thereby eliminating the need to repetitive procedures.

The proposed system architecture drastically reduces the Medicare spending for imaging services. The sharing of medical records, specifically radiology imaging databases, will drastically reduce medical redundancies, and exposure to radiations. Total national healthcare spending is in excess of $2.6 Trillion or about 17% of our Gross Domestic Product. The proposed architecture would significantly contribute the reduction in national healthcare spending by eliminating the repetition of procedure due to unavailability of medical records. In year 2006 itself various medical imaging services accounted for 58% of Medicare's physician office spending. In order to control this spending on medical imaging, the "Deficit Reduction Act" (DRA) was created in year 2005 to reduce medical imaging spending by $2.8 Billion by 2011. This project will allow various CDOs to share the medical records and imaging thereby, eliminating the need to repeat the procedures during a defined time period thereby, serving the objectives of Deficit Reduction Act. Please note with the current technology radiology imaging can be shared within a CDO, however not among various CDOs. The development of a lossless accelerated presentation layer would allow to access all radiology images residing on a cloud based distributed database in a lossless manner through a web-based DICOM viewer. This layer would provide a seamless access to all radiology imaging from any location in real-time thereby increasing the efficiency of overall medical record systems.

Centralizing medical records can also create new and more intelligent perspectives in medicine. Such database with medical information will be extremely valuable for advanced data mining in clinical research. This will have potential to analytically evaluate and innovate new disease information and test methods that will improve the health care delivery and lead the exploration of new preventive treatment. In addition, the proposed project also serves the criteria of national Health Information Technology agenda.

24.3 Background and Related Work

System Integration has been always the most critical issue for the development of information systems in healthcare industry. Medical Information Systems (MIS) are heterogeneous in nature and therefore pose a severe challenge in their interoperability (Hersh, 2002; Lu, Duan, Li, Zhao, & An, 2005). A large number of healthcare applications are isolated and do not communicate with each other. Therefore, the integration of existing information systems represents one of the most urgent priorities of healthcare information systems (Saito & Shapiro, 2005). A 2005 RAND study estimates that the U.S.A. could save $81 billion annually and help to improve the quality of care through the adoption of high quality and integrated MISs (Weber-Jahnke & Price, 2007). At the end of 2003, the Medicare Prescription Drug Improvement and Modernization Act (MMA) was signed that required the establishment of a Commission on Systemic Interoperability to provide a road map for interoperability standards in order for the majority of the U.S. citizens to have interoperable electronic health records within ten years (Katehakis, Sfakianakis, Kavlentakis, Anthoulakis, & Tsiknakis, 2007).

Many efforts have been made on integrating the heterogeneous systems in hospitals (Haux, 2006). Healthcare industry has developed several standards through which relevant data can be transferred among different information systems

(Hasselbring, 2000). These standards are Health Language Seven (HL7), Electronic Data Interchange (EDI) X12 Version 4010, Health Insurance Portability And Accountability (HIPAA), Digital Image Communication in Medicine (DICOM), Integrating Healthcare Enterprise (IHE) among others (Lenz, Beyer, & Kuhn, 2007). All these standards are currently being widely used in healthcare industry. According to Open Source Clinical Research Group HL7 is the most widely used messaging standard in health, not only in North America, but also around the world. Further, a 1998 survey found the HL7 standard in use in more than 95 percent of hospitals with more than 400 beds. Overall, more than 80 percent of the respondents in that study reported using HL7 in their information system departments with another 13.5% planning to do so. The proposed solution will be able to integrate all medical information systems that are in compliance with HL7 standard.

Standardized interfaces are available to many healthcare "Object Oriented Services" such as CORBAmed (Common Object Request Broker Architecture in Medicine), which realizes the share of common functionalities like access control among different systems. Others, like DICOM, HL7 (Health Level Seven) and the initiative of IHE (Integrating the Healthcare Enterprise) (Carr & Moore, 2003), specify the guidelines or standards for exchanging messages among different systems, which make the different systems work in harmony and implement the workflow integration (Lu, Duan, Li, Zhao, An, 2005; Wang, Wang, & Zhu, 2005). Broker, an embedded device facilitated the communication between HIS, RIS and Picture Archiving and Communication System (PACS) by integrating HL7 with DICOM. Broker accepted HL7 messages from RIS, translated and mapped the data to produce DICOM messages for transmission to PACS. However, the Broker system posed a challenge since it allowed RIS information to flow only in one direction resulting in the duplication of databases. IHE initiative, jointly established by Hospital Information Management Systems Society (HIMSS) and the Radiological Society of North America (RSNA), later addressed the issue by allowing the integration of clinical information within a healthcare delivery network. Later a consolidated solution with RIS/PACS/HIS integration was offered by healthcare companies. This was a major step towards the successful integration of patient records within a network (Boochever, 2004).

There have been some efforts to share the patient information within a limited group of in order to facilitate the teamwork and effective healthcare management. The Medical Informatics Network Tool (MTNL). The software included an intelligent engine that was used for treating only Schizophrenia, a chronic brain disorder (Young, Mintz, Cohen, & Chiman, 2004). The software allowed collecting useful information about the patient and facilitating the communication among all the team members in addition to other providing other useful information related to Schizophrenia. This system was very helpful in taking meaningful decisions quickly and therefore had a positive impact on the patient healthcare. Similarly, the proposed solution would provide all information regarding a patient in a centralized location. Aggregation of such information will have profound impact on the overall patient healthcare and would reduce the probability of making wrong decisions such as prescribing conflicting prescriptions. In 2006 Al- Busaidi et al.

researched on introducing a personalized patient information that was extracted from a single patient database (Al-Busaidi, Gray, & Fiddian, 2006). This research was more focused on web mining and intelligent information retrieval from web that can provide a simplified and meaningful description to the problem that a patient might be experiencing. Project was more focused towards analyzing the information based on conceptual integration of ontology. However in this project our focus is to integrate several patient record systems (Radiology Information System (RIS), Electronic Medical Records (EMR), Hospital Information System (HIS), Patient Health Record (PHR) and Clinical Information System (CIS) among others) and provide a cloud computing based patient centric interface for all patient records.

Recently service oriented architecture type IT platforms are emerging as solutions for clinical enterprises (Strähle et al., 2007). Web Services (WS) provide an open and standardized way to achieve interoperation between different software applications, running on a variety of platforms and/or frameworks. Therefore, they constitute an important technological tool toward the incremental delivery of advanced inter-enterprise services. Significant advantages of using WS on top of any existing middleware solution is location transparency, language and platform independence, together with their embracement by big vendors and the acceptance they enjoy between the users. WS, with their extensible markup language (XML) roots, open the door for the next generation, loosely coupled, coarse-grained, document-oriented architectures. The term "loosely coupled" is used to characterize services where the assumptions made by the communicating parties about each other's implementation, internal structure, and the communication itself are kept to a minimum. In Wang, Wang, and Zhu (2005) researchers have proposed the use of a SOA for combining few workflows for integrating various health information systems. Although the workflows are not complete however it is an important contribution towards integrating various informatics systems.

Harvard Medical School CIO John Halamka quoted, "Putting servers and exchanges into doctors' offices is not going to work". He suggested a better model is using regional health-care information technology centers that use cloud computing systems to work with doctors (Gallagher, 2009). Computing done at cloud scale allows users to access virtual supercomputer-level (Hayes, 2008). Cloud computing's aim is to deliver tens of trillions of computations per second to problems such as delivering medical information in a way that users can tap through the Web. When implemented correctly, cloud computing allows the development and deployment of applications that can easily grow capacity, deliver needed performance, and have a high-degree of fault tolerance, all without any concern as to the nature and location of the underlying infrastructure (Buyya, Yeo, Venugopal, Broberg, Brandic, 2009). IBM announced that American Occupational Network and HyGen Pharmaceuticals are improving patient care by digitizing health records and streamlining their business operations using cloud-based software. GoogleHealth and Microsoft Vault health solutions are commercial steps in the direction of aggregating patient records in a unified environment. However, the major issue with their solution is the inability of CDOs to upload patient health records in a central data repository. In both systems patients must upload all the records, which require patients to first gain access to their medical history. It is important to note that EMR is the basic building block, the source of information that feeds the Electronic Health Record (EHR). The EHR is the longitudinal record made possible by Regional Health Information organizations (RHIO's). While he Patient Health Record (PHR) is the record owned, accessed, and managed by the consumer. The interdependencies among them are very clear. Without linking (interfacing) a EMR with a PHR, the consumer will have to manually input vital data, like laboratory results. Without an EHR, the PHR cannot accept information from multiple providers (McDonald, 1997). This is the case in both solutions offered by Microsoft and Google. Often it takes several weeks before one can gain access to their medical records from a hospital thereby, limiting its usage. In addition, both Google and Microsoft health solutions do not provide a lossless solution to the imaging services, which are important component of the overall patient centric system. In this chapter, we discuss a framework that will allow us to interface all medical records systems those are HL7 compliant and store the data in a multi-cloud based distributed database system.

24.4 Brief Discussion of Medical Standards

Healthcare industry currently has several standards through which relevant data is transferred between different information systems, these standards are HL7, EDI X12 Version 4010 (EDI X12), HIPAA, DICOM, IHE among others (Spyrou et al., 2002). A brief discussion of these standards is discussed below:

Health Language 7: aims at enabling communication between applications provided by different vendors, using different platforms, operating systems, application environments (e.g. programming languages, tools). In principle, HL7 enables communication between any systems regardless of their architectural basis and their history. Therefore, HL7 supports communication between real-world systems, newly developed or legacy. This is achieved through syntactically and semantically standardized messages. HL7 interfaces realize the request/service procedure by sending and receiving these standardized messages. HL7 functional areas include typical health-care (clinical) domains as Admission Discharge and Transfers (ADT), Patient Registration, Orders, Results, Financial and Master files. More recent versions of HL7 also include Non-ASCII character sets, Query language support, Medical documents, Clinical trials, Immunization reporting, Ad6erse drug, Reactions, Scheduling, Referrals, and Problems and goals. An example of an HL7 transaction set is shown below:

```
MSH|^~\&|GHH LAB|ELAB-3|GHH OE|BLDG4|200202150930||ORU^R01|CNTRL-3456|P|2.4<cr>
PID|||555-44-4444||EVERYWOMAN^EVE^E^^^^L|JONES||19620320|F|||153 FERNWOOD DR.^
^STATESVILLE^OH^35292||(206)3345232|(206)752-121||||AC555444444||67-A4335^OH^20030520<cr>
OBR|1|845439^GHH OE|1045813^GHH LAB||15545^GLUCOSE|||200202150730||||||
555-55-5555^PRIMARY^PATRICIA P^^^^MD^^|||||||F|||||444-44-4444^HIPPOCRATES^HOWARD H^^^^MD<cr>
OBX|1|SN|1554-5^GLUCOSE^POST 12H CFST:MCNC:PT:SER/PLAS:QN||^182|mg/dl|70_105|H|||F<cr>
```

Electronic Data Interchange: is a data format based on ASC (Accredited Standards Committee) X12 standards. It is used to exchange specific data between two or more trading partners. Term "trading partner" may represent organization, group of organizations or some other entity. EDI X12 is governed by standards released by ASC X12. Each release contains set of message types like invoice, purchase order, healthcare claim, etc. Each message type has specific number assigned to it instead of name. For example: an invoice is 810, purchase order is 850 and healthcare claim is 837. Some key EDI transactions are:

* 837: Medical claims with subtypes for professional, institutional, and dental varieties
* 820: Payroll deducted and other group premium payment for insurance products
* 834: Benefits enrollment and maintenance
* 835: Electronic remittances
* 270/271: Eligibility inquiry and response
* 276/277: Claim status inquiry and response
* 278: Health services review request and reply

Health Insurance Portability and Accountability: regulation impacts those in healthcare that exchange patient information electronically. HIPAA regulations were established to protect the integrity and security of health information, including protecting against unauthorized use or disclosure of the information. HIPAA states that a security management process must exist in order to protect against "attempted or successful unauthorized access, use, disclosure, modification, or interference with system operations". In allows monitoring, reporting and sounding alert on attempted or successful access to systems and applications that contain sensitive patient information. Current version of HIPAA is X12 4010 and recently a new guideline X12 5010 was released and mandated by the Department of Health to be complied with and implemented by all care givers before January 2013.

Digital Imaging and Communication in Medicine: is the industry standard for storing and transferring all radiology images. The standard ensures the interoperability of system and can be use to produce, display, send, query, store, process, retrieve, and print DICOM objects. Patterned after the Open System Interconnection of the International Standards Organization, DICOM enables digital communication between diagnostic and therapeutic equipment and systems from various manufacturers. The DICOM 3.0 standard, developed by the American College of Radiology (ACR) and National Electrical Manufacturers Association (NEMA), evolved from versions 1.0 and 2.0 which evolved in 1985 and 1988 respectively.

In addition to these standards it is must that the designed medical informatics system is in compliance with the following requirements.

Patient Safety: One of the most important requirements of any medical informatics system is the availability of consistent and correct information. At no point should the system show inaccurate, incomplete and unintended information that may jeopardize the safety of a patient.

Disaster Recovery: Since the proposed system is cloud computing based (hosted on Internet) there must be provision for backing up the entire system data incase of a system failure. A method must be included that would prevent the data from being corrupted or lost. If not done so this may lead to major crisis in terms of patient safety.

Accuracy, Availability and Accessibility: It is must that health informatics system must achieve the availability target of above 99% since the system stores critical information. The data stored must be accurate, available and always accessible from any location.

Integration: As discussed above, for the system to serve as a global medical record that will include all patient records, all medical standards must be correctly and carefully integrated into the system. Several of these standards have already been discussed above.

Ease of use and Customer Satisfaction: It is expected that the system will be widely used by all patients, doctors, nurses and other entities involved in healthcare. Therefore, the system must provide a simple user interface for all entities (users) involved. Inability to achieve this may prohibit users in using the system thereby reducing the potential impact of the proposed informatics system.

Government Compliance: The most important system requirement is security and the HIPAA compliance. The system must support both. Each workflow must be carefully designed such that it meets the HIPAA standard.

24.5 Architecture Description and Research Methods

Currently clinical data, in standardized format, is distributed among Care Delivery Organizations (CDO) such as hospitals, pharmacies, insurance companies among others. Creating and authorized & secured sharing of this data repository on cloud(s) based distributed database is the goal of proposed project. In this section we provide an architecture description of each layer of our proposed medical information system as shown in Fig. 24.2 above in section "Introduction". The project is focused on several critical research components such as architecture of distributed database, load balancing algorithms for traffic management on cloud computing system, system integration, development of lossless presentation layer for image viewing and performance evaluation.

24.5.1 Objective 1: A Service Oriented Architecture for Interfacing Medical Messages

This section discusses the approach that has been adopted for interfacing several medical systems in order to centralize all patient records. The proposed solution does not focus on the developing yet another standard which would try and enforce other organizations to comply with. Rather the federated approach was selected, based on a set of already existing healthcare industry standards through which medical messages are transferred between different information systems. Currently medical messaging standards enable data transfer between systems in a request – service manner, where in the data is sent from one system to another directly or via a modulator like an EDI VAN service provider. Such data transfer is occurs only upon the request and is not based on the occurrence of an event such as admission of patient.

Through a service oriented architecture various medical information systems can be integrated by collecting standardized data on a cloud based distributed database repository. In this architecture Web-services will be hosted securely on a cloud while the Web service clients serving as agents will be running on various health-information-systems. In order to facilitate a seamless integration of various medical information databases the schemas of the distributed database residing on cloud(s) will imitate the existing schema of healthcare standards like HL7, EDI and DICOM. During initial setup of the web-service clients, the schema of the existing HIS or EMR systems will be mapped to the proposed cloud based distributed database schema. This would allow the agent to periodically query the client databases through established connections which will facilitate the transfer of data to the cloud(s) over secure HTTP connection. Figure 24.3 shows the proposed approach as discussed above.

Fig. 24.3 Layered view of the proposed service oriented architecture for interfacing medical messages

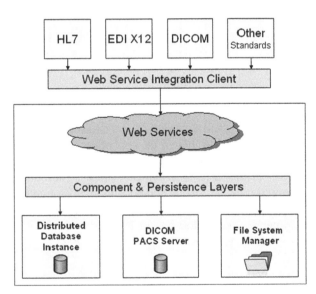

Web-Services (WS) is major, service-oriented, connection technologies which is specification based and mostly open. In addition to its open source development potential in a technology neutral environment, major vendors are embracing World Wide Web Consortium (W3C) and the Internet Engineering Task Force (IETF) efforts. Significant advantages of using WS on top of any existing middleware solution are location transparency, language and platform independence, in addition to their adoption by big vendors and wide acceptance. WS, with their XML roots, open the door for the next generation, loosely coupled, coarse-grained, document-oriented architectures. Security should not be considered an afterthought but it should be built into the communication platform itself. WS were originally considered as an easy way to do business across the Internet since it allows tunneling through the hypertext transfer protocol that usually bypasses corporate firewalls. The use of transport layer security may not be enough to provide the desired levels of authentication, authorization, and trust. The use of technologies like XML-Signature, XML-Encryption, and WS-Security should be mandatory in order to achieve the necessary quality of protection for message integrity and confidentiality. Additional efforts such as WS-Trust, WS-Policy, and WS-Secure Conversation must be consideration as well. Currently, the most common technological tool to cover various security aspects is the Public Key Infrastructure (PKI). PKI is used to describe the processes, policies, and standards that govern the issuance, maintenance, and revocation of the certificates, public, and private keys that the encryption and signing operations require. PKI incorporates the necessary techniques to enable two entities that do not know each other to securely exchange information using an insecure network such as the Internet.

24.5.2 Objective 2: Lossless Accelerated Presentation Layer for Viewing DICOM Objects on Cloud

A key requirement for DICOM viewers is lossless image coding; users accessing DICOM images should receive lossless image to rule out any compression artifacts. Figure 24.4 shows the proposed architecture for the imaging sub-system. When users open a DICOM image, a DICOM viewer is executed in the cloud. The views rendered by the DICOM viewer have to be communicated to the users remotely accessing the image. Commercial remote access tools such as Citrix use lossy compression for remote viewing and hence are not suitable for medical imaging application. A few hospitals have used such solution for making the DICOM objects available outside the hospital network. However, such use of lossy compression may not be an acceptable solution under several medical conditions. For instance, such a lossy compression may provide wrong information about the size of a cancer cells that may be growing in any part of a body. Since the stage of a

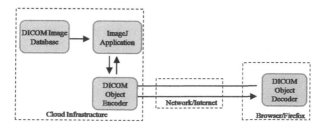

Fig. 24.4 The proposed architecture for the imaging sub-system

cancer is determined by the volume of the cancer cells; a lossy image may show a reduced volume by removing some pixels.

Our proposed solution is to use the open source remote control software TightVNC as a basis and modify the image coding engine to support lossless images. The complexity of encoding and the compression achieved varies with algorithms. One can easily evaluate and measure the compression performance for lossless JPEG-2000 and JPEG. High performing compression algorithms can then be selected to get maximum performance of the imaging sub-system.

24.5.3 Objective 3: Web Based Interface for Patient Health Records

Patient health records stored in a centralized data repository over the distributed cloud(s) can be instantly viewed by any authorized user connected to this system through a web-based interface designed as part of the proposed system. The data can be accessed by existing health information systems with the help of remote calls to cloud hosted web services as shown in Fig. 24.5. The proposed SOA would allow various medical information systems to interface with these web-services through their interfacing clients. All clients will go through a standard layer of authentication and authorization through public key encryptions standards.

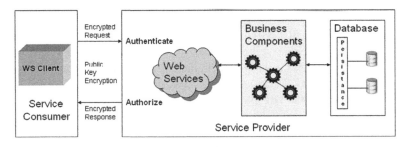

Fig. 24.5 Service oriented architecture for presentation layer

Since the data is stored in the standardized format (HL7 or EDI) on the cloud based GHIS database, we must present the data in a readable format. The web portal can provide all users with the ability to search for a patient's identity given a set of demographic criteria and retrieval of all the related health and medical information pertaining to the patient under consideration. Additional filtering of patient data will be possible if the consumer of the service is only restricted to view some parts of the patient's medical records. This will be accomplished by the use of user roles and access grants. Secured logins to access patient records for authorized users such as physicians, radiologists, laboratory technicians among others can be created using the existing methods like one-time passwords (OTP). OTP methods can be facilitated through the use of a standard medical hardware device such as a "Token" that would generate a time synchronized one time password to allow the access to patient database.

A web-based ImageJ interface can be easily made available through this web-portal system for viewing DICOM objects. ImageJ is a public domain Java image processing program inspired by NIH ImageJ for the Macintosh. It was designed with an open architecture that provides extensibility via Java plugins. ImageJ will be integrated with a PACS server on the cloud to read DICOM objects residing on the distributed database. The user interface of ImageJ viewer application would depend upon the role of the accessing user. For instance, a radiologist will have the permission to alter the DICOM object that will be stored as a new version in form of a separate image layer. A web-browser presents the remote "desktop" from which an authorized user may launch ImageJ to open DICOM objects. Users interact with ImageJ directly using the controls provided by ImageJ. As the views of ImageJ change, a view encoder based on the TightVNC server compresses the "desktop" and transmits this to the user. TightVNC uses the standard Remote Frame Buffering (RFB) protocol for desktop sharing and control. Since lossless compression will be used in the View Encoder, the users will see images that are identical to the images rendered by ImageJ.

24.5.4 Objective 4: A Globally Distributed Dynamically Scalable Cloud Based Application Architecture

The proposed medical information system concept is for patients, doctors and other care providers to have immediate access to medical records, images and other digital resources. Once connected to information system, the services available to a consumer will be filtered depending upon the consumer's role, type or responsibility. Figure 24.6 shows high level layered architecture for a proposed globally distributed dynamically scalable cloud that will be used for storing all medical data. Every tier in the Fig. 24.6 includes multiple instances in the local and geographically distributed clouds.

Tier 1 (Security Tier) of each application partition would include firewall with VPN, traffic filtering, statistical reporting and balancing functionalities and capabilities. In order increase efficiency of the tier, one would explore combining few of these services together on the same host or device, though cross-cloud Virtual Private Network (VPN) services will reside on isolated hosts. Additionally, due to the sensitive nature of the Internet Protocol Security (IPSEC) hosts and to ensure data security, this separation is considered necessary.

Tier 2 (Presentation Tier) represents a web server for serving of http clients. Ultimately each instance on this tier should be able to detect a failed node on its tier and take over the load in order to provide fault tolerance in our overall proposed system. This can be accomplished by using the "Linux-Ha" clustering software in an active/active configuration (http://www.linux-ha.org/GettingStarted/TwoApaches). Secure Socket Layer (SSL)/Transport Layer Security (TLS) and http proxy services are somewhat compute intensive therefore their impact on overall performance is

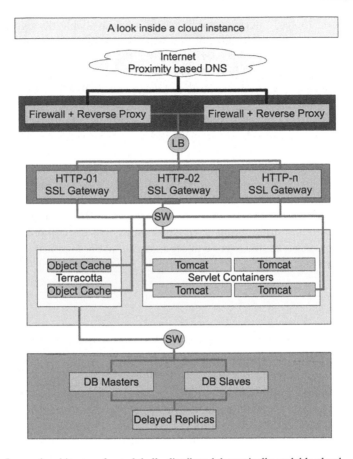

Fig. 24.6 Layered architecture for a globally distributed dynamically scalable cloud

close to linear. Thus, resources on this layer can be estimated based on number of connections.

Tier 3 (Application Tier) is the actual application/business logic tier. The primary platforms in this tier are Apache Tomcat and Sun Java. Since the load balancing will be done mostly on the outer perimeter and http tier, high availability becomes the primary concern at this level. It being a healthcare domain, one of our prime objectives is to ensure the application availability all the time. Being data critical, healthcare applications cannot afford to lose the connection even during the major hardware outage of x-1 nodes (where x is the number of nodes serving the application via Apache Tomcat), the layer would ensure the constant availability of all applications.

The final Tier, Tier 4 (Database Tier) of the server systems is the database systems. We discuss a brief architecture of our proposed system.

24.5.4.1 Distributed Data Consistency Across Clouds

One can easily carry out a detailed performance evaluation and benchmarking of various database storage methodologies such as traditional relational database management systems (RDBMS) Object Oriented Database Systems and Distributed Key-Value Persistence. The performing database architecture could then be implemented on a single-cloud in a standard master-slave topology with distributed reads and master-only writes.

A special replication server can be configured to replicate the data from the master database after every 20–60 min. Such configuration would allow us to preserve a consistent data state which is lagged by 20–60 min. In case of a system failure the data from this state can be recovered. All data-backups can be scheduled to run from the special replication server such as to avoid affecting read or write performance. The existence of multiple databases scattered over multiple clouds will pose a data consistence issue (Holden, et al., 2009; Saito & Shapiro, 2005). Cross-cloud architecture can be developed to handle this issue safely and efficiently. Figure 24.7 shows the proposed method for distributed master database synchronization technique.

The proposed solution is to perform offline synchronization on a schedule. On the system being replicated to, we can develop an agent to stop and reroute new connections, pause all automated maintenance agents flush all the caches on each node of each system and then perform cross replication from a replica of the online system to its own master.

The agents responsible for this will also communicate amongst each other to ensure that it would be performed in a rolling pattern, where no more than 1/3 to $\frac{1}{2}$ of the individual global cloud instances are unavailable at any given time. This will eliminate perceived service interruption. Since this data is ultra sensitive and must be protected at all costs, an industry standard IPSEC VPN must be implemented to facilitate this cross-cloud replication or synchronization.

24.5.4.2 Higher availability and application scalability

We propose the architecture of a global medical information system that may have millions of users accessing the system for accessing personal health records. Therefore the system must ensure high scalability in order to serve increasing number of users on the system. Further, it will be imperative to ensure the persistence and integrity of the information store while maintaining high performance.

One can easily explore and benchmark the methods for distributed HTTP serving (Zegura, Ammar, Fei, Bhattacharjee, 2000). One method distributed HTTP serving, referred to as Geographic Load Balancing, is controversial as to its effectiveness yet being quite heavily used among large web presences like Google, Inc. or Amazon.com. The premise of Geographic Load Balancing method is that any host with a public IP address can be cross referenced with the IP address block assignments on a per country basis (Colajanni, Yu, 1997) (See Fig. 24.8).

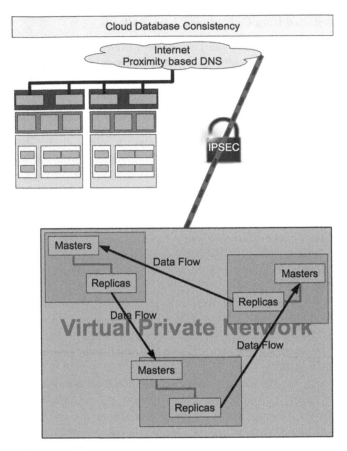

Fig. 24.7 Proposed method for distributed master database synchronization

For the application tier, we propose to implement load balancing using the JK connector from the http layer in a weighted round robin load balancing scheme. The application cluster software stack will include Apache Tomcat on Sun Java [TM] and a JVM heap clustering suite, Terracotta (See Fig. 24.9) (Eeckhout, Georges, & De Bosschere, 2003; Taboada, Touriño, & Doallo, 2009).

On cloud computing clusters one can spawn new computing resources, virtual machines, dynamically. We propose to that one should develop a method that would allow us to effectively allocate/deallocate a new application instance in a timely manner. Such a method would further interface with the JK connector(s) in order to dynamically alter connection weights and notify the HTTP layer of a new resource against which it can balance (Arzuaga & Kaeli, 2010). The proposed system would include four application partitions: Core System Services, Hospital Information Web Services, PACS System and Accelerated DICOM Presentation Services. The agent will know which of these services needs more resources. We will develop

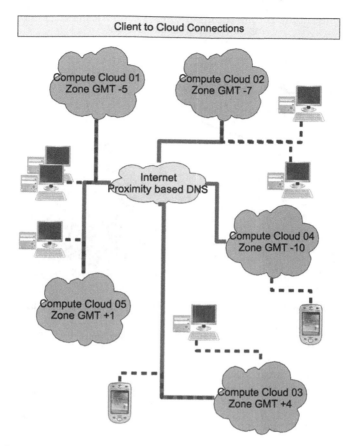

Fig. 24.8 Geographic proximity cloud selection

an algorithm to detect rate of load increase based on the special needs of each subsystem.

24.5.4.3 Concerning Low Level Security

Although user authentication and authorization will reside in the application and integration services, the GMIS Infrastructure must be developed in such a way so as to ensure trustworthy use of the cloud systems and networks. GMIS security components and layers will be enforced on any internet capable platform. The existing security methods such as use of firewalls with a minimum necessary access policy and Public key infrastructure will be deployed in order to ensure secured access to the healthcare cloud. Further, one must aim at certifying all clients by GMIS Root Certificate Authority which in-turn may be certified by a third party. Figure 24.10 shows a simplified view of trusted client connection to the healthcare cloud.

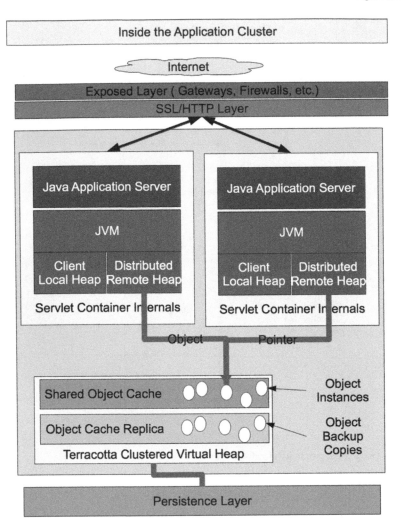

Fig. 24.9 Application cluster

The public key infrastructure should be used for accessing the services and applications using Transport Layer Security (TLS) and require the servers to provide their credentials to the client. Additionally, by requiring the client also to present their security credentials (or certificate) we can easily establish a low level trust and assume that both parties are very likely to be who they claim. One must further configure the SSL/TLS processing servers with an HTTP based reverse proxy and Intrusion Detection suite.

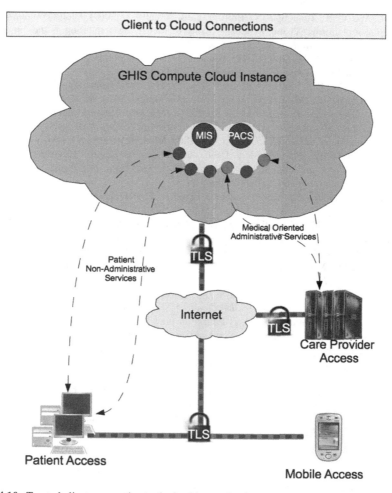

Fig. 24.10 Trusted client connection to the healthcare cloud

References

Arzuaga, E., & Kaeli, D. R. (2010). *Quantifying Load Imbalance on Virtualized Enterprise Servers* (pp. 235–242). ACM Proceedings joint WOSP/SIPEW International Conference on Performance Engineering, New York, NY.

Al-Busaidi, A., Gray, A., & Fiddian, N. (2006). Personalizing web information for patients: Linking patient medical data with the web via a patient personal knowledge base. *Health Informatics Journal, 12*(17), 27–39.

Boochever, S. S. (2004). HIS/RIS/PACS Integration: Getting to the Gold Standard. *Radiology Management, 26*(3), 16–24.

Buyya, R., Yeo, C. S., Venugopal, S., Broberg, J., & Brandic, I. (2009). Cloud computing and emerging it platforms: Vision, hype, and reality for delivering computing as the 5th utility. *Elsevier Future Generation Computer Systems, 25*, 599–616.

Carr, C. D., & Moore, S. M. (2003). IHE: A model for driving adoption of standards. *Journal of Computerized Medical Imaging and Graphics, 27*, 137–147.

Colajanni, M., & Yu, P. S. (1997). Adaptive TTL schemes for load balancing of distributed web servers. *ACM SIGMETRICS Performance Evaluation Review, 25*(2), 36–42.

Eeckhout, L., Georges, A., & De Bosschere, K. (2003). *How Java Programs Interact With Virtual Machines at the Microarchitectural Level. ACM Conference on Object Oriented Programming Systems Languages and Applications* (pp. 169–186). ACM, New York, NY.

Gallagher, J. (2009). Harvard Med Dean Sees Long Road to Electronic Records. *Triangle Business Journal*, Retrieved April 23, 2009, from http://triangle.bizjournals. com/triangle/stories/2009/04/20/daily53.html

Hasselbring, W. (2000). Information system integration. *Communications of the ACM, 43*(6), 32–38.

Haux, R. (2006). Health information systems – past, present, future. *International Journal of Medical Informatics, 75*, 268–281.

Hayes, B. (2008). Cloud computing. *Communications of the ACM, 51*(7), 9–11.

Hersh, W. R. (2002) Medical informatics improving health care through information. *Journal of American Medical Association, 288*(16), 1955–1958.

Holden, E. P., Kang, J. W., Bills, D. P., Ilyassov, M., Holden, E. P., Kang, J. W., Bills, D. P., & Ilyassov, M. (2009). *Databases in the Cloud: A Work in Progress* (pp. 138–143). ACM Conference on Information Technology Education. New York: ACM.

Katehakis, D. G., Sfakianakis, S. G., Kavlentakis, G., Anthoulakis, D. N., Tsiknakis M. (2007) Delivering a lifelong integrated electronic health record based on a service oriented architecture. *IEEE Transactions on Information Technology in Biomedicine, 11*(6), 639–650.

Lenz, R., Beyer, M., Kuhn, K. A. (2007). Semantic integration in healthcare networks. *International Journal of Medical Informatics, 7*(6), 201–207.

Lu, X., Duan, H., Li, H., Zhao, C., An, J. (2005). The architecture of enterprise hospital information system. *27th Annual Conference Proceedings of IEEE Engineering in Medicine and Biology, 7*, 6957–6960.

McDonald, C. J. (1997) The barriers to electronic medical record systems and how to overcome them. *Journal of the American Medical Informatics Association, 4*(3), 213–221.

Saito, Y., & Shapiro, M. (2005). Optimistic replication. *Journal of ACM Computing Surveys, 37*(1), 42–81.

Spyrou, S. S., Bamidis, P., Chouvarda, I., Gogou, G., Tryfon, S. M., & Maglaveras, N. (2002) Healthcare information standards: Comparison of the approaches. *Health Informatics Journal, 8*, 14–19.

Strähle, M., Ehlbeck, M., Prapavat, V., Kück, K., Franz, F., & Meyer, J.-U. (2007). Towards a service-oriented architecture for interconnecting medical devices and applications (pp. 153–155). IEEE Workshop on High Confidence Medical Devices, Software, and Systems and Medical Device Plug-and-Play Interoperability.

Taboada, G. L., Touriño, J., & Doallo R. (2009). Java for high performance computing: assessment of current research and practice (pp. 30–39). ACM Proceedings of the 7th International Conference on Principles and Practice of Programming in Java, New York, NY.

Wang, W., Wang, M., & Zhu, S. (2005). Healthcare Information System Integration: A Service Oriented Approach. *IEEE International Conference on Services Systems and Services Management, 2*, 1475–1480.

Weber-Jahnke, J. H., Price, M. (2007). *Engineering Medical Information Systems: Architecture, Data and Usability & Security* (pp. 188–189). IEEE International Conference on Software Engineering,.

Young, A. S., Mintz, J., Cohen, A. N., Chiman, M. J. (2004). Network-based system to improve care for schizophrenia: The Medical Informatics Network Tool (MINT). *Journal of American Medical Informatics, 11*(5), 358–367

Zegura, E. W., Ammar, M. H., Fei, Z., Bhattacharjee, S. (2000). Application-Layer Anycasting: A Server Selection Architecture and Use in a Replicated Web Service. *IEEE/ACM Transactions on Networking, 8*(4), 455–466.

Chapter 25
Cloud@Home: A New Enhanced Computing Paradigm

Salvatore Distefano, Vincenzo D. Cunsolo, Antonio Puliafito, and Marco Scarpa

25.1 Introduction

Cloud computing is a distributed computing paradigm that mixes aspects of *Grid computing*, ("... hardware and software infrastructure that provides dependable, consistent, pervasive, and inexpensive access to high-end computational capabilities" (Foster, 2002)) *Internet Computing* ("...a computing platform geographically distributed across the Internet" (Milenkovic et al., 2003)), *Utility computing* ("a collection of technologies and business practices that enables computing to be delivered seamlessly and reliably across multiple computers, ... available as needed and billed according to usage, much like water and electricity are today" (Ross & Westerman, 2004)) *Autonomic computing* ("computing systems that can manage themselves given high-level objectives from administrators" (Kephart & Chess, 2003)), *Edge computing* ("... provides a generic template facility for any type of application to spread its execution across a dedicated grid, balancing the load ..." Davis, Parikh, & Weihl, 2004) and *Green computing* (a new frontier of Ethical computing[1] starting from the assumption that in next future energy costs will be related to the environment pollution).

The development and the success of Cloud computing is due to the maturity reached by both hardware and software, in particular referring to virtualization and Web technologies. These factors made realistic the L. Kleinrock outlook of computing as the 5th utility (Kleinrock, 2005), like gas, water, electricity and telephone.

Cloud computing is derived from the *service-centric perspective* that is quickly and widely spreading on the IT world. From this perspective, all capabilities and

---

S. Distefano (✉), V.D. Cunsolo, A. Puliafito, and M. Scarpa
University of Messina, Contrada di Dio, S. Agata, Messina, Italy
e-mails: {sdistefano; vdcunsolo; apuliafito; mscarpa}@unime.it

[1] *Ethical computing* puts into practice the principles of *computer ethics*. Computer ethics is a branch of practical philosophy which deals with how computing professionals should make decisions regarding professional and social conduct. Computer ethics is a very important topic in computer applications, whose interests are quickly rising in the last period due to the increase of raw materials and basic commodities prices.

B. Furht, A. Escalante (eds.), *Handbook of Cloud Computing*,
DOI 10.1007/978-1-4419-6524-0_25, © Springer Science+Business Media, LLC 2010

resources of a Cloud (usually geographically distributed) are provided to users *as a service*, to be accessed through the Internet without any specific knowledge of, expertise with, or control over the underlying technology infrastructure that supports them.

Cloud computing is strictly related to *service oriented science* (Foster, 2005), *service computing* (Zhang, 2008) and *IT as a service* (ITAAS) (Foster & Tuecke, 2005), a generic term that includes: platform AAS, software AAS, infrastructure AAS, data AAS , security AAS, business process management AAS and so on. It offers a user-centric interface that acts as a unique, user friendly, point of access for users' needs and requirements.

Moreover, Cloud computing provides *on-demand service provision*, *QoS guaranteed offer*, and *autonomous system* for managing hardware, software and data transparently to users (Wang et al., 2008).

In order to achieve such goals it is necessary to implement a level of abstraction of physical resources, uniforming their interfaces and providing means for their management, adaptively to user requirements. This is done through *virtualizations*, *service mashups* (Web 2.0) and *service oriented architectures* (SOA).

Virtualization allows to execute a software version of a hardware machine into a host system, in an isolated way. It "homogenizes" resources: problems of compatibility are overcome by providing heterogeneous hosts of a distributed computing environment (the Cloud) with the same virtual machine. The software implementing virtualization is named *hypervisor*.[2]

The Web 2.0[3] provides an interesting way to interface Cloud services, implementing service mashup. It is mainly based on an evolution of JavaScript with improved language constructs (late binding, clousers, lambda functions, etc) and AJAX interactions.

The *Service Oriented Architecture* (SOA) is a paradigm for organizing and utilizing distributed capabilities that may be under the control of different ownership domains (MacKenzie et al., 2006). In SOA, *services* are the mechanism by which needs and capabilities are brought together. SOA defines standard interfaces and protocols that allow developers to encapsulate information tools as services that clients can access without knowledge of, or control over, their internal workings (Foster, 2005).

As pictorially described in Fig. 25.1[4] a great interest on Cloud computing has been manifested from both academic and private research centers, and numerous projects from industry and academia have been proposed. In commercial contexts, among the others we highlight: Amazon Elastic Compute Cloud (http://aws.amazon.com/ec2), IBM's Blue Cloud (http://www-03.ibm.com/press/us/en/pressrelease/22613.wss/), Sun Microsystems Network.com

---

[2] According to the ties between the host and the guest OS, two kinds of virtualization techniques are available [19]: *full-virtualization* (completely decoupled OS such as QEMU, VirtualBox, VMWare, etc) and *para-virtualization* (guest OS partially depends on host OS such as XEN).

[3] http://www.oreillynet.com/pub/a/oreilly/tim/news/2005/09/30/what-is-web-20.html

[4] http://peterlaird.blogspot.com/2008/09/visual-map-of-cloud-computingsaaspaas.html

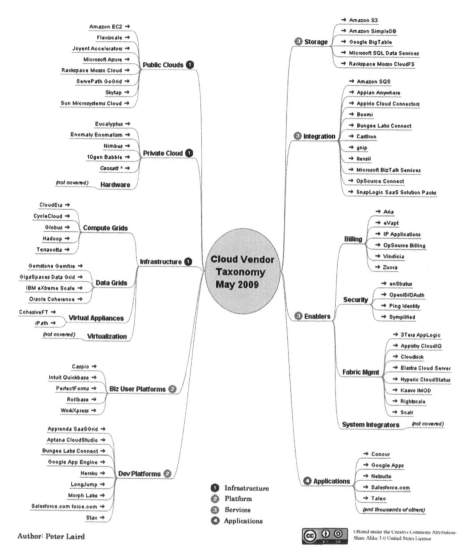

Fig. 25.1 Cloud computing taxonomy

(http://www.network.com), Microsoft Azure Services Platform (http://www.
microsoft.com/azure/default.mspx), Dell Cloud computing solutions (http://www.
dell.com/cloudcomputing). There are also several scientific activities, such as:
Reservoir (http://www-03.ibm.com/press/us/en/pressrelease/23448.wss/), Nimbus-
Stratus-Wispy-Kupa[5] and OpenNEbula (http://www.opennebula.org/). All of them

---

[5]Respectively: http://workspace.globus.org/clouds/nimbus.html/,   http://www.acis.ufl.edu/vws/,
http://www.rcac.purdue.edu/teragrid/resources/#wispy, http://meta.cesnet.cz/cms/opencms/en/docs/
clouds

support and provide an on-demand computing paradigm, in the sense that a user submits his/her requests to the Cloud that remotely, in a distributed fashion, processes them and gives back the results. This client-server model well fits aims and scopes of commercial Clouds: the business. But, on the other hand, it represents a restriction for scientific Clouds, that have a view closer to *Volunteer computing*.

Volunteer computing (also called *Peer-to-Peer computing*, *Global computing* or *Public computing*) uses computers volunteered by their owners, as a source of computing power and storage to provide distributed scientific computing (Anderson & Fedak, 2006).

It is behind the "@home"[6] philosophy of sharing/donating network connected resources for supporting distributed scientific computing.

We believe the Cloud computing paradigm is applicable also at lower scales, from the single contributing user, that shares his/her desktop, to research groups, public administrations, social communities, small and medium enterprises, which make available their distributed computing resources to the Cloud. Both free sharing and pay-per-use models can be easily adopted in such scenarios.

From the utility point of view, the rise of the "techno-utility complex" and the corresponding increase of computing resources demand, in some cases growing dramatically faster than Moore's Law as predicted by the Sun CTO Greg Papadopoulos in the *red shift theory* for IT (Martin, 2007) could bring, in a close future, towards an *oligarchy*, a lobby or a trust of few big companies controlling the whole computing resources market. In this sense: "vested economic and political interests could conspire together to build huge technology-based utility industries that preserve and reinforce their power bases".[7]

To avoid such pessimistic but achievable scenario, we suggest to address the problem in a different way: instead of building costly private *data centers*, that the Google CEO Eric Schmidt likes to compare to the prohibitively expensive cyclotrons (Baker, 2008), we propose a more "democratic" form of Cloud computing, in which the computing resources of single users accessing the Cloud can be shared with the others, in order to contribute to the elaboration of complex problems.

Since this paradigm is very similar to the Volunteer computing one, it can be named *Cloud@Home*. Both hardware and software compatibility limitations and

---

[6]Examples of projects on the topic are: SETI@home (Search for Extra-Terrestrial Intelligence http://setiathome.berkeley.edu/) - looks for radio evidence of extraterrestrial life; FOLDING@home (http://folding.stanford.edu/) Predictor@home (http://predictor.chem.lsa.umich.edu/) - investigate protein-related diseases; Einstein@home (http://einstein.phys.uwm.edu/) - looks for gravitational signals coming from pulsars; LHC@home (http://lhcathome.cern.ch/) - improves the design of LHC particles accelerator; AQUA@home (http://aqua.dwavesys.com/) - predicts the performance of superconducting adiabatic quantum computers on a variety of hard problems.

[7]http://www.roughtype.com/archives/2007/12/the_technoutili.php

restrictions of Volunteer computing can be solved in Cloud computing environments, allowing to share both hardware and software resources or *services*.

The Cloud@Home paradigm could be also applied to commercial Clouds, establishing an *open computing-utility market* where users can both buy and sell their services. Since the computing power can be described by a "long-tailed" distribution, in which a high-amplitude population (Cloud providers and commercial data centers) is followed by a low-amplitude population (small data centers and private users) which gradually "tails off" asymptotically, Cloud@Home can catch the *Long Tail* effect (Anderson, 2006), providing similar or higher computing capabilities than commercial providers' data centers, by grouping small computing resources from many single contributors.

25.2 Why Cloud@Home?

The necessity of such a new computing paradigm is strictly related to the limits of existing Cloud solutions. For years the Grid computing paradigm has been considered as the solution for all the computing problems: a secure, reliable, performing platform for safely managing geographically distributed resources. But the Grid computing has some drawbacks: it is sensitive to hardware or software differences or incompatibility; it is not possible to dynamically extend a Virtual Organization by on-line enrolling resources, and consequently is not possible to share local resources, if they are not initially enrolled in the VO; it often does not face QoS and billing problems; it mainly implements *data parallelism* against *task parallelism*, making difficult the composition of services; a user needs to have knowledge of both the distributed system and the application requirements in order to submit and manage jobs.

These lacks have been partially faced and solved in Utility and Cloud computing, implementing service oriented paradigms with higher level user friendly interfaces. Utility and Cloud implement on-demand computing paradigms: users commission their computing, pay and get the results. Since they are mainly thought for commercial applications, QoS and business policies have to be carefully addressed. Utility and Cloud computing lack of an open, free viewpoint: as in the Grid computing, it is not possible to enroll resources or services, as also to build custom data centers by dynamically aggregating resources and services not conceived with this purpose.

Moreover, each Cloud has its own interface and services, therefore it cannot communicate or interoperate with the other Clouds. Another important issue is the *customizability*, i.e. the capability of expressing a custom application by means of services.

On the other hand the Volunteer computing paradigm is born for supporting the philosophy of open computing. It implements an open distributed environment in which resources (not services as in the Cloud) can be shared. But it manifests the same problem of Grid with regard to the compatibility among resources. Moreover, due to its purpose, it also does not implement any QoS and billing policy.

25.2.1 Aims and Goals

Ian Foster summarizes the computing paradigm of the future as follows[8]: "... we will need to support on-demand provisioning and configuration of integrated 'virtual systems' providing the precise capabilities needed by an end-user. We will need to define protocols that allow users and service providers to discover and hand off demands to other providers, to monitor and manage their reservations, and arrange payment. We will need tools for managing both the underlying resources and the resulting distributed computations. We will need the centralized scale of today's Cloud utilities, and the distribution and interoperability of today's Grid facilities."

We share all these requirements, but in a slightly different perspective: we want to actively involve users into such a new form of computing, allowing to create their own interoperable Clouds. In other words, we believe that it is possible to export, apply and adapt the "@home" philosophy to the Cloud computing paradigm. By merging Volunteer and Cloud computing, a new paradigm is created: *Cloud@Home*. This new computing paradigm gives back the power and the control to users, who can decide how to manage their resources/services in a global, geographically distributed context. They can voluntarily sustain scientific projects by voluntarily providing their resources to scientific research centres for free, or they can earn money by selling their resources to Cloud computing providers in a pay per use/share context.

In this way, the focus is moved from Cloud providers to users: Cloud@Home can be a Cloud computing framework that take as main goal user's needs. Thus, in such perspective, both the commercial/business and the volunteer/scientific viewpoints coexist: in the former case the end-user orientation of Cloud is extended to a collaborative two-way Cloud in which users can buy and/or sell their resources/services; in the latter case, the Grid philosophy of few but large computing requests is extended and enhanced to *open* Virtual Organizations. In both cases QoS requirements could be specified, introducing in the Grid and Volunteer philosophy (*best effort*) the concept of quality.

Cloud@Home can be also considered as a generalization and a maturation of the @home philosophy: a context in which users voluntarily share their resources without any compatibility problem. This allows to knock down both hardware (processor bits, endianness, architecture, network) and software (operating systems, libraries, compilers, applications, middlewares) barriers of Grid and Volunteer computing. Moreover, in Cloud@Home the term resources must be interpreted in the more general Cloud sense of services. This means that Cloud@Home allows users to share not only physical resources, as in @home projects or in Grid environments, but any kind of service. The *flexibility* and the *extendibility* of Cloud@Home could allow to easily arrange, manage and make available (for free or paying) significant computing resources (greater than in Clouds, Grids and/or @home environments) to everyone that owns a computer.

---

[8]http://ianfoster.typepad.com/blog/2008/01/theres-grid-in.html

From the other hand, Cloud@Home can be considered as the enhancement of the Grid-Utility vision of Cloud computing. In this new paradigm, users' hosts are not passive interfaces to Cloud services, but they can be actively involved in computing. Single nodes and services can be enrolled by the Cloud@Home middleware, in order to build own-private Cloud infrastructures that can (for free or paying) interact with other Clouds. This allows to customize Cloud applications with own special purpose services.

The Cloud@Home motto is: *heterogeneous hardware for homogeneous Clouds.* Thus, the scenario we prefigure is composed of several coexisting and interoperable Clouds, as pictorially depicted in Fig. 25.2. *Open Clouds* (yellow) identify open VO operating for free Volunteer computing; *Commercial Clouds* (blue) characterize entities or companies selling their computing resources for business; *Hybrid Clouds* (green) can both sell or give for free their services. Both Open and Hybrid Clouds can interoperate with any other Clouds, also Commercial, while these latter can interoperate each other if and only if the Commercial Clouds are mutually *recognized*. In this way it is possible to make *federations* of Clouds working together on the same project.

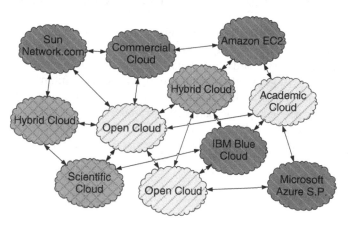

Fig. 25.2 Cloud@Home scenario

This can give to users the possibility to *choose* the best provider that matches their requirements. In such *flexible* context, Cloud providers can establish business relationships, agreements and strategies to achieve the best market performance, reducing costs and maximizing revenues.

The overall infrastructure must deal with the high dynamism of its nodes/resources, allowing to move and reallocate data, tasks and jobs. It is therefore necessary to implement a lightweight middleware, specifically designed to optimize migrations. The choice of developing such middleware on existing technologies (as done in Nimbus-Stratus starting from Globus) could be limitative, inefficient or not adequate from this point of view. This represents another significant point in favor of Could@home against Grid: a lightweight middleware allows to involve limited resources' devices into the Cloud, implementing some specific (light) services. Moreover, the Cloud@Home middleware does not influence the code writing as Grid and Volunteer computing paradigms do.

Another important goal of Cloud@Home is the *security*. Volunteer computing has some lacks in security concerns, while the Grid paradigm implements complex security mechanisms. The virtualization in Clouds implements the isolation of the services, but does not provide any protection from local access. With regards security, the specific goal of Cloud@Home is to extend the security mechanisms of Clouds to the protection of data from local access. Since Cloud@Home is composed of an amount of resources potentially larger than commercial or proprietary Cloud solutions, its *reliability* can be compared to the Grid or the Volunteer computing one, and it should be greater than other Clouds.

Last but not least, *interoperability* is one of the most important goal of Cloud@Home. This is an open problem in Grid, Volunteer and Cloud computing that we want to adequately face in Cloud@Home. In Grid environment interoperability is a very tough issue, many people tried to address it for many years and still we are far away to solve the problem. Interoperability in Cloud contexts is easier, since virtualization avoids the major architectural, physical, hardware and software problems.

New standards and interfaces enabling enhanced portability and flexibility of virtualized applications have to be implemented. Up to now, significant discussion has occurred around open standards for Cloud computing. In this context, the "Open Cloud Manifesto" (www.opencloudmanifesto.org) provides a minimal set of principles that will form a basis for initial agreements as the Cloud community develops standards for this new computing paradigm. Moreover, problems of compatibility among different virtual machines (VM) monitors can arise and therefore must be adequately faced, as the Open Virtualization Format (OVF) group is trying to do.

25.2.2 Application Scenarios

Several possible application scenarios can be imagined for Cloud@Home:

- Scientific research centers, communities – the Volunteer computing inspiration of Cloud@Home provides means for the creation of open, interoperable Clouds for supporting scientific purposes, overcoming the portability and compatibility problems highlighted by the @home projects. Similar benefits could be experienced in public administrations and open communities (social network, peer-to-peer, cloud gaming, etc). Through Cloud@Home it could be possible to implement resources and services management policies with QoS requirements (characterizing the scientific project importance) and specifications (QoS classification of resources and services available). A new deal for Volunteer computing that does not take into consideration such aspect, following a best effort approach.

- Enterprises - planting a Cloud@Home computing infrastructure in business/commercial locations can bring considerable benefits, especially in small and medium but also in big enterprises. It could be possible to implement own data center with local, existing, off the shelf, resources: usually in any enterprise there exist a capital of stand-alone computing resources for office automation, monitoring, designing and so on. Since such resources are only (partially) used in office hours, by Internet connecting them altogether it becomes possible to build up a Cloud@Home data center, in which allocate the shared services (web server, file server, archive, database, etc) without compatibility constraints or problems. The interoperability among Clouds allows to buy computing resources from commercial Cloud providers if needed or, otherwise, to sell the local Cloud computing resources to the same providers. This allows to reduce and optimize business costs according to QoS/SLA policies, improving performances and reliability. For example, this paradigm allows to deal with the *flow peaks economy*: data centers could be sized for the medium case, and worst cases (peaks) could be managed by buying computing resources from Cloud providers. Moreover, Cloud@Home drives towards a resources rationalization: all the business processes can be securely managed by web, allocating resources and services where needed. In particular this fact can improve marketing and trading (E-commerce), making available to sellers and customers a lot of customizable services. The interoperability could also point out another scenario, in which private companies buy computing resources in order to resell them.
- Ad-hoc networks, wireless sensor networks, home automation - the Cloud computing approach, where both the software and the computing resources are owned and managed by the service providers, eases the programmers' efforts in facing the device heterogeneity and prevents application downloads. Mobile application designers should start to consider that their applications, besides to be usable on a small device, will need to interact with the Cloud. Service discovery, brokering, and reliability are important, and services are usually designed to interoperate (The Programmable Web. http://www.programmableweb.com). In order to consider the arising consequences related to the access of mobile users to service-oriented grid architecture, researchers have proposed new concepts such as the one of a mobile dynamic virtual organization (Waldburger & Stiller, 2006). New distributed infrastructures have been designed to facilitate the extension of Clouds to the wireless edge of the Internet. Among them, the Mobile Service Clouds enables dynamic instantiation, composition, configuration, and reconfiguration of services on an overlay network to support mobile computing (Samimi, McKinley, & Sadjadi, 2006).

A still open research issue is whether or not a mobile device should be considered as a service provider of the Cloud itself. The use of modern mobile terminals such as smart-phones not just as Web Service requestors, but also as mobile hosts that can themselves offer services in a true mobile peer-to-peer setting is discussed in (Srirama, Jarke, & Prinz, 2006). Context aware operations involving control and monitoring, data sharing, synchronization, etc, could be implemented and exposed as Cloud@Home Web services involving wireless and Bluetooth devices, laptop, Ipod, cellphone, household appliances, and so on. Cloud@Home could be a way for implementing *Ubiquitous* and *Pervasive* computing: many computational devices and systems can be engaged simultaneously for performing ordinary activities, and may not necessarily be aware that they are doing so.

25.3 Cloud@Home Overview

Our basic idea is to reuse "domestic" computing resources to build voluntary contributors' Clouds. With Cloud@Home, anyone can experience the power of Cloud computing, both actively providing his/her own resources and services, and passively submitting his/her applications.

25.3.1 Issues, Challenges and Open Problems

In order to implement such a form of computing the following issues should be taken into consideration:

- Resources and Services management – a mechanism for managing resources and services offered by Clouds is mandatory. This must be able to enroll, discovery, index, assign and reassign, monitor and coordinate resources and services. A problem to face at this level is the compatibility among resources and services and their portability.
- Frontend – abstraction is needed in order to provide users with a high level service oriented point of view of the computing system. The frontend provides a unique, uniform access point to the Cloud. It must allow users to submit functional computing requests only providing requirements and specifications, without any knowledge of the system resources deployment. The system evaluates such requirements and specifications and translates them into physical resources' demand, deploying the elaboration process. Another aspect concerning the frontend is the capability of customizing Cloud services and applications.
- Security – effective mechanisms are required to provide: authentication, resources and data protection, data confidentiality and integrity.
- Resource and service accessibility, reliability and data consistency – it is necessary to implement redundancy of resources and services, and hosts' recovery policies since users voluntarily contribute to the computing, and therefore they can asynchronously, at any time, log out or disconnect from the Cloud.
- Interoperability among Clouds – it should be possible for Clouds to interoperate each other.
- Business models – for selling Cloud computing it is mandatory to provide QoS and SLA management for both commercial and open volunteer Clouds (traditionally best effort), in order to discriminate among the applications to be run.

25.3.2 Basic Architecture

In order to accomplish the introduced issues, we recur to virtualization. This technology provides solution to the problem of incompatibility among resources, implements an adequate level of abstraction and guarantees the isolation of services and resources, i.e. the security protection.

A possible Cloud@Home architecture is shown in Fig. 25.3, identifying three hierarchical layers: *frontend*, *virtual* and *physical*.

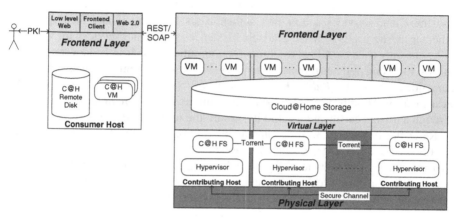

Fig. 25.3 Basic architecture of Cloud@Home

According to this point of view the Cloud is composed of several *contributing hosts* that share their resources. A user can interact with the Cloud through the *consumer host* after authenticating him/herself into the system. The main enhancement of Cloud@Home is that a host can be at the same time both contributing and consumer host, establishing a symbiotic mutual interaction with the Cloud.

25.3.3 Frontend Layer

The Cloud@Home frontend layer is responsible for the resources and services management (enrolling, discovery, allocation, coordination, monitoring, scheduling, etc.) from the global Cloud system's perspective. The frontend layer provides tools for translating end-user requirements into physical resources' demand, also considering QoS/SLA constraints, if specified by the user. Moreover, in commercial Clouds, it must be able to negotiate the QoS policy to be applied (SLA), therefore monitoring for its fulfillment and, in case of unsatisfactory results, adapting the computing workflow to such QoS requirements.

If the available Cloud's resources and services can not satisfy the requirements, the frontend layer provides mechanisms for requesting further resources and services to other Clouds, both open and/or commercial. In other words, the Cloud@Home frontend layer implements the interoperability among Clouds, also checking for services' reliability and availability. In order to improve reliability and availability of services and resources, especially if QoS policies and constraints have been specified, it is necessary introduce redundancy.

The frontend layer is split into two parts, as shown in Fig. 25.3: the server side, implementing the resources management and related problems, and the *light* client side, only providing mechanisms and tools for authenticating, accessing and interacting with the Cloud.

In a widely distributed system, globally spread around the world, the knowledge of resources' accesses and uses assumes great importance. To access and use the Cloud services a user first authenticates him/herself and then specifies whether he/she wants to make available his/her resources and services for sharing, or he/she only uses the Cloud resources for computing. The frontend layer provides means, tools and policies for managing users. The best mechanism to achieve secure authentications is the *Public Keys Infrastructure* (PKI) (Tuecke, Welch, Engert, Pearlman, & Thompson, 2004), better if combined with smartcard devices that, through a trusted certification authority, ensure the user identification.

Referring to Fig. 25.3, three alternative solutions can be offered by the frontend layer for accessing a Cloud: (a) Cloud@Home frontend client, (b) Web 2.0 user interface and (c) low level Web interface (directly specifying REST or SOAP queries). These also provide mechanisms for customizing user applications by composing services (service mashup and SOA) and submitting own services.

25.3.4 Virtual Layer

The virtualization of physical resources offers end-users a homogeneous view of Cloud's services and resources. Two basic services are provided by the virtual layer to the frontend layer and, consequently, to the end-user: *execution* and *storage* services.

The execution service is the tool provided by the virtual layer for creating and managing virtual machines. A user, sharing his/her resources within a Cloud@Home, allows the other users of the Cloud to execute and manage virtual machines locally at his/her node, according to policies and constraints negotiated and monitored at the frontend layer. In this way, a Cloud of virtual machine's executors is established, where virtual machines can migrate or can be replicated in order to achieve reliability, availability and QoS targets. As shown in Fig. 25.3, from the end-user point of view an execution Cloud is seen as a set of virtual machines available and ready-to-use. The virtual machines' *isolation* implements protection and therefore security. This security is ensured by the hypervisor that runs the virtual machine's code in an isolated scope, similarly to a sandbox, without affecting the local host environment.

The storage service implements a storage system distributed across the storage hardware resources composing the Cloud, highly independent of them since data and files are replicated according to QoS policies and requirements to be satisfied.

From the end-user point of view, a storage Cloud appears as a locally mounted remote disk, similarly to a Network File System or a Network Storage. The tools, libraries and API for interfacing end-user and storage Clouds are provided to user by the frontend client, but are implemented at virtual and physical layers.

In a distributed environment where any users can host part of private data, it is necessary to protect such data from unauthorized accesses (data security). A way to obtain data *confidentiality* and *integrity* could be the cryptography, as better explained in the physical layer description.

25.3.5 Physical Layer

The physical layer is composed of a "cloud" of generic nodes and/or devices geographically distributed across the Internet. They provide to the upper virtual layer both physical resources for implementing execution and storage services and mechanisms and tools for locally managing such resources.

Cloud@Home negotiates with users that want to join a Cloud about his/her contribution. This mechanism involves the physical layer that provides tools for reserving physical execution and/or storage resources for the Cloud, and monitors these resources, such that constraints, requirements and policies thus specified are not violated. This ensures reliability and availability of physical resources, avoiding to overload the local system and therefore reducing the risk of crashes.

To implement the execution service in a generic device or to enroll it into an execution Cloud, the device must have a hypervisor ready to allocate and run virtual machines, as shown in Fig. 25.3. If a storage service is installed into the device, a portion of the local storage system must be dedicated for hosting the Cloud data. In such cases, the Cloud@Home file system is installed into the devices' shared storage space.

At physical layer it is necessary to implement data security (integrity and confidentiality) also ensuring that stored data cannot be accessed by who physically hosts them. We propose an approach that combines the inviolability of the Public Key Infrastructure asymmetric cryptography and the speed of the symmetric cryptography. Data are firstly encrypted by the symmetric key, and then stored into the selected host with the symmetric key encrypted by the user private key. This ensures that only authorized users can decrypt the symmetric key and consequently can access data.

SSH, TLS, IPSEC and other similar transmission protocols could be used to manage the connection among nodes. However, since the data stored in a Cloud@Home storage are encrypted, it is not necessary to use a secure channel for data transfers, more performant protocol, such as BitTorrent[9], can be used. The secure channel is required for sending and receiving non-encrypted messages and data to/from remote hosts.

---

[9]http://www.bittorrent.org/beps/bep_0003.html

Once the functional architecture of Cloud@Home has been introduced, it is necessary to characterize the blocks implementing the functions thus identified. These blocks have been depicted in the layered model of Fig. 25.4, that reports the core structure of the overall system implementing the Cloud@Home server-side, subdivided into *management* and *resource subsystems*.

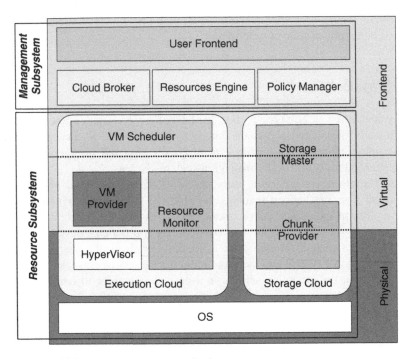

Fig. 25.4 Cloud@home core structure organization

25.3.6 Management Subsystem

In order to enroll and manage the distributed resources and services of a Cloud, providing a unique point of access them, it is necessary to adopt a centralized approach that is implemented by the management subsystem. It is composed of four parts: the *user frontend* (UF), the *Cloud broker*, the *resource engine* and the *policy manager*.

The user frontend provides tools for Cloud@Home-User interactions. It collects and manages the users' requests issued by different types of clients (frontend client, Web 2.0 and low level SOAP/REST Web interface). All such requests are transferred to the blocks composing the underlying layer (resource engine, Cloud broker and policy manager) for processing.

The Cloud broker collects and manages information about the available Clouds and the services they provide (both *functional* and *non-functional* parameters, such as QoS, costs, reliability, *request formats' specifications* for Cloud@Home-foreign Clouds translations, etc).

The policy manager provides and implements the Cloud's access facilities. This task falls into the security scope of identification, authentication and permission management. To achieve this target, the policy manager uses an infrastructure based on PKI, smartcard devices and Certification Authority. The policy manager also manages the information about users' QoS policies and requirements.

The resource engine is the hearth of Cloud@Home. It is responsible for the resources' management, the equivalent of a Grid *resource broker* in a broader Cloud environment. To meet this goal, the resource engine applies a hierarchical policy. It operates at higher level, in a centralized way, indexing all the resources of the Cloud. Incoming requests are delegated to *VM schedulers* or *storage masters* that, in a distributed fashion, manage the computing or storage resources respectively, coordinated by the resource engine.

In order to manage QoS policies and to perform the resources discovery, the resource engine collaborates with both Cloud broker and policy manager, as depicted in Fig. 25.5 showing the step-by-step interactions among such blocks. After authenticating into the system (steps 1 and 2), an end-user specifies his/her requirements (step 3), saved by the policy manager (step 4). Then, a negotiation between the two parties is triggered (step 5), iteratively interacting with the end-user till an agreement is met (SLA). This task is split into two parallel subtasks: the former (step 5a), performed by the policy manager under the supervision of the resource engine, estimates and evaluates the QoS requirements of the request; the latter (step 5b), performed by the resource engine, discovers resources and services to be used. Both subtasks can require the collaboration of the Cloud broker, that looks for other Clouds able to provide resources and services to satisfy SLA/QoS requirements.

Figure 25.6 shows the interaction between a contributing user, that wants to provide his/her resources to a Cloud, and the Cloud@Home management system. A user, authenticated by the Cloud's policy manager (steps 1 and 2), sends a request

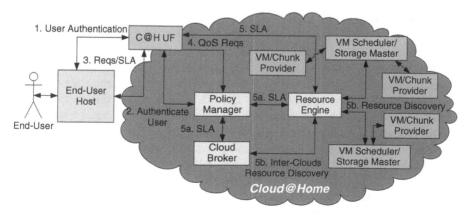

Fig. 25.5 Cloud@Home end-user negotiation

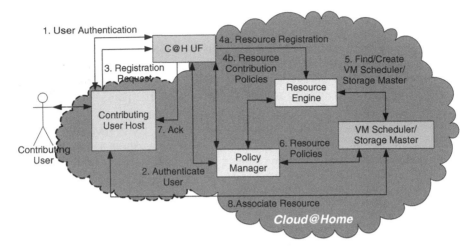

Fig. 25.6 Cloud@Home resource setup

for registering resources and services to the user frontend (step 3), also specifying policies for using them. It sorts the request at the resource engine (step 4a), and constraints and policies at the policy manager (step 4b). After that, the resource engine searches for a VM scheduler or a storage master to which such resources/services have to be assigned (step 5), collaborating with the policy manager. It can also create a new VM scheduler/storage master if the search results obtained do not satisfy the requirements.

Once the scheduler/master is identified the policy manager contacts it for exchanging policies and specifications of the resources. Then the resource engine sends the acknowledgement and the scheduler/master reference to the contributing host (step 7), that signals its availability and actual status (step 8).

25.3.7 Resource Subsystem

The resource subsystem contains all the blocks implementing the local and distributed management functionalities of Cloud@Home. This subsystem can be logically split into two parts offering different services over the same resources: the *execution Cloud* and the *storage Cloud*. The management subsystem merges them providing a unique Cloud that can offer both execution and/or storage services.

The execution Cloud provides tools for managing virtual machines according to users' requests and requirements coming from the management subsystem. It is composed of four blocks: *VM scheduler, VM provider, resource monitor* and *hypervisor.*

The VM Scheduler is a peripheral resource broker of the Cloud@Home infrastructure, to which the resource engine delegates the management of computing/execution resources and services of the Cloud. It establishes which, what, where and when allocate a VM, moreover it is responsible for moving and managing VM services.

From the end-user point of view a VM is allocated somewhere on the Cloud, therefore its migration is transparent for the end-user that is not aware of any VM migration mechanism. The association between resources and scheduler is made locally, as shown in Fig. 25.6. Since a scheduler can become a bottleneck if the system grows, to avoid the congestion further decentralized and distributed scheduling algorithms can be implemented. Possible strategies and tricks for facing the problem are:

- implementing a hierarchy of schedulers with geographic characterization (local, zone, area, region, etc);
- replicating schedulers, which can communicate each other for synchronization;
- autonomic scheduling.

The VM provider, the resource monitor and the hypervisor are responsible for managing a VM locally to a physical resource. A VM provider exports functions for allocating, managing, migrating and destroying a virtual machine on the corresponding host.

The resource monitor allows to take under control the local computing resources, according to requirements and constraints negotiated in the setup phase with the contributing user. If during a virtual machine execution the local resources crash or become insufficient to keep running the virtual machine, the resource monitor asks the scheduler to migrate the VM elsewhere.

Figure 25.7 depicts the process of requesting and allocating computing resources in Cloud@Home environments. The overall process is coordinated by the resource engine that estimates requests and requirements submitted by the end-user (steps 0,1 and 2), previously authenticated, and therefore evaluates and selects proper schedulers (step 3). Each of such schedulers, in its turn, allocates the physical resources

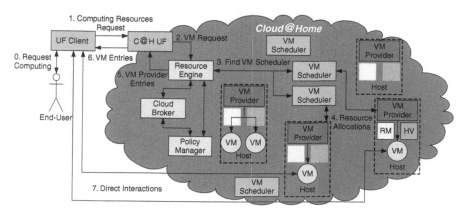

Fig. 25.7 User computing request processing

that will host the VM (step 4). The access points of such resources are then fed back to the end-user (steps 5 and 6), and consequently the two parties get connected and can directly interact (step 7).

In order to implement the storage Cloud, we specify the *Cloud@Home file system* (FS), adopting an approach similar to the Google FS one (Ghemawat, Gobioff, & Leung, 2003). The Cloud@Home FS splits data and files into *chunks* of fixed or variable size, depending on the storage resource available. The architecture of such file system is hierarchical: data chunks are physically stored on *chunk providers* and corresponding *storage masters* index the chunks through specific *file indexes* (FI).

The storage master is the directory server, indexing the data stored in the associated chunk providers, It directly interfaces with the resource engine to discover the resources storing data. In this context the resource engine can be considered as the directory server indexing all the storage masters. To improve the storage Cloud reliability, storage masters must be replicated. Moreover, a chunk provider can be associated to more than one storage master.

In order to avoid a storage master becoming a bottleneck, once the chunk providers have been located, data transfers are implemented by directly connecting end-users and chunk providers. Similar techniques to the ones discussed about VM schedulers can be applied to storage masters for improving performance and reliability of the storage Clouds.

Chunk providers physically store the data, that, as introduced above, are encrypted in order to achieve the confidentiality goal.

Data reliability can be improved by replicating data chunks and chunk providers, consequently updating the corresponding storage masters. In this way, a corrupted data chunk can be automatically recovered and restored through the storage masters, without involving the end-user.

Similarly to the execution Cloud, the storage Cloud can be implemented as shown in Fig. 25.8: an end-user data I/O request to the Cloud (steps 0 and 1) is delivered to the resource engine (step 2), that locates the storage masters managing the chunk

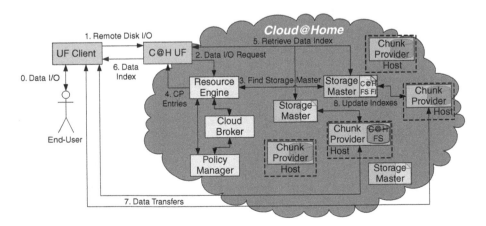

Fig. 25.8 User remote disk I/O request processing

providers where data are stored or will be stored (step 3), and feeds back the list of chunk providers and data indexes to the end-user (step 4, 5 and 6). In this way the end user can directly interact with the assigned chunk providers storing his/her data (step 7).

25.4 Ready for Cloud@Home?

In this paper we proposed an innovatory computing paradigm merging volunteer contributing and Cloud approaches into Cloud@Home. This proposal represents a solution for building Clouds, starting from heterogeneous and independent nodes, not specifically conceived for this purpose. This can implement a generalization of both Volunteer and Cloud computing by aggregating the computational poten- tialities of many small, low power systems, exploiting the long tail effect of computing.

In this way Cloud@Home opens the Cloud computing world to scientific and academic research centers, as well as to communities or single users: anyone can voluntarily support projects by sharing his/her resources. On the other hand, it opens the utility computing market to the single user that wants to sell his/her computing resources. To realize this broader vision, several issues must be ade- quately taken into account: reliability, security, portability of resources and services, interoperability among Clouds, QoS/SLA and business models and policies.

It is necessary a common understanding, an ontology that fixes metrics and concepts such as resources, services and also overall Clouds functional and non- functional parameters (QoS, SLA, exposition format, and so on), that must be translated into specific interoperability standards. Fundamental aspects to take into account are reliability and availability: in a heterogeneous Cloud we can have resources highly reliable and available, such as NAS and/or computing servers, and barely reliable and available, such as temporary contributors connected only few hours. Cloud@Home must consider such parameters, specifying adequate policies for optimizing their management.

References

Anderson, C. (2006, July). The Long Tail: How Endless Choice is Creating Unlimited Demand. Random House Business Books, London.

Anderson, D. P., & Fedak, G. (2006). The Computational and Storage Potential of Volunteer Computing. *Proceedings of the Sixth IEEE international Symposium on Cluster Computing and the Grid (May 16–19, 2006), CCGRID. IEEE Computer Society, Washington, DC,* 73–80.

Baker, S. (2008, December, 24). *Google and the Wisdom of Clouds, BusinessWeek.* Retrieved December, 2008, from http://www.businessweek.com/magazine/content/07_52/b406404 8925836.htm.

Davis, A., Parikh, J., & Weihl, W. E. (2004). *Edgecomputing: Extending Enterprise Applications to the Edge of the Internet* (pp. 180–187). WWW Alt.'04: Proceedings of the 13th international World Wide Web Conference on Alternate Track Papers & Posters. New York: ACM.

Foster, I. (2002, July) "What is the grid? – A three point checklist. *GRIDtoday, 1*(6).

Foster, I. (May 2005). Service-oriented science. *Science, 308*(5723).

Foster, I., Tuecke, S. (2005). Describing the elephant: The different faces of IT as service. *Queue,* 3 (6) 26–29.

Ghemawat, S., Gobioff, H., & Leung, S.-T. (2003, December). The Google File System. *SIGOPS Operating Systems Review, 37* (5), pp. 29–43.

Kephart, J. O., & Chess, D. M. (2003). The vision of autonomic computing. *Computer, 36,* (1) 41–50.

Kleinrock, L. (2005, November) A vision for the internet. *ST Journal of Research, 2,*(1), 4–5.

MacKenzie, C. M., Laskey, K., McCabe, F., Brown, P. F., Metz, R., & Hamilton, B. A. (2006). Reference Model for Service Oriented Architecture 1.0," OASIS SOA Reference Model Technical Committee, http://docs.oasis-open.org/soa-rm/v1.0/.

Martin, R. The Red Shift Theory," InformationWeek, Retrieved August 20, 2007, from http://www.informationweek.com/news/hardware/showArticle.jhtml? articleID=201800873.

Milenkovic, M., Robinson, S. H., Knauerhase, R. C., Barkai, D., Garg, S., Tewari, A., Anderson, T. A., & Bowman, M. (2003, May). Toward internet distributed computing. *Computer, 36*(5), 38–46.

Ross, J. W., & Westerman, G. (2004). *Preparing for Utility Computing: The Role of it Architecture and Relationship Management. IBM System Journal, 43*(1), 5–19.

Samimi, F. A., McKinley, P. K., & Sadjadi, S. M. (2006). *Mobile Service Clouds: A Self-Managing Infrastructure for Autonomic Mobile Computing Services.* LCNS 3996, Heidelberg: Springer.

Srirama, S. N., Jarke, M., & Prinz, W. (2006). *Mobile Web Service Provisioning," AICT-ICIW'06: Proceedings of the Advanced Int'l Conference on Telecommunications and Int'l Conference on Internet and Web Applications and Services* (p. 120). Washington, DC: IEEE Computer Society.

Tuecke, S., Welch, V., Engert, D., Pearlman, L., & Thompson, M. (2004, June). *Internet X.509 Public Key Infrastructure (PKI) Proxy Certificate Profile.* RFC 3820 (Proposed Standard).

VMWare, Understanding Full Virtualization, Paravirtualization, and Hardware Assist, 2007. White Paper.

Waldburger, M., & Stiller, B. (2006). *Toward the Mobile Grid: Service Provisioning in a Mobile Dynamic Virtual Organization* (pp. 579–583). IEEE International Conference on Computer Systems and Applications. Washington, DC: IEEE Computer Society.

Wang, L., Tao, J., Kunze, M., Castellanos, A., Kramer, C. D., & Karl, W. (2008). *Scientific Cloud Computing: Early Definition and Experience* (pp. 825–830). HPCC'08, IEEE Computer Society. Washington, DC: IEEE Computer Society.

Zhang, L.-J. (2008, April–June). *EIC Editorial: Introduction to the Body of Knowledge Areas of Services Computing. IEEE Transactions on Services Computing,1*(2), 62–74.

Chapter 26
Using Hybrid Grid/Cloud Computing Technologies for Environmental Data Elastic Storage, Processing, and Provisioning

Raffaele Montella and Ian Foster

26.1 Introduction

High-resolution climate and weather forecast models, and regional and global sensor networks, are producing ever-larger quantities of multidimensional environmental data. To be useful, this data must be stored, managed, and made available to a global community of researchers, policymakers, and others.

The usual approach to addressing these problems is to operate dedicated data storage and distribution facilities. For example, the Earth System Grid (ESG) (Bernholdt et al., 2005) comprises data systems at several US laboratories, each with large quantities of storage and a high-end server configured to support requests from many remote users. Distributed services such as replica and metadata catalogs integrate these different components into a single distributed system.

As both environmental data volumes and demand for that data grows, servers within systems such as ESG can easily become overloaded. Larger datasets also lead to consumers wanting to execute analysis pipelines "in place" rather than downloading data for local analysis – further increasing load on data servers. Thus, operators are faced with important decisions as to how best to configure systems to meet rapidly growing, often highly time-varying, loads.

The emergence of commercial "cloud" or *infrastructure on demand* providers (Mell & Tim, 2009) – operators of large storage and computing farms supporting quasi-instantaneous on-demand access and leveraging economics of scale to reduce cost – provides a potential alternative to the servers operated by systems such as ESG. Hosting environmental data on cloud storage (e.g., Amazon S3) and running analysis pipelines on cloud computers (e.g., Amazon EC2) has the potential to reduce costs and/or improve the quality of delivered services, especially when responding to access peaks (Armbrust, Fox, Griffith, Joseph, Katz et al., 2009).

R. Montella (✉)
Department of Applied Science, University of Napoli Parthenope, Napoli, Italy
e-mail: raffaele.montella@uniparthenope.it

I. Foster
Argonne National Laboratory, Argonne, IL, USA; The University of Chicago, Chicago, IL, USA
e-mail: foster@anl.gov

B. Furht, A. Escalante (eds.), *Handbook of Cloud Computing*,
DOI 10.1007/978-1-4419-6524-0_26, © Springer Science+Business Media, LLC 2010

In this chapter, we present the results of a study that aims to determine whether it is indeed feasible and cost-effective to apply cloud services to the hosting and delivery of environmental data. We approach this question from the twin perspectives of architecture and cost. We first examine and present the design, development, and evaluation of a cloud-based software infrastructure that leverages some grid computing services and dedicated to the storage, processing, and delivery of multidimensional environmental data. In our design we used the Amazon EC2/S3 cloud computing APIs in order to provide an elastic hosting and processing facilities for data, and the Globus Toolkit v4 (GT4) to federate data elements in a wider grid application context. The scalability is ensured by a hybrid virtual/real aggregation of computing resources.

The rest of this chapter is as follows. In Section 26.2, we introduce the application context, describing important characteristics of environmental data and what scientists need to accelerate research. In Section 26.3, we describe the current status of elastic allocated storage and the tools we developed in order to realize our goals, stressing on the new capabilities produced benefitting the computer scientist community. A GT4 (Foster, 2006) web service providing environmental multidimensional datasets using a grid/cloud hybrid approach is described in Section 26.4, while Section 26.5 presents experiments that characterize the performance and cost of our approach; these results can help identify the best deployment scenario in a real world operational applications. Finally, in Section 26.6 we present some conclusions and outline future work.

26.2 Distributing Multidimensional Environmental Data

ESG is built upon the Globus Toolkit and other related technologies. ESG continues to expand its hosted data and data processing services, leveraging an environment that addresses authentication, authorization for data access, large-scale data transport and management, services and abstractions for high-performance remote data access, mechanisms for scalable data replication, cataloging with rich semantic and syntactic information, data discovery, distributed monitoring, and Web-based portals for using the system. Current work aims to expand the scope of ESG to address the need for federation of many data sources, as will be required for the next phase of the Intergovernmental Panel on Climate Change (IPCC) assessment process.

In the world of environmental computational science, data catalogues are implemented, managed, and stored using community developed file standards such as Network Common Data File (NetCDF), mainly used for storage and (parallel) high performance retrieval, and the Gridded Binary format (GriB), usually used for data transfer. The most widely used data transfer protocol is OpenDAP (the Open source Project for a Network Data Access Protocol), formerly DODS (Distributed Oceanographic Data System). OpenDAP supports a set of standard features for requesting and transporting data across the web (Gallagher, Potter, & Sgouros, 2004). The current OpenDAP Data Access Protocol (DAP) uses HTTP for requests and responses. The Grid Analysis and Display System (GrADS) (Doty & Kinter, 1995) is a free and open source interactive desktop tool used for easy access, manipulation, and visualization of earth science data stored in various formats such as binary, GRIB, NetCDF, and HDF-SDS. The GrADS-DODS Server (GDS) combines GrADS and OPeNDAP to create an open-source solution for serving distributed scientific data (Wielgosz & Doty, 2003).

In previous work, Montella et al. developed the GrADS Data Distribution Service (GDDS) (Montella, Giunta, & Riccio, 2007) service, a GT4-based web service for publishing and serving environmental data. The use of GT4 mechanisms enables the integration of advanced authentication and authorization protocols and the convenient publication of service metadata, which is published automatically to a GT4 index service. This latter feature enables resource brokering involving both data and other grid resources such as CPUs, storage, and instruments – useful, for example, when seeking data sources that also support data processing.

Hyrax is a data server that combines the efforts at UCAR/HAO to build a high performance DAP-compliant data server based on software developed by OpenDAP (Gallagher, Potter, West, Garcia, & Fox, 2006). A servlet frontend formulates a query to a second backend server that reads data from the data stores and returns DAP-compliant responses to the frontend. The frontend may then either pass responses back to the requestor, perhaps with modifications, or it may use them to build more complex responses.

Montella et al. turned to a Hyrax-based software architecture when they developed the Five Dimensional Data Distribution Service (FDDDS), leveraging the Hyrax OpenDAP server to achieve a better integration within a web/grid service ecosystem. To implement FDDDS, they extended the client OpenDAP class APIs to separate them from the frontend and to provide the needed interfaces in order to provide services to the grid. FDDDS inherits most GDDS features included the automatic index service advertisement of available metadata. This GT4-based environmental data delivery service operates on local data storage. Thus, cloud deployment is possible in a fully virtualized context with no kind of cloud-specific optimization (Giunta, Laccetti, & Montella, 2008).

26.3 Environmental Data Storage on Elastic Resources

Our goal in this work is to explore whether it is feasible to leverage Amazon cloud services to host environmental data. In other words, we want to determine the difficulty, performance, and economic cost of operating a service like FDDDS with data (and perhaps processing as well) hosted not on local resources but on cloud resources provided by Amazon. This service, like FDDDS, should allow remote users to request both entire datasets and subsets of datasets, and ultimately also to perform analysis on datasets. Whether the service is deployed in a native grid environment or in a grid on cloud fashion should be transparent to the consumer. Ideally, the service will inherit the security and standard connection interface from grid computing and achieve scalability and availability thanks to the elastic power of the cloud.

In conducting this study, we focus in particular on performance issues. The use of dynamically allocated cloud resources has the potential for poor performance, due to the virtualized environment, internal details of cloud storage behavior, and extra cloud/intra-cloud network communication. We anticipate that it will be desirable to move as much processing work (subsetting and data analysis) as possible into the cloud, so as to minimize the need for cloud-to-outside world data transfer. This approach can also help to reduce costs, given that Amazon charges for the movement of data between its cloud storage and the outside world.

26.3.1 Amazon Cloud Services

We summarize important characteristics of the Amazon EC2, S3, and EBS services that we use in this work.

The *Elastic Compute Cloud* (EC2) service allows clients to request the creation of one or more virtual machine (VM) instances, each configured to run a VM image supplied by the client. The user is charged only for the time (rounded up to the nearest full hour) a EC2 instance is up and running. Different instance types are supported with different configurations (number of virtual cores, amount of memory, etc.) and costing different amounts per hour. An EC2 user can configure multiple VM images and run multiple instances of each to instantiate complex distributed scenarios that incorporate different functional blocks such as web servers, application servers, and database servers. EC2 provides tools, web user interface and APIs in many languages that make it straightforward to create and manage images and instances. A global image library offers a starting point from which to begin image setup and configuration.

The *Simple Storage Service* (S3) provides a simple web service interface that can be used to store and retrieve data objects (up to 5 GB in size) at any time and from anywhere on the web. Only write, read, and delete operations are allowed. The number of objects that can be created is effectively unlimited. The object name space is flat (there is no hierarchical file system): each data object is stored in a bucket and is retrieved via a unique key assigned by the developer. The S3 service replicates each object to enhance availability and reliability. The physical location of objects is invisible to the user, except that the user can choose the geographical *zone* in which to create the object (currently, US West, US East, and Europe). Unless objects are explicitly transferred, they never leave the region in which they are created.

S3 users can control who can access data or alternatively can make objects available to all. Data is accessed via REST and SOAP interfaces designed to work with any Internet development toolkit. S3 users are charged for storage and for transfers between S3 and the outside world. The default download protocol is HTTP; a BitTorrent protocol interface is provided to lower costs for large-scale distribution. Large quantities of data (e.g., a large environmental dataset) can be moved into S3 by using an import/export service based on physical delivery of portable storage units, which is more rapid and less expensive than Internet upload.

The *Elastic Block Store* (EBS) provides block-level storage volumes that can be attached to a EC2 instance as a device, but that persist independently from the life of any particular instance. This service is useful for applications that require a database, file system, or access to raw block-level storage, as in the case of NetCDF file storage. EBS volumes can range in size from one to 1,000 GB. Multiple volumes can be mounted to the same instance, allowing for data striping. Storage volumes behave like raw, unformatted block devices, with user-supplied device names and a block device interface. Instances and volumes must be located in the same zone. Volumes are automatically replicated. EBS uses S3 to store volume snapshots in order to protect data for long-term durability and to instantiate as many volumes as needed by the user. EBS performance can vary because the needed network access and the S3 snapshot interfacing and are deeply related to the specific application, so benchmark are needed for each case. The user is charged only for the data stored in the volume plus how much S3 consumed for snapshots (storage space and I/O operations).

From the developer's perspective, the use of EBS is completely transparent because volumes are seen as block devices attached to EC2 instances. In contrast, S3 features require the use of web service APIs. In order to interface S3 with NetCDF we choose the Java implementation freely available in source code.

26.3.2 Multidimensional Environmental Data Standard File Format

As we deal here with data in NetCDF format, we describe that data format briefly. NetCDF is a data format and abstraction, implemented by a software library, for storing and retrieving environmental multidimensional data. Developed by UCAR in the early 90s for meteorological data management, NetCDF has become a widely used data format for a wide range of environmental computing science applications. The NetCDF implementation provides a self-describing and machine-independent format for representing multidimensional scientific data: the abstraction, the access library, and the data format support the creation, access, and sharing of scientific information.

The NetCDF data abstraction models a scientific data set as a collection of named multidimensional variables (scalars and arrays of bytes, characters, integers, and floating-point numbers) along with their coordinate systems and some of their named auxiliary properties. Each variable has a type, a shape specified by a list of named dimensions, and a set of other properties described by attribute pairs. The

NetCDF interface allows data to be accessed by providing a variable name and a specification for what part of the data associated with that variable is to be read or written, rather than by sequential access and individual reads and writes. A dimension is a named integer used to specify the shape of one or more variables, and usually represents a real physical dimension, such as time, latitude, longitude, or atmospheric level. A variable is an array of values of the same type, and is characterized by a name, a data type, and a shape described by a list of dimensions. Attributes may be a single value or a vector of values. One dimension may be unbounded. A variable with a shape that includes an unbounded dimension can grow to any length along that dimension. The unbounded dimension is like a record number in conventional files; it allows us to append data to variables.

NetCDF software interface implementations are available in C, Fortran, Java, and MatLab, among others. We work here with the NetCDF-Java library, a 100% Java framework for reading NetCDF and other file formats into the Common Data Model (CDM), a generalization of the NetCDF, OpenDAP and HDF5 data models, and for writing to the NetCDF file format. The NetCDF-Java library also implements NcML, which allows the developer to add metadata to CDM datasets, as well as to create virtual datasets through aggregation. This library implementation permits access to NetCDF files via network protocols such as HTTP and, via a plug in architecture, enables the development of different data reader.

26.3.3 Enhancing the S3 APIs

S3 has two important limitations that complicate its use for large environmental multidimensional data sets. The first is the five gigabyte maximum object size, which is too small for environmental applications. For example, a real time weather forecasting application developed at DSA/uniParthenope in 2003 and still running today produces each day an 11 gigabyte NetCDF dataset just from runs performed with the Weather Research and Forecast (WRF) model (Ascione, Giunta, Montella, Mariani, & Riccio, 2006). The second is the requirement that an object must be read or written in its entirety. The most basic operation performed on multidimensional environmental datasets is subsetting: extraction of scalars, arrays, and matrices along one or more dimensions. This operation requires random access to stored objects. Thus, simply storing each NetCDF file as an S3 object is inefficient and (for larger files) also infeasible.

On the positive side, S3 uses a highly reliable replica and location service completely transparent to the user: each object is named by a unique resource identifier and accessed by an URL that can be made public. Multiple concurrent accesses to different objects belonging to the same bucket are possible without an evident loss in performance.

The design of our S3-enhanced Java API seeks to overcome S3's limitations while preserving S3's high performance and availability (Figs. 26.1 and 26.2). As

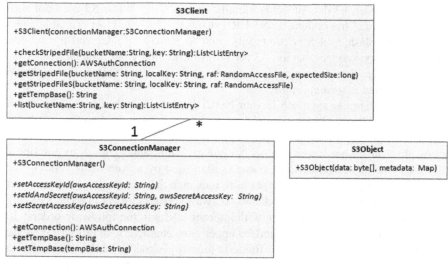

Fig. 26.1 Part of the enhanced S3 Java interface class diagram

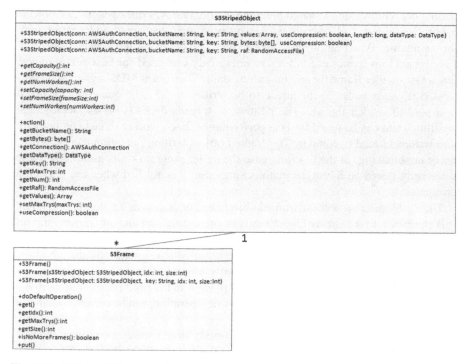

Fig. 26.2 S3StripedObject class diagram

noted earlier, each S3 data object is identified via a URL. While S3 does not support a hierarchical file system, an S3 URL string can include most printable chars including the slash, which is commonly used to structure folder and file names. Internal Amazon services that use S3 for storage commonly partition large datasets (e.g., VM images) across multiple S3objects, with the set of objects that make up the dataset listed in a *manifest file*.

We adopt a similar approach for our NetCDF files. Specifically, we implement a self-managed "framed object," a virtual, large (perhaps more than five gigabyte) object, identified by its name (no manifest file is needed) and partitioned across a set of physical frames stored one per S3 object, with names chosen to describe a subfolder-like organization. Each frame is identified by a "sub-name" coding the number of the frame, the total number of frames, and the frame size. Because object names are limited in size in 255 characters, we encode numbers in base 62, using all printable characters compatible with internet URLs in the following orders: 10 digits, 26 lower case characters, and 26 upper case characters. In order to increase the reliability of the service, each frame is digitally signed by MD5. Each time a frame is retrieved it is checked against the signature and re-requested if an error is detected.

Because saving space in S3 can both reduce costs (for storage and data transfer) and improve performance, we compress each frame using the zip algorithm. S3 supports high performance concurrent access, so we implement all writing and reading operations using a shared blocking queue. Our API manages framed objects in both write and read operations. A developer sees each read and write operation as atomic. When an object is to be stored on S3 using the framed approach, it is divided into frames each of a size previously evaluated for best performance, plus a last smaller frame for any remaining data. Then, each MD5-signed and compressed frame is added to the queue to be written to S3. The queue is consumed by a pool of worker threads. The number of threads depends on the deployment conditions and can be tuned for best performance. Each worker thread can perform both write and read operations. The framed object writing operation can be blocking or nonblocking. In the blocking case the caller program waits until each frame is correctly stored on S3; in the nonblocking case, it is notified when the operation terminates.

The developer also sees the framed object read operation as an atomic operation. This operation first extracts object features (size, total amount of frames, size of each frame plus the rest frame size) from the name of a stored frame. Then the object is allocated in client memory and the required read operations are submitted to the queue. The worker threads concurrently check for frame integrity using the MD5 signature, uncompress the frame, and place it in the right position in memory. The framed object read operation, like the write operation, can be either blocking or nonblocking (Fig. 26.2).

The S3-enhanced Java API that we developed can be used as a middleware interface to S3, on which we can then build higher-level software such as our S3-enabled NetCDF Java interface.

26.3.4 Enabling the NetCDF Java Interface to S3

The NetCDF-Java library supports access to environmental datasets stored both locally, via file system operations, and remotely, via protocols such as HTTP and OpenDAP. Data access is performed via an input/output service provider (IOSP) interface. The developer can build custom IOSPs that implement methods for file validity checking, file opening and closing, and data reading, in each case specifying the variable name and the section to read. An IOSP generates requests to a low-level random access file (RAF) component, a buffered drop-in replacement for the homonymous component present in the Java input/output package. The use of RAF provides substantial speed increases through the use of buffering. Unfortunately, RAFs are not structured in the NetCDF-Java library as a customizable service provider, so developers cannot build and register their own random access file component.

The NetCDF file format is self describing: data contain embedded metadata. The NetCDF Markup Language (NcML) is an XML representation of netCDF metadata. NcML is similar to the netCDF network Common data form Description Language (CDL), but uses XML syntax. A NetCDF file can be seen as a folder in which each variable can be considered a file. Thus, a NcML file description can be considered as an aggregator for variables, dimensions and attributes.

In our S3-enabled NetCDF Java Interface we use the NcML file representation as a manifest file, the NetCDF file name as a folder name, and each variable as a framed object: basically a subfolder in which each variable is stored in frames. Thus, if an S3-stored NetCDF file is named s3://bucketname/path/filename.ncml, then its data is stored in s3://bucketname/path/filename/varname/framename objects.

Our S3IOServiceProvider *open method* interacts with our S3-enhanced Java package to retrieve the NcML file representation and create an empty local image of the remote netCDF file using the NcMLReader component. Thus, metadata are available locally, but no variable data are actually downloaded from S3. In order to implement this behavior we create a custom RAF, which defines three methods: canOpen returns true if the passed filename can be considered as a valid file; getName returns the name of the RAF; and getRaf returns the underlying random access file component. We also provide an abstract RAF provider component that defines the RAF reference and implements the getRaf method. Finally, we implemented the S3RandomAccessFileProvider component, which returns the string "S3RAFProvider" as name and verifies that the filename begins with the "s3:" string. The canOpen method creates an instance of the S3RandomAccessFile and performs the needed initialization.

S3RandomAccessFile (S3RAF) is the key component in our S3-enabled NetCDF service. It implements the low-level interaction with our S3 Java API. When a variable section is read, S3RAF retrieves only the needed data frames stored on S3 and writes them to a temporary NetCDF local file. If more than one frame is needed, our S3 Java interface uses the framed object component to read them concurrently. A caching mechanism ensures that we do not download the same frame twice. This

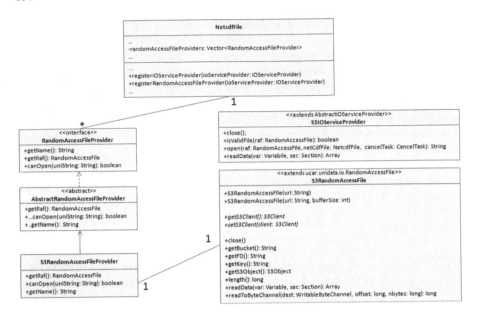

Fig. 26.3 The S3-enabled NetCDF Java interface class diagram

feature minimizes download time and reduces the number of S3 get operations and consequently the overall cost (Fig. 26.3).

The following code reads a variable section from a NetCDF file stored on S3:

```
1: String fileName = "s3://12WREKPN1ZEX2RN4SZG2/
     wrfout_d01_2009-02-10_00-00-00.nc_1_test";
2: S3ConnectionManager.setIdAndSecret
     ("12WREKPN1ZEX2RN4SZG2", "...");
3: NetcdfFile.registerIOProvider("ucar.unidata.io.s3.
     S3IOServiceProvider");
4: NetcdfFile.registerRAFProvider("ucar.unidata.io.s3.
     S3RandomAccessFileProvider");
5: NetcdfFile ncfile = NetcdfFile.open(testFileName);
6: String section = "1:1:1,1:19:,1:190:1,1:254:1";
7: Array arrayResult = ncfile.findVariable("U").read(range);
```

In line 1, we define the file name. The string "s3://" identifies the protocol, allowing the S3RAFProvider component to recognize that the location can be accessed by the S3RAF and that it must create an instance of this object. The string immediately following the protocol name is the bucket name. (Because bucket names must be unique, a good strategy is to use the S3 user id.) The last part of the string is the actual file name.

In line 2, we use the S3ConnectionManager component to set the user id and password required by the S3 infrastructure. The S3ConnectionManager component also allows the developer to configure deployment and performance details, such as the temporary file path, the size of the read and write queues, and the number of worker threads.

In line 3, we register the S3IOServiceProvider using the standard NetCDF Java interface, while in line 4 we register the S3RAFProvider, which as stated above improves the standard NetCDF-Java interface to implement completely transparent access to S3-stored datasets.

In line 5, we open the file, using the same syntax as for a local operation. Line 6 specifies that we wish to read just one time step of the whole two-dimensional variable. We read variable in line 7, retrieving an Array object reference to data in the same manner as for a local data access. Observe that the underlying cloud complexity is completely hidden.

26.4 Cloud and Grid Hybridization: The NetCDF Service

The NetCDF service developed by Montella, Agrillo, Mastrangelo, and Menna (2008) is a GT4-based web service. It leverages useful GT4 features and captures much previous experience in environmental data delivery using grid tools. The service integrates multiple data sources and data server interaction modes, interfaces to an index service to permit discovery, and supports embedded data processing. Last but not least, it is designed to work in a hybrid cloud/grid environment.

26.4.1 The NetCDF Service Architecture

The NetCDF service provides its clients with access to *resources*: an abstracted representations of data objects that are completely disjoint from the underlying data storage associated with the data objects. A *connector* links the NetCDF service resource to a specific underlying data storage system. Available connectors include the *NetCDF file connector*, which using our S3-enhanced NetCDF Java interface can serve local files, DODS-served files, HTTP-served files and S3-stored files; the *GDS connector* that can serve Grib and GrADS files served by a Grads Data Server; and the *Hyrax connector* for OpenDAP Hyrax-based servers. We are also developing an *instrument connector* as a direct interface to data acquisition instruments (Montella, Agrillo, Mastrangelo, & Menna, 2008) based on our Abstract Instrument Framework (Montella, Agrillo, Di Lauro, 2008).

The primary purpose of a connector is to dispatch requests to different data servers and to convert all responses into NetCDF datasets (Fig. 26.4).

Once a requested subset of a dataset is delivered by the data connector and stored locally, the user can process that subset using local software. This feature is implemented using another full customizable plug in. Thanks to the factory/instance approach, each web service consumer deals with its own data in a temporary private storage area physically close to the web service. The processor connector component mission is to interface different out of the process NetCDF dataset processors

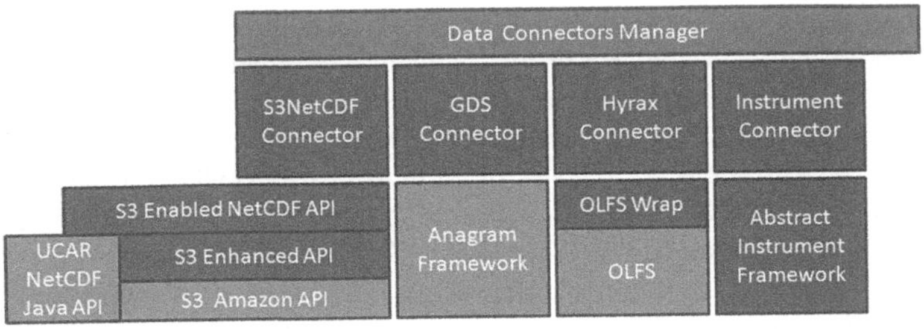

Data Connectors Manager
S3NetCDF Connector
GDS Connector
Hyrax Connector
Instrument Connector
S3 Enabled NetCDF API
Anagram Framework
OLFS Wrap
Abstract Instrument Framework
UCAR NetCDF Java API
S3 Enhanced API
S3 Amazon API
OLFS

Fig. 26.4 The NetCDF service data connector architecture

caring out a standard way to perform data input, processing job submission and data output (Fig. 26.5).

Two processor connectors are available. The *GrADS processor connector* provides access to NetCDF data as gridded data. A consumer can send to the processor connector complex command sequences using a common Java interface or directly with GrADS scripts. The GrADS processor connector is build on top the GrADSj Java interface we developed in previous work (Montella & Agrillo, 2009a).

The *NetCDF Operator connector* is a Java interface to the homonymous software suite. The netCDF Operators, or NCO, are a suite of standalone, command-line programs that each take netCDF files as input, operate on those files (e.g., derive new data, compute averages, extract hyperslabs, manipulate metadata), and produce a netCDF output file. NCO primarily aids manipulation and analysis of gridded scientific data. The single-command style of NCO allows users to manipulate and analyze files interactively, with simple scripts that avoid some overhead (and power) of higher level programming environments. As in the case of the GrADS processor connector, the web service consumer interacts with the NCO processor connector using a simple Java interface or directly with shell-like scripts. As internally GrADSj, the NCO processor connector leverages on the

Processing Connectors Manager
GrADS Connector
GrADSj
NCO Connector
Abstract Execution Framework

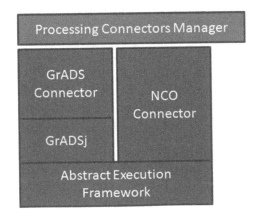

Fig. 26.5 The NetCDF service processing connector architecture

AbstractExecutionFramework (AEF) we developed in order to manage the execution of out of the process software from the Java environment is a standard high level fashion (Montella & Agrillo, 2009b).

Once a data subset is extracted and processed, the main actor in the data pipeline is the data transfer connector. This plug-in-like component permits the developer to customize how the selected data is made available to the user. All transfer connectors share the cache management system and the automatic publishing of data descriptions into an index service. (These features are applied automatically only when there are no privacy issues.) In general, subset data from a public available dataset are still public, but the user can set a subsetting result as private; in contrast, a processing result is private by default, but the user can declare it as public (Fig. 26.6).

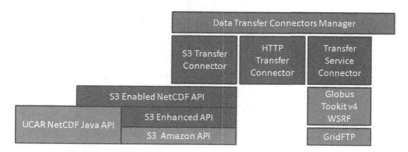

Fig. 26.6 The NetCDF service transfer connector architecture

The caching and auto publication processes work on the two kinds of datasets in the same way. Because each NetCDF file is completely described by its metadata, each dataset can be uniquely identified by a MD5 hash of that metadata. Thus, we sign each dataset generated by a subsetting or processing operation and copy it to a caching area. Then, each time a NetCDF file is requested, we first evaluate the MD5 metadata signature and check the cache. In the case of a cache miss, the dataset is requested for subsetting or submitted for processing. Following a cache hit the dataset is copied from the cache area to the requestor's temporary storage area, and in addition the NetCDF dataset cache manager component increase the usage index of the selected dataset. If this index overcomes a chosen threshold, the dataset is promoted to be a stored dataset.

The web service resource manager explores resources and advertises them automatically on the index service. The NetCDF dataset cache manager component periodically explores the cache, decreasing the usage count of each cached NetCDF dataset. A dataset is deleted if its count reaches zero, in order to save local storage space. This usage checking process is performed even on promoted datasets. The actual storage device priority sequence, ordered from the most frequently accessed to the least is: local file system, EBS, S3 and then deleted. The policy that is used to manage data movement of these storage devices is fully configurable and based on the number of requests per unit of time multiplied by the space needs for the specific device.

The transfer connector act as the main data delivery component implementing how the web service result can be accessed by the user. The default transfer connector is the TransferServiceConnector. In this way we represent the result by an End Point Reference (EPR) to a temporary resource managed by a TransferService. This service is a wrapper over the GridFTP service. The TransferService uses the Grid Security Infrastructure and works in an efficient and effective way for secure data transfer. Other available transfer connectors include the HTTPTransferConnector and the DODSTransferConnetor suitable for publically available result datasets. Finally the S3TransferConnector stores the results on S3 and then the user can access them directly (Fig. 26.7).

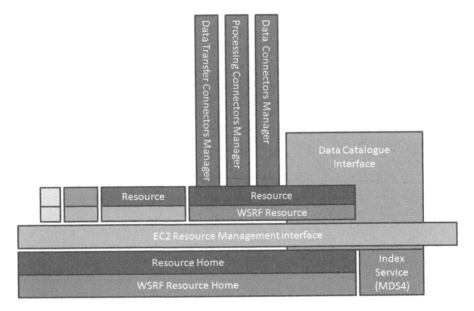

Fig. 26.7 NetCDF service architecture: the big picture

26.4.2 NetCDF Service Deployment Scenarios

The NetCDF Service represents the grid aggregator component for cloud hosted multidimensional environmental data resources. We explore three different deployment scenarios, in which the NetCDF service is deployed variously (see Fig. 26.8):

1. on a computer outside the cloud (*Grid+S3*);
2. on an EC2 instance (*Cloud*); or
3. in a proxy-like configuration in which the service runs on a computer outside the cloud and automatically runs one or more EC2 instances to manage operations on datasets (*Hybrid*).

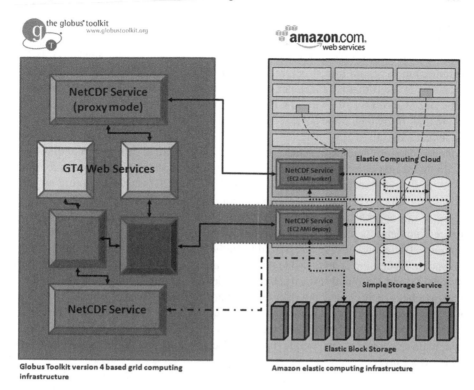

Fig. 26.8 The three NetCDF service deployment scenarios: Grid+S3 (*bottom left*), Cloud (the lower NetCDF service on the *right*), and Hybrid (*top left*)

In the *Grid+S3* deployment scenario, a standalone NetCDF service runs on a server external to the cloud. This server must be powerful enough and have enough storage space to support subsetting and processing operations. Data can be hosted locally, by DODS servers on secured private networks, and/or on S3-based cloud services. Thanks to the AbstractExecutionFramework component, the processing (GrADS and NCO) software can work as jobs submitted to a local queue and execute on a high performance computing cluster. In this scenario, we use the cloud provider for data storage as in the case of locally stored large datasets and S3 stored automatically promoted cached datasets.

In the *Cloud* scenario, the NetCDF service is deployed on an EC2 instance, directly accessible from outside the cloud using Amazon's Elastic IP service. Data can be stored on EBS and accessed locally by the instance, while automatically promoted cached datasets can be stored on EBS and/or S3, depending on frequency of use. The EC2 instance has to provide enough computing power to satisfy dataset processor needs. The advantage of this kind of deployment is high-speed access to elastically stored datasets using EBS (faster) and S3 (slower, but accessible from outside the cloud). The main drawback to this approach is that the user is charged for the cost of a continuously running EC2 instance and the need for an assigned Elastic IP.

The *Hybrid* scenario has the highest level of grid/cloud hybridization. Here, we deploy the NetCDF service on a real computer outside the cloud infrastructure. This service acts as a proxy for another instance of the service running (as an EC2 instance) in the cloud infrastructure. When the NetCDF service receives a request, it checks to see if that EC2 instance is already running. If there is not, or if the running one is too busy (the virtual CPU's load exceeds a threshold), a new instance is created. If the data to be accessed is hosted by EBS, and an active EC2 instance is connected to the EBS where the data is hosted. Because an EBS volume can be connected to only one EC2 instance a time, the request must be directed to that.

In this scenario, a consumer interacts directly only with the NetCDF service running on the real machine; the interaction with the NetCDF service running on the elastically allocated computing resource is completely transparent. The main advantages of this approach are that (a) the user is charged for running AMIs only when processing datasets and (b) there is no need for assigned Elastic IP. Moreover, the amount of data that must be transferred from the cloud to the grid is limited because most data processing is done inside the cloud. In addition, the elasticity provides for scalability and the cloud infrastructure provides for predictable quality of service, data availability, backups, and disaster recovery. Finally, the grid machine serving the NetCDF service does not need huge computing power or storage space because it acts only as a proxy providing a transparent interface to the cloud infrastructure. The main drawback is the complexity of the deployment and the increased number of interface layers.

26.5 Performance Evaluation

All software components developed in this work are high-quality prototypes ready for real world application test beds and even production use. To evaluate the performance of the different deployment methods, we measured performance in three different scenarios: reading and writing data from S3 using the framed object implemented in our Java interface; reading NetCDF data from S3 comparing intra-cloud and extra-cloud performance; and NetCDF dataset serving comparing EBS and S3 storages.

26.5.1 Parameter Selection for the S3-Enhanced Java Interface

We must evaluate the I/O performance of the S3-enhanced Java interface in order to identify an optimal framework configuration. The two developer-controllable parameters are the frame size and number of concurrent threads used as workers by the framed object components. (A third parameter, the size of the blocking queue, was empirically evaluated as four times the number of worker threads.) The choice of frame size is critical because, once set, the stored framed object must be re-uploaded in order to change it. In contrast, the number of worker threads can vary even while the application is running and can potentially be adjusted automatically based on observed network congestion and machine CPU usage.

We used a 130 MB NetCDF file to evaluate upload and download performance for a range of frame sizes (1, 2, 4, and 8 MB) and thread counts (1–64 threads). In each case, we ran the experiment 100 times and averaged the results. We used a worst-case approach, storing the file in a US zone and accessing it from Europe (Fig. 26.9).

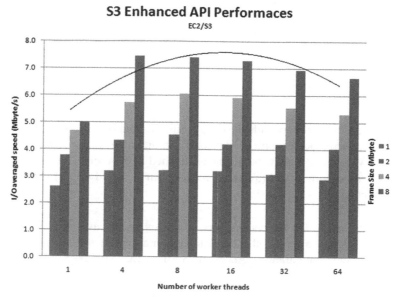

Fig. 26.9 S3-enhanced Java interface tuning

The best performance for both reading and writing is achieved when using 8 MB frames. The best number of concurrent threads vary depends on the operation: for writing (uploading) the best performance is achieved 8–16 worker threads, while for reading (downloading) the maximum number of tested worker threads, 64, was the best. We have applied these lessons in an improved version of our S3-enhanced Java interface that uses a live performance evaluator to vary the number of worker threads over time.

26.5.2 Evaluation of S3- and EBS-Enabled NetCDF Java Interfaces

Our second set of experiments compare the performance of our S3-enabled NetCDF Java interface with analogous operations using EBS. We use a four-dimensional NetCDF file of size ~11 GB produced by the WRF model on a grid with 256 × 192 cells and 28 vertical levels. This file contains six days (144 h) of 111 variables: 14 one-dimensional, varying in time; 10 two-dimensional, varying in time and level; 72 three-dimensional, varying in time, latitude, and longitude; and 15 four-dimensional, varying in time, level, latitude, and longitude.

We define five separate tests, corresponding to reading the first one (5.5 MB), 24 (132 MB), 48 (264 MB), 72 (396 MB), and 144 h (792 MB), respectively, of a single four-dimensional variable. The subset variable is the west to east component of the wind speed $U(X, Y, Z, T)$, where X, Y, and Z are the spatial dimensions and T is the temporal dimension, in the order T, Z, Y, X, meaning that the disk file created for that variable contains data for all time periods for $(X = 0, Y = 0, Z = 0)$, then data for all time periods for $(X = 0, Y = 0, Z = 1)$, and so on. Thus, a request for 144 h of the variable involves a single contiguous sequence of bytes, while a request for 1 h of the variable involves multiple small reads. As before, we run each test 100 times and average the results.

We ran the test suite in four configurations, in which the subsetting operation is performed variously on:

1. a computer in Europe, outside the cloud, with the data on S3 in the US;
2. an EC2 virtual machine with the data on S3 in the same zone;
3. an EC2 virtual machine with data on an attached EBS volume; and
4. a server outside the cloud (a quad-core Xeon Linux based server) with the data on that computer's local disk.

As shown in Fig. 26.10, performance increases in all cases with subset time period and thus read size. We also see that performance varies considerably across the four different configurations. Accessing S3 from EC2 (EC2←S3) is faster than

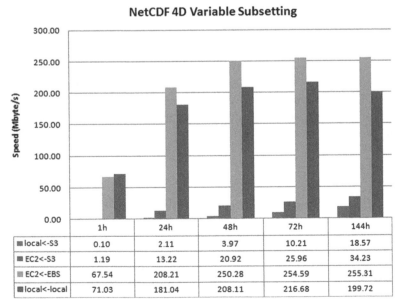

Fig. 26.10 Performance in MB/s of the S3-enabled NetCDF Java interface for different subset time periods and different data transfer scenarios

when accessing S3 from outside the cloud (local←S3), with the difference being proportionally larger for smaller reads. This result suggests that there may be advantages to performing subsetting operations in the cloud and then moving only the subset data to a remote client. We also see that EBS performance (EC2←EBS) is far superior to S3 performance – indeed, superior to accessing local disk (local←local). We might conclude from this result that data should always be located on EBS. However, there are other subtle issues to consider when choosing between S3 and EBS storage. If data are to be accessed only from an EC2 virtual machine, EBS is the best choice. On the other hand, S3 permits data to be accessed directly (for example, using our S3-enhanced Java NetCDF interface) from outside the cloud, and permits multiple accesses to proceed concurrently. Cost must also be considered: running instances on EC2 machines may or may not be cost effective, depending on workload. Creating virtual machine instances dynamically, only when needed for data subsetting, analysis, and transfer purposes, could be a winning choice. We return to these issues below.

26.5.3 Evaluation of NetCDF Service Performance

Our final experiments are designed to evaluate our NetCDF service from the perspectives of both cost and performance. For these studies, we developed an Amazon Web Service Simulator (AWSS), a Java framework that simulates the behavior of the Amazon EC2, EBS, and S3 components. Given specified data access and pricing profiles, this simulator generates a performance and cost forecast. To evaluate the simulator's accuracy, we compared its predicted cost against results obtained in which we ran data hosting and virtual machines instances without any kind of cloud application. We obtained cost profiles comparable with the Amazon Simple Monthly Simulator.

We evaluated the three scenarios described in Section 26.4.1. In *Grid+S3*, the NetCDF service is hosted on a computer outside the cloud and accesses data hosted on S3 via the Amazon HTTP-based S3 protocol.

In *Cloud*, the NetCDF service is hosted on a "standard.extralarge" EC2 instance (15 GB of memory, 8 EC2 Compute Units, 1690 GB of local instance storage, 64-bit platform) that runs continuously for the entire month. Data are hosted on 15 one terabyte EBS volumes attached to this EC2 instance.

In *Hybrid*, the NetCDF service is hosted on an EC2 instance with the same configuration as in *Cloud*. However, this instance is not run continuously but instead is created when a request arrives and then shut down after two hours unless an additional request arrives during that period.

Because *Cloud* and *Hybrid* generate subset data on EC2 instances, that data must be transferred from the cloud to the outside world. We measured performance for three different protocols – HTTP ($\sim$2.5 MB/s), gridFTP ($\sim$16 MB/s), and scp ($\sim$0.56 MB/s) – and use those numbers in our simulations.

We present results for the following configuration and workload. We assume 15 Tbyte of environmental data, already stored in the cloud; thus no data upload costs are incurred. We consider a 30-day duration (during which time data storage costs are incurred), with requests for varying subsets of the same four-dimensional dataset arriving as follows:

- Days 0–2: No requests.
- Days 3–5: 144 requests, each for a 1 h subset (i.e., 5.5 MB).
- Days 6–8: 144 requests, each for a 24 h subset (132 MB).
- Days 9–11: 144 requests, each for a 48 h subset (264 MB).
- Days 12–14: 144 requests, each for a 72 h subset (396 MB).
- Days 15–17: 144 requests, each for a 144 h subset (792 MB).
- Days 18–29: No requests.

Requests arrive according to a normal distribution with a mean of 0.5 h; they total 280 GB over the 30 days. We assume a worst-case situation with a machine in Europe accessing AWS resources hosted in the US West zone. We use Amazon costs as of March 2010, as summarized in Table 26.1.

Table 26.2 shows the cost predicted by our simulator for the three scenarios over the simulated month (the Hybrid has been evaluated using EBS or S3 storages), while Fig. 26.11 shows predicted time *per request* for the different combinations of scenario, subset size, and transfer protocol.

Grid+S3 is the most expensive, due to its use of S3 and the associated storage and data transfer costs. Its performance is good, especially for larger subsets, because the S3 Enhanced Java interface that is used to move data outside the cloud performs well with large data selections.

Table 26.1 Amazon Web Services costs as of March 2010

Component	Cost ($US)
EC2 standard.extralarge, US East, per hour	0.68
Data transfer from outside to EC2, per Gbyte	0.15
S3 storage, per Gbyte per month	0.15
Data movement out of S3, per Gbyte	0.15
S3 10.000 get operations	0.01
EBS storage, per Gbyte per month	0.10
EBS million of I/O operations	0.10

Table 26.2 Simulated monthly costs for each approach

Approach	Storage ($US)	Computing ($US)	Networking ($US)	Total $US)
Grid+S3	2211	0	50	2261
Cloud EBS	1472	482	50	2004
Hybrid EBS	1474	231	49	1754
Hybrid S3	2216	233	49	2498

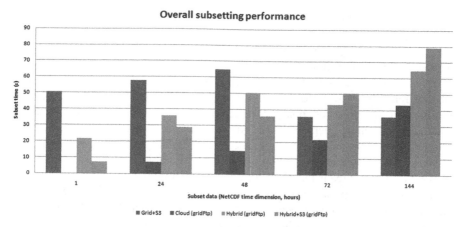

Fig. 26.11 Modelled overall subsetting performance for each deployment scenario

Cloud is cheaper than *Grid+S3* and has better performance than *Grid+S3* (when using GridFTP for external transfers), except in the 144 h (no subsetting) case.

Hybrid is the most cost effective because of its use of EBS to host data and its creation of EC2 instances only when required. When using GridFTP to transfer data outside the cloud, is more economical as problem size (storing, subsetting and transfer data) increases. *Grid+S3* is expensive and slow for small subsets, but becomes competitive when the subset data grow.

One factor not addressed in these results is what happens when multiple remote clients request data at the same time. In that situation, *Grid+S3* may become more competitive, as the fact that an EBS volume can be attached to only one EC2 instance at a time limits the performance of *Hybrid*.

It is also instructive to compare these results with what we know of ESG. As of early 2010, ESG systems hold around 150 Terabytes of climate simulation data. During 2009, one of the two major servers operating at that time (at NCAR) delivered a total of 112 Tbytes in response to 170,886 requests, for an average of 20 requests per hour and 654 MB/file. Our experimental configuration had 15 Tbytes (~1/10th of the total size) and returned a total of 280 GB in response to 720 requests, for an average of 1 request per hour and 388 MB/file. We need to repeat our simulations with an actual ESG workload, but an estimate based on Fig. 26.11 suggests a cost of roughly $20,000 per month to host ESG on Amazon.

26.6 Conclusions and Future Directions

We have sought to answer in this chapter the question of whether it is feasible, cost-effective, and efficient to use Amazon cloud services to host large environmental datasets. We believe that the answer to this question is "yes," albeit with some caveats.

To evaluate feasibility, we have developed a NetCDF service that uses a combination of external and cloud-based services to enable remote access to data hosted on Amazon S3 storage. Underpinning this service is an S3-enabled NetCDF Java API that uses a framed object abstraction to enable near random access read and write access to NetCDF datasets stored in S3 objects. MD5 signatures and compression enhance reliability and performance. The S3-enabled NetCDF-Java interface permits the developer to access both local files and S3-stored (or EBS-stored) NetCDF data in an identical manner. The use of Amazon storage is transparent with the only difference being the need to provide access credentials. Overall, this experience leads us to conclude that an Amazon-based ESG is feasible.

The NetCDF service is the product of our considerable previous work on (GT4-based) environmental data provider web services. This service is highly modular and based on a plug in architecture. Multidimensional environmental datasets are exposed as resources that furthermore are advertised automatically via an index service. The web service consumer interacts with a private, dynamically allocated instance of the service resource. The operation provider permits data selection, subsetting, processing, and transfer. Each feature is implemented by a provider component that allows for improvements, expansion, and customization. The service can be deployed in three main ways: stand alone on a machine belonging to a computing grid to distribute data hosted locally, on secured custom servers, or using the S3 service; stand alone on a virtual machine instance running in the EC2 ecosystem providing data stored on EBS or S3; and in a proxy mode running on an external computer that acts as a proxy to service instances managing resources on dynamically allocated EC2 instances.

To evaluate performance, we have conducted detailed experimental studies of data access performance for our NetCDF service in these different configurations. To evaluate cost, we developed and applied a simple cloud simulator. Based on these results, we believe that the hybrid solution represented by the third NetCDF Service deployment scenario, *Hybrid*, provides the best balance between external and cloud hosted resources. This strategy minimizes costs because EC2 instances are run only when required. Another cost reduction is achieved because there is no need for Elastic IP: the service instance running on the virtual machine connects to the one on the real machine identified by a fully qualified domain name. Two other sources of costs are the put and get data operations on S3 and the data transferred over the cloud infrastructure boundary. In *Hybrid*, most data processing is performed inside the cloud, and thus a consumer who requests subsets or analysis results retrieves less data. Less data transfer means also better performance.

The use of EBS rather than S3 provides increased data access performance. While this approach permits to use the S3 as an effective way to carry out the cloud large datasets in an effective and efficient way leveraging on concurrent S3 object access with the Java API we developed. Finally, as the need for storage is scaled by the elastic computing approach typical for the cloud infrastructure, even the processing computing power scales with the needs thanks to the possibility of instancing as much EC2 instances hosting NetCDF Services as are needed.

A more detailed comparison of cost and performance for ESG awaits the availability of detailed ESG access logs. However, it seems that the cost of hosting the current ESG holdings on Amazon, and responding to current workloads, might be ~$20,000 month. Determining ESG's current data hosting costs would be difficult, but when all costs are considered, it may not be too different.

ESG is now preparing for the next generation of climate models that will generate petabytes of output. A detailed performance vs. costs comparison will be needed to evaluate the suitability of commercial cloud providers such as Amazon for such datasets – a task for which our simulator is well prepared. Unfortunately, this comparison is difficult or impossible to perform at present because of lack of data on future cloud configurations and costs, and on future client workloads. However, we do note that server-side analysis is expected to become increasingly important as data sizes grow, and infrastructures such as Amazon are well designed for compute-intensive analysis of large quantities of data.

This work is the result of a year of prototyping and experiments in the use of hybrid grid and cloud computing technologies for environmental data elastic storing, processing, and delivery. The NetCDF service integrates multiple software layers to identify a convenient hybrid deployment scenario. In an immediate next step, we will apply this technology to more realistic weather forecast and climate simulation scenarios. In the process, we will improve the stability of the individual components and develop further features identified by experience.

References

Armbrust, M., Fox, A., Griffith, R., Joseph, A. D., Katz, R. H., et al. (2009). Above the clouds: A berkeley view of cloud computing, Electrical Engineering and Computer Sciences University of California, Berkeley, (Technical Report No. UCB/EECS-2009-28). http://www.eecs.berkeley.edu/Pubs/TechRpts/2009/EECS-2009-28.html, February 10, 2009.

Ascione, I., Giunta, G., Montella, R., Mariani, P., & Riccio, A. (2006). A grid computing based virtual laboratory for environmental simulations. In W. E. Nagel, W. V. Nagel & W. Lehner (Eds.), *Euro-par 2006 parallel processing* (pp. 1085–1094). LNCS 4128, Springer, Heidelberg.

Allcock, B., Bester, J., Bresnahan, J., Chervenak, A. L., Foster, I., Kesselman, C., Meder, S., Nefedova, V., Quesnal, D., & Tuecke, S. (May 2002). Data management and transfer in high performance computational grid environments. *Parallel Computing Journal, 28* (5), 749–771.

Bernholdt, D. et al. (March 2005). The earth system grid: Supporting the next generation of climate modeling research. *Proceedings of the IEEE,93*(3).

Buyya, R., Yeo, C. S., & Venugopal, S. (2008) *Market-oriented cloud computing: Vision, hype, and reality for delivering it services as computing utilities.* Proceedings the of 10th IEEE International Conference on High Performance Computing and Communications HPCC '08, Dalian, China.

Doty, B. E., Kinter III, J. L. (1995) Geophysical data analysis and visualization using GrADS. In E. P. Szuszczewicz & J. H. Bredekamp (Eds.), *Visualization techniques in space and atmospheric sciences* (pp. 209–219). NASA,Washington, DC.

Foster, I. (July 2002). What is the Grid? A Three Point Checklist.

Foster, I. (2006). Globus toolkit version 4: Software for service-oriented systems. *Journal of Computational Science and Technology, 21*, 513–520.

Foster, I., Zhao, Y., Raicu, I., & Lu, S. (2008). Cloud computing and grid computing 360-degree compared. *Proceedings of Grid Computing Environments Workshop*, GCE '08, Austin, TX.

Giunta, G., Laccetti, G., & Montella, R. (2008). Five dimension environmental data resource brokering on computational grids and scientific clouds (pp. 81–88), APSCC, IEEE Asia-Pacific Services Computing Conference.

Gallagher, J., Potter, N., Sgouros, T. (2004). DAP Data Model Specification DRAFT, Rev.: 1.68, www.opendap.org. November 6, 2004.

Gallagher, J., Potter, N., West, P., Garcia, J., & Fox, P. (2006). OPeNDAP's Server4: Building a High Performance Data Server for the DAP Using Existing Software, AGU Meeting in San Francisco.

Mell, P., Tim, G. (July 2009). The NIST Definition of Cloud Computing, National Institute of Standards and Technology, Version 15. Information Technology Laboratory.

Montella, R., Agrillo, G., & Di Lauro, R. (April 2008). Abstract Instrument Framework: Java Interface for Instrument Abstraction," (DSA Technical Report. Napoli).

Montella, R., Agrillo, G., Mastrangelo, D., & Menna, M. (June 2008). A globus toolkit 4 based instrument service for environmental data acquisition and distribution. Proceedings of Upgrade Content Workshop HPDC2008. Boston, MA.

Montella, R., Giunta, G., & Riccio, A. (June 2007). Using grid computing based components in on demand environmental data delivery. *Proceedings of upgrade content Workshop HPDC2007*. Monterey Bay.

Montella, R., & Agrillo, G. (April 2009). *GrADSj: A GrADS Java interface* (DSA Technical Report, Napoli).

Montella, R., & Agrillo, G. (June 2009). Abstract Execution Framework: Java Interface for Out of the Process Execution, (DSA Technical Report, Napoli).

Rew, R. K., & Davis, G. P. (July 1990). NetCDF: An interface for scientific data access. *IEEE Computer Graphics and Applications, 10*(4), 76–82.

Sotomayor, B., Keahey, K., & Foster, I. (June 2008). Combining batch execution and leasing using virtual machines. ACM/IEEE International Symposium on High Performance Distributed Computing 2008 (HPDC 2008), Boston, MA.

Wielgosz, J. & Doty, J. A. B. (2003). The Grads-Dods Server: An Open-Source Tool for Distributed Data Access and Analysis, 19th International Conference on Interactive Information and Processing Systems (IIPS) for Meteorology, Oceanography, and Hydrology.

Wielgosz, J. (2004). Anagram – A modular java framework for high-performance scientific data servers. 20th International Conference on Interactive Information and Processing Systems (IIPS) for Meteorology, Oceanography, and Hydrology.

Index

B. Furht, A. Escalante (eds.), *Handbook of Cloud Computing*,
DOI 10.1007/978-1-4419-6524-0, © Springer Science+Business Media, LLC 2010